Environmental Radiobiology

This book focuses on the impacts of anthropogenic radiation on wildlife and ecosystems and provides an in-depth look at the approaches and available tools we can use to gain information about biological effects of radiation in the environment.

The nuclear accidents in Chornobyl in 1986 and Fukushima in 2011 focused the attention of the world on the vulnerability of ecosystems to radiation. In Chornobyl, there still remains an exclusion zone where levels are considered to be too high for people and impacts on terrestrial and aquatic ecosystems can still be measured 35 years later. In the area impacted by the Fukushima disaster, intense remediation is still under way at tremendous cost and causing widespread disruption to the environment. That accident impacted the terrestrial and marine ecosystems. In both accidents, it became obvious that a radiation protection framework focusing on protection of "humans" (a single species) and using evacuation as a key strategy was not sufficient to protect the natural environment. The complexity of ecosystems makes developing a protection framework very challenging, but in order to even start the process it is vital to gather information about likely impacts of low dose exposures on wildlife and to develop monitoring tools to measure changes over time. This book contains reviews and original research aimed at filling our knowledge gaps about these important areas.

Environmental Radiobiology will be a key resource for academics, researchers, and advanced students of Radiobiology, Radioecology, Biology, Ecology, Biomedicine, and Research Methods. The chapters included in this book were originally published as a special issue of *International Journal of Radiation Biology.*

Paul N. Schofield is the University Professor of Biomedical Informatics at the University of Cambridge, UK, and an Adjunct Professor at The Jackson Laboratory, USA. His research focus is on experimental and informatics approaches to understanding human disease using model organisms. He has a long-standing interest in the biological effects of low dose ionizing radiation and FAIR data in radiobiology.

Carmel E. Mothersill is an Environmental Radiobiologist at McMaster University, Hamilton, Canada, where she studies low dose radiation effects and development of biomarkers for wildlife. Her main goal is to identify ecosystem level bioindicators of environmental impact.

Environmental Radiobiology

Edited by Paul N. Schofield and Carmel E. Mothersill

CRC Press
Taylor & Francis Group
Boca Raton London New York

CRC Press is an imprint of the
Taylor & Francis Group, an **informa** business

First edition published 2024
by CRC Press
4 Park Square, Milton Park, Abingdon, Oxon, OX14 4RN

and by CRC Press
6000 Broken Sound Parkway NW, Suite 300, Boca Raton, FL 33487-2742

British Library Cataloguing-in-Publication Data
A catalogue record for this book is available from the British Library

ISBN13: 978-1-032-55768-7 (hbk)
ISBN13: 978-1-032-55769-4 (pbk)
ISBN13: 978-1-003-43213-5 (ebk)

DOI: 10.1201/9781003432135

Typeset in Minion Pro
by codeMantra

Publisher's Note
The publisher accepts responsibility for any inconsistencies that may have arisen during the conversion of this book from journal articles to book chapters, namely the inclusion of journal terminology.

Disclaimer
Every effort has been made to contact copyright holders for their permission to reprint material in this book. The publishers would be grateful to hear from any copyright holder who is not here acknowledged and will undertake to rectify any errors or omissions in future editions of this book.

This volume is dedicated to the memory of Professor Nick Beresford, a colleague, a friend, and a remarkable radioecologist, whose passion for the environment and dedication to scientific research has left a lasting mark on the world. His contributions to radioecology, particularly that of Chornobyl, have not only enriched our scientific understanding but also influenced environmental policies and conservation efforts worldwide. His work fostered a deeper understanding of the complex interplay between radiation and the natural world and continues to guide us as we strive to protect and preserve the beauty of our planet for generations to come.

Contents

Citation Information

The chapters in this book were originally published in the *International Journal of Radiation Biology*, volume 98, issue 6 (2022). When citing this material, please use the original page numbering for each article, as follows:

Introduction
Paul Schofield and Carmel Mothersill
International Journal of Radiation Biology, volume 98, issue 6 (2022) pp. 997–998

Chapter 1
Neglecting the ecosystemic dimension of life hinders efficient environmental protection from radiation and other hazards
François Bréchignac
International Journal of Radiation Biology, volume 98, issue 6 (2022) pp. 999–1007

Chapter 2
What makes radiation protection so challenging?
Lawrence A. Kapustka
International Journal of Radiation Biology, volume 98, issue 6 (2022) pp. 1008–1011

Chapter 3
Insight into the evolutionary profile of radio-resistance among insects having intrinsically evolved defence against radiation toxicity
Jagdish Gopal Paithankar, Tanhaji Sandu Ghodke, and Rajashekhar K. Patil
International Journal of Radiation Biology, volume 98, issue 6 (2022) pp. 1012–1024

Chapter 4
The sine qua non *of the fish invitrome today and tomorrow in environmental radiobiology*
Nguyen T. K. Vo
International Journal of Radiation Biology, volume 98, issue 6 (2022) pp. 1025–1033

Chapter 5
Environmental radiobiology of amphibians – knowledge gaps to be filled using cell lines
Nguyen T. K. Vo
International Journal of Radiation Biology, volume 98, issue 6 (2022) pp. 1034–1046

Chapter 6
Radiation-induced transgenerational effects in animals
Yuri E. Dubrova and Elena I. Sarapultseva
International Journal of Radiation Biology, volume 98, issue 6 (2022) pp. 1047–1053

Chapter 7

A review of the impact on the ecosystem after ionizing irradiation: wildlife population
Georgetta Cannon and Juliann G. Kiang
International Journal of Radiation Biology, volume 98, issue 6 (2022) pp. 1054–1062

Chapter 8

Impact of nonionizing electromagnetic radiation on male infertility: an assessment of the mechanism and consequences
Rohit Gautam, Eepsita Priyadarshini, JayPrakash Nirala, and Paulraj Rajamani
International Journal of Radiation Biology, volume 98, issue 6 (2022) pp. 1063–1073

Chapter 9

Environmental and industrial developments in radiation cataractogenesis
Lukasz Laskowski, David Williams, Colin Seymour, and Carmel Mothersill
International Journal of Radiation Biology, volume 98, issue 6 (2022) pp. 1074–1082

Chapter 10

Bio-acoustic signaling; exploring the potential of sound as a mediator of low-dose radiation and stress responses in the environment
Bruno F. E. Matarèse, Jigar Lad, Colin Seymour, Paul N. Schofield, and Carmel Mothersill
International Journal of Radiation Biology, volume 98, issue 6 (2022) pp. 1083–1097

Chapter 11

Radiation induces stress and transgenerational impacts in the cricket, Acheta domesticus
Xiaobing Li and C. D. Rollo
International Journal of Radiation Biology, volume 98, issue 6 (2022) pp. 1098–1105

Chapter 12

Evaluation of interactive effects of phosphorus-32 and copper on marine and freshwater bivalve mollusks
Emily L. Vernon, Michael N. Moore, Tim P. Bean, and Awadhesh N. Jha
International Journal of Radiation Biology, volume 98, issue 6 (2022) pp. 1106–1119

Chapter 13

Combined action of gamma radiation and exposure to copper ions on Lemna minor L
Irina S. Bodnar and Evgenia V. Cheban
International Journal of Radiation Biology, volume 98, issue 6 (2022) pp. 1120–1129

Chapter 14

Copper uptake in adult rainbow trout irradiated during early life stages and in non-irradiated bystander trout which swam with the irradiated fish
Richard Smith, Sunita Nadella, Richard Moccia, Colin Seymour, and Carmel Mothersill
International Journal of Radiation Biology, volume 98, issue 6 (2022) pp. 1130–1138

Chapter 15

A pilot study of radiation-induced bystander effect in radio-adapting frogs at a radiologically contaminated site located on the chalk river laboratories property
Nguyen T. K. Vo, Harleen Singh, Marilyne Stuart, Colin B. Seymour, and Carmel E. Mothersill
International Journal of Radiation Biology, volume 98, issue 6 (2022) pp. 1139–1146

Chapter 16

Environmental radiation on large Japanese field mice in Fukushima reduced colony forming potential in hematopoietic progenitor cells without inducing genomic instability
Kentaro Ariyoshi, Tomisato Miura, Kosuke Kasai, Valerie Swee Ting Goh, Yohei Fujishima, Akifumi Nakata, Atsushi Takahashi, Yoshinaka Shimizu, Hisashi Shinoda, Hideaki Yamashiro, Colin Seymour, Carmel Mothersill, and Mitsuaki A. Yoshida
International Journal of Radiation Biology, volume 98, issue 6 (2022) pp. 1147–1158

Chapter 17

*Assessment of chromosome aberrations in large Japanese field mice (*Apodemus speciosus*) in Namie Town, Fukushima*
Yohei Fujishima, Akifumi Nakata, Risa Ujiie, Kosuke Kasai, Kentaro Ariyoshi, Valerie Swee Ting Goh, Ko-jun Suzuki, Hirofumi Tazoe, Masatoshi Yamada, Mitsuaki A. Yoshida, and Tomisato Miura
International Journal of Radiation Biology, volume 98, issue 6 (2022) pp. 1159–1167

Chapter 18

Isolation of the effects of alpha-related components from total effects of radium at low doses
Chandula Fernando, Soo Hyun Byun, Xiaopei Shi, Colin B. Seymour, and Carmel E. Mothersill
International Journal of Radiation Biology, volume 98, issue 6 (2022) pp. 1168–1175

Chapter 19

Investigation of Radon Sources, Health Hazard and Risks assessment for children using analytical and geospatial techniques in District Bannu (Pakistan)
Huma Shakoor, Noor Jehan, Sardar Khan, and Nimat Ullah Khattak
International Journal of Radiation Biology, volume 98, issue 6 (2022) pp. 1176–1184

Chapter 20

From tangled banks to toxic bunnies; a reflection on the issues involved in developing an ecosystem approach for environmental radiation protection
Carmel E. Mothersill, Deborah H. Oughton, Paul N. Schofield, Michael Abend, Christelle Adam-Guillermin, Kentaro Ariyoshi, Nicholas A. Beresford, Andrea Bonisoli-Alquati, Jason Cohen, Yuri Dubrova, Stanislav A. Geras'kin, Tanya Helena Hevrøy, Kathryn A. Higley, Nele Horemans, Awadhesh N. Jha, Lawrence A. Kapustka, Juliann G. Kiang, Balázs G. Madas, Gibin Powathil, Elena I. Sarapultseva, Colin B. Seymour, Nguyen T. K. Vo, and Michael D. Wood
International Journal of Radiation Biology, volume 98, issue 6 (2022) pp. 1185–1200

Notes on Contributors

Michael Abend, Bundeswehr Institute of Radiobiology, Munich, Germany.

Christelle Adam-Guillermin, Institut de Radioprotection et de Surete Nucleaire, PSE-SANTE/SDOS/LMDN, Cadarache, France.

Kentaro Ariyoshi, Integrated Center for Science and Humanities, Fukushima Medical University, Fukushima City, Japan.

Tim P. Bean, Cefas Weymouth Laboratory, UK.

Nicholas A. Beresford, UK Centre for Ecology & Hydrology, Lancaster Environment Centre, Lancaster, UK.

Irina S. Bodnar, Institute of Biology of the Komi Science, Center of the Ural Branch of the Russian Academy of Sciences, Syktyvkar, Russia.

Andrea Bonisoli-Alquati, Department of Biological Sciences, California State Polytechnic University, Pomona, USA.

François Bréchignac, Institute for Radioprotection and Nuclear Safety (IRSN) & International Union of Radioecology (IUR), Fontenay aux Roses, France.

Soo Hyun Byun, Department of Physics and Astronomy, McMaster University, Hamilton, Canada.

Georgetta Cannon, Radiation Combined Injury Program, Scientific Research Department, Armed Forces Radiobiology Research Institute, Uniformed Services University of the Health Sciences, Bethesda, USA.

Evgenia V. Cheban, Institute of Biology of the Komi Science, Center of the Ural Branch of the Russian Academy of Sciences, Syktyvkar, Russia.

Jason Cohen, Department of Biology; Department of Physics and Astronomy, McMaster University, Hamilton, Canada.

Yuri E. Dubrova, Department of Genetics and Genome Biology, University of Leicester, UK.

Chandula Fernando, Radiation Sciences Graduate Program, McMaster University, Hamilton, Canada.

Yohei Fujishima, Department of Radiation Biology, Tohoku University School of Medicine, Sendai, Japan.

Rohit Gautam, School of Environmental Sciences, Jawaharlal Nehru University, New Delhi, India.

Stanislav A. Geras'kin, Russian Institute of Radiology and Agroecology, Obninsk, Russia.

Tanhaji Sandu Ghodke, Centre for Applications of Radioisotopes and Radiation Technology (CARRT); Department of Applied Zoology, Mangalore University, Mangalore, India.

Valerie Swee Ting Goh, Department of Bioscience and Laboratory Medicine, Hirosaki University Graduate School of Health Sciences, Japan.

Tanya Helena Hevrøy, Norwegian Radiation and Nuclear Safety Auhtority, Norway.

Kathryn A. Higley, School of Nuclear Science and Engineering, Oregon State University, Corvallis, USA.

Nele Horemans, Belgian Nuclear Research Centre (SCK CEN), Belgium.

Noor Jehan, Department of Environmental Sciences, University of Peshawar, Pakistan.

Awadhesh N. Jha, School of Biological and Marine Sciences, University of Plymouth, UK.

Lawrence A. Kapustka, LK Consultancy, Turner Valley, Alberta, Canada.

Kosuke Kasai, Department of Bioscience and Laboratory Medicine, Hirosaki University Graduate School of Health Sciences, Japan.

Sardar Khan, Department of Environmental Sciences, University of Peshawar, Pakistan.

Nimat Ullah Khattak, Department of Environmental Sciences, University of Peshawar, Pakistan.

Juliann G. Kiang, Radiation Combined Injury Program, Scientific Research Department, Armed Forces Radiobiology Research Institute; Department of Pharmacology and Molecular Therapeutics; Department of Medicine, Uniformed Services University of the Health Sciences, Bethesda, USA.

Jigar Lad, Department of Physics and Astronomy, McMaster University, Hamilton, Canada.

Lukasz Laskowskia, Department of Physics and Astronomy, McMaster University, Hamilton, Canada.

Xiaobing Li, Department of Biology, McMaster University, Hamilton, Canada.

Balázs G. Madas, Environmental Physics Department, Centre for Energy Research, Budapest, Hungary.

Bruno F. E. Matarèse, Department of Haematology; Department of Physics, University of Cambridge, UK.

Tomisato Miura, Department of Radiation Biology, Institute of Radiation Emergency Medicine, Hirosaki University, Japan.

Richard Moccia, Department of Animal Biosciences, University of Guelph, Canada.

Michael N. Moore, School of Biological and Marine Sciences, University of Plymouth, UK; European Centre for Environment and Human Health (ECEHH), University of Exeter Medical School, Truro, UK; Plymouth Marine Laboratory, UK.

Carmel E. Mothersill, Department of Biology, McMaster University, Hamilton, Ontario, Canada.

Sunita Nadella, Department of Biology, McMaster University, Hamilton, Canada.

Akifumi Nakata, Department of Pharmacy, Faculty of Pharmaceutical Science, Hokkaido University of Science, Sapporo, Japan.

JayPrakash Nirala, School of Environmental Sciences, Jawaharlal Nehru University, New Delhi, India.

Deborah H. Oughton, CERAD, Norwegian University of Life Sciences, Aas, Norway.

Jagdish Gopal Paithankar, Division of Environmental Health and Toxicology, Nitte University Centre for Science Education and Research (NUCSER), Nitte (Deemed to be University), Mangalore, India.

Rajashekhar K. Patil, Department of Applied Zoology, Mangalore University, Mangalore, India.

Gibin Powathil, Department of Mathematics, Computational Foundry, Swansea University, Swansea, UK.

Eepsita Priyadarshini, School of Environmental Sciences, Jawaharlal Nehru University, New Delhi, India.

Paulraj Rajamani, School of Environmental Sciences, Jawaharlal Nehru University, New Delhi, India.

C. D. Rollo, Department of Biology, McMaster University, Hamilton, Canada.

Elena I. Sarapultseva, Department of Biotechnology, National Research Nuclear University MEPhI, Moscow, Russian Federation; A. Tsyb Medical Radiological Research Centre, National Medical Research Radiological Center of the Ministry of Health of the Russian Federation, Obninsk, Russian Federation.

Paul N. Schofield, Department of Physiology; Development and Neuroscience, University of Cambridge, Cambridge, UK.

Colin B. Seymour, Department of Biology, McMaster University, Hamilton, Canada.

Huma Shakoor, Department of Environmental Sciences, University of Peshawar, Peshawar, Pakistan.

Xiaopei Shi, Department of Biology, McMaster University, Hamilton, Canada.

Yoshinaka Shimizu, Graduate School of Dentistry, Tohoku University, Sendai, Japan.

Hisashi Shinoda, Graduate School of Dentistry, Tohoku University, Sendai, Japan.

Harleen Singh, Department of Medical Physics and Applied Radiation Sciences, Hamilton, Canada; Buffalo General Hospital, Buffalo, NY, USA.

Richard Smith, Department of Animal Biosciences, University of Guelph, Guelph, Canada.

Marilyne Stuart, Canadian Nuclear Laboratories, Chalk River, Canada.

Kojun Suzuki, Michinoku Fauna Research, Hirosaki, Japan.

Atsushi Takahashi, Tohoku University Hospital, Tohoku University, Sendai, Japan.

Hirofumi Tazoe, Department of International Cooperation and Collaborative Research, Institute of Radiation Emergency Medicine, Hirosaki University, Japan.

Risa Ujiie, Department of Radiation Medical Sciences, Atomic Bomb Disease Institute, Nagasaki University, Japan.

Emily L. Vernon, School of Biological and Marine Sciences, University of Plymouth, UK.

Nguyen T. K. Vo, Department of Biology, McMaster University, Hamilton, Canada.

David Williams, Department of Veterinary Medicine, University of Cambridge, UK.

Michael D. Wood, School of Science, Engineering & Environment, University of Salford, UK.

Masatoshi Yamada, Central Laboratory, Marine Ecology Research Institute, Chiba, Japan.

Hideaki Yamashiro, Graduate School of Science and Technology, Niigata University, Nishiku, Japan.

Mitsuaki A. Yoshida, Department of Radiation Biology, Institute of Radiation Emergency Medicine, Hirosaki University, Japan.

INTRODUCTION

The development and implementation of a universal, evidence-based radiation protection policy is one of the greatest challenges that we face. While different contexts and unique situations make the implementation of a global policy complex, the idea that there might be a commonly accepted set of principles around which regulation, legislation and monitoring might be built has long been accepted as desirable, if not even essential. Recent developments in radiation protection and in radiobiology have started to suggest that the current, anthropocentric approach may underestimate the impact of radiation on the environment. This, through ecosystem perturbations, may prove deleterious to humans as well as other species in a way not captured by assessing human cancer risk as the main endpoint under consideration. Our evolving understanding of other human endpoints such as cardiovascular disease and the impact of multiple stressors such as pesticides and heavy metals have started to broaden our concept of an endpoint for human radiation protection and in doing so start to draw in the importance of the non-radioactive anthropogenic and natural environments. The questions raised by considering non-medical radiation protection as primarily an issue of environmental protection, stimulated a recent discussion and accompanying paper in IJRB (Mothersill et al. 2020) and were the stimulus for this special issue.

Climate change and the recent terrible events in Ukraine, have focused world attention on the vulnerability of the environment and the accompanying risk to human health and safety. Logically this brings the need to reduce dependence on oil and gas into sharp focus and emphasizes the need to work to secure energy supplies which are carbon neutral. Among the alternatives being discussed is the expansion of national nuclear programmes and the proliferation of small modular nuclear reactors (SMR) to serve local needs. This option is particularly favored in remote areas where transportation of energy is difficult and expensive. Implicit in this option is the fact that remote areas are generally environmentally important e.g., wetlands, forests or wilderness where siting an SMR could compromise the ecosystem and bring with it a risk of radiation release into a pristine environment.

Unfortunately while we have considerable knowledge about impacts of medium to high doses of radiation on humans due to its use in radiotherapy and medical diagnostics, we have little information about impacts on non-humans and virtually no information on impacts on ecosystems. In fact we do not even yet know how to assess damage at the ecosystem level.

In this special issue papers cover the resulting environmental and human impacts from the Fukushima TEPCO reactor disaster, Chalk River leakage incident in 2008, and the impact of low level radiation exposure on the marine environment, and invertebrate species. The consequences of low level exposure for children and transgenerational effects as well as those that influence evolution and adaptation are of critical importance if the consequences of exposure might be seen two three or more generations afterwards. This changes completely the concept of when to measure radiation effects and when to set endpoints. Cataractogensis, both in the natural and the industrial environment, also provides instances where the mechanisms of radiation action can be investigated in a defined timeframe. We include several papers on the implications of our understanding, or in many cases the realization of our lack of understanding, for public policy and we look to the future where we can fill in some of these gaps in our knowledge from experiment and epidemiology and come to a better understanding of the risks of ionizing and non-ionizing radiations. Whether these challenges result in a mitigation of our concerns, or a wake-up call to consider the wisdom of increasing our reliance on nuclear energy, the consequences for humanity are profound and it is only be questioning our current wisdom in the light of evidence that we can understand where the sustainable boundaries of our activities on the planet might reasonably lie.

This special issue is therefore critically important at this time, to start the process of documenting impacts of radiation on animals, plants and ecosystems. We hope it will serve as a state of the art collection of papers in this field, and that it will stimulate further research and interest in environmental impacts of radiation as we move to alternative sources of electricity generation.

Knowledge is empowering and without objective and unbiased knowledge accumulation and critical assessment, we cannot make informed decisions about risks and benefits of energy alternatives.

ORCID

Paul Schofield http://orcid.org/0000-0002-5111-7263

Reference

Mothersill CE, Oughton DH, Schofield PN, Abend M, Adam-Guillermin C, Ariyoshi K, Beresford NA, Bonisoli-Alquati A, Cohen J, Dubrova Y, et al. 2020. From tangled banks to toxic bunnies; a reflection on the issues involved in developing an ecosystem approach for environmental radiation protection. Int J Radiat Biol. [2020 Jul 21]:[16 p.]. doi:10.1080/09553002.2020.1793022.

Carmel Mothersill

Paul Schofield

Neglecting the ecosystemic dimension of life hinders efficient environmental protection from radiation and other hazards

François Bréchignac

1. Introduction

During the past decades, and notably after the Chernobyl accident, radioecology lost the vision and skills of its founders who were much inspired and involved in ecology and ecosystem science (Rhodes et al. 2020). In Russia, V. I. Vernadsky started off his carreer by studying radionuclides environmental behavior and their impact on biocenoses, and this led him to conceive the most advanced planetary concepts in ecology. Later, a similar development happened in USA with the Odum brothers who emerged as prominent ecologists by demonstrating the vast potential of using radionuclides as tracers of biogeochemical cycling in ecosystems (Odum 1953, 1962). Similar historical trends can be traced in other nuclearized countries. However, after a while within the radioecology community, consideration of the ecosystem concept, which was central at the very beginning, started to vanish out under the pressure for pragmatic human radioprotection needs arising from the industrial growth of civil nuclear energy (Bréchignac et al. 2008; Rhodes et al. 2020). This is why the recent efforts undertaken by the radioecology community to protect the environment against radiation have largely ignored the ecosystem concept. Instead, it has simply transferred to non human biota the concepts and methods developped for human radioprotection. This oversimplification, rooted within a narrow anthropocentric view dominating radiation protection, urgently needs to be corrected.

Moreover, it must be strengthened that neglecting the ecosystem concept is not an issue restricted to radiation protection alone, but it is a generic problem affecting most current environmental assessments of stressors impact (UNEP-CBD-COP-7-11 2004). Indeed, present methods for environmental/ecological risk assessment are still largely based upon traditional ecotoxicological conceptualizations which rely exclusively on responses to stress of individual organisms. We know today that this is far too much limited to reach an appropriate understanding of critical ongoing issues of environmental concern such as collapsing biodiversity and loss of ecosystem services (Forbes and Calow 2013), which radiation might contribute to.

One paramount issue to realize is that under pressure for change, as elicited by stressors, ecosystem responses are specific to this level of organization and differ from responses of individual organisms. This is because an ecosystem is basically a system of interactions between species populations that are highly symbiotic, interactions featuring numerous feed-backs, promoting self-organization together with new "emergent properties" (Nielsen and Müller 2000; Ponge 2005). As a consequence, understanding overall ecological impacts cannot ignore the ecosystem level processes at work behind these emergent properties.

This paper does not intend to review once more the recent literature on low dose effects on biota and ecosystems which has already been covered recently, especially for what can be learnt from the Fukushima accident (Fukumoto 2020; Strand et al. 2017). Instead, it aims at highlighting how current discrepancies in experimental results and associated controversies, which go far beyond pure radiation issues, could usefully be unraveled when adopting an ecosystemic perspective. The complexity of ecosystems which is to be faced should not be used as a argument not to embark on this research track. Today, in front of systemic environmental situations that prove to be critical (global change, biodiversity decline, widespread pollution …), it is a moral duty for environmental scientists, incuding radioecologists, to more seriously tackle the issue.

2. Justifications for an ecosystem approach in radiation protection

The first recommendation from scientists to design an ecosystem approach for protection of the environment from radiation started to be expressed in concomitance with the early international brainstorming engaged by the International Commission on Radiological Protection (ICRP) at the very beginnng of this century (Bréchignac

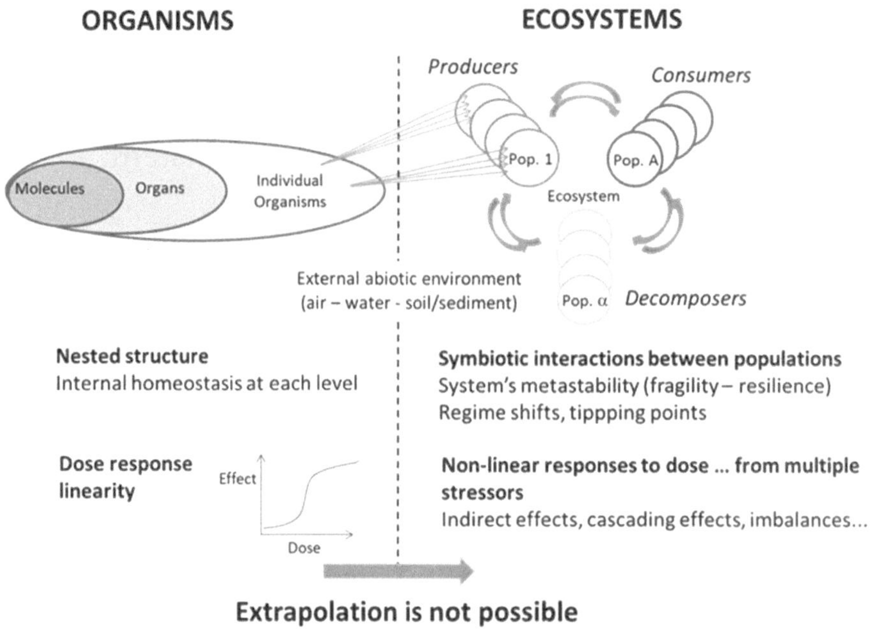

Figure 1. Organisms versus ecosystems: a different structural organization featuring different responses to stressors. As a consequence, extrapolation from organisms to ecosystems is very uncertain, if at all possible.

2003). Next, it has been repeatedly justified and conceptualized, especially under the International Union of Radioecology (IUR) the full board of which elaborated a consensus statement about "protection of the environment in the 21st century: radiological protection of the biosphere including humankind" together with recommendations (Bréchignac et al. 2003).

Since then, a continuous effort of reflexion evolved up to now through dedicated task groups, seminars and consensus conferences and further developed the issue (Bréchignac et al. 2003; Bréchignac and Doi 2009; IUR 2012; Bradshaw et al. 2014; Bréchignac 2017; Bréchignac et al. 2016; Kapustka and Bréchignac 2016; Mothersill et al. 2018, 2019, 2020; Rhodes et al. 2020) especially widening the perspective toward neighbor scientific communities called to contribute. Meanwhile, it has gained continuously growing interest with the recognition that major environmental issues today, such as climate change, biodiversity decline and loss of ecosystem services, are all ecosystem-centered.

A brief review of the main justifications for designing an ecosystem approach in radiation protection of the environment is synthesized in the following.

2.1. The current methodology does not match protection goals

The first justification to found environmental protection upon an ecosystem approach is arising from a mismatch between the assigned protection goals - which target essentially at population and ecosystem level - and the methodology proposed to achieve them - which is exclusively based upon radiotoxicological data obtained at individual organism level (IUR 2012). This initial methodology currently recommended at international level has been designed by the International Committee on Radiological Protection (ICRP) along its 12 Reference Animals and Plants (RAPs) conceptualization (ICRP 2003, 2007, 2008) following on previous IUR recommendations (IUR 2002). Being exclusively based upon radiation effects data obtained at individual organism level, the RAPs method is therefore hardly applicable to truly demonstrate protection of populations and ecosystems of a large variety of species. Strictly speaking, the only situation where protection of individual organisms would be of pertinence would be when the last individual members of an endangered species are at risk of eradication. This is far from the overall objective of environmental protection, and one must acknowledge that when species become endangered, this is precisely because the system of protection has failed.

Extrapolation from organisms to ecosystems, a strategy commonly used by regulating bodies when the scientific knowldege is missing, and which has often been proposed to solve this mismatch, will not work without promoting very large uncertainties, and finally a much questionable evaluation of risk. Here, the widely used linear representation of the scale of biological organization is higly misleading. It erronneously suggests that extrapolation can be used all along the scale. In fact, there is a major discontinuity in the structural and functional organization of life between the levels of individuals and populations (Figure 1). This discontinuity actually translates in different processes of response to impacting perturbations (Bréchignac 2017). In such a context, relying on extrapolation without knowing more about the processes at work at the ecosystem level would be totally blind. Altogether, this calls to stimulate stronger focus on the ecosystem concept.

What is suggested here is not to go back to old experimental strategies based upon irradiator studies in the field of the previous century, but instead to use model ecosystems in micro- and mescosms allowing to link organism and ecosystem level effects and understand how these combine to yield overall ecological impacts (Haanes et al. 2020; Hevrøy et al. 2019).

2.2. Diverging effects when observed in situ or in laboratory controlled conditions

The ICRP proposed RAPs methodology is not only relying exclusively on dose-response of individual organisms, but these data further came to a large extent from laboratory controlled conditions widely different from real nature. In laboratory controlled conditions, organisms live and grow in a reconstructed ideal situation without any limiting factor apart from variation of radiation. This is purposedly very theoretical and artificial as, in nature, populations of organisms are embeded within their ecosystem where they undergo a series of natural limitations, identical to stress, such as food limitation (because other species are competing for it) and predation pressure from predators species, etc…

In addition, species *in situ* are most often exposed to a series of pollutants, promoting additional source of stress. Altogether, this certainly contributes to the divergence of the observed apparent responses: the radiation dose rate to elicit the same effect being about 10 times lower in populations of organisms *in situ* than in laboratory-based organisms (Garnier-Laplace et al. 2013). If similar trends have already been reported (Woodwell and Sparrow 1963), the more recent application of species sensitivity distributions (SSD), would it be correctly used in this context, critically puzzles its use as a baseline for defining regulatory benchmarks, or limit dose rate values, that has been suggested in the context of environmental protection.

Crucially, some of the few studies that used an ecosystem level approach by analyzing leaf litter decomposition (an ecosystem function) also reported divergent conclusions (Mousseau et al. 2014; Bonzom et al. 2016), strengthening the need to better unravel ecosystem level impacts.

2.3. Environmental protection still hindered by anthropocentrism

Human domination over nature has been a leading precept within human history up to today's practice (Messerli et al. 2000). Perhaps, this is best illustrated by the widespread development of intensive agriculture which literally submit species of agronomic interest (animal and plants) to the exclusive service of human beings (essentially to meet their food requirements). In order to favor species of agronomic value, the intensive production methods make large use of fertilizers and pesticides to alleviate competition from other "useless" species (weeds, "pest" animals, insects and fungi, …). We realize today that most environmental compartments - soils, oceans, rivers and lakes and even the atmosphere – are now widespreadly contaminated with pesticides specially designed to kill what we have long considered as "useless" species for humans (not part of the human food chain). We see today that scientists also critically question the role of these production methods in the dramatic biodiversity decline that has been repeatedly reported (Grime 2002; Inger 2015; Ceballos et al. 2017; Hallmann et al. 2017; Ripple et al. 2017). It is like a two shots gun: on a local scale, pesticides are designed to kill all natural species that may outcompete the species of agronomic interest in the cultivated field, but also, on a broad scale, due to their persistence and lack of selectivity, they have largely spread through almost all planetary habitats and consequently expanded their destructive effect on biodiversity at large, even ultimately on *Homo sapiens* himself. Saying it in brief, anthropocentrism has led to the erroneous thinking that humans would be protected by killing large sets of other species of life seen exclusively as competitors. This is wrong because ignoring that other species also interact with the human species on cooperative and symbiotic grounds.

Anthropocentrism has been much influenced by medical approaches and practice where the natural focus is to understand and protect physiological functionalities of organisms at individual level. It is therefore not surprising that when ICRP, an international expert group born in the medical field, started to address radiation protection of the environment, it restricted its understanding to a simple collection of individual organisms of "non-human biota," and addressed the issue exclusively at this level. ICRP therefore proposed a methodology transposed from humans, i.e. fully based upon individual organisms, meant to demonstrate that non-human biota are also protected from radiation (ICRP 2003, 2008). The demonstration however proves to be weak: ignoring the multiple interactions which interlink all species one with another (Morin 2016), including *Homo sapiens*, it cannot address the ecological dimension of the environment, i.e. at population and ecosystem levels. It should be recalled here that the goals assigned to environmental protection are set at population and ecosystem level, rarely at individual organisms level.

One recent and popular issue within the radiation protection community is the objective to design an integrated system of protection that would address man and the environment at the same time, hopefully with a similar methodology and line of reasoning. The idea is indeed attractive as it would allow to simplify greatly how both issues can be regulated in practice. This line of reasoning however is purely methodology-driven as it reduces the environment to individual organisms of non-human biota because this allows for deploying a similar methodology. The environmental objects of protection however are generally not individual organisms, but rather populations forming communities and ecosystems. What is lacking therefore is a global conceptual view describing how the human population, as a species, is interacting with other species populations (including agronomical and wild species) and what are their respective ecological positions and relationships. Such an ecosystem conceptualization gives access to describing how humans and other species are interacting (Figure 2). This is a prerequisite to start with when expecting to develop an integrated system of protection.

Finally, it appears that the initial statement from ICRP which triggered the development of a methodology for environmental protection against radiation – "protecting humans will *de facto* also protect non-human biota" (because humans are more radiosensitive) – is quite

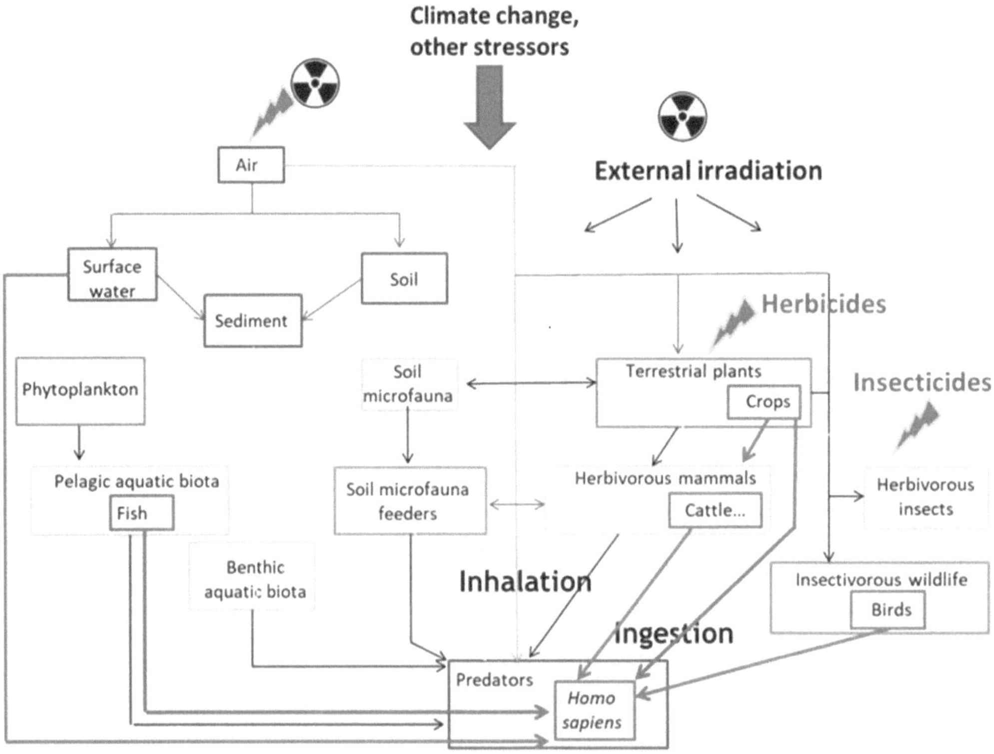

Figure 2. Conceptual model for a scenario of release of radioactivity to the air, integrating the various ecosystem components – abiotic media (air, water, sediment/soil), various functional categories of wildlife (animal and plant) – together with the Homo *sapiens* species and its companion agricultural species (inner boxes inserted within larger categories), in order to describe radionuclides impact on wildlife and man (*Homo sapiens*). Arrows represent physico-chemical transfers of radionuclides from abiotic compartments, and trophic transfers, including those (from inner boxes) which are directed at feeding the *Homo sapiens* species.

symptomatic of a narrow, anthropocentric and ultimately erroneous view. The correct statement would rather be reversed, saying "protecting the environment (i.e. the ecosystem) will make sure that the human population, and others, are not at risk".

2.4. The multipollution concern and the low dose context both call for ecocentrism

A general consensus has emerged with respect to the importance of considering exposure to multiple stressors and their consequences (also called "cocktail effect") (Hunting et al. 2019). This is first because real situations of environmental contamination are rarely single contaminant, but involve multiple concomitant contaminants (radiation, chemical stressors, temperature, etc…). Hence, unraveling effects from radiation might be biased, masked or exacerbated by the presence of other stressors.

But this is also a consequence of the "low dose" very context, where radiation no longer constitutes a dominant signal which can be discriminated out from the background of multiple signals. Since different species exhibit various sensitivities to a given stressor, it is hard to address the issue as it is traditionally done, i.e. at the organism level of one single species. What is on scope here is the overall multispecies ecosystemic response from multiple stressors. Hence, one understands why this calls on the ecosystem concept to be addressed through building more ecocentrism in environmental radiation protection.

3. The ecosystem, the neglected dimension of life

From the radiation protection perspective (above discussion), but also in the broader context of all environmental stressors, consideration of the ecosystem concept appears to be central to any system of environmental protection. Obviously, this is not yet the case although biodiversity impoverishment, climate change, loss of ecosystem services, and ecological impact of pollutions, are all acknowledged to be ecosystem level problems. In order to address these problems more appropriately, we need to upgrade conceptualizations from individual organisms up to ecosystems. Indeed, the ecosystem approach is a concept which first emerged from the international convention on biodiversity (UNEP-CBD-COP-7-11 2004).

More ecosystem science is needed to build appropriate ecological risk assessments. Such assessments would recognize the value of the ecosystem concept by means of which all interacting species including *Homo sapiens* are maintained alive in the long term. The following briefly discusses what are the new ecosystem-related features that ought to be considered when building methods for real ecological risk assessment.

3.1. Ecosystems feature symbiotic interspecies relationships

If biology describes processes of life at molecular up to organisms level whithin a time scale coherent with organisms, ecology expands this knowledge to the larger space and time scales of populations and ecosystems level. One new process of interaction that emerges at the ecosystem level is symbiosis. Since the beginning of multicellular life evolution, living organisms could evolve only as part of ecosystems because their populations all rely upon symbiotic relationships for their long-term maintenance. Ecosystems actually drive the sustainability of life through symbiotic interspecies relationships. In order to be appropriate, concepts for environmental protection and risk assessment in the context of pollutants do require a much better understanding of ecosystem attibutes, processes and functioning, in addition to pure toxicological understanding. Following this reasoning, environment protection, therefore, is an issue of ecology (ecosystems) as much as an issue of biology (organisms), as it is concerned with the maintenance of these high level symbiotic relationships.

Perhaps, the easiest understandable example of collaborative interaction betwen species in ecosystems is the trophic interdependance which promotes bioregeneration: autotrophic photosynthtic species transform inorganic (CO_2, minerals, …) into organic matter (mostly carbohydrates) that is used as a food source by heterotrophic animal species, the autotrophs also regenerate the O_2, required by the heterotrophs respiration, from the CO_2 these later produce whislt oxydising the organic matter that they ingest as food. Thus, the ecosystem features a bioregeneration capacity through an autotrophs-heteroptrophs cycling where the by-products from autotrophs are used by heterotrophs as ressources and vice versa.

Symbiosis is therefore a key feature of ecosystems. The human species is but one of the heterotophs. As such, he cannot escape from this general symbiotic-like mode of functionning and he is not capable to survive disconnected from his ecosystem. This is actually what drove to the concept of "ecosystem services" (ecosystems do provide services to the human species, as well as for the other companion species). Humans still need to breethe O_2 produced by plants, to ingest plant food that supports their physiological needs, and to be cleared by microoganisms from their organic wastes accumulation that would otherwise quickly choke them. Symbiosis is a high level fascinating feature of ecosystems which can be traced back to the first times of life, when monocellular organisms started to collaborate with some ingested bacteria instead of simply digesting them as food, these bacteria being recognized nowadays as the ancestors of chloroplasts and mitochondria of today's cellular life (Margulis and Sagan 1997).

3.2. Ecosystems are self-organised complex systems

Complex system theory explains that any system whose constituents interact in a collaborative way (with many retroaction loops) will, under the pressure from a constraint, react in a way that cannot be simply deduced from the behavior of its constituents. This can be transposed to the life kingdom in saying that if organisms (the system's constituents) linearly respond to the dose of a toxic substance – usually, the effect grows together with the dose – the ecosystem gathering these organisms (i.e. the complex system formed by the constituents assembly), in turn, does no longer respond linearly - the effect does not necessarily increase with dose: large doses may appear to be ineffective (ecological effects are "masked," as per Polikarpov's wise terminology; Polikarpov 1998), or alternatively, small doses may well promote large effects. Complex systems are self-organised, they have an internal organization which automatically increases without being driven by an external source, leading them to show "emergent properties" that differ from those of their constituents (Nielsen and Müller 2000; Quan-Xing 2013).

As compared to organisms, another novelty with ecosystems is therefore the appearance of "ermergent properties" which include the looped circulation of energy and nutrients, or biodiversity and resilience. These appear at ecosystem level, but do not exist at organism level. The differences of behavior at the constituents scale (organisms) versus the system scale (ecosystem) result from the different processes, acting at these two scales when responding to a stressor, which do not obey the same laws. Acknowledging that an ecosystem is a complex system as described above, it is quite illusory to pretend that protecting some organisms against stressors (it will never be possible to protect them all …) will ensure that the ecosystem is protected. Though, most current methodologies for protection of the environment are still based upon this postulate.

Today, we know rather well about the laws of toxicological response of organisms, but we know little about laws of ecosystem reponses submitted to an accumulation of toxicants. We often remain restricted to naturalist descriptive observations. For example, we observe that ecosystems react abruptly and often non-linearly to a constraint accumulation, such as the sudden death of thousands of fishes in Australia last year upon high temperatures (AAS 2019) or within the eutrophication process in aquatic media which is now well understood (Figure 3).

In this last example, the microalgae density (measured as turbidity) abruptly shifts to very high levels ("blooms") when nutrients inputs are passing a threshold. Therory further indicates that complex systems are also characterized by one (or several) "attraction bassin(s)" surrounded by "tipping points". An attraction bassin represents a metastable functioning zone where apparent stability is maintained provided that any perturbation does not reach a threshold, i.e. "the tipping point". Nothing appears to change below the tipping point. This is obviously linked to the concept of ecosystem resilience which allows ecosystems, when they have not accumulated too much stress, to resist against the pressure of change. Alternatively, if the pressure to change passes the threshold, the ecosystem abruptly shifts to a different zone of apparent stability that may be quite different from the initial one.

Such behaviors, dominated by regime shifts through attractors and tipping points, are typical of ecosystems as we observe them today, and they cannot be predicted from the pure toxicological understanding gathered for individul organisms. This is probably very much relevant to the significance of the dramatic biodivesity decline that we are

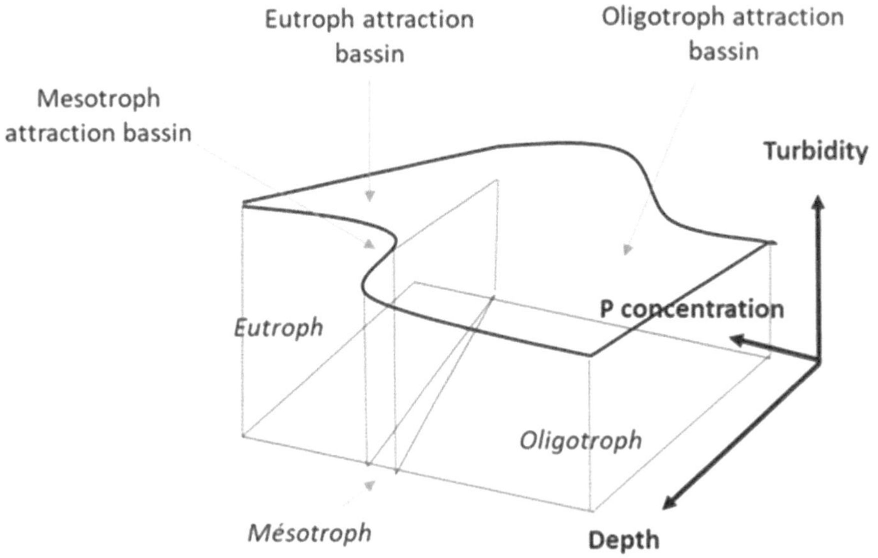

Figure 3. Lake eutrophication process: organization in two alternative resilient domains of stability, pelagic eutroph and bentic oligotroph (Bréchignac 2003, adapted from Kay 2000).

observing today worldwide. Toward which new attraction bassin are we going to? Will it still be compatible with *Homo sapiens* and other species living requirements?

3.3. Ecosystem resilience and ecological risk misinterpretations

There is one further aspect of complexity pertaining to ecosystems: resilience. By virtue of the many species that are inter-related, healthy ecosystems include functional redundancy. This means that several species run approximately the same function (photosynthtic autotrophy, for example) such that if one species is removed, the overall ecosystem stability, and its associated emergent properties, will not be much affected. Stability results from the steady-state dynamic equilibrium that is achieved between the various ecosystem components (i.e. species interacting one with each other). Ecosystem stability is strongly linked to resilience. Ecological resilience provides an ecosystem with the capacity to absorb disturbance and reorganize while undergoing change so as to still retain essentially the same function, structure, identity and feed-backs (Holling 1973; Walker et al. 2004). Only disturbances of sufficient magnitude or duration can profundly affect the ecosystem and may force it to reach a threshold (tipping point) beyond which a different regime of processes and structures predominates. Resilience is one of the emergent properties of ecosystems (Oliver et al. 2015; Truchy et al. 2015).

Ecosystem resilience may therefore mask on-going impact for quite a while, as long as the stressor promoting this impact remains below the threshold/tipping point. Given that, one understands how irresponsible it would be to wait for the manifestation of impacts before acting (this is actually one foundation of the precautionary principle). Though, most regulators usually follow a rule stipulating that the need to regulate is to be restricted to situations of observed and demonstrated impact detriment. Waiting for observed ecological detriment before acting is a major mistake because of the strong protracted effects at ecosystem level that are a manifestation of their resilience. This is a looser strategy, always acting when it is too late for alleviating the detriment… We unfortunately have learnt this in practice from too long inaction in the face of climate change. The level of resilience of an ecosystem would be an extremely pertinent ecosystem level endpoint to support ecological risk assessment.

There are quite interesting consequences to hypothezise from resilience in the context of an ecosystem exposure to low doses of a stressor. At organism level of one single species, low doses will promote small physiological/genetic effects which are small, subtle, non lethal, and hard to discriminate from other stressors effects. But the same low doses would be capable of triggering perturbation at multispecies ecosystem level, especially when the ecosystem is poorly resilient (as would result from permanent long-term exposure to low doses of contaminants). Here, the perturbation would not be much mediated by dose itself (which is small and producing only subtle effects in organisms) but driven instead by differences in sensitivity to toxicants among interacting species which would alter the previously established interaction balance and overall stability.

At low doses, the biological responses in individuals constitute a triggering mechanism for ecosystem-level effects which are not proportional to the dose, puzzling the assertion: "the dose makes the poison". Indeed, this is a popular statement shared among toxicologists working at organism level. But at ecosystem level, the poison would rather depend on a combination of the accumulated dose together with the receptor system's configuration and associated level of resilience. An ecosystem having a much degraded level of resilience could be easily pushed out from its domain of stability, and therefore exhibit a large effect response, by a small further accumulation of dose.

4. Implications for ecological risk assessment

Adopting an ecosystem approach in addition to pure toxicological knowledge is a prerequisite if one wants to reach a real and pertinent capacity of performing ecological risk assessment, i.e. at population and ecosystem level. This leads to several new considerations which also point at novel paradigms to be grasped.

Tackling assessment of ecosystem condition in relation to stressors requires the construction of appropriate ecosystem-level endpoints (possibly integrating as well population and community level ones) in order to capture ecosystem features such as biodiversity, network connectivity and/or number of nodes, resilience, stability, and threshold tipping points. Significant progresses of knowledge on these features are necessary, like exploiting theoretical models associated to experiments on simplified ecosystem models (based on microcosms and mesocosms), and ultimately testing *in situ* the functioning laws they shoud allow to hypothesize. For example, both top-down and bottom-up mechanisms have been shown to control ecosystem functioning (Schuldt et al. 2017). Recent efforts along these lines have now started (Haanes et al. in press; Hevrøy et al. 2019) and it is worth mentioning the temptative application of ecological network analysis to radioactively and chemically contaminated ecosystems which appears very promising (Golz, personnal communication).

Tackling assessment of ecosystem condition in relation to stressors requires to integrate toxicological knowledge together with ecosystem understanding, such as to be able to derive how ecosystem structure and functions can be best protected from degradation triggered by stressors. One example that can be mentioned is the need to reconsider the concept of dose as promoting detriment to an ecosystem. The concept of dose has been conceived for analytical mechanistic understanding of stressor interaction at moleclar and cellular level, in view of deriving dose-response curves for organisms. Such dose-response curves have founded the methodology of protection of organisms by exploting them to derive threshold values below which risk is considered negligible (NOEC, LD_{10}, etc…). But given the larger space and time scales of ecosystems, and the uneven repartition of stressors across their constituents, an average dose is probably meaningless and the principle of linearity between dose and reponse becomes also much questionable. A new paradigm needs to be constructed acknowledging that if stressor dose will promote effects in cells and organisms, this will not be the only cause responsible for the magnitude of the ecosystem response, but a further combination with other system level endpoints to be considered in addition.

Tackling assessment of ecosystem condition in relation to stressors is an integrated issue which no longer can be addressed stressor by stressor, species by species, or discipline by discipline. In order to design an effective environment protection system that is protective enough of ecosystems, the methodology should evolve by considering all potential stressors at a time, calling therefore for a wide integrated collaboration to design a generic methodology (and not several methodologies for each kind of stressor). This is quite a challenge because this means disrupting long established hierarchies and habits within the scientific community. Current borders existing between disciplines according to specific stressors (chemical, physical, radiation, etc…) should be abolished to form one single multifacet and target-oriented community dealing with ecosystem reaction upon all kinds of stressors. With this view in mind, it would be much desirable to set up an international and independent advisory committee of experts to construct common principles, and orient international research work in accordance.

5. Conclusion

Finally, it is worthwile pinpointing that most ecological risk assessment methodologies currently in use are obsolete, and not fulfiling, or only partially, the goals assigned to them (Kapustka 2008; Tannenbaum 2010). For example, with the today's methodologies of ecological risk assessment, we have failed to predict the dramatic decline of biodiversity observed today. Furthermore, the media regularly report on late discoveries of toxic effects from pesticides or pharmacological substances once they have prompted large health diseases and associated medical costs. In the one hand, many pollutants derive more or less directly from very profitable industrial activities, which have not considered environmental impact during a long time. On the other hand, there is now a pressing demand from society to live in a clean environment, further calling for up-to-date regulations designed to make sure that this is indeed the case. So far, because of this antagonism, the course to profit has generally, and quite unhappily, outcompeted the goal of maintaining a clean environment and healthy ecosystems. This explains why efforts dedicated to develop up-to-date regulations of these toxicants prove to have rarely been sufficiently supported. The negative evidences which we are now facing should drive us to rebalancing the situation. Reluctances from the concerned scientific communities to adopt ecosystem approaches urgently need to be overcome. If changing paradigms have always proven to be hard and slow, requesting to spend much energy to fight conservatisms, it is now a scientific duty to move forward.

Acknowledgements

The author warmly thanks many worldwide scientists for their continuous support through many years of collaborative brainstorming, and especially during two decades of work conducted by the IUR "Ecosystem approach" task group that has forged the emergence of many findings expressed in this paper.

Disclosure statement

No potential conflict of interest was reported by the author(s).

References

Australian Academy of Sciences. 2019. Investigation of the causes of mass fish kills in the Menindee region NSW over the summer 2019–2019. Canberra; p. 162.

Bonzom J-M, Hättenschwiler S, Lecomte-Pradines C, Chauvet E, Gaschak S, Beaugelin-Seiller K, Della-Vedova C, Dubourg N, Maksimenko A, Garnier-Laplace J, et al. 2016. Effects of radionuclide contamination on leaf litter decomposition in the Cher nobyl exclusion zone. Sci Total Environ. 562:596–603.

Bradshaw C, Kapustka L, Barnthouse L, Brown J, Ciffroy P, Forbes V, Geras'kin S, Kautsky U, Bréchignac F. 2014. Using an ecosystem approach to complement protection schemes based on organism-level endpoints. J Environ Radioact. 136:98–104.

Bréchignac F. 2003. Protection of the environment: how to position radioprotection in an ecological risk assessment perspective. Sci Total Environ. 307:37–54.

Bréchignac F. 2017. Assessing ecological risk from radiation requires an ecosystem approach. In: Korogodina VL, Mothersill CE, Inge-Vechtomov SG, Seymour CB, editors. Genetics, evolution and radiation. Cham, Switzeland: Springer; p. 207–223.

Bréchignac F, Oughton D, Mays C, Barnthouse L, Beasley JC, Bonisoli-Alquati A, Bradshaw C, Brown J, Dray S, Geras'kin S, et al. 2016. Addressing ecological effects of radiation on populations and ecosystems to improve protection of the environment against radiation: agreed statements from a Consensus Symposium. J Environ Radioact. 158–159:21–29.

Bréchignac F, Alexakhin R, Godoy JM, Oughton D, Sheppard S, Strand P. 2008. Integrating environment protection, a new challenge: strategy of the International Union of Radioecology. Radioprotection. 43(3):339–356.

Brèchignac F, Doi M. 2009. Challenging the current strategy of radiological protection of the environment: arguments for an ecosystem approach. J Environ Radioact. 100(12):1125–1134.

Bréchignac F, Polikarpov G, Oughton DH, Hunter G, Alexakhin R, Zhu YG, Hilton J, Strand P, International Union of Radioecology Board of Council. 2003. Protection of the environment in the 21st century: radiation protection of the biosphere including humankind. Statement of the International Union of Radioecology. J Environ Radioact. 70(3):155–159.

Ceballos G, Ehrlich PR, Dirzo R. 2017. Biological annihilation via the ongoing sixth mass extinction signaled by vertebrate population losses and declines. Proc Natl Acad Sci USA. 114(30):E6089–E6096.

Forbes VE, Calow P. 2013. Use of the ecosystem services concept in ecological risk assessment of chemicals. Integr Environ Assess Manag. 9(2):269–275.

Fukumoto M, editor. 2020. Low-dose radiation effects on animals and ecosystems; long-term study on the Fukushima nuclear accident. Gateway East, Singapore: Springer.

Garnier-Laplace J, Geras'kin S, Della-Vedova C, Beaugelin-Seiller K, Hinton TG, Real A, Oudalova A. 2013. Are radiosensitivity data derived from natural field conditions consistent with data from controlled exposures? A case study of Chernobyl wildlife chronically exposed to low dose rates. J Environ Radioact. 121:12–21.

Grime JP. 2002. Declining plant diversity: empty niches or functional shifts? J Veg Sci. 13(4):457–460.

Haanes H, Hansen EL, Hevrøy TL, Kiel Jensen L, Gjelsvik R, Jaworska A, Bradshaw C. 2020. Realism and usefulness of multispecies experiment designs with regard to application in radioecology: a review. Sci Total Environ. 718:134485.

Hallmann CA, Sorg M, Jongejans E, Siepel H, Hofland N, Schwan H, Stenmans W, Müller A, Sumser H, Hörren T, et al. 2017. More than 75 percent decline over 27 years in total flying insect biomass in protected areas. PLoS One. 12(10):e0185809.

Hevrøy TH, Golz A-L, Hansen EL, Xie L, Bradshaw C. 2019. Radiation effects and ecological processes in a freshwater microcosm. J Environ Radioact. 203:71–83.

Holling CS. 1973. Resilience and stability of ecological systems. Annu Rev Ecol Syst. 4(1):1–23.

Hunting ER, Bundschuh M, Schrama M. 2019. Editorial: multiple stressors across ecosystem boundaries. Front Environ Sci. 7:98.

ICRP. 2003. A Framework for assessing the impact of ionizing radiation on non-human species. Ann ICRP. 33(3):266.

ICRP. 2007. The 2007 recommendations of the International Commisssion on Radiological Protection. Ann ICRP. 37(2–4):332.

ICRP. 2008. Environmental protection: the concept and use of reference animals and plants. Ann ICRP. 38(4–6):242.

Inger. 2015. Common European birds are declining rapidly while less abundant species' numbers are rising. Ecol Lett. 18:28–36.

IUR. 2002. Protection of the environment: current status and future work. IUR Rep. 3:23. (www.iur-uir.org)

IUR. 2012. Towards an ecosystem approach for environment protection with emphasis on radiological hazards. IUR Rep. 7:82. (www.iur-uir.org)

Kapustka LA. 2008. Limitations of the current practices used to perform ecological risk assessment. Integr Environ Assess Manag. 4(3): 290–298.

Kapustka L, Bréchignac F. 2016. Advancing research and policy in the field of environmental radiation safety. Integr Environ Assess Manag. 12(4):611–612.

Kay J. 2000. Ecosystems as self-organizing holarchic open systems: narratives and the second law of thermodynamics. In: Jorgensen SE, Muller F, editors. Handbook of ecosystems theories and management. London: CRC Press - Lewis Publishers; p. 135–160.

Margulis L, Sagan D. 1997. Microcosmos: four billion years of microbial evolution from our microbial ancestors. Berkeley (CA): University of California Press; p. 304.

Messerli B, Grosjean M, Hofer T, Núñez L, Pfister C. 2000. From nature-dominated to human-dominated environmental changes. Quat Sci Rev. 19(1–5):459–479.

Morin E. 2016. Penser global: l'homme et son univers. Paris: Flammarion éd; p. 144.

Mothersill C, Abend M, Bréchignac F, Iliakis G, Impens N, Kadhim M, Møller AP, Oughton D, Powathil G, Saenen E, et al. 2018. When a duck is not a duck; a new interdisciplinary synthesis for environmental radiation protection. Environ Res. 162:318–324.

Mothersill C, Abend M, Bréchignac F, Copplestone D, Geras'kin S, Goodman J, Horemans N, Jeggo P, McBride W, Mousseau TA, et al. 2019. The tubercular badger and the uncertain curve: the need for a multiple stressor approach in environmental radiation protection. Environ Res. 168:130–140.

Mothersill C, Oughton DH, Schofield PN, Abend M, Ariyoshi K, Beresford NA, Bonisoli-Alquati A, Dubrova Y, Geras'kin SA, Hevrøy TH, et al 2020. From tangled webs to toxic bunnies; a reflection on the issues involved in developing an ecosystem approach for environmental radiation protection. Int J Rad Biol.

Mousseau TA, Milinevsky G, Kenney-Hunt J, Møller AP. 2014. Highly reduced mass loss rates and increased litter layer in radioactively contaminated areas. Oecologia. 175(1):429–437.

Nielsen SN, Müller F. 2000. Emergent properties of ecosystems. In: Joergensen SE, Müller F, editors. Handbook of ecosystem theories and management. Boca Raton (FL): Lewis Publishers; p. 195–216.

Odum EP. 1953. Fundamentals of ecology. Saunders WB, editor. Philadelphia (PA): W. B. Saunders Company.

Odum EP. 1962. Relationships between structure and function in the ecosystem. Jpn J Ecol. 12:108–118.

Oliver TH, Heard MS, Isaac NJB, Roy DB, Procter D, Eigenbrod F, Freckleton R, Hector A, Orme CDL, Petchey OL, et al. 2015. Biodiversity and resilience of ecosystem functions. Trends Ecol Evol (Amst). 30(11):673–684.

Polikarpov G. 1998. Conceptual model of responses of organisms, populations and ecosystems to all possible dose rates of ionising radiation in the environment. Radiat Prot Dosimetry. 75(1):181–185.

Ponge J-F. 2005. Emergent properties from organisms to ecosystems: towards a realistic approach. Biol Rev Camb Philos Soc. 80(3): 403–411.

Quan-Xing L. 2013. The emergent properties of spatial self-organization: a study of patterned mussels beds. Gröningen: Gröningen University; p. 121–128.

Rhodes OE, Bréchignac F, Bradshaw C, Hinton TG, Mothersill C, Arnone JA, Aubrey DP, Barnthouse LW, Beasley JC, Bonisoli-Alquati A, et al. 2020. Integration of ecosystem science into radioecology: a consensus perspective. Sci Total Environ. 740:140031.

Ripple WJ, Wolf C, Newsome TM, Galetti M, Alamgir M, Crist E, Mahmoud MI, Laurance WF, 15,364 scientist signatories from 184 countries. 2017. World scientists' warning to humanity: a second notice. BioScience. 67(12):1026–1028.

Schuldt A, Bruelheide H, Buscot F, Assmann T, Erfmeier A, Klein A-M, Ma K, Scholten T, Staab M, Wirth C, et al. 2017. Belowground top-down and aboveground bottom-up effects structure multitrophic community relationships in a biodiverse forest. Sci Rep. 7(1):4222.

Strand P, Sundell-Bergman S, Brown JE, Dowdall M. 2017. On the divergences in assessment of environmental impacts from ionising radiation following the Fukushima accident. J Environ Radioact. 169–170:159–173.

Tannenbaum LV. 2010. Moving beyond obsolete ecological risk assessments. Environ Sci Technol. 44(9):3211–3212.

Truchy A, Angeler DG, Sponseller RA, Johnson RK, McKie BG. 2015. Linking biodiversity, ecosystem functioning and services, and ecological resilience: towards an integrative framework for improved management. Adv Ecol Res. 53:55–96.

UNEP-CBD-COP-7-11 2004. Convention on Biological Diversity, COP 7 Decision VII/11, Ecosystem approach. Kuala Lumpur, Malaysia, 9–20 February.

Walker B, Holling CS, Carpenter SR, Kinzig A. 2004. Resilience, adaptability and transformability in social-ecological systems. E&S. 9(2):5.

Woodwell GM, Sparrow AH. 1963. Predicted and observed affects of chronic gamma irradiation on a near-climax forest ecosystem. Radiat Bot. 3(3):231–237.

What makes radiation protection so challenging?

Lawrence A. Kapustka

ABSTRACT

This commentary explores the challenges in radiation safety that derives from the inherent complexity of social–ecological systems. The framework needed to address the challenges acknowledges the characteristics of wicked problems in this era of postnormal science. My objective for this piece is to summarize relevant characteristics of social–ecological systems that underscore the importance, even the necessity, of adopting a holistic approach to radiation safety. This work builds on several publications that have come out of the ecosystems approach working group of the International Union of Radioecology. The nature of wicked problems is that they require meaningful engagement among diverse groups of affected stakeholders so that negotiated consensus regarding assessment and management for radiation safety can be achieved. I conclude by stating that this approach is complementary to the reference animal and plant approach, that it is consistent with the views for postnormal science, and it conforms with the United Nations Sustainable Development Goals (SDGS) that were adopted in 2015.

A work group of the International Union of Radioecology analyzed the reference animal and plant (RAP) approach to radiation safety and recommended using a complementary ecosystem approach to the improve the relevance and alignment with the agreed protection goals (Brechignac et al. 2011). Subsequent publications of workshop products have elaborated the benefits and challenges of using and ecosystem approach (Mothersill et al. 2020; Rhodes et al. 2020; Bradshaw et al. 2014; Kapustka and Brechignac 2016; Stark et al. 2017). Despite what seem to be obvious benefits to shift to an ecosystem approach widespread acceptance remains elusive.

There are underlying reasons for the slow pace of adopting an ecosystem approach broadly. Central among them are engrained views of ecology that have not kept pace with the advances in science over the past few decades. For many years, biology taught in schools and universities introduced the concept of organizational hierarchy, that is the progressive complexity building from atoms to molecules to cells and onward up a linear scale ending with the biosphere. The implication, often stated explicitly, was that these building blocks were neatly added to yield the next level up. Missing from this linear progression was the understanding of systems, specifically that emergent properties are revealed at different levels. In a colloquial sense, this can be stated that the system is more than the sum of its parts.

Reductionism served as a useful approach to take in search of certain mechanisms of action, how the different parts of an organism function. The basis of experimentation is to examine the role that one part has by comparing directly in controlled settings how something works with versus without a part. Much of what we know about how the world works was derived from reductionist approaches. But there are limitations to what reductionism can reveal (Mazzocchi 2011).

The basis for this limitation is found in part by understanding emergent properties. We know for example that ecological systems are highly dimensional (that is, with connectivity spanning wide spatial and temporal scales) and self-organizing; that dynamic collections of organisms and abiotic structures potentially interact with a multitude of responses modulated by positive (or reinforcing) and negative (or dampening) feedback loops. Matthews et al. (1996) detailed five features of ecological systems in what they called community conditioning: (1) systems are complex, multidimensional, and dynamic; (2) true equilibrium is never attained; (3) historical events determine current and future structures; (4) past conditions cannot be repeated; and (5) forecasting future states of systems is tenuous at best.

From a landscape perspective, we know that temporal and spatial scales are paramount. Organisms interact with their environment across different dimensions spanning several orders of magnitude. For example, microorganisms may inhabit a few cubic centimeters of soil and have a life span measured in minutes to days, whereas mammals traverse a few meters to a 100 square kilometers and live for days to many decades. Migratory animals range across continents even hemispheres. When conducting an environmental assessment or incorporating a management plan, the relevant scales need to be matched with the life histories of the valued ecological receptors.

Clearly, the complexity observed in ecological systems should call into question any attempt to manage radiation safety relying solely on generalized responses of a limited suite of species types. In recent years, species sensitivity distributions (Posthuma et al. 2001), adverse outcome pathways (Ankley et al. 2010), and various 'omics (Buesen et al. 2017) have been offered as potential breakthroughs to improve our collective responses to environmental stressors. Each suffers deficiencies in that they are organism-based and lack incorporation of emergent properties known to operate at population, community, and system levels.

The goals of radiation protection are couched in the concepts of sustainability and resilience of ecological systems (Brechignac et al. 2011; Bradshaw et al. 2014; Bréchignac et al. 2016; and references therein). Sustainability science builds from foundational work in Systems Ecology and calls for a postnormal science that is based on the recognition that societies and ecologies are fundamentally nonlinear in their behavior due in large part to multiple nested feedback loops (some reinforcing or positive; some dampening or negative) that operate on different spatial and temporal scales. There is also the recognition that the welfare of societies is ultimately dependent on reasonably constant flows of ecological goods and services. Establishing a framework for sustainability, McCormick et al. (2013) and Kapustka et al. (2013) pose four questions that can be used to engage a meaningful dialogue, namely (1) what do you want to sustain? (2) for how long? (3) who benefits, and (4) who pays.

Biggs et al. (2015) identified three ecological parameters and four sociological parameters that they offer as principles for building resilience in social–ecological systems. Those on the ecological side are (1) maintain diversity and redundancy, (2) manage connectivity across landscapes, and (3) manage slow variables and feedback loops. The sociological aspects that are needed are (1) foster an understanding of social–ecological systems as complex and adaptive, (2) encourage learning and experimentation, (3) encourage broad participation, and (4) promote polycentric governance systems.

Radiation protection occurs in and across social–ecological systems. Whether dealing with safe practices for the extraction of radioactive ores, naturally occurring radon gas, routine operation of nuclear facilities, disposal of radioactive waste (energy plants or medical supplies), or coping with accidents, there are sociological and ecological considerations. These of course vary markedly in different biomes and in different economic, cultural, and ethnic settings. This complexity further underscores the limitation of the RAP species that were selected to represent northern temperate climes.

The concept of wicked problems originated in the 1960s and was described in a publication by Rittel and Webber (1973). Although initially conceived to explain the difficulties of social planning and the policy arena (Table 1), it is applicable to most environmental challenges including radiation safety, in large part because these challenges interface multiple publics and diverse ecological systems.

Table 1. Characteristics of wicked problems (adapted from Rittel and Webber 1973).

- Those that cannot be defined so all agree on the problem to solve
- Require complex judgment about the level of abstraction at which to define the problem
- Have no clear stopping rules
- Have no right/wrong answer; just better/worse conditions
- Have no objective measure of success
- Require iteration—every trial counts
- Have no given alternative solutions—these must be discovered
- Often have strong moral, political, or professional dimensions

Rittel and Webber (1973) began by stating that wicked problems are those that cannot be defined in a way that all affected stakeholders can agree on what is to be solved. They further state that complex judgment regarding the level of abstraction to be applied to the situation is needed. Perhaps most unsettling to applied scientists and engineers is that there are no clear stopping rules; there is no equation to solve in a way that one knows they have achieved the answer. Moreover, there are no right or wrong answers because that which works for one group of stakeholders is often unacceptable to another group. A corollary to this is that there is no objective measure of success. In working toward an acceptable solution every attempt or trial counts and provides useful information to be considered by the stakeholders. Each situation is unique and as such solutions must be discovered anew. Finally, wicked problems, including those associated with radiation safety, have strong moral, political, or professional dimensions.

Most scientists, especially in the applied fields and engineers are predisposed to finding definitive answers to problems they take on. Typically, they pursue problems in a linear fashion moving closer and closer to a workable solution that can be evaluated objectively. Yet as Funtowicz and Ravetz (1994) argued for a shift to postnormal science, no one can claim 'truth' from their results and uncertainty cannot be banished. Due to ethical complexities, inclusion of affected stakeholders is needed. They describe the realm of postnormal science as the situations in which predicting the outcome of complex systems with high uncertainty is tenuous if not impossible, especially over long time horizons. Accommodating the diverse values of stakeholders raises the decision stakes. In combination, uncertainty and decision stakes demand iterative dialogue that is characteristic of wicked problems.

Ionizing radiation comes in different forms. Some is encountered as background dosages. Some has been elevated through mining, processing, use in nuclear power facilities, in disposal of waste (medical and energy), and accidents. Each of these present varying scenarios ranging from solely human exposure concerns to complex environmental settings involving humans and various ecological receptors. The simplistic standard human and RAP approaches work for the simplest scenarios in that the level of effort that goes into an assessment matches the magnitude of the issue—it is reasonable and acceptable to err on the side of protectiveness. But as the scenarios become more complex more effort is needed to match the magnitude of the challenge. Moreover, if there is markedly excess protectionism, there can be adverse consequences in terms of opportunity costs that effect societies.

Radiation concerns are associated with legacy uranium mining sites that exist in Canada, Australia, Niger, Russia, Kazakhstan, Namibia as well as with coal mines and metal mining operations throughout the world. Think for a moment about the diversity in human societies in these regions and layer on top of that the biodiversity in these varied ecoregions. Any framework for assessing the risks to humans and ecological receptors across such diverse settings clearly requires flexibility to take into account how these systems work. The values expressed by the Indigenous Peoples, nonindigenous residents and workers, industries, and governments present unique challenges across each of these regions. Clearly, what works for the communities in the boreal forest region of Saskatchewan, Canada will not transfer to northern Australia or Namibia. Each locale requires that the affected stakeholders have opportunities to express their values, their needs, their expectations; some will express strong anthropocentric values, others will champion biocentric views. And this is squarely within the realm of wicked problems.

Similarly, if we think about radon exposures in urban settings the challenges and potential mitigation are different across the spectrum of poor to affluent communities. Those in affluent communities can readily modify structures to prevent accumulation of radon in their homes, but the poor are likely to lack the resources to do so. Moreover, we know that added stress associated with being poor (increased hypertension, inadequate nutrition, and other health issues) increases sensitivity to environmental stressors (Jensen et al. 2017). Managing risks associated with radon becomes a wicked problem.

Finally, the complexity of social–ecological systems that contain nuclear reactors and research facilities from the Cold War era present the greatest challenges. Contrast the differences between the Chernobyl Exclusion Zone and the Prefectures around Fukushima in terms of human cultural norms and ecological systems. Contrast the social–ecological system differences among the sites at Savannah River, Georgia, Fernald, Ohio, Hanford, Washington, in the United States and Mayak, Russia. Tick through the characteristics of wicked problems articulated by Rittel and Webber (1973) and it should be obvious that these sites individually and collectively indeed pose wicked problems.

This analysis of how we address radiation safety is not intended to dismiss the powerful investigative tools that are used to refine our ability to measure dose or dose rate and to interpret the effects at cellular, organismal, or population endpoints. Nor is it dismissing the important screening-level work that has been done using the RAP approach. Rather, it is a call to expand our perspective that aims for a more holistic assessment of risk and subsequent management of risks, one that honors the varied perspectives of diverse human communities and embraces the complexities of social–ecological systems. This plea is consistent with the views for a postnormal science expressed by Funtowicz and Ravetz (1994). And it conforms with the United Nations Sustainable Development Goals (SDGS) that were adopted in 2015 (United Nations Development Programme 2015).

Acknowledgement

The author thanks Andrej Rusin for his assistance in reviewing and formatting the manuscript prior to submission.

Disclosure statement

The author reports no conflict of interest.

References

Ankley GT, Bennett RS, Erickson RJ, Hoff DJ, Hornung MW, Johnson RD, Mount DR, Nichols JW, Russom CL, Schmieder PK, et al. 2010. Adverse outcome pathways: a conceptual framework to support ecotoxicology research and risk assessment. Environ Toxicol Chem. 29(3):730–741.

Biggs R, Schlüter M, Schoon ML. 2015. Principles for building resilience: sustaining ecosystem services in social-ecological systems. Cambridge (UK): Cambridge University Press.

Bradshaw C, Kapustka L, Barnthouse L, Brown J, Ciffroy P, Forbes V, Geras'kin S, Kautsky U, Bréchignac F. 2014. Using an ecosystem approach to complement protection schemes based on organism-level endpoints. J Environ Radioact. 136:98–104.

Brechignac F, Bradshaw C, Carroll S, Jaworska A, Kapustka L, Monte L, Oughton D. 2011. Recommendations from the International Union of Radioecology to improve guidance on radiation protection. Integr Environ Assess Manag. 7(3):411–413.

Bréchignac F, Oughton D, Mays C, Barnthouse L, Beasley JC, Bonisoli-Alquati A, Bradshaw C, Brown J, Dray S, Geras'kin S, et al. 2016. Addressing ecological effects of radiation on populations and ecosystems to improve protection of the environment against radiation: agreed statements from a Consensus Symposium. J Environ Radioact. 158-159:21–29.

Buesen R, Chorley BN, da Silva Lima B, Daston G, Deferme L, Ebbels T, Gant TW, Goetz A, Greally J, Gribaldo L, et al. 2017. Applying 'omics technologies in chemicals risk assessment: report of an ECETOC workshop. Regul Toxicol Pharmacol. 91(Suppl 1):S3–S13.

Funtowicz SO, Ravetz JR. 1994. The worth of a songbird: ecological economics as a post-normal science. Ecol Econ. 10(3):197–207.

Jensen SKG, Berens AE, Nelson CA. 3rd 2017. Effects of poverty on interacting biological systems underlying child development. Lancet Child Adolesc Heal. 1(3):225–239.

Kapustka L, Brechignac F. 2016. Advancing research and policy in the field of environmental radiation safety. Integr Environ Assess Manag. 12(4):611–612.

Kapustka L, McCormick R, Stahl C. 2013. Informing the sustainability dialogue. Integr Environ Assess Manag. 9(3):355–356.

Matthews RA, Landis WG, Matthews GB. 1996. The community conditioning hypothesis and its application to environmental toxicology. Environ Toxicol Chem. 15(4):597–603.

Mazzocchi F. 2011. The limits of reductionism in biology: what alternatives? E-Logos Electron J Philos. 11:1–19. DOI:10.18267/j.e-logos.301

McCormick R, Kapustka L, Stahl C, Fava J, Lavoie E, Robertson C, Sanderson H, Scott H, Seager T, Vigon B. 2013. Exploring SETAC's roles in the global dialogue on sustainability-an opening debate. Integr Environ Assess Manag. 9(1):7–11.

Mothersill CE, Oughton DH, Schofield PN. 2020. From tangled banks to toxic bunnies; a reflection on the issues involved in developing an Ecosystem approach for environmental radiation protection. Int J Radiat Biol. DOI:10.1080/09553002.2020.1793022

Posthuma L, Suter GW, Traas TP. 2001. Species sensitivity distributions in ecotoxicology. Boca Raton (FL): CRC Press.

Rhodes O, Bryan A, Ward A. 2020. Integration of ecosystem science into radioecology: a consensus perspective. STOTEN. 740. DOI:10.1016/j.scitotenv.2020.140031

Rittel HWJ, Webber MM. 1973. Dilemmas in a general theory of planning. Policy Sci. 4(2):155–169.

Stark K, Gomez-Ros JM, Vives I, Batlle J, Lindbo Hansen E, Beaugelin-Seiller K, Kapustka LA, Wood MD, Bradshaw C, Real A, et al. 2017. Dose assessment in environmental radiological protection: state of the art and perspectives. J Environ Radioact. 175-176:105–114.

United Nations Development Programme. 2015. United Nations Sustainable Development Goals. https://www.undp.org/content/undp/en/home/sustainable-development-goals.html.

Insight into the evolutionary profile of radio-resistance among insects having intrinsically evolved defence against radiation toxicity

Jagdish Gopal Paithankar, Tanhaji Sandu Ghodke, and Rajashekhar K. Patil

ABSTRACT

Ionizing radiation (IR) has wide-ranging applications in various fields. In agriculture, pest control is one of the important applications, because insect pests have become a threat to the global agriculture industry. IR are used routinely to prevent crop loss and to protect stored food commodities. Radio-sterilization and disinfestation treatments are commonly used procedures for insect pest control. From various studies on insect radio-sterilization and disinfestation, it has been established that compared to vertebrates' insects have high levels of radiation resistance. Therefore, to achieve adequate radio-sterilization/disinfestation; exposure to high doses of IR is necessary. However, studies over decades made a presumption that radiation resistance is general among insects. Recent studies have shown that some insect orders are having high IR resistance and some insect orders are sensitive to IR. These studies have clarified that radiation resistance is not uniform throughout insect class. The present review is an attempt to insight at the evolutionary profile of insect species studied for radio-sterilization and disinfestation treatment and are having the trait of radio-resistance. From various studies on insect radiation resistance and after phylogenetic analysis of insect species it appears that the evolutionary near species have drastically different levels of radio-resistance and trait of radiation resistance appears to be independent of insect evolution.

Abbreviations: IR: Ionizing radiation; ROS: Reactive oxygen species; SIT: Sterile insect technique; SMT: Sterile male technique; Gy: Gray

Introduction

Annually insects destroy one-fifth of the agricultural commodities produced. For example, the Diamondback moth (*Plutella xylostella*) species damages crops all over the world and this leads to a loss of 4 to 5 billion US dollars every year (Alphey and Bonsall 2018). To prevent crop loss, the overall cost estimated for controlling insect pests was approximately 55.921 billion US dollars for the year 2012 (Atwood and Paisley-Jones 2017). According to the study, a larger part of this money (consecutively increasing) was spent on the use of 600 million pounds of pesticides (Atwood and Paisley-Jones 2017). Indiscriminate use of chemical compounds is heavily contributing towards environmental toxicity and posing threat to the ecosystem (Lu et al. 2015). Among insect pest control measures, ionizing radiation (IR) appears significant in terms of effectiveness, economics and low environmental impacts (Alphey and Bonsall 2018, Marec and Vreysen 2019). Therefore, several studies have been carried out to control insect populations by using IR. However, these studies reported varying responses of insects against radiation toxicities (Bakri et al. 2005; Hallman and Blackburn 2016).

Radiation

Radiation is a form of an energy emitted by the nuclei of unstable elements. There are two kinds of radiations namely, IR and non-ionizing radiations. IR are high-energy radiation and are capable of generating ions in a biological tissue, whereas non-ionizing radiation are low energy radiation (Maurya and Devasagayam 2011). As, IR are capable of producing ions and induce toxicities in living organisms, therefore IR are of more concern. IR induces toxicities by either imparting its energy to biomolecules or by knocking off electrons from an atom of water molecules, which further lead to the series of cascading reactions. These cascading reactions may lead to alterations in the structure and function of various biomolecules (Casarett 1968). This in turn causes impairments in metabolic functions, can lead to genomic damages and may alter biological responses. Short- and long-term adaptive effects following exposure to IR have proposed many applications in agriculture for producing genetically modified crops (Shin et al. 2011), post-harvest exposure for disinfestation (Hallman and Blackburn 2016), pre-harvest pest management through radio-sterilization (Hallman et al. 2013). All these developments facilitated to develop radiobiology as a new field of research and drew the attention of the scientific community. However, the radiation-induced toxicities in cellular and molecular processes remain less explored until recent past (Paithankar et al. 2018a; Paithankar et al. 2020).

Direct and indirect effects are responsible for IR induced toxicities in a biological system (Figure 1). The direct IR effects include absorption of radiation energy by the bio-molecules directly, resulting in the loss of electron/s (followed by molecular damage). Indirect IR effects are triggered by free radical generation/reactive oxygen species (ROS) through radiolysis of water (Casarett 1968). ROS are the oxygen containing highly reactive species, the high reactivity is due to the presence of lone pair of an electron (superoxide – $O_2^{\bullet -}$ and hydroxyl radicals – $OH^{\bullet}$) (Li et al. 2016). These free radicals mainly formed through the radiolysis of water followed by Fenton reaction. The free radicals are responsible for the greater extent of IR induced toxicities and it is well established that following IR exposure one-third radiation effects are caused by direct radiation absorption and two-third of the effects are caused indirectly through ROS (Michael and O'Neill 2000).

In living organisms, apart from direct and indirect modes of IR-induced toxicities, several other factors need emphasis. These factors decide the fate of an organism following induction of radiation toxicities. Subsequent to the IR exposure, cell activates pro-survival genes and promote DNA damage repair pathways, however, if the damage is irreparable cell activates either of the cell death pathway; mitotic catastrophe, apoptosis, necrosis autophagy, and senescence (Figure 1) (Eriksson and Stigbrand 2010; Sia et al. 2020). Onset of the apoptotic pathway is influenced by several factors including type of IR, dose of IR, exposure time, cell type, cell cycle phase, DNA damage repair capacity and p53 status (Casarett 1968; Eriksson and Stigbrand 2010). Subsequent to radiation exposure, depending on the extent of damage cell facets are decided. Mitotic catastrophe starts in cells that are unable of further replication following IR exposure. Activation of an apoptosis is dependent on the balance between pro- and anti-apoptotic factors. Necrotic pathway reported to be activated by cellular micro-environment, pH, ionic imbalance and energy loss following radiation exposure. Excessive autophagic responses were reported to activate an autophagic cell death pathway. The

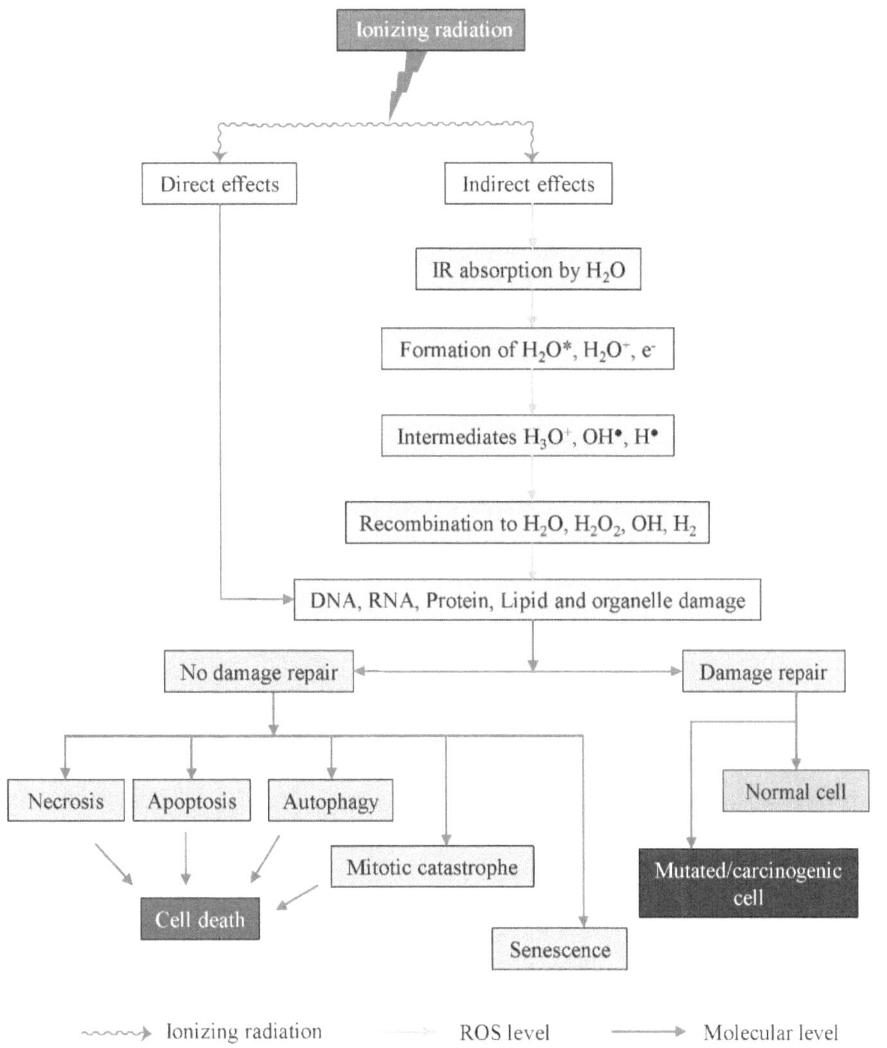

Figure 1. Modes of interaction of ionizing radiation with the biological tissue. In direct effect biomolecule receives energy of ionizing radiation directly whereas, indirect effects include radiolysis of water and formation of the reactive oxygen species that further reacts with biomolecules.

autophagy-related genes are activated following irradiation and they subsequently activate the autophagy pathway (Sia et al. 2020).

As, a larger part of the biological tissues are made-up of water therefore, radiation exposure generates a huge number of ROS. Consequently, antioxidant molecules that are scavengers of free radicals grabbed the attention for several years. Antioxidant molecules are considered as key players in radiation resistance. Because of this, enzymatic antioxidant molecules were at the center of the research for several decades (Sun et al. 1998; Ozden et al. 2011). Only in the last two decades, several studies reported the presence of non-enzymatic antioxidant molecules playing significant role in radiation resistance (Daly et al. 2004; Paithankar et al. 2018a, 2018b, 2020). These studies questioned the very presumption of enzymatic antioxidants are solely responsible for radiation resistance among different organisms.

Radiation resistance

The mechanisms of radiation toxicities, signaling pathways and organisms' responses to IR are yet to unfold completely. Findings about high radiation resistance among microorganisms and some other animals have raised many questions. It is indeed a challenge in cell biology to know, how some cells of some organisms can survive at high doses of radiation. Based on responses of living organisms to IR, they can be categorized into two groups; one is radiosensitive and another is radio-resistant. Among living organisms, the prokaryotes reported to have distinctly high radiation resistance compared to eukaryotes (Cox and Battista 2005; Daly 2009; Krisko and Radman 2013, Ranawat and Rawat 2017). Microorganisms are known for their extraordinary capacity to withstand high doses of IR. *Deinococcus radiodurans* and *Deinococcus geothermalis*, were found to survive up to 12,000 Gy of gamma radiation, which places this genus at top in the list of radio-resistant organisms (Daly 2012, Ranawat and Rawat 2017). Among vertebrates, lower animals are relatively more radio-resistant than higher animals. For instance, mice and bats have higher radiation resistance than humans (Harrison and Anderson1996; Bakri et al. 2005). Among eukaryotes, the invertebrate animals showed considerably higher radiation resistance than vertebrate animals. Tardigrade (water bears or moss piglets) a water dwelling animal found in extreme environmental conditions was reported to have significantly high radiation resistance (Jonsson et al. 2008). Other invertebrate group known for high radiation resistance is insect. Insects are reported to have high radiation resistance compared to vertebrates (Paithankar et al. 2017). Insects are widely studied for radiation effects and for inducing reproductive sterilization and disinfestation. Comprehensively, radio-biological studies in insects were carried out to control insect populations, aiming to reduce harmful impact on agriculture products and associated commodities (Bakri et al. 2005; Hallman et al. 2010, Hallman and Blackburn 2016; Dohino et al. 2017). Insects are responsible for larger share of annual loss of agricultural products because they are the principal agricultural pest. Insect pests are responsible for 10.8% loss of the total agricultural production in the world. This leads to global loss of 36 billion US dollar every year. In India, this loss is 15.7% of total agricultural production (Dhaliwal et al. 2015).

Excellent reviews have been written to explain the phenomenon of radiation resistance among microbes several reviews were written; with special emphasis on *Deinococcus* (Cox and Battista 2005; Daly 2009; Krisko and Radman 2013; Ranawat and Rawat 2017). A review on radiation effects in higher animals following total body irradiation was published in 1984 (Broerse and Macvittie 1984). By which the review summarized studies comprising radiation responses amid higher animals. The review discussed studies following accidental nuclear exposures and studies on the Japanese population exposed to IR after Hiroshima–Nagasaki bombings. Hallman (1999) wrote a detailed review on radiation doses required for the treatment of fruit flies, known to affect agricultural commodities. In review, the author specifically focused on effective treatment measures against the tephritidae family, with special emphasis on Oriental and Mediterranean fruit flies. In 2005, Bakri and colleagues reviewed the radiation treatments and levels of radiation resistance in insects. They summarized data available on a web portal; International Database on Insect Disinfestation and Sterilization (IDIDAS) (Bakri et al. 2005).

In the present review, we consolidated the current knowledge on radiation doses required for radio-sterilization and disinfestation in different insect groups (Tables 1 and 2). Subsequently, we carried out phylogenetic analysis for these insects, to understand the presence of any evolutionary relationship between the insect groups and trait of radiation resistance. To ascertain, whether the trait of radiation resistance has evolved along with insects or it is independent of insect evolution.

Insects as pest in agriculture

Insects are one of the most successful life forms throughout the history of terrestrial life on our planet. This has been supported by several observations, such as species diversity among insects, an ecological impact of insect populations, their distribution throughout the planet and their existence since beginning of the terrestrial life until today (Bradley et al. 2009). There are many factors, which are responsible for the evolution of the insects and making them a successful life form. Insects show numerous physiological and metabolic adaptations. Some insects have evolved with specialized physiological mechanisms to detoxify potent toxins and some can withstand against toxicants (Denlinger and Lee 2010). Some of the insects reported to thrive in extreme environmental conditions of cold and heat. Two species of *Chironomidae* family found at Antarctica are the southernmost living insects known currently (Oliver 1971). Chironomids are reported to thrive in radioactively contaminated areas of accident sites in Chernobyl, Ukraine (Williams et al. 2001), this environmental adaptation

Table 1. Radiation doses required for radio-sterilization in insect species of different orders.

Insect order	Insect species	Life stage	Radiation effects	Radiation dose (Gy)	Reference
Coleoptera	*Aethina tumida*	Adults	Sterility	75 Gy	Downey et al. 2015
	Cryptorhynchus mangiferae	Emerging adults	Sterility	100 Gy	Follett 2001
	Diaprepes abbreviates	Eggs	Hatching inhibition	50 Gy	Gould and Hallman 2004
	Hypothenemus hampei	Adults	Sterility	150 Gy	Follett 2018
	Sitophilus oryzae	Adults	Sterility	120 Gy	Follett et al. 2013a
Beetles	*Callosobruchus maculatus*	Pupae & adults	Hatching inhibition & sterility	650 Gy	Hammad et al. 2020
Diptera	*Anastrepha ludens*	Late larval instar	Emergence inhibition	60 Gy	Hallman and Loaharanu 2002
	Anastrepha obliqua	Late larval instar	Emergence inhibition	100 Gy	Bustos et al. 1992
	Anastrepha serpentine	Late larval instar	Emergence inhibition	100 Gy	Bustos et al. 1992
	Anastrepha suspensa	Late larval instar	Mortality	50 Gy	Gould and Windeguth 1991
	Bactrocera correcta	Adults	Sterility	30 Gy	Puanmanee et al. 2010
	Bactrocera cucurbitae	Late larval instar	Mortality	125 Gy	Follett and Armstrong 2004
	Bactrocera dorsalis	Late larval instar	Mortality	150 Gy	Follett and Armstrong 2004; Srimartpirom et al. 2018
	Bactrocera jarvisi	Eggs & larvae	Emergence inhibition	101 Gy	Heather et al. 1991
	Bactrocera tryoni	Adult male	Sterility	50 Gy	Dominiak et al. 2014
	Bactrocera zonata (Saunders)	Adult male	Sterility	90 Gy	Draz et al. 2016
	Ceratitis capitata	Late larval instar	Mortality	100 Gy	Follett et al. 2013b
	Dacus ciliatus	Adult male	Sterility	140 Gy	Nemny-Lavy et al. 2016
	Drosophila suzukii	Larvae & pupae	Emergence inhibition	80 Gy	Follett et al. 2014
	Drosophila melanogaster	Adults	Sterility	200 Gy	Bakri et al. 2005
	Toxotrypana curvicauda	Late larval instar	Emergence inhibition	150 Gy	Gould and Hallman 2004
Hemiptera	*Bemisia tabaci*	a. Eggs b. Late instar larvae c. Adults	a. Hatching inhibition b. Emergence inhibition c. Sterility	a. 50 Gy b. 150 Gy c. 100 Gy	Cho et al. 2019
	Trialeurodes vaporariorum	a. Eggs b. Late instar larvae c. Adults	a. Hatching inhibition b. Emergence inhibition c. Sterility	a. 50 Gy b. 150 Gy c. 100 Gy	Cho et al. 2019
Hymenoptera	*Solenopsis invicta*	Adults	Sterility	150 Gy	Follett et al. 2016
	Wasmannia auropunctata	Adults	Sterility	$\geq$ 70 Gy	Calcaterra et al. 2012
Lepidoptera	*Amorbia emigratella* Busck	Late pupae	Emergence inhibition	150 Gy	Follett 2008
	Epiphyas postvittana	Egg & larvae	Hatching & emergence inhibition	250 Gy	Follett and Snook 2012
	Lobesia botrana	Late larval instar	Mortality	250 Gy	Nadel et al. 2018
	Leucoptera coffeella Guer	Adult female	Sterility	700 Gy	Katiyar and Ferrer 1968
Orthoptera	*Schistocerca gregaria*	Male	Sterility	4 Gy	Dushimirimana et al. 2012
Thrips	*Frankliniella occidentalis*	Adults	Sterility	250 Gy	Nicholas and Follett 2018
Mean dose for radio-sterilization of above species				151.5	

*Sterility- following irradiation of parents (either male/female/both), the eggs in F1 generation will not hatch/will not survive.

enhanced radio-resistance of these insects. Some species of chironomidae are reported to accumulate uranium in their body and these species are thriving in uranium-contaminated areas (Dias et al. 2008).

Due to their omnipresence, insect species are causing huge crop loss all over the world. For example, *Bactrocera dorsalis* (Handel), of order dipteran is a well-known plant pest. The United States Department of Agriculture–Animal and Plant Health Inspection Service (USDA-APHIS) reported that fruit fly *B. dorsalis,* infests on approximately 450 plant species (Dohino et al. 2017). *B. dorsalis* infests on fruits of the plant, female lay eggs on fruits and leads to huge loss in fruit produce. Insects were reported to consume 5–20% of major grain crops produced all over the world (Deutsch et al. 2018). The agricultural loss caused by insects is also depends on the environmental factors, one such factor is a temperature. The amount of damage to crops due to insect pests was reported to increase along with the increasing environmental temperature. 10–25% increase in loss of total agriculture yield of major grains with 1-degree increase in temperature has been reported (Deutsch et al. 2018). The rise in environmental temperature results in accelerated metabolic rates of insects and consequently rate of food consumption among insects goes up (Deutsch et al. 2018). The increasing environmental temperature causing increased crop loss all over the world therefore, radio-sterilization and disinfestation treatments become significant. Over the years, insects are being treated using IR to protect crops and prevent loss of related agricultural commodities (Hallman and Blackburn 2016). To increase shelf-life of stored food grains, vegetables, fruits and spices radiation treatment was recommended (Arvanitoyannis et al. 2009).

Table 2. Radiation doses required for disinfestation treatment in insect species of different orders.

Insect order	Insect species	Life stage	Radiation effects	Radiation dose (Gy)	Reference
Blattodea	*Blattella germanica*	Nymph	Mortality	700 Gy	Cole et al. 1959
	Cimex lectularius	Adult male	Mortality	1500 Gy	
	Periplaneta americana	Adult male	Mortality	500 Gy	
Coleoptera	*Aethina tumida*	Adults	Mortality	80 Gy	Downey et al. 2015
	Cryptorhynchus mangiferae	Developing stages	Emergence inhibition	300 Gy	Follett 2001
	Dermestes maculatus	Early pupae	Emergence inhibition	100 Gy	Seal and Tilton 1986
	Diaprepes abbreviates	Late larval instar	Emergence inhibition	150 Gy	Gould and Hallman 2004
	Hypothenemus hampei	Adults	Mortality	3200 Gy	Kiran et al. 2019
	Lasioderma serricorne	Adults	Mortality	>1000 Gy	Tilton et al. 1966
	Rhyzopertha dominica	Pupae	Emergence inhibition	>1000 Gy	
	Sitophilus oryzae	Pupae	Emergence inhibition	>1000 Gy	
	Tribolium confusum	Pupae	Emergence inhibition	>1000 Gy	
	Sitophilus granaries	Pupae	Emergence inhibition	300 Gy	Brown et al. 1972
	Sitophilus zeamais	Pupae	Emergence inhibition	1000 Gy	
	Trogoderma granarium	Adults	Mortality	80 Gy	Downey et al. 2015
Diptera	*Anastrepha obliqua*	Larvae	Emergence inhibition	600 Gy	Arthur and Wiendl 1994
	Chironomus ramosus	Adults	Mortality	3500 Gy	Datkhile et al. 2009
	Drosophila melanogaster	Late larval instar	Mortality	1595 Gy	Paithankar et al. 2017
	Musca domestica	Pupae	Emergence inhibition	300 Gy	Dauer et al. 1965
	Polypedilum vanderplanki	Larvae	Mortality	2397 Gy	Watanabe et al. 2006
Hymenoptera	*Dolichoderus quadripunctatus*	Adults	Mortality	1500 Gy	Torossian and Causse 1968
	Monomorium pharaonis	Adults	Mortality	1900 Gy	Cole et al. 1959
Lepidoptera	*Bombyx mori*	Eggs	Hatching inhibition	320 Gy	Murakami and Miki 1972
	Cadra cautella (Ephestia cautella)	Eggs	Hatching inhibition	300 Gy	Ozyardimci et al. 2006
	Cryptophlebia illepida	Pupae	Emergence inhibition	250 Gy	Follett and Lower 2000
	Dendrolimus spectabilis	Late larval instar	Emergence inhibition	4000 Gy	Lee et al. 1968
	Leucoptera coffeella Guer	Pupae & male	Emergence inhibition	> 600 Gy	Katiyar and Ferrer 1968
	Plodia interpunctella	a. Eggs b. Adults	a. Hatching inhibition b. Mortality inhibition	450 Gy	Ayvaz et al. 2008; Ozyardimci et al. 2006
Louse	*Pediculus humanus humanus*	Adult female	Mortality inhibition	1800 Gy	Cole et al. 1959
Orthoptera	*Schistocerca gregaria*	Eggs	Mortality	10 Gy	IDIDAS 2020
Zygentoma	*Thermobia domestica*	Adults	Mortality inhibition	980 Gy	Cole et al. 1959
Mean dose for disinfestation of above species				1035 Gy	

Radiation treatments for insect control

In recent years, insect radiation treatments have been performed as a routine practice in post-harvest processing. This has been achieved by irradiating agricultural produce and related commodities to high doses of IR (Arvanitoyannis et al. 2009; Hallman and Blackburn 2016). It is generally accepted that insects are radio-resistant organisms, due to the reason they have higher radiation tolerance than mammals (Harrison and Anderson 1996; Bakri et al. 2005; Paithankar et al. 2017). Though insects are radio-resistant organisms, their exposure to IR is associated with radiation-induced toxicities, such as sterility, early onset of ageing, reduced life span, decreased fecundity and many others (Hooper 1975). The radio-sterilization and disinfestation treatments are considered as pre- and post-harvest pest control measures and include IR exposure of grains/produce before harvest and prior to harvest before storage. In radio-sterilization treatment, the grains are irradiated to low doses of IR to induce sterility. Whereas, during disinfestation treatment, the food products and harvested grains are exposed to high doses of IR that lead to the death of the entire insect population. These lethal doses are higher than doses required for induction of sterility in organisms (Hallman and Blackburn 2016). The radio-sterilization doses and the disinfestation doses may vary significantly for the same species. For example, dose required to induce sterility in *D. melanogaster* was reported to be 200 Gy whereas, dose required for killing adult flies of *D. melanogaster* was reported to be greater than 1000 Gy (Bakri et al. 2005; Paithankar et al. 2017).

The post-harvest processing includes irradiation of food grains and other commodities at high doses of IR. The high dose exposure is to disinfests all possible pest in grains before long term storage. The doses required for disinfestation treatments are several folds higher than doses required for radio-sterilization techniques (Bakri et al. 2005; Paithankar et al. 2017).

Besides, the pre-harvest measure includes the sterile insect technique (SIT), in which sterile males are introduced in the natural insect population. This is achieved by irradiating pupae/larvae/adults (before achieving sexual maturity- in case of locust) (Dushimirimana et al. 2012), at the desired dose followed by collection and release of emerged F1 offspring into the natural population. The dose of IR will be selected in such a way, the emerged insects will be sterile but equally competent to unirradiated males. Introduction of sterile males into the natural insect population will reduce the number of fertile eggs laid by females (Marec and Vreysen 2019). Therefore, the progeny will be non-viable, which leads to reduction in the insect population. This technique is also known as the sterile male technique (SMT) (Aldryhim and Adam 1999; Hallman 1999; Bakri et al. 2005; Marec and Vreysen 2019). The use of sterile insects helped in decreasing both the attack intensity and insect population density. In general, these studies keep check on insect population and reduce impact on harvested agriculture products and related commodities. However, while accomplishing these goals, the remarkable property of insects to withstand high doses of IR has been overlooked.

During radio-sterilization treatment, the lepidopteran insects were found to be more radio-resistant compared to other groups, whereas radio-sterilization in orthopterans achieved at relatively low doses (Bakri et al. 2005). Besides, species-to-species variation in radiation resistance was observed and several factors were reported to affect radiation resistance in an organism. One such factor is oxygen levels in the environment, under hypoxic conditions radiation resistance was found to be increased in the Cabbage looper moth (*Trichoplusia ni*), compared to oxygenated environments (Condon et al. 2016). The study by Bell et al. (2007) has explained the role of hypoxia-inducible factor (HIF) in life span extension under hypoxic conditions. The study highlighted the crucial role of mitochondrial ROS in the activation of HIF and its downstream signaling. Besides, the extension of life-span, the HIF signaling might be helping in the radiation resistance of an organism. Similarly, in *Drosophila suzukii* the radiation resistance was found to be increased with reduction in oxygen levels (Follett et al. 2018).

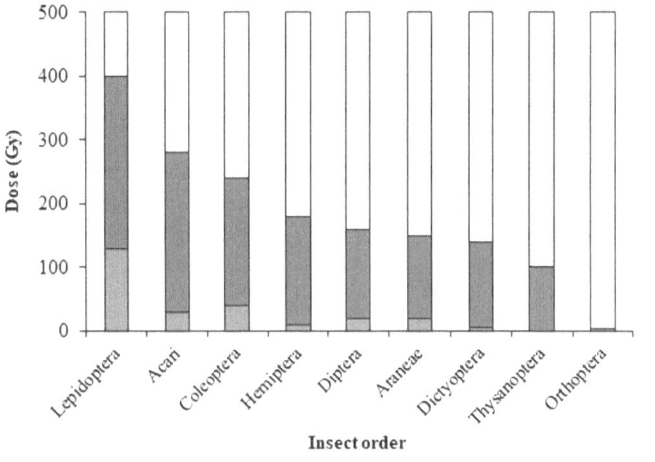

Figure 2. Doses of ionizing radiation required for quarantine treatment of insects of different orders. X- axis represents radiation dose (Gy) of ionizing radiation, Y- axis represents insect orders. The red color represents the doses recommended for radio-sterilization studies; green color represents doses of radiation inefficient to induce radio-sterilization. Among insect orders reported for radio-sterilization treatment, lepidoptera needs high doses of radiation from 120 to 400 Gy and orthopteran needs low dose about 4 Gy.

Radio-sterilization treatment

Following decades of studies in the area of radiation toxicities and its effects on biological systems, IR are widely used to control insect populations known for infesting stored food grains, fruits, and vegetables (Condon et al. 2016). The aim of radio-sterilization is to control insect population by inducting sterility in adults. Different insect groups were reported to need different doses of radiation for sterility induction. The existing information about radiation doses required for sterility induction are from different studies, in different insect species comprising various stages of life i.e. eggs, larvae, instars, pupae or sub-adults/adults. The mean doses required for the radio-sterilization of various insect orders were found to be in the range of 130–400 Gy for Lepidoptera (Hallman et al. 2013), 30–280 Gy for Acari, 40–200 Gy for Coleoptera, 10–180 Gy for Hemiptera, 20–160 Gy for Diptera, 20–150 Gy for Araneae (spiders), 5–140 Gy for Dictyoptera, 100 Gy for Thysanoptera, and the dose was 4 Gy for order Orthoptera (Figure 2) (Bakri et al. 2005). Among these insect orders some are considered as radiation resistant namely, Lepidoptera (moths and butterflies), Acari (Mites), Coleoptera (Beetles), Hemiptera (true bugs) and Diptera (true flies). Some of the lepidopteran species require doses up to 400 Gy to attain reproductive sterilization whereas, the order orthoptera need very low dose of radiation to attain sterility (Willard and Cherry 1975; Bakri et al. 2005). Table 1, highlights the level of radiation doses required for either induction of sterility or prevention of adult emergence in insects of different orders. In some insect species irradiation of females carried out for induction of sterility. In case of *Aethina tumida* it was reported that the males and females have equal radio-sensitivity. However, irradiation of females only, lowered the reproduction effectively than irradiation of only males (Downey et al. 2015). In lepidopteran, at adult stage separation of males and females is difficult, therefore the method of inherited sterility (IS) or F1 sterility has been used. In IS technique, males and females (i.e. parents) both were irradiated to IR; they show reduced reproduction capability however, F1 generation exhibits elevated sterility compared to both the parents (Marec and Vreysen 2019). During radio-sterilization treatments females found to be radiosensitive compared to males. Therefore, irradiation to doses capable of inducing female sterility will induce partial sterility in males. The partially sterile males produce F1 offspring with more sterility without affecting competitiveness (Marec and Vreysen 2019). Therefore, selection of dose capable of inducing sterility in males are more suitable for SIT (Robinson 2005).

Disinfestation treatment

The level of radiation resistance in insects is not uniform throughout their life cycle, it is found to vary from one stage to another (Balock et al. 1963; Murakami and Miki 1972; Datkhile et al. 2009; Paithankar et al. 2017). Prior to transformation from one life stage to another, insects undergo enormous cell division. Subsequent to the phase of enormous cell division, insects transform into different forms and this process is called metamorphosis (Truman 2019). Interestingly, the process of metamorphosis is not uniform in insect class, there are some insect groups that undergo complete metamorphosis and are called holometabolous insects, some insect groups undergo partial metamorphosis called hemimetabolous insects, and insect groups which do not undergo metamorphosis are called ametabolous insects. During the process of metamorphosis new tissue formation and organogenesis takes place (Truman 2019). Subsequent to metamorphosis, there is a change in feeding pattern/preferences, behavior and physiology of insects (Mitra 2013). In insects, the requisite cell divisions usually happen during metamorphosis and therefore the adults are largely post-mitotic (Parashar et al. 2008). Due to presence of non-dividing cells in large number (amitotic nature) the adults appear to have more resistance against IR compared to stage having rapidly dividing cells (Paithankar et al. 2017). Besides, the other physiological differences, variation in metabolism, cell division rates might be the reason behind variation in levels of radiation resistance amid different life stages. Compared to holometabolous insects ametabolous and hemimetabolous insects show different trend. Among these insects the radiation resistance increases along with developmental stages and adults showed high radiation resistance compared to juveniles (Cole et al. 1959). Among various insect groups, some life stages during development were reported as radio-sensitive and some reported as radio-resistant (Datkhile et al. 2009; Paithankar et al. 2017). From various studies in different holometabolous insects such as fruit flies, chironomids, silk worm, and so on it appears that, the late instar larvae are most radio-resistant whereas, the early pupae appear least radio-resistant (more radiation sensitive) (Balock et al. 1963; Murakami and Miki 1972; Paithankar et al. 2017). For example, there was a specific trend observed in levels of radiation resistance of Mexican fruit fly (*Anastrepha ludens*) and West Indian fruit fly (*Anastrepha obliqua*). The study reported that early-pupal stage was more radio-sensitive than later stages, it was also reported as the most radio-sensitive stage among all life stages (Hallman and Worley 1999). Similar trend of radiation resistance was reported in Australian fruit fly *Bactocera jarvisi* and *Bactocera tryoni*. First and second instar larvae of these flies found to be radiosensitive than third instar larvae. The radiation resistance in late third instar larvae was found to be significantly greater than early third instar larvae. In short, for *B. jarvisi* and *B. tryoni*, late instar larvae were more radio-tolerant compared to other life stages (Heather et al. 1991). From the studies of disinfestation treatment of holometabolous insects, it appears that late larval instars are significantly resistant to radiation induced toxicities compared to other life stages. Table 2 highlights, different holometabolous, ametabolous and hemimetabolous insect species from different orders reported for damaging agriculture produce. The table highlights previously reported disinfestation doses in different stages with radiation induced effects in them.

Evolutionary relationship between insect species reported for radiation resistance

The insect species considered for phylogenetic analysis are having economic importance and are well documented for their effects on agriculture products. Phylogenetic analysis was carried out using cytochrome oxidase (Cyt Oxi) gene sequences for insect species mentioned in Tables 1 and 2. Cyt Oxi is well known for its role in respiratory electron transport chain and is known for regulating oxidative phosphorylation and ROS generation (Li et al. 2016). Cyt Oxi gene sequences for different organisms are available on the database of the National Center for Biotechnology Information (NCBI). Mega-7 free online software was used for analysis. As insect species show significant differences in their radio-sterilization doses and disinfestation doses. Two separate phylogenetic trees were plotted, (1) Insects studied for radio-sterilization treatments (Figure 3), and (2) Insects studied for disinfestation treatments (Figure 4).

Figures 3 and 4 are highlighting random distribution of radiation resistance among insect species. From phylogenetic analysis, it appears that the phenotype of radiation resistance probably not essential for the survival of these insect groups therefore; some species have this trait and others lack. Previously, studies have explained that the phenotype of radiation resistance is a by-product of desiccation tolerance (May et al. 1964; Jonsson 2003). However, recent study on water bear reported that, desiccation tolerance and radiation resistance are two independent traits in *Hypsibius dujardini* (Beltrán-Pardo et al. 2015). In insects, it is yet to establish whether, the radiation resistance is a by-product of desiccation stress tolerance or it is an independent trait acquired as vestige from ancestors. Microbes and tardigrades have developed radiation resistance over a period following continuous exposure to environmental stress such as heat and desiccation (Jonsson et al. 2016; Ranawat and Rawat 2017). From phylogenetic analysis carried out in present study it appears that, species with low levels of radiation resistance either lost this trait as it was not necessary for their survival or radiation resistance was evolved in some species (species with high radiation tolerance) as a by-product of other stress tolerance mechanisms. From the phylogenetic analysis it appears that, there is no clear correlation between radiation resistance and insect evolution.

Radiation resistance: an outline

Microbes are having high levels of radiation resistance compared to eukaryotes (Cox and Battista 2005; Daly 2009; Krisko and Radman 2013). Among eukaryotes, invertebrate animals such as tardigrades and insects are significantly resistant to IR compared to vertebrate animals such as mammals (Jonsson et al. 2008). Some insects belong to a category of major agricultural pests therefore; studies on insect radiation responses have unique importance. Insects and their radiation responses were studied widely to control their population for protecting agricultural commodities (Bakri et al. 2005). The packed food commodities are processed using radiation before long-term storage; this disinfests the commodities and increases their shelf-life (Arvanitoyannis et al. 2009). During this process of disinfestation treatment, the insects of orders Lepidoptera and thrips (Thysanoptera) appear to be most resistant to radiation treatment compared to insects of other orders. To inhibit emergence and induce sterility the dose of 400 Gy was recommended as a generic dose for treatment of lepidopteran insects (Bakri et al. 2005, Nicholas and Follett 2018).

Radiation resistance in lepidopteran insects was reported to increase in hypoxic conditions and similar results were observed previously for fruit flies (Balock et al. 1963; Condon et al. 2016, Follett et al. 2018). Therefore, it is

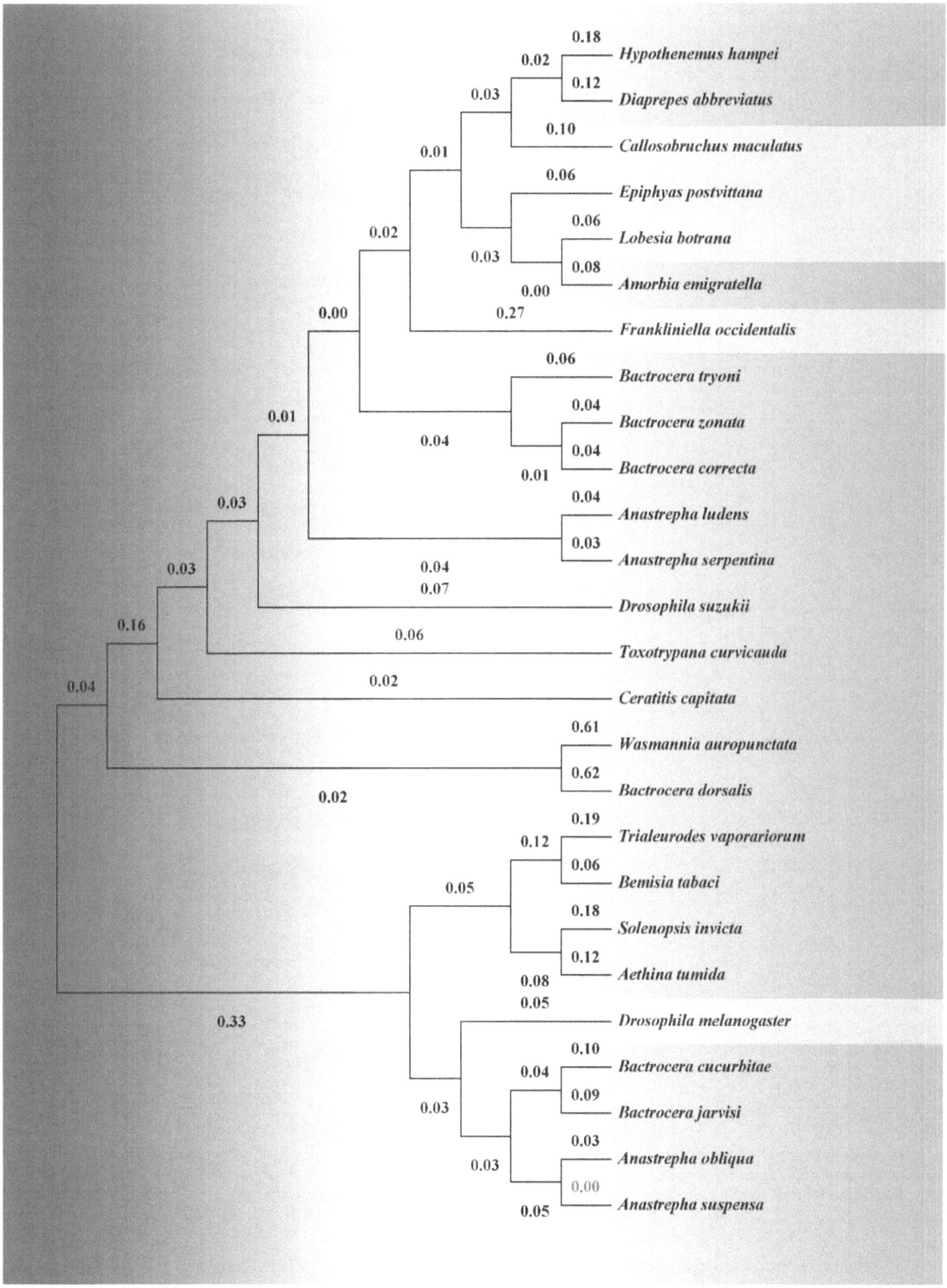

Figure 3. Cytochrome oxidase based phylogenetic analysis of insect groups reported for radio-sterilization treatment. When, dose required for radio-sterilization treatment of a species is more than mean dose for species mentioned in Table 1, such species are considered as radio-tolerant and highlighted in green color whereas, species requiring dose less than mean for radio-sterilization are considered as radio-sensitive species and highlighted in red color.

essential to recheck the doses suggested for radio-sterilization treatments of insects, whether these doses are actually for insects or they are due to hypoxic conditions generated because of packaging. The packing of grains or fruits can create hypoxic conditions in containers/packets; this can affect the radio-sterilization treatment. Sometimes the actual gamma dose required for radio-sterilization or disinfestation can be more than the suggested dose.

Studies on different insect groups have clarified that the radiation resistance increases with the progress in the developmental stages of insects. However, increment in radiation resistance has shown a particular trend amid

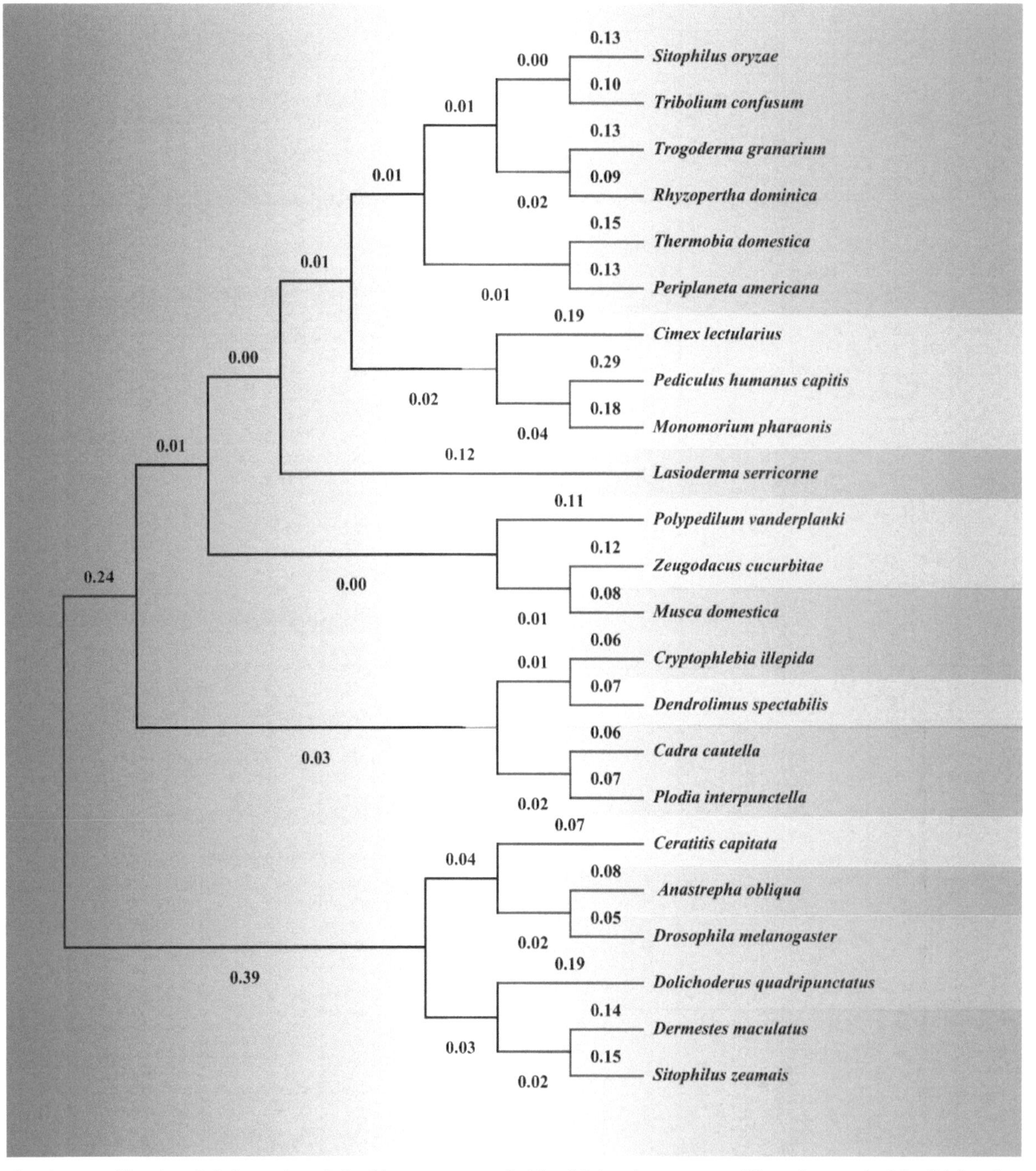

Figure 4. Cytochrome oxidase based phylogenetic analysis of insect groups studied for disinfestation treatment. When, dose required for disinfestation treatment of a species is more than mean dose for species mentioned in Table 2, such species are considered as radio-tolerant and highlighted in green color whereas, species requiring dose less than mean for disinfestation are considered as radio-sensitive species and highlighted in red color.

holometabolous insects. From reported studies, it appears that late larval instars (non-dividing stage) are resistant to radiation, whereas the early pupal stage (highly dividing stage) is radiation sensitive. (Balock et al. 1963; Murakami and Miki 1972; Datkhile et al. 2009; Paithankar et al. 2017). Whereas, in the case of the ametabolous and hemimetabolous insects the radiation resistance increases along with the development stage but it does not come down in adults as that of holometabolous insects (Cole et al. 1959). The ups and downs in the levels of radiation resistance among holometabolous insects may be due to the life stages of holometabolous insects are drastically different in terms of their physiology, metabolic activity, cell division rates and so on (Mitra 2013). From these studies, it is evident that radiation resistance varies in different life stages. Therefore, following disinfestation studies, the highest dose required for disinfestation of the most resistant stage can be selected for future disinfestations as a generic dose. On careful observation of the doses known for disinfestation of various life stages of different insects, a specific pattern of ups and downs in levels of radiation resistance can be observed (Figure 5). Different studies have proposed different rationales for high and varying radiation resistance. These studies explained that non-enzymatic antioxidants such as; trehalose and uric

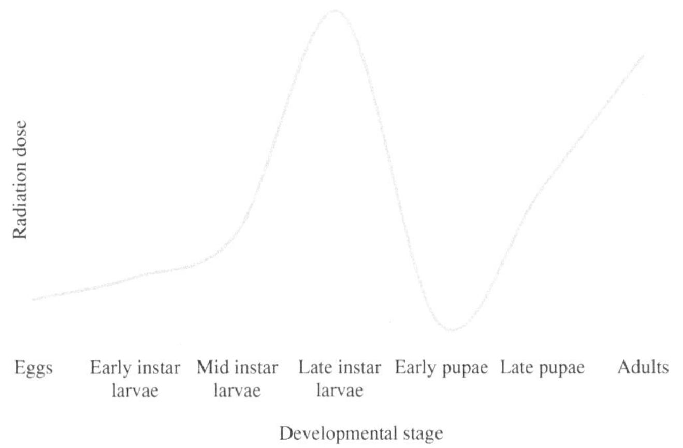

Figure 5. General trend in radiation resistance of different developmental stages of holometabolous insect species. X- axis is showing developmental stages and Y-axis is showing lethal dose of ionizing radiation (arbitrary scale). (Balock et al. 1963; Murakami and Miki 1972; Datkhile et al. 2009; Paithankar et al. 2017).

acid are playing important role in imparting radiation resistance to the insects (Paithankar et al. 2018a, 2020). Besides, studies in *Deinococcus* reported the efficient DNA repair mechanism responsible for high radiation tolerance (Daly 2009). The study also reported the accumulation of Manganese ions (Mn^{2+}) playing crucial role in radiation resistance. The Mn^{2+} forms complexes with proteins, these complexes are protecting proteome of this bacterium against radiation induced damage and also these Mn^{2+} complexes reported to reduce cellular ROS. Apart from presence of non-enzymatic antioxidants and efficient DNA repair, the recent study on arthropods showed that, organisms with holocentric chromosomes are having high radiation resistance compared to organisms with monocentric chromosomes (Zedek and Bureš 2019).

Subsequent to the analysis of evolutionary tree, it appears that the trait of radiation resistance and insect evolution are independent (Figures 3 and 4). Evolutionarily near insect species appeared to exhibit significantly varying levels of radiation resistance. Apart from evolutionary basis; the physiological parameters and adaptive responses appear to play a key role in determining radiation resistance of insects. Similarly, various environmental factors may be playing role in attributing characteristics of high radiation resistance to some species (Condon et al. 2016; Follett et al. 2018). Further, additional studies are required to understand mechanisms of radiation resistance to elucidate random occurrence of the trait of radiation resistance among insect groups.

Conclusion

IR treatment has been used for several decades to prevent agricultural products and reduce crop loss from insect pests. Among insects, IR doses required to induce toxicity and achieve radio-sterilization/disinfestation vary dramatically. Insects show high levels of radiation resistance compared to higher animals; however, in largely dividing stages the radiation resistance was found to be considerably low. The robust cell division during these stages appears to cause a significant fall in radiation resistance. Therefore, these radiation-sensitive stages can be used as a target stage for effective radio-sterilization and for SIT/SMT. From phylogenetic analysis carried out in the present study, it is evident that levels of radiation resistance vary even in evolutionarily related and near species. The phylogenetic analysis clarifies that there is no correlation between evolution of insects and the trait of radiation resistance. Therefore, we conclude that the trait of radiation resistance is independent of insect evolution. Further, during this study, we observed that even after decades of studies several insect species are merely reported for radio-sterilization treatments and we do not know about IR doses required for their disinfestation. There is also a need for stage-specific studies because not all the species are reported for radiation effects in all the life stages. The knowledge of species-specific and stage-specific radiation doses will help to protect agricultural commodities in a better way. However, systematic analysis of stage-specific radiation doses is required for appropriate induction of sterility and disinfestation in insect pests.

Acknowledgement

JGP sincerely thanks the Council of Scientific and Industrial Research (CSIR), India, for the research associate fellowship (Award no. 09/1257(0001)/2019-EMR-I).

Author contributions

JGP Conceptualized the work, wrote the original draft, prepared figures and tables. TSG carried out phylogenetic analysis RKP reviewed the manuscript. All authors read the manuscript and approved for submission.

Disclosure statement

No potential conflict of interest was reported by the author(s).

ORCID

Jagdish Gopal Paithankar http://orcid.org/0000-0001-8064-1649
Tanhaji Sandu Ghodke http://orcid.org/0000-0002-9952-7990
Rajashekhar K. Patil http://orcid.org/0000-0002-5362-0479

References

Aldryhim YN, Adam EE, 1999. Efficacy of gamma irradiation against *Sitophilus granarius* (L.) (Coleoptera: Curculionidae). J Stored Prod Resh. 35(3):225–232.

Alphey N, Bonsall MB, 2018. Genetics-based methods for agricultural insect pest management. Agric For Entomol. 20(2):131–140.

Arthur V, Wiendl FM, 1994. Desinfestação de *Averrhoa carambola* infestada por anastrepha obliqua (macquart, 1835) (diptera - tephritidae) através de radiação gama. Sci Agric. 51(2):216–221.

Arvanitoyannis IS, Stratakos AC, Tsarouhas P, 2009. Irradiation applications in vegetables and fruits: a review. Crit Rev Food Sci Nutr. 49(5):427–462.

Atwood D, Paisley-Jones C, 2017. Pesticides industry sales and usage: 2008–2012 Market Estimates. U.S. Environmental Protection Agency, Washington, DC 20460.

Ayvaz A, Albayrak S, Karaborklu S, 2008. Gamma radiation sensitivity of the eggs, larvae and pupae of Indian meal moth *Plodia interpunctella* (Hübner) (Lepidoptera: Pyralidae). Pest Manage Sci. 64: 505–512.

Bakri A, Heather N, Hendrichs J, Ferris I, 2005. Fifty years of radiation biology in entomology: lessons learned from IDIDAS. Ann Entomol Soc Am. 98(1):1–12.2.0.CO;2]

Balock JW, Burditt AK, Christenson LD, 1963. Effects of gamma radiation on various stages of three fruit fly species. J Econ Entomol. 56(1):42–46.

Bell EL, Klimova TA, Eisenbart J, Schumacker PT, Chandel NS, 2007. Mitochondrial reactive oxygen species trigger hypoxia-inducible factor-dependent extension of the replicative life span during hypoxia. Mol Cell Biol. 27(16):5737–5745.

Beltrán-Pardo E, Jönsson KI, Harms-Ringdahl M, Haghdoost S, Wojcik A, 2015. Tolerance to gamma radiation in the tardigrade *Hypsibius dujardini* from embryo to adult correlate inversely with cellular proliferation. PLoS One. 10(7):e0133658.

Bradley TJ, Briscoe AD, Brady SG, Contreras HL, Danforth BN, Dudley R, Grimaldi D, Harrison JF, Kaiser JA, Merlin C, Reppert SM, et al. 2009. Episodes in insect evolution. Integr Comp Biol. 49(5):590–606.

Broerse JJ, Macvittie A. 1984. Response of different species to total body irradiation: series in radiology. Netherlands: Springer. Vol. 10: p. 1.

Brown GA, Brower JH, Tilton EW, 1972. Gamma radiation effects on *Sitophilus zeamais* and *S. granarius*. J Econ Entomol. 65:203–205.

Bustos ME, Enkerlin W, Toledo J, Reyes J, Casimiro A, 1992. Irradiation of mangoes as a quarantine treatment. In: Use of irradiation as a quarantine treatment of food and agricultural commodities. Vienna: International Atomic Energy Agency. p. 77–90.

Calcaterra LA, Coulin C, Briano JA, Follett PA, 2012. Acute exposure to low-dose radiation disrupts reproduction and shortens survival of *Wasmannia auropunctata* (Hymenoptera: Formicidae) queens. J Econ Entomol. 105(3):817–822.

Casarett AP. 1968. Radiation biology. Englewood Cliffs (NJ): Prentice-Hall.

Cho SR, Koo HN, Shin S, Kim HK, Park JH, Yoon YS, Kim GH, 2019. Gamma-ray irradiation control of whiteflies *Bemisia tabaci* (Hemiptera: Aleyrodidae) and *Trialeurodes vaporariorum* in the exportation of fresh strawberries. J Econ Entomol. 112(4): 1611–1617.

Cole MM, LaBrecque GC, Burden GS, 1959. Effects of gamma radiation on some insects affecting man. J Econ Entomol. 52(3): 448–450.

Condon CH, White S, Meagher RL, Jeffers LA, Bailey WD, Hahn DA, 2016. Effects of low-oxygen environments on the radiation tolerance of the Cabbage Looper Moth (Lepidoptera: Noctuidae). J Econ Entomol. 110(1):80–86.

Cox MM, Battista JR, 2005. *Deinococcus radiodurans*—the consummate survivor. Nat Rev Microbiol. 3:882–892.

Daly MJ, 2009. a new perspective on radiation resistance based on *Deinococcus radiodurans*. Nat Rev Microbiol. 7:237–245.

Daly MJ, 2012. Death by protein damage in irradiated cells. DNA Repair. 11(1):12–21.

Daly MJ, Gaidamakova EK, Matrosova VY, Vasilenko A, Zhai M, Venkateswaran A, Hess M, Omelchenko MV, Kostandarithes HM, Makarova KS, et al. 2004. Accumulation of Mn (II) in *Deinicoccus radiodurans* facilitates gamma-radiation resistance. Science. 306(5698):1025–1028.

Draz KA, Tabikha RM, El-Aw MA, Darwish HF, 2016. Impact of gamma radiation doses on sperm competitiveness, fecundity and morphometric characters of peach fruit fly *Bactrocera zonata* (Saunders) (Diptera: Tephiritidae). J Radiat Res Appl Sci. 9(3): 352–362.

Datkhile KD, Dongre TK, Mukhopadhyaya R, Nath BB, 2009. Gamma radiation tolerance of a tropical species of midge, *Chironomus ramosus* Chaudhuri (Diptera: Chironomidae). Int J Radiat Biol. 85(6): 495–503.

Dauer M, Bhatnagar PL, Rockstein M, 1965. X-irradiation of pupae of the house fly, *Musca domestica* L, and male survival. J Gerontol. 20(2):219–223.

Denlinger DL, Lee RE. jr, 2010. Low temperature biology of insects. Cambridge (UK): Camridge University Press.

Deutsch CA, Tewksbury JJ, Tigchelaar M, Battisti DS, Merrill SC, Huey RB, Naylo RL, 2018. Increase in crop losses to insect pests in a warming climate. Science. 361(6405):916–919.

Dhaliwal GS, Jindal V, Mohindru B, 2015. Crop losses due to insect pests: global and Indian scenario. Ind Jour of Entomol. 77(2): 165–168.

Dias V, Bonzom VC, Marc J, 2008. Exposure of *Chironomus riparius* larvae to uranium: effects on survival, development time, growth, and mouthpart deformities. Chemosphere. 71:574–581.

Dohino T, Hallman GJ, Grout TG, Clarke AR, Follett PA, Cugala DR, Tu DM, Murdita W, Hernandez E, Pereira R, et al. 2017. Phytosanitary treatments against *Bactrocera dorsalis* (Diptera: Tephritidae): current situation and future prospects. J Econ Entomol. 110(1):67–79.

Dominiak BC, Sundaralingam SL, Jiang L, Fanson BG, Collins SR, Banos C, Davies JB, Taylor PW, 2014. Evaluating irradiation dose for sterility induction and quality control of mass-produced fruit fly *Bactrocera tryoni* (Diptera: Tephritidae). J Econ Entomol. 107(3): 1172–1178.

Downey D, Chun S, Follett P, 2015. Radiobiology of small hive beetle (Coleoptera: Nitidulidae) and prospects for management using sterile insect releases. J Econ Entomol. 108(3):868–872.

Dushimirimana S, Hance T, Damiens D, 2012. Comparison of reproductive traits of regular and irradiated male desert locust *Schistocerca gregaria* (Orthoptera: Acrididae): evidence of last-male sperm precedence. Biol Open. 1(3):232–236.

Eriksson D, Stigbrand T, 2010. Radiation-induced cell death mechanisms. Tumor Biol. 31:363–372.

Follett PA, Lower RA, 2000. Irradiation to ensure quarantine security for Cryptophlebia spp. (Lepidoptera: Tortricidae) in sapindaceous fruits from Hawaii. J Econ Entomo. 93(6):1848–1854.

Follett PA, 2001. Irradiation as a quarantine treatment for mango seed weevil. Proc Hawaii Entomol Soc. 35:85–90.

Follett PA, Armstrong JW, 2004. Revised irradiation doses to control melon fly, mediterranean fruit fly, and oriental fruit fly (Diptera: Tephritidae) and a generic dose for tephritid fruit flies. J Econ Entomol. 97(4):1254–1262.

Follett PA, 2008. Effect of irradiation on Mexican leafroller (Lepidoptera: Tortricidae) development and reproduction. J Econ Entomol. 101(3):710–715.

Follett PA, Snook K, 2012. Irradiation for quarantine control of the invasive light brown apple moth (Lepidoptera: Tortricidae) and a generic dose for Tortricid eggs and larvae. J Econ Entomol. 105: 1971–1978.

Follett PA, Snook K, Janson A, Antonio B, Haruki A, Okamura M, Bisel J, 2013a. Irradiation quarantine treatment for control of *Sitophilus oryzae* (Coleoptera: curculionidae) in rice. J Stored Prod Res. 52:63–67.

Follett PA, Wall M, Bailey W, 2013b. Influence of modified atmosphere packaging on radiation tolerance in the phytosanitary pest melon fly (Diptera: Tephritidae). J Econ Entomol. 106(5):2020–2026.

Follett PA, Swedman S, Price DK, 2014. Postharvest irradiation treatment for quarantine control of *Drosophila suzukii* (Diptera: Drosophilidae) in fresh commodities. J Econ Entomol. 107(3): 964–969.

Follett PA, Porcel S, Calcaterra LA, 2016. Effect of irradiation on queen survivorship and reproduction in the invasive fire ant *Solenopsis invicta* (Hymenoptera: Formicidae) and a proposed phytosanitary irradiation treatment for ants. J Econ Entomol. 109(6):2348–2354.

Follett PA, 2018. Irradiation for quarantine control of coffee berry borer, *Hypothenemus hampei* (Coleoptera: Curculionidae: Scolytinae) in coffee and a proposed generic dose for snout beetles (Coleoptera: Curculionoidea). J Econ Entomol. 111(4):1633–1637.

Follett PA, Swedman A, Mackey B, 2018. Effect of low-oxygen conditions created by modified atmosphere packaging on radiation tolerance in *Drosophila suzukii* (Diptera: Drosophilidae) in sweet cherries. J Econ Entomol. 111(1):141–145.

Gould WP, Windeguth DL, 1991. Gamma irradiation as a quarantine treatment for carambolas infested with Caribbean fruit flies. Fla Entomol. 74(2):297–300. –

Gould WP, Hallman GJ, 2004. Irradiation disinfestation of Diaprepes root weevil (Coleoptera: Curculionidae) and papaya fruit fly (Diptera: Tephritidae). Fla Entomol. 87(3):391–392

Hallman GJ, 1999. Ionizing radiation quarantine treatment against Tephritid fruit flies. Postharvest Biol Technol. 16(2):93–106.

Hallman GJ, Worley JW, 1999. Gamma radiation doses to prevent adult emergence from Mexican and West Indian fruit fly (Diptera: Tephritidae) immatures. J Econ Entomol. 92(4):967–973.

Hallman GJ, Loaharanu P, 2002. Generic ionizing radiation quarantine treatments against fruit flies (Diptera: Tephritidae) proposed. J Econ Entomol. 95:893–901.

Hallman GJ, Levang-Brilz NM, Zettler JL, Winborne IC, 2010. Factors affecting ionizing radiation phytosanitary treatments, and implications for research and generic treatments. J Econ Entomol. 103(6): 1950–1963.

Hallman GJ, Parker AG, Blackburn CM, 2013. The case for a generic phytosanitary irradiation dose of 400 gy for lepidoptera that infest shipped commodities as pupae. J Econ Entomol. 106(2):525–532.

Hallman GJ, Blackburn CM, 2016. Phytosanitary irradiation. Foods. 5(4):81–10.

Hammad A, Gabarty A, Zinhoum R, 2020. Assessment irradiation effects on different development stages of *Callosobruchus maculatus* and on chemical, physical and microbiological quality of cowpea seeds. Bull Entomol Res. 110(4):497–499.

Harrison F, Anderson S. 1996. Taxonomic and developmental aspects of radiosensitivity. Proceedings of the symposium: Ionizing radiation, the Swedish Radiation Protection Institute (SSI) and The Atomic Energy Control Board (AECB) of Canada, Stockholm, Sweden, May 1996. p. 65–88.

Heather NW, Corcoran RJ, Banos C, 1991. Disinfestation of mangoes with gamma irradiation against two Australian fruit flies (Diptera: Tephritidae). J Econ Entomol. 84(4):1304–1307.

Hooper GHS, 1975. Sterilization of *Dacus cucumis french* (Diptera: Tephritidae) by gamma radiation I. Effect of dose on fertility, survival and competitiveness. Aust J Entomol. 14(1):81–87.

IDIDAS. [Internet]. 2020; [accessed 2020 Oct 25]. http://www-ididas.iaea.org/ididas/.

Jonsson KI, 2003. Causes & consequences of excess resistance in cryptobiotic metazoans. Physiol Biochem Zool. 76:429–435.

Jonsson KI, Hygum TL, Andersen KN, Clausen LKB, Mobjerg N, 2016. Tolerance to gamma radiation in the marine heterotardigrade, *Echiniscoides sigismundi*. PLoS One. 11(12):e0168884.

Jonsson KI, Rabbow E, Schill RO, Ringdahl MH, Rettberg P, 2008. Tardigrades survive exposure to space in low earth orbit. Curr Biol. 18(17):R729–31.

Katiyar KP, Ferrer F, 1968. Proceedings of a symposium on the use of isotopes and radiation in entomology. International Atomic Energy Agency. Rearing technique, biology and sterilization of the coffee leaf miner, Leucoptera coffeella Guer. (Lepidoptera: Lyonetiidae), Vienna. p. 165–173.

Kiran R, Shenoy KB, Venkatesha MG, 2019. Effect of gamma radiation as a post-harvest disinfestation treatment against life stages of the coffee berry borer, *Hypothenemus hampei* (Ferrari) (Coleoptera: Curculionidae). Int J Radiat Biol. 95(9):1301–1308.

Krisko A, Radman M, 2013. Biology of extreme radiation resistance: the way of *Deinococcus radiodurans*. Cold Spring Harbor Perspect Biol. 5(7):a012765–11.

Lee KB, Hwang LK, Lee KS, 1968. Exploratory studies on the eradication of the Korean pine caterpillar Dendrolimus spectabilis butler by means of radiation. Proceedings of a Symposium on the Use of Isotopes and Radiation in Entomology. International Atomic Energy Agency. Vienna. p. 273–286.

Li R, Jia Z, Trush MA, 2016. Defining ROS in biology and medicine. Reactive Oxygen Species. 1(1):9–21.

Lu Y, Song S, Wang R, Liu Z, Meng J, Sweetman AJ, Jenkins A, Ferrier RC, Li H, Luo W, et al. 2015. Impacts of soil and water pollution on food safety and health risks in China. Environ Int. 77: 5–15.

Marec F, Vreysen MJB, 2019. Advances and challenges of using the sterile insect technique for the management of pest lepidoptera. Insects. 10(11):371–326.

Maurya DK, Devasagayam TPA. 2011. Role of radioprotectors in the inhibition of DNA damage and modulation of DNA repair after exposure to gamma-radiation. Selected Topics in DNA Repair. Prof Clark Chen (Ed.), ISBN: 978-953-307-606-5, InTech.

May RM, Maria M, Guimard J, 1964. Actions différentielles des rayonset ultraviolets sur le tardigrade *Macrobiotus areolatus*, a'l' e'tat actif et desse'che'. Bulletin Biologique de la France et de la Belgique. 98: 349–367.

Michael BD, O'Neill P, 2000. Molecular biology: a sting in the tail of electron tracks. Science. 287:1603–1604.

Mitra A, 2013. Cinderella's new shoes – how and why insects remodel their bodies between life stages. Curr Sci. 104(8):1028–1036.

Murakami A, Miki M, 1972. Age-dependent changes of radio-sensitivity in embryo of *Bombyx mori*. J Radiat Res. 13:183–192.

Nadel H, Follett PA, Perry CL, Mack RG, 2018. Postharvest irradiation treatment for quarantine control of the invasive *Lobesia botrana* (Lepidoptera: Tortricidae). J Econ Entomol. 111(1):127–134.

Nemny-Lavy E, Nestel D, Rempoulakis P, 2016. Laboratory longevity and competitiveness of *Dacus ciliatus* Loew (Diptera: Tephritidae) following sub-sterilizing gamma irradiation. Bull Entomol Res. 106(3):415–419.

Nicholas AH, Follett PA, 2018. Postharvest irradiation treatment for quarantine control of western flower thrips (Thysanoptera: Thripidae). J Econ Entomol. 111(3):1185–1189.

Oliver DR, 1971. Life story of the chironomidae. Annu Rev Entomol. 16(1):211–230.

Ozden O, Park SH, Kim HS, Jiang H, Coleman MC, Spitz DR, Gius D, 2011. Acetylation of MnSOD directs enzymatic activity responding to cellular nutrient status or oxidative stress. Aging. 3(2):102–107.

Ozyardimci BÃ, Cetinkaya N, Denli E, Ic E, Alabay M, 2006. Inhibition of egg and larval development of the Indian meal moth *Plodia interpunctella* (Hubner) and almond moth *Ephestia cautella* (Walker) by gamma radiation in decorticated hazelnuts. J Stored Prod Res. 42(2):183–196.

Paithankar JG, Deeksha K, Patil RK, 2017. Gamma radiation tolerance in different life stages of the fruit fly *Drosophila melanogaster*. Int J Radiat Biol. 93(4):440–448.

Paithankar JG, Raghu SV, Patil RK, 2018a. Concomitant changes in radiation resistance and trehalose levels during life stages of *Drosophila melanogaster* suggest radio-protective function of trehalose. Int J Radiat Biol. 94(6):576–589.

Paithankar JG, Raghu SV, Patil RK, 2018b. Levels and fluxes in enzymatic antioxidants following gamma irradiation are inadequate to confer radiation resistance in *Drosophila melanogaster*. Mol Biol Rep. 45(5):1175–1186.

Paithankar JG, Kudva AK, Raghu SV, Patil RK, 2020. Radio-protective role of uric acid: evidences commencing with studies on Drosophila and Human dermal fibroblast cells. Mol Biol Rep. 47(4):2427–2436.

Parashar V, Frankel S, Alan GL, Blanka R, 2008. The effects of age on radiation resistance and oxidative stress in adult *Drosophila melanogaster*. Radiat Res. 169:707–711.

Puanmanee K, Wongpiyasatid A, Sutantawong M, Hormchan P, 2010. Gamma irradiation effect on guava fruit fly, Bactrocera correcta (Bezzi) (Diptera: Tephritidae), Kasetsart. J Nat Sci. 44:830–836.

Ranawat P, Rawat S, 2017. Radiation resistance in thermophiles: mechanisms and applications. World J Microbiol Biotechnol. 33:112.

Robinson AS. 2005. Genetic basis of the sterile insect technique. In Hendrichs, J, Dyck VA, Robinson AS, editors, Sterile insect technique principles and practice in area-wide integrated pest management. Berlin: Springer. p. 95–114.

Sia J, Szmyd R, Hau E, Gee HE, 2020. Molecular mechanisms of radiation-induced cancer cell death: a primer. Front Cell Dev Biol. 8(41):1–8.

Seal DR, Tilton EW, 1986. Effect of gamma radiation on the metamorphic stages of *Dermestes maculatus* DeGeer (Coleoptera: Dermestidae). Appl Radiat Isot. 37(6):531–535.

Shin JM, Kim BK, Seo SG, Jeon SB, Kim JS, Jun B, Kang SY, Lee JS, Chung MN, Kim SH, 2011. Mutation breeding of sweet potato by gamma-ray radiation. Afr J Agric Res. 6(6):1447–1454.

Srimartpirom M, Burikam I, Limohpasmanee W, Kongratarporn T, Thannarin T, Bunsiri A, Follett PA, 2018. Low-dose irradiation with modified atmosphere packaging for mango against the oriental fruit fly (Diptera: Tephritidae). J Econ Entomol. 111(1):135–140.

Sun J, Chen Y, Li M, Ge Z, 1998. Role of antioxidant enzymes on ionizing radiation resistance. Free Radic Biol Med. 24(4):586–593.

Tilton EW, Burkholder WE, Cogburn RR, 1966. Effects of gamma radiation on *Rhyzopertha dominica*, *Sitophilus oryzae*, *Tribolium confusum*, and *Lasioderma serricorne*. J Econ Entomol. 59:1363–1368.

Torossian C, Causse R, 1968. The effect of gamma radiation on the fertility and longevity of *Dolichoderus quadripunctatus* (Hymenoptera: *Formicoidea dolichoderidae*); effets des radiations gamma sur la fertilite et la longevite des colonies de *Dolichoderus quadripunctatus* (Hymenoptere: *Formicoidea dolichoderidae*). International Atomic Energy Agency, Vienna (Austria). p. 155–162.

Truman JW, 2019. The evolution of insect metamorphosis. Current Biology. 29(23):R1252–R1268.

Watanabe M, Tetsuya S, Akihiko F, Takahiro K, Yuichi N, Nobuyuki H, Daiki DH, Wada S, Funyama T, Yasuhiko K, et al. 2006. Estimation of radiation tolerance to high LET heavy ions in an anhydrobiotic insect, *Polypedilum vanderplanki*. Int J Radiat Biol. 82(12):835–842.

Willard WK, Cherry DS, 1975. Comparative radiosensitivity in the class Insecta. J Theor Biol. 52:149–158.

Williams DD, Nesterovitch AI, Tavares AF, Muzzatti EG, 2001. Morphological deformities occurring in Belarusian Chironomids (Diptera; Chironomidae) subsequent to the Chernobyl nuclear disaster. Freshw Biol. 46(4):503–512.

Zedek F, Bureš P, 2019. Pest arthropods with holocentric chromosomes are more resistant to sterilizing ionizing radiation. Radiat Res. 191(3):255–261.

The *sine qua non* of the fish invitrome today and tomorrow in environmental radiobiology

Nguyen T. K. Vo

ABSTRACT

Fish cell lines, collectively referred to as the fish invitrome, are useful diagnostic tools to study radiation impacts on aquatic health and elucidate radiation mechanisms in fish. This paper will highlight the advantages, discuss the challenges, and propose possible future directions for uses of the fish invitrome in the field of environmental radiobiology. The fish invitrome contains at least 714 fish cell lines. However, only a few of these cell lines have been used to study radiation biology in fish and they represent only 10 fish species. The fish invitrome is clearly not yet explored for its full potential in radiation biology. Evidence suggests that they are useful and, in some cases, irreplaceable in making underlying theories and fundamental concepts in radiation responses in fish. The debate of whether environmental radiation is harmful, presents risks, has no effect on health, or is beneficial is on-going and is one that fish cell lines can help address in a time-effective fashion. Any information obtained with fish cell lines is useful in the framework of environment radiation risk assessments. Radiation threats to aquatic health will continue due to the very likely rise of nuclear energy and medicine in the future. The fish invitrome, in theory, lives forever and can meet new challenges at any given time to provide diagnostic risk analyses pertaining to aquatic health and environmental radiation protection.

1. Introduction

Threats to aquatic health have been and will continue to be an important focus of monitoring, analysis, reassessment, and refinement of water quality and management due to various uncertainties and unknown variables in regional and global water security. Radiological contamination, either from natural or man-made sources, heightens these issues and can pose serious concerns, especially in light of potential multiple stressor and transgenerational effects. The Fukushima Dai-ichi nuclear power plant meltdown accident in 2011 is the most recent example of a catastrophic release of 7000 PBq of radioactive materials into the environment and open water (Imanaka et al. 2015). In fact, it was only the 2nd case after the Chernobyl nuclear disaster in 1986 to receive the highest level 7 classification for 'widespread health and environmental effects' on the International Nuclear and Radiological Event Scale by the International Atomic Energy Agency (IAEA 2013). Chernobyl and Fukushima radiation effects on aquatic animal health have been reported (Belova et al. 2007; Suzuki 2015). Radiological contamination can also come from nuclear weapon tests and waste discharge from industrial and medical nuclear facilities. It is recognized that radiation exposure types and levels are geographically disproportionate and species sensitivity in radiation responses is varied. Therefore, the ability to access radiobiological effects in a species-specific, time-efficient, labor- and cost-effective is essential.

Fish are considered to be the most radiosensitive animal group for aquatic poikilotherms (Sazykina and Kryshev 2003). As of today, 35,423 fish species have been identified, accounting for 49% of all vertebrates (IUCN 2020). It is perhaps most fitting to study aquatic health using fish species because of their complete dependence on water for the entire or a vital part of their life. Fish also are an important component of the food web and radiological contamination can affect food safety for human consumption. As the biology and radiosensitivity of mammals and fish are vastly different (Whicker and Schultz 1982), using human/mammalian data to predict what could happen to fish is inappropriate. To date, radiobiological effects have been studied in only a small number of fish species, thus limiting how we understand various adaptive mechanisms fish species use to respond to stressors. It is a challenge to examine so many fish species in a well-planned-out and well-thought-out studies due to time, labor, funding and technical constraints. There is nevertheless a need to expand studies with more fish species in both acute and chronic exposure to different radiation types. Such a need seeks an approach that is more accessible while still generating important information.

Continuous cell lines derived from fish, collectively referred to as the fish invitrome (Bols et al. 2017), are powerful biotechnological creations. For decades, the fish invitrome has revolutionized aquatic health sciences and

Table 1. List of fish cell lines that were used to study environmental radiation biology.

Species	Cell line	Tissue, cell type	References	Radiation type, exposure type	Evaluation endpoint(s)
American eel (*Anguilla rostrata*)	eelB	Brain, endothelial	Vo et al. (2017a, 2017b, 2019a, 2019b)	Gamma, acute Beta, chronic	clonogenic cell survival, mitochondrial functions
	PBLE	Peripheral blood, fibroblastic	O'Neill-Mehlenbacher et al. (2007)	Gamma, acute	Clonogenic cell survival
Channel catfish (*Ictalurus punctatus*)	3B11	B lymphoblast	Stuart et al. (2016)	Beta, chronic	Cellular functions, cell survival, genomic aberrations
Chinook salmon (*Oncorhynchus tshawytscha*)	CHSE-214	Embryo, epithelial	O'Neill-Mehlenbacher et al. (2007); Cassidy et al. (2007); Shi et al. (2016a, 2016b, 2018)	Alpha, chronic Gamma, acute	clonogenic cell survival; mitochondrial functions
Fathead minnow (*Pimephales promelas*)	EPC	Skin, epithelial	O'Reilly and Mothersill (1997)	UV, acute Gamma, acute	Clonogenic cell survival
	FHM	Connective tissues, epithelial	Belisheva et al. (2012)	Secondary solar cosmic rays, acute	Genomic aberrations
	FHMT-W1	Testis, epithelial	Stuart et al. (2016)	Beta, chronic	Cellular functions, cell survival, genomic aberrations
Goldfish (*Carassius auratus*)	CAF-MM1	Fin, unknown	Mitani et al. (1982); Mitani (1983)	UV, acute Gamma, acute	clonogenic cell survival
	GEM 199	Erythrophoroma, unknown	Mitani (1983)	Gamma, acute	clonogenic cell survival
	RBCF-1	Fin, fibroblast	Shima and Setlow (1984)	UV, acute	clonogenic cell survival; DNA lesions
Medaka (*Oryzias latipes*)	CAB-2	Fin, fibroblast	Cassidy et al. (2007)	Gamma, acute	Genomic aberrations
	Ol-Hdr R-e3	Embryo, fibroblast	Cassidy et al. (2007)	Gamma, acute	Genomic aberrations
Mudminnow (*Umbra limi*)	ULF-23	Fin, fibroblast	Kurihara et al. (1992)	X-rays, acute	Chromosomal aberrations
Rainbow trout (*Oncorhynchus mykiss*)	RTG-2	Gonads, fibroblastic	O'Neill-Mehlenbacher et al. (2007); Cassidy et al. (2007)	Gamma, acute	Clonogenic survival, genomic aberrations
	RTgill-W1	Gill, epithelial	O'Neill-Mehlenbacher et al. (2007)	Gamma, acute	Clonogenic cell survival
	RTS-34st	Spleen, stromal	O'Neill-Mehlenbacher et al. (2007); Xing et al. (2009)	Gamma, acute	Clonogenic cell survival, cell arrest
	RTS-pBk+	Spleen, unknown	O'Neill-Mehlenbacher et al. (2007)	Gamma, acute	Clonogenic cell survival
Tilapia (*Tilapia mossainbica* x *T. nilotica*)	TK	Kidney, epithelial	Yew and Chang (1984)	UV, acute	DNA synthesis and repair, cell survival
Zebrafish (*Danio rerio*)	ZEB2J	Embryo, epithelial	O'Neill-Mehlenbacher et al. (2007)	Gamma, acute	Clonogenic cell survival
	ZF4	Embryo, fibroblast	Pereira et al. (2011)	Gamma, acute and chronic	DNA damage
	ZF-L	Liver, epithelial	Costa et al. (2019)	UV, acute	DNA repair, gene expression
	ZSSJ	Spleen, stromal	Xing et al. (2009)	Gamma, acute	Cell arrest

served as useful diagnostic tools in both basic research and clinical applications. Since the establishment of the first fish cell line RTG-2 (Wolf and Quimby 1962), at least 714 fish cell lines have been developed according to the Cellosaurus database (Bairoch 2018). There are more that have been developed and some were never formally described in the literature or public domains so could have been lost. Surprisingly, only a small fraction of the reported fish cell lines has been used in radiation biology and they represent only 10 fish species (Table 1).

Exploiting and expanding the fish invitrome to study radiation biology in teleosts presents promising opportunities and, more importantly, is genuinely practical. Fish cell lines are species- and tissue-specific and therefore have the genetic information, cellular physiological characteristics, and biochemical functions of the species/tissue they represent. Compared to the mammalian/human invitrome, they require significantly less care so are more cost-effective. They grow indefinitely and enable experimentations in a time-effective manner without fear of losing the experimental materials for good. Additionally, fish cell lines have consistent baseline levels of measured endpoints and increases/decreases above and below the baseline would be caused by radiation effects. This makes result comparisons between studies and research groups more feasible and reliable. Fish cell lines can be used to study how radiation exposure affects mutation loads, chromosomal aberrations, micronucleus number, mitochondrial functions, differential gene and protein expressions, to name a few. For large-size fish species, required space and infrastructure to keep them can be a limiting factor. Using cell lines eliminates this challenge. In contrast to mammalian cell lines, fish cell lines are known to withstand a wide range of temperatures and oxygen fluctuations (Vo et al. 2015a; Bols et al. 2017; Semple et al. 2017). This would make it possible to bring fish cells enclosed in sterile vessels on research expedition trips without fear of loss in cell viability of function (Curtis et al. 2013; Vo et al. 2016). In this manner, they can be used as species-specific and tissue-specific biodosimeters in locales of increased radioactivity levels to measure real-time induced damage levels.

This paper will highlight the advantages, discuss the challenges, and propose possible future directions for uses of the fish invitrome in the field of environmental radiobiology.

2. Use of fish cell lines in discovering new radiation response mechanisms in fish

Many fundamental concepts and underlying theories in radiation biology have been established based on the mammalian invitrome. While some concepts are rudimentary and its validity is, without question, firm, others are worthwhile to be reexamined or explored with the fish invitrome. Some unique features of fish biology are described below; however, this is not an exhaustive list that perhaps may become more definitive in another paper or a different special issue elsewhere.

First, fish undergo a number of whole-genome duplication events (Glasauer and Neuhauss 2014) throughout evolution. Different gene variants could impact radiation responsive pathways as some do in other molecular and biological functions in fish (Sever et al., 2014, 2018).

Second, fish are cold-blooded. Temperature tolerance ranges vary drastically from one species to another. Biochemical functions, cellular activities, and physiological performances in fish change when water temperature changes. These changes are more pronounced when the temperature shift is large and sudden as well as at the temperature extremes; in some cases biological functions are suppressed whereas in other cases they are maintained normally regardless of temperature shifts. Radiation responses are influenced by thermal energies. The level of interactions between temperature-sensitive expressions and functions of biomolecules with radiation responses is not well understood in fish. Concepts of hypothermia and hyperthermia in radiation responses that are well-studied in radiation oncology initially seem peculiar for fish. In fish physiology, they are important in our understanding in fish health in the winter months and due to global warming, respectively, when radiation exposure is involved.

Third, fish respiratory system, in contrast to mammals whose lungs are internal, is externally exposed to the environment, making it a vital organ susceptible to both dense and sparse ionizing radiation.

Fourth, fish develop adaptative mechanisms to oxygen tensions (Gattuso et al. 2018; Montgomery et al. 2019), raising the question of the extent of oxygen effect's contribution to their radiation responses.

Fifth, their cellular senescence program operates differently from the mammalian counterparts (Hartmann et al. 2009; Graf et al. 2013; Vo et al. 2015b, 2018).

Sixth, the tumor suppressor protein p53 is central to the DNA damage response and differences in its function and regulation have already been noted (Krause et al. 1997; Chen et al. 2001; Rau Embry et al. 2006; Liu et al. 2011; Zeng et al. 2016).

These data suggest that some aspects of radiation responses could be different in fish when compared to mammalian biology, but the degree of how different is unknown. It is surprising how little has been done with fish cell lines to study radiation-associated molecular and biochemical mechanisms in fish. Taking advantage of the large fish invitrome would be a catalyst in elucidating potentially new radiation response mechanisms.

Use of the fish invitrome would be a more ethical approach when it concerns endangered, threatened and at-risk species. Thanks to recent advances in *in vitro* animal cell sciences, generating continuous cell lines from fish is now more robust than ever and even possible with non-lethal approaches (for example, use of fin clippings). Novel radiation responses in those fish at the cellular level could be studied without further threatening their survival.

3. Use of fish cell lines in clonogenic survival assays in radiation biology

Clonogenic survival assays or colony-forming assays are one of the most important *in vitro* techniques in radiation biology. The beginning of the assay development exploited cell lines of the human invitrome (Puck and Marcus 1956). The assay and the invitrome have been instrumental to the establishment of many underlying theories and concepts in radiation biology such as linear-quadratic cell survival models, radiosensitivity, relative biological effectiveness, fractionated radiation, oxygen effects, radiation carcinogenesis, low-dose hypersensitivity, radiation hormesis, and tumor vs. normal cell responses to radiation (Hall and Giaccia 2006).

Clonogenic assays evaluate the potential of single cells to survive and replicate (or self-clone) indefinitely when placed into tissue culture vessels at very low cell densities and free from close interactions with neighbor cells (Puck and Marcus 1956). Unlike normal differentiated cells, continuous cell lines, which *in vitro* are thought to be representative of stem cells, tissue progenitor cells, cancer cells, or cancer stem cells, have high cloning efficiency (CE). If cultured normal differentiated cells increase their CE after irradiation, this suggests that the cells are transformed and become tumourigenic and information learned would be useful in assessing radiation-associated cancer risks. Irradiated cells can reduce their CE due to reproductive cell death (RCD) when damage is beyond repair. RCD is the impairment of cellular reproductive integrity and encompasses several cell fates such as programmed cell death, senescence, and failure of clonal expansion (Hall and Giaccia 2006). Ultimately, clonogenic survival assays enable the quantitative measurement of radiation-induced RCD. This concept is particularly useful in radiation oncology in the interest of finding dose the effective dose to eradicate the tumor or in rehabilitation therapy for knowing the tissue progenitor cells' repopulation capacity and rate after radiation treatment. Typically, cell survival curves are constructed after clonogenic surviving fraction is scored to determine the radiobiological response. One can learn a great deal about the nature and consequence of radiation interactions with cells based on the shape and the shoulder of the cell survival curve and mathematically derived values. For example, survival curves for gamma-irradiated fish cells have a much wider shoulder and a slope that opens the curve more broadly as compared to those for mammalian cells (Mitani et al. 1982; Mitani 1983; Vo et al. 2017a). For fish cells, the survival curve is generally linear at 5 Gy or higher, which is the dose range that would be in the quadratic region for mammalian cells (Mitani et al. 1982; Mitani 1983; Dowling et al. 2005; Ryan et al. 2008; Vo et al. 2017a, 2017b). Reproductive lethality does not occur at <0.5 Gy direct irradiation (Mitani et al. 1982; Mitani 1983; Ryan et al. 2008; Vo et al. 2017a, 2017b). n (extrapolation number), D_o (mean lethal dose), and D_q (threshold dose) values can be calculated from cell survival curves. They represent relative radiosensitivity of cells. For eelB brain endothelial cells, n = 4.42 Gy, Do = 2.16 Gy, and Dq =3.21 Gy (Vo et al. 2017a). These values are greater than 3.2 Gy for goldfish cells (Mitani et al. 1982; Mitani 1983). These numbers suggest that eelB cells are more radiosensitive than goldfish cells and are more capable of recovering from sub-lethal damages. Together, these data suggest that just like *in vivo* fish cells are more radioresistant than mammalian cells.

The concept of clonogenic survival assays can be applied to fish invitrome to study various aspects of radiation biology pertaining to fish or aquatic health. Questions about how *in vitro* results with a few fish cells in a culture flask and changes in their reproductive integrity enable the accurate data interpretation at the whole organism response are raised. Studies using the invitrome have never been meant to replace works in animals but rather complement or enhance them. However, in some cases, the invitrome is the *sine qua non* in science; for example, cell lines are used to make vaccines to control the epidemics and prevent/cure/eradicate infectious diseases. In radiation biology and pharmaceutical sciences, it is crucial that this concept be worth re-emphasis: regardless of species under study, the clonogenic survival assay, in principle, enables the quantitative measurement of radiation-induced RCD and is the only assay that can do this. Significant findings with the clonogenic survival assay provide a necessary radiation risk assessment to perform comprehensive studies to establish a link with possible adverse health effects.

In other for clonogenic assays to work, fish cells need to be clonogenic. Although there are at least 714 fish cell lines, it appears that only a small number of the fish cell lines are clonogenic. These include, but are not limited to, the embryonic CHSE-214 cell line from Chinook salmon (Shi et al. 2016a, 2016b), the embryonic ZEB-2J cell line from zebrafish (O'Neill-Mehlenbacher et al. 2007), the brain endothelial eelB cell line from American eel (Vo et al. 2017a, 2017b), the EPC cell line from fathead minnow (O'Reilly and Mothersill 1997), the fin CAF-MM2 and RBCF-1 cell lines from goldfish (Mitani et al. 1982), the erythrophoroma GEM 199 cell line from goldfish (Mitani 1983; Shima and Setlow 1984), the gonadal RTG-2 and gill RTgill-W1 cell lines from rainbow trout (O'Neill-Mehlenbacher et al. 2007), the eight head kidney cell lines from Siberian sturgeon (Ryu et al. 2018), the caudal trunk BF-2 cell line from bluegill (Babich et al. 1986), the retinal epithelial cell line WE-rpe from walleye (Vo and Bols 2016), the heart endothelial ASHe cell line from Atlantic salmon (Pham et al. 2017), the embryonic MES-1 cell line from medaka (Hong et al. 2011), and the embryonic TES1 cell line from tilapia (Fan et al. 2017). At the present time, uncertainties remain as to why some fish cell lines are clonogenic and many others are not. Perhaps there are fish-specific clonogen-promoting factors that are lacking in commercial animal serum supplements in media. Lysophosphatidic acid, a type of phospholipid mitogen, is one example as it was found to promote clonogenic potential in fish heart endothelial cells (Pham et al. 2017).

Due to the limitation of few clonogenic fish cell lines, only several radiation cell survival curves have been constructed following gamma irradiation and some of the important radiobiological values determined (Mitani et al. 1982, Mitani 1983; Vo et al. 2017a, 2017b). One general conclusion is that fish cells are more radioresistant than mammalian cells and this trend holds true *in vivo* as well. This is evident by the broad shoulders of fish cell survival curves (Mitani et al. 1982, Mitani 1983; Vo et al. 2017b) and larger threshold dose values indicative of more efficient recovery from sub-lethal damages (Hall and Giaccia 2006; Vo et al. 2017b). Further use of clonogenic fish cell lines in examining other aspects of radiation biology is a promising research direction.

4. Use of fish cell lines in evaluating radiation risks in fish health

As responses to environmental radiation exposure are typically chronic and low-dose/low-dose-rate biology, the topic of radiation risks is of special concerns. Radiation risks can be broadly categorized into two groups: cancer risk and non-cancer risk.

Environmental radiation-associated cancer risks in fish are not a frequently reported subject in the literature. In the most recent study by Lerebours et al. (2018), it was found that fish from the heavily radioactive-contaminated lakes in the Chernobyl Exclusion Zone do not develop any sign of tumors. Given that the Chernobyl nuclear disaster was over 30 years ago, their results suggest that cancer risk in fish exposed to radiation in natural environments are probably very low to nil. Monitoring cancer morbidity in the environment can be difficult if routine surveillance with the right expertise is not performed. Additionally, there are other stressors in the natural environment that can contribute to the stochastic effects as well. As diagnostic tools, fish cell lines can aid in determining if one radiation exposure type, singularly or in combination with other stressors may lead to carcinogenesis, by using cell transformation assays. They can also aid in identifying sensitive tumourigenic biomarkers for exposure to environmental radiation by using omics technologies.

Environmental radiation-associated non-cancer risks are mostly hereditary and thus contribute to transgenerational effects (Hurem et al. 2018; Kamstra et al. 2018; Hancock et al. 2019a, 2019b, 2020). Understanding of how low of a radiation dose or dose rate can affect germline cells and the effects carried on to the offspring is pertinent to environmental radiation protection programs. As some species have a long and complex lifestyle, it may be a challenge for studies in transgenerational effects. Cell lines derived from those fish can be useful in tracking genetic and epigenetic changes in progenies of radiation surviving cells over a long period of time in the laboratory. This would give useful insights into what could occur in natural environments.

5. Use of fish cell lines in studying non-targeted effects in fish

The paradigm of radiation damage has shifted from one that focuses on DNA as the direct target of radiation and cause of disease to another that expands on the contribution of non-targeted effects (NTEs) (Mothersill and Seymour 2019). NTEs occur when cells or organisms exhibit changes due to factors other than direct products of radiation-induced DNA aberrations. NTEs include genomic instability, delayed lethal mutations, bystander effects, increased carcinogenesis, and radio-adaptive response (Mothersill et al. 2017). In the environment, all types of NTEs except for radio-adaptive response are considered to be harmful or at least presenting greater risks to holistic health. Research has shown that NTEs are governed by cell-to-cell communication and epigenetic mechanisms (Mothersill and Seymour 2019). NTEs predominate at low doses, which are more common in environmental radiation exposures and thus of pressing concerns (Mothersill et al. 2017, 2020) in environmental radiation protection.

Genomic instability refers to an increase in mutation loads and varied phenotypic expressions after subsequent generations. Delayed lethal mutations can develop as a result of genomic instability, which can lead to the diminishment of stem cell pools in the body and embryonic lethality. Genomic instability and delayed lethal mutations have been previously demonstrated in animals both kept in the laboratory and sampled in the wild (Shimada and Shima 2004; Hurem et al. 2018; Kamstra et al. 2018; Hancock et al. 2019a, 2019b, 2020). This concept can be conveniently demonstrated with fish cell lines. Recent demonstrations were with CHSE-214 and eelB cell lines following chronic low dose alpha and acute gamma irradiations (Shi et al. 2016b, 2018; Vo et al. 2017b, 2019b). Chronic irradiation with environmentally relevant doses of alpha-emitting Ra-226 induces genomic instability and delayed lethal mutations in CHSE-214 salmon embryonic cells (Shi et al. 2018). Acute gamma irradiation induces genomic instability and delayed lethal mutations in eelB brain endothelial cells but not in CHSE-214 cells (Vo et al. 2017b, 2019b; Shi et al. 2018). Inheritance of radiation-induced DNA methylation patterns from radiation parental survivors is strongly linked to genomic instability in fish (Kamstra et al. 2018). Studying DNA methylation mechanisms with fish cell lines is both convenient and high throughput.

Radiation-induced bystander effect (RIBE) is a phenomenon in which non-irradiated cells or organisms exhibit cellular aberrations as if they have been irradiated (Mothersill et al. 2017). Several mechanisms have been elucidated and appear to involve both chemical and physical means. As demonstrated in mammalian models, chemical means are mediated via gap junctions or exchange of damaging secretable signals via the extracellular medium (Azzam et al. 2001; Lyng et al. 2002; Le et al. 2017). RIBE has been demonstrated *in vivo* in fish (Mothersill et al. 2006; Smith et al. 2013a) and even at environmentally relevant concentrations of radium-226 (an alpha emitter) (Smith et al. 2013b). Irradiated fish were shown to release bystander signals into the water and affected non-irradiated bystander fish. Physical means involve at least UV biophotons emitted by irradiated cells (Le et al. 2015; Cohen et al. 2020). Many RIBE discoveries were made using the human invitrome and this can certainly be extended to the fish invitrome. Some fish cell lines do not appear to respond to RIBE signals (O'Neill-Mehlenbacher et al. 2007; Vo et al. 2017a) but they can produce them (O'Neill-Mehlenbacher et al. 2007; Vo et al. 2019b). This shows that bystander signal production and response are operated independently. It seems that fish cells produce and secrete molecules that protect from radiation damage (O'Neill-Mehlenbacher et al. 2007; Pereira et al. 2014; Vo et al. 2017a). Exosome profiling in irradiated cultured fish cells in combination bioinformatic analysis can be done to tease out what specific molecules in the exosomes contribute to damage induction. Among signaling molecules that mediate RIBE is serotonin, which is ubiquitously produced in all animals. Serotonin exacerbates the bystander response in human cells (Curtis et al. 2020) and in live fish (Saroya et al. 2009). Inhibitors of serotonin receptors effectively block RIBE (Saroya et al. 2009; Curtis et al. 2020). All of this information would be important in radiation risk assessment in environmental health.

Radio-adaptive response (RAR) is a protective response in which cells or organisms develop enhanced radioresistance. RAR is inducible by low doses including those found in environmental exposures. It works by activating various defense mechanisms including DNA repair machineries and oxidative stress scavengers, subsequently priming the system to defend against detrimental radiation attacks more effectively (Cohen et al. 2019). RAR was first demonstrated in goldfish cells (Kurihara et al. 1992) and occurs in fish *in vivo* as well (Choi et al. 2010) in acute exposures. By contrast, Shi et al. (2016a) showed that chronic very low-dose radium-226 exposure did not induce an RAR in Chinook salmon CHSE-214 cells. The molecular mechanisms of RAR in fish are largely not well studied. Fish cell lines can be used to elucidate the mechanisms at play in the future.

6. Use of fish cell lines in studying multiple stressor effects in fish

Radioecological evidence suggests that organisms in wildlife are more radiosensitive than when they are studied in controlled environments in the laboratory (Garnier-Laplace et al. 2013). This is likely due to their exposure to other environmental stressors besides radiation that act as radiosensitizers. Complex mixtures of stressors can be difficult and time-consuming to separate one by one in the field. Conversely, by combining known or suspected stressors one by one with radiation challenge, multiple-stressor studies can be performed more quickly with fish cell lines in the laboratory. The end result is to determine possible outcomes and yield novel observations that can substantiate what is reported in the field. Two recent demonstrations of how fish cell lines can be used in multiple-stressor studies involving radiation were with the RTG-2 and eelB cell lines from rainbow trout and American eel, respectively. RTG-2 was used to study how environmental stressors such as heat shock and a wastewater contaminant can influence radiation responses (Sreetharan et al. 2018). eelB was used to show that the common field lampricide used to kill larval sea lampreys is a potential radiosensitizer for fish (Vo et al. 2019a). Using fish cell lines to study many classes of chemical and physical stressors that have not been systematically investigated in combination with radiation can be a research direction that is worth exploring.

Additionally, infections can add to stress burdens in susceptible animals, along with activated immune responses, can exacerbate radiation responses. The extent of how much infections and immune responses interact with radiation responses in fish is not known. This is possibly due to the lack of expertise crossing multiple specialized disciplines such as radiation biology, radioecology, fish biology, fish immunology, fish virology, and fish microbiology. Perhaps fish cell lines can be the first approach to make useful observations into what can take place.

7. Use of fish cell lines in studying the biology of p53 in radiation responses in fish

The p53 protein is regarded to as the 'guardian of the genome' and is critical in regulating DNA damage repair as well as NTEs (Lane 1992; Mothersill et al. 2011; Cohen et al. 2019). About 50% of human cancers have the *TP53* gene encoding p53 mutated (Hollstein et al. 1991), accentuating the role of p53 in genome stability and cancer development. Acquired p53 mutants dictate whether cells are more and less sensitive to radiation responses (Lee and Bernstein 1993; Ribeiro et al. 1997; Murley et al. 2017). Immortalization of mammalian cell lines often results in the *TP53* gene mutated and this has an impact on how cell lines respond to radiation (Harvey and Levine 1991; Lehman et al. 1993). Presently, we do not know if the gene is mutated during the immortalization of fish cell lines as well as the p53 status of the established fish cell lines. Finding this out and making correlations with sensitivity of the cell lines' radiation responses would provide us valuable information about fish genome biology and p53 biology and their associated changes in response to environmental radiation.

8. Concluding remarks

The fish invitrome is large; however, its potential in radiation biology is currently not fully explored. While the mammalian/human invitrome is extensively used to study various aspects of radiation for human health and medicine, the fish invitrome can certainly help to elucidate radiation mechanisms in fish and associated risks for aquatic health.

There are several unique cell lines in the fish invitrome that can be further explored just like the mammalian/human counterparts. I am briefly discussing one example with the rainbow trout lens epithelial RTLens cell line (Li 2019) in the study of radiation cataractogenesis in fish. Radiation cataractogenesis can occur in response to low dose radiation and has recently been suspected to be due to NTEs rather than deterministic effects (Laskowski et al. 2020). Fish in the wild or in domestication/hatcheries succumb to cataract development (Laycock et al. 2000; Bjerkås et al. 2004; Björnsson 2004). The involvement of environmental radiation on cataract formation in fish is not well studied and studying RTLens cells would a first effort in improving this poor literature.

As the 3Rs (Replacement, Reduction, and Refinement) initiative promoting less animal use in research is spearheaded in Europe (Scholz et al. 2013) and gains worldwide support including from Canada (by the establishment of the Canadian Center for Alternatives to Animal Methods in Windsor, Ontario), fish cell lines offer a unique ethical solution while ensuring the integrity and rigor of scientific discoveries. The fish invitrome can be a game changer in its profound contribution to environmental radiobiology. What it needs now are the expertise, the mind, the effort, and the open dialogue to push it forward.

Disclosure statement

The author reports no conflict of interest.

ORCID

Nguyen T. K. Vo http://orcid.org/0000-0002-4709-3078

References

Azzam EI, de Toledo SM, Little JB. 2001. Direct evidence for the participation of gap junction-mediated intercellular communication in the transmission of damage signals from alpha-particle irradiated to nonirradiated cells. Proc Natl Acad Sci U S A. 98:473–478.

Babich H, Shopsis C, Borenfreund E. 1986. In vitro cytotoxicity testing of aquatic pollutants (cadmium, copper, zinc, nickel) using established fish cell lines. Ecotoxicol Environ Saf. 11:91–99.

Bairoch A. 2018. The cellosaurus, a cell-line knowledge resource. Version 2. J Biomol Tech. 29:25–38.

Belisheva NK, Lammer H, Biernat HK, Vashenuyk EV. 2012. The effect of cosmic rays on biological systems – an investigation during GLE events. Astrophys Space Sci Trans. 8:7–17.

Belova NV, Emel'yanova NG, Makeeva AP, Ryabov IN. 2007. The state of the reproductive system of several fish species from water bodies polluted with radionuclides during the Chernobyl catastrophe. J Ichthyol. 47:366–384.

Bjerkås E, Holst JC, BjerkåS I, 2004. Cataract in farmed and wild Atlantic salmon (*Salmo salar* L.). Anim Eye Res. 23:3–13.

Björnsson B. 2004. Can UV-treated seawater cause cataract in juvenile cod (*Gadus morhua* L.)? Aquaculture. 240:187–199.

Bols NC, Pham PH, Dayeh VR, Lee LEJ. 2017. Invitromatics, invitrome, and invitroomics: introduction of three new terms for in vitro biology and illustration of their use with the cell lines from rainbow trout. In Vitro Cell Dev Biol Anim. 53:383–405.

Cassidy CL, Lemon JA, Boreham DR. 2007. Impacts of low-dose gamma-radiation on genotoxic risk in aquatic ecosystems. Dose Response. 5:323–332.

Chen S, Hong Y, Scherer SJ, Schartl M. 2001. Lack of ultraviolet-light inducibility of the medakafish (*Oryzias latipes*) tumor suppressor gene p53. Gene. 264:197–203.

Choi VW, Konishi T, Oikawa M, Iso H, Cheng SH, Yu KN. 2010. Adaptive response in zebrafish embryos induced using microbeam protons as priming dose and X-ray photons as challenging dose. J Radiat Res. 51:657–664.

Cohen J, Vo NTK, Chettle D, McNeill FE, Seymour CB, Mothersill CE. 2020. Quantifying biophoton emissions from human cells directly exposed to low-dose gamma radiation. Dose Response. 18: 1559325820926763.

Cohen J, Vo NTK, Seymour CB, Mothersill CE. 2019. Parallel comparison of pre-conditioning and post-conditioning effects in human cancers and keratinocytes upon acute gamma irradiation. Int J Radiat Biol. 95:170–178.

Costa SR, Velasques RR, Hoff MLM, Souza MM, Sandrini JZ. 2019. Characterization of different DNA repair pathways in hepatic cells of Zebrafish (*Danio rerio*). DNA Repair (Amst). 83:102695.

Curtis JJ, Vo NTK, Seymour CB, Mothersill CE. 2020. 5-HT_{2A} and 5-HT_3 receptors contribute to the exacerbation of targeted and non-targeted effects of ionizing radiation-induced cell death in human colon carcinoma cells. Int J Radiat Biol. 96:482–490.

Curtis TM, Collins AM, Gerlach BD, Brennan LM, Widder MW, van der Schalie WH, Vo NTK, Bols NC. 2013. Suitability of invertebrate and vertebrate cells in a portable impedance-based toxicity sensor: temperature mediated impacts on long-term survival. Toxicol In Vitro. 27:2061–2066.

Dowling K, Seymour C, Mothersill C. 2005. Delayed cell death and bystander effects in the progeny of Chinook salmon embryo cells exposed to radiation and a range of aquatic pollutants. Int J Radiat Biol. 81:89–96.

Fan Z, Liu L, Huang X, Zhao Y, Zhou L, Wang D, Wei J. 2017. Establishment and growth responses of Nile tilapia embryonic stem-like cell lines under feeder-free condition. Dev Growth Differ. 59: 83–93.

Garnier-Laplace J, Geras'kin S, Della-Vedova C, Beaugelin-Seiller K, Hinton TG, Real A, Oudalova A. 2013. Are radiosensitivity data derived from natural field conditions consistent with data from controlled exposures? A case study of Chernobyl wildlife chronically exposed to low dose rates. J Environ Radioact. 121:12–21.

Gattuso A, Garofalo F, Cerra MC, Imbrogno S. 2018. Hypoxia tolerance in teleosts: implications of cardiac nitrosative signals. Front Physiol. 9:366.

Glasauer SM, Neuhauss SC. 2014. Whole-genome duplication in teleost fishes and its evolutionary consequences. Mol Genet Genomics. 289: 1045–1060.

Graf M, Hartmann N, Reichwald K, Englert C. 2013. Absence of replicative senescence in cultured cells from the short-lived killifish *Nothobranchius furzeri*. Exp Gerontol. 48:17–28.

Hall EJ, Giaccia AJ. 2006. Radiobiology for the radiologist. 6th ed. Philadelphia (PA): Lippincott Williams & Wilkins.

Hancock S, Vo NTK, Byun SH, Zainullin VG, Seymour CB, Mothersill CE. 2019a. Effects of historic radiation dose on the frequency of sex-linked recessive lethals in Drosophila populations following the Chernobyl nuclear accident. Environ Res. 172:333–337.

Hancock S, Vo NTK, Goncharova RI, Seymour CB, Byun SH, Mothersill CE. 2020. One-decade-spanning transgenerational effects of historic radiation dose in wild populations of bank voles exposed to radioactive contamination following the Chernobyl nuclear disaster. Environ Res. 180:108816.

Hancock S, Vo NTK, Omar-Nazir L, Batlle JVI, Otaki JM, Hiyama A, Byun SH, Seymour CB, Mothersill C. 2019b. Transgenerational effects of historic radiation dose in pale grass blue butterflies around Fukushima following the Fukushima Dai-ichi nuclear power plant meltdown accident. Environ Res. 168:230–240.

Hartmann N, Reichwald K, Lechel A, Graf M, Kirschner J, Dorn A, Terzibasi E, Wellner J, Platzer M, Rudolph KL, et al. 2009. Telomeres shorten while Tert expression increases during ageing of the short-lived fish *Nothobranchius furzeri*. Mech Ageing Dev. 130: 290–296.

Harvey DM, Levine AJ. 1991. p53 alteration is a common event in the spontaneous immortalization of primary BALB/c murine embryo fibroblasts. Genes Dev. 5:2375–2385.

Hollstein M, Sidransky D, Vogelstein B, Harris CC. 1991. p53 mutations in human cancers. Science. 253:49–53.

Hong N, Li Z, Hong Y. 2011. Fish stem cell cultures. Int J Biol Sci. 7: 392–402.

Hurem S, Martín LM, Lindeman L, Brede DA, Salbu B, Lyche JL, Aleström P, Kamstra JH. 2018. Parental exposure to gamma radiation causes progressively altered transcriptomes linked to adverse effects in zebrafish offspring. Environ Pollut. 234:855–863.

Imanaka T, Hayashi G, Endo S. 2015. Comparison of the accident process, radioactivity release and ground contamination between Chernobyl and Fukushima-1. J Radiat Res. 56:i56–i61.

International Atomic Energy Agency (IAEA). 2013. INES: The International Nuclear and Radiological Event Scale User's Manual. [accessed 2020 May 23]. https://www-pub.iaea.org/MTCD/Publications/PDF/INES2013web.pdf

International Union for Conservation of Nature and Natural Resources (IUCN). 2020. The IUCN Red List of Threatened Species. Version 2020-1. Table 1a "Number of species evaluated in relation to the overall number of described species, and numbers of threatened species by major groups of organisms". [accessed 2020 May 24]. https://www.iucnredlist.org/resources/summary-statistics#Summary%20Tables

Kamstra JH, Hurem S, Martin LM, Lindeman LC, Legler J, Oughton D, Salbu B, Brede DA, Lyche JL, Aleström P. 2018. Ionizing radiation induces transgenerational effects of DNA methylation in zebrafish. Sci Rep. 8:15373.

Krause MK, Rhodes LD, Van Beneden RJ. 1997. Cloning of the p53 tumor suppressor gene from the Japanese medaka (*Oryzias latipes*) and evaluation of mutational hotspots in MNNG-exposed fish. Gene. 189:101–106.

Kurihara Y, Rienkjkarn M, Etoh H. 1992. Cytogenetic adaptive response of cultured fish cells to low doses of X-rays. J Radiat Res. 33:267–274.

Lane DP. 1992. Cancer. p53, guardian of the genome. Nature. 358: 15–16.

Laskowski L, Williams D, Seymour C, Mothersill C. 2020. Environmental and industrial developments in radiation cataractogenesis. Int J Radiat Biol.

Laycock NLC, Schirmer K, Bols NC, Sivak JG. 2000. Optical properties of rainbow trout lenses after in vitro exposure to polycyclic aromatic hydrocarbons in the presence or absence of ultraviolet radiation. Exp Eye Res. 70:205–214.

Le M, Fernandez-Palomo C, McNeill FE, Seymour CB, Rainbow AJ, Mothersill CE. 2017. Exosomes are released by bystander cells exposed to radiation-induced biophoton signals: reconciling the mechanisms mediating the bystander effect. PLoS One. 12:e0173685.

Le M, McNeill FE, Seymour C, Rainbow AJ, Mothersill CE. 2015. An observed effect of ultraviolet radiation emitted from beta-irradiated HaCaT cells upon non-beta-irradiated bystander cells. Radiat Res. 183:279–290.

Lee JM, Bernstein A. 1993. p53 mutations increase resistance to ionizing radiation. Proc Natl Acad Sci U S A. 90:5742–5746.

Lehman TA, Modali R, Boukamp P, Stanek J, Bennett WP, Welsh JA, Metcalf RA, Stampfer MR, Fusenig N, Rogan EM. 1993. p53 mutations in human immortalized epithelial cell lines. Carcinogenesis. 14: 833–839.

Lerebours A, Gudkov D, Nagorskaya L, Kaglyan A, Rizewski V, Leshchenko A, Bailey EH, Bakir A, Ovsyanikova S, Laptev G, et al. 2018. Impact of environmental radiation on the health and reproductive status of fish from chernobyl. Environ Sci Technol. 52: 9442–9450.

Li Z. 2019. Effects of temperature on the proliferation of human and fish lens epithelial cells [master's thesis]. Waterloo (ON): University of Waterloo.

Liu M, Tee C, Zeng F, Sherry JP, Dixon B, Bols NC, Duncker BP. 2011. Characterization of p53 expression in rainbow trout. Comp Biochem Physiol C Toxicol Pharmacol. 154:326–332.

Lyng FM, Seymour CB, Mothersill C. 2002. Early events in the apoptotic cascade initiated in cells treated with medium from the progeny of irradiated cells. Radiat Prot Dosimetry. 99:169–172.

Mitani H. 1983. Lethal and mutagenic effects of radiation and chemicals on cultured fish cells derived from the erythrophoroma of goldfish (*Carassius auratus*). Mutat Res. 107:279–288.

Mitani H, Etoh H, Egami N. 1982. Resistance of a cultured fish cell line (CAF-MM1) to γ irradiation. Radiat Res. 89:334–347.

Montgomery DW, Simpson SD, Engelhard GH, Birchenough SNR, Wilson RW. 2019. Rising CO_2 enhances hypoxia tolerance in a marine fish. Sci Rep. 9:15152.

Mothersill C, Bristow RG, Harding SM, Smith RW, Mersov A, Seymour CB. 2011. A role for p53 in the response of bystander cells to receipt of medium borne signals from irradiated cells. Int J Radiat Biol. 87:1120–1125.

Mothersill C, Bucking C, Smith RW, Agnihotri N, Oneill A, Kilemade M, Seymour CB. 2006. Communication of radiation-induced stress or bystander signals between fish in vivo. Environ Sci Technol. 40: 6859–6864.

Mothersill C, Oughton DH, Schofield Pn, Abend M, Adam-Guillermin C, Ariyoshi K, Beresford NA, Bonisoli-Alquati A, Cohen J, Dubrova Y. 2020. From tangled banks to toxic bunnies; a reflection on the issues involved in developing an ecosystem approach for environmental radiation protection. Int J Radiat Biol. DOI: 10.1080/09553002.2020.1793022

Mothersill C, Rusin A, Seymour C. 2017. Low doses and non-targeted effects in environmental radiation protection; where are we now and where should we go? Environ Res. 159:484–490.

Mothersill C, Seymour C. 2019. Targets, pools, shoulders, and communication - a reflection on the evolution of low-dose radiobiology. Int J Radiat Biol. 95:851–860.

Murley J, Miller R, Weichselbaum R, Grdina D. 2017. TP53 mutational status and ROS effect the expression of the survivin-associated radio-adaptive response. Radiat Res. 188:659–670.

O'Neill-Mehlenbacher A, Kilemade M, Elliott A, Mothersill C, Seymour C. 2007. Comparison of direct and bystander effects induced by ionizing radiation in eight fish cell lines. Int J Radiat Biol. 83:593–602.

O'Reilly JP, Mothersill C. 1997. Comparative effects of UV A and UV B on clonogenic survival and delayed cell death in skin cell lines from humans and fish. Int J Radiat Biol. 72:111–119.

Pereira S, Bourrachot S, Cavalie I, Plaire D, Dutilleul M, Gilbin R, Adam-Guillermin C. 2011. Genotoxicity of acute and chronic gamma-irradiation on zebrafish cells and consequences for embryo development. Environ Toxicol Chem. 30:2831–2837.

Pereira S, Malard V, Ravanat JL, Davin AH, Armengaud J, Foray N, Adam-Guillermin C. 2014. Low doses of gamma-irradiation induce an early bystander effect in zebrafish cells which is sufficient to radioprotect cells. PLoS One. 9:e92974.

Pham PH, Vo NTK, Tan EJH, Russell S, Jones G, Lumsden JS, Bols NC. 2017. Development of an Atlantic salmon heart endothelial cell line (ASHe) that responds to lysophosphatidic acid (LPA). In Vitro Cell Dev Biol Anim. 53:20–32.

Puck TT, Marcus PI. 1956. Action of X-rays on mammalian cells. J Exp Med. 103:653–666.

Rau Embry M, Billiard SM, Di Giulio RT. 2006. Lack of p53 induction in fish cells by model chemotherapeutics. Oncogene. 25:2004–2010.

Ribeiro JC, Barnetson AR, Fisher RJ, Mameghan H, Russell PJ. 1997. Relationship between radiation response and p53 status in human bladder cancer cells. Int J Radiat Biol. 72:11–20.

Ryan LA, Seymour CB, O'Neill-Mehlenbacher A, Mothersill CE. 2008. Radiationinduced adaptive response in fish cell lines. J Environ Radioact. 99:739–747.

Ryu JH, Kim MS, Kang JH, Kim DH, Nam YK, Gong SP. 2018. Derivation of the clonal-cell lines from Siberian sturgeon (*Acipenser baerii*) head-kidney cell lines and its applicability to foreign gene expression and virus culture. J Fish Biol. 92:1273–1289.

Saroya R, Smith R, Seymour C, Mothersill C. 2009. Injection of resperpine into zebrafish, prevents fish to fish communication of radiation-induced bystander signals: confirmation in vivo of a role for serotonin in the mechanism. Dose Response. 8:317–330.

Sazykina TG, Kryshev AI. 2003. EPIC database on the effects of chronic radiation in fish: Russian/FSU data. J Environ Radioact. 68: 65–87.

Scholz S, Sela E, Blaha L, Braunbeck T, Galay-Burgos M, García-Franco M, Guinea J, Klüver N, Schirmer K, Tanneberger K, et al. 2013. A European perspective on alternatives to animal testing for environmental hazard identification and risk assessment. Regul Toxicol Pharmacol. 67:506–530.

Semple SL, Vo NTK, Li AR, Pham PH, Bols NC, Dixon B. 2017. Development and use of an Arctic charr cell line to study antiviral responses at extremely low temperatures. J Fish Dis. 40:1423–1439.

Sever L, Vo NTK, Bols NC, Dixon B. 2014. The rainbow trout (*Oncorhynchus mykiss*) contain two calnexin genes which encode distinct proteins. Dev Comp Immunol. 42:211–219.

Sever L, Vo NTK, Bols NC, Dixon B. 2018. Tapasin's protein interactions in the rainbow trout peptide-loading complex. Dev Comp Immunol. 81:262–270.

Shi X, Mothersill C, Seymour C. 2016a. No adaptive response is induced by chronic low-dose radiation from Ra-226 in the CHSE/F fish embryonic cell line and the HaCaT human epithelial cell line. Environ Res. 151:537–546.

Shi X, Seymour C, Mothersill C. 2016b. The effects of chronic, low doses of Ra-226 on cultured fish and human cells. Environ Res. 148: 303–309.

Shi X, Seymour C, Mothersill C. 2018. Change of cell growth and mitochondrial membrane polarization in the progeny of cells surviving low-dose high-LET irradiation from Ra-226. Environ Res. 167: 51–65.

Shima A, Setlow RB. 1984. Survival and pyrimidine dimers in cultured fish cells exposed to concurrent sun lamp ultraviolet and photoreactivating radiations. Photochem Photobiol. 39:49–56.

Shimada A, Shima A. 2004. Transgenerational genomic instability as revealed by a somatic mutation assay using the medaka fish. Mutat Res. 552:119–124.

Smith RW, Seymour CB, Moccia RD, Hinton TG, Mothersill CE. 2013a. The induction of a radiation-induced bystander effect in fish transcends taxonomic group and trophic level. Int J Radiat Biol. 89: 225–233.

Smith RW, Seymour CB, Mothersill CE. 2013b. Short and long term bystander effect induction by fathead minnows (*Pimephales promelas*, Rafinesque, 1820) injected with environmentally relevant whole body doses of 226-Ra. J Environ Radioact. 126:133–136.

Sreetharan S, Thome C, Tsang KK, Somers CM, Manzon RG, Boreham DR, Wilson JY. 2018. Micronuclei formation in rainbow trout cells exposed to multiple stressors: Morpholine, heat shock, and ionizing radiation. Toxicol In Vitro. 47:38–47.

Stuart M, Festarini A, Schleicher K, Tan E, Kim SB, Wen K, Gawlik J, Ulsh B. 2016. Biological effects of tritium on fish cells in the concentration range of international drinking water standards. Int J Radiat Biol. 92:563–571.

Suzuki Y. 2015. Influences of radiation on carp from farm ponds in Fukushima. J Radiat Res. 56:i19–i23.

Vo NTK, Bender AW, Lee LEJ, Lumsden JS, Lorenzen N, Dixon B, Bols NC. 2015a. Development of a walleye cell line and use to study the effects of temperature on infection by viral haemorrhagic septicaemia virus group IVb. J Fish Dis. 38:121–136.

Vo NTK, Bender AW, Lumsden JS, Dixon B, Bols NC. 2016. Differential viral haemorrhagic septicaemia virus genotype IVb infection in fin fibroblast and epithelial cell lines from walleye, *Sander vitreus* (Mitchill), at cold temperatures. J Fish Dis. 39: 175–188.

Vo NTK, Bols NC. 2016. Demonstration of primary cilia and acetylated α-tubulin in fish endothelial, epithelial and fibroblast cell lines. Fish Physiol Biochem. 42:29–38.

Vo NTK, Mikhaeil MS, Lee LEJ, Pham PH, Bols NC. 2015b. Senescence-associated β-galactosidase staining in fish cell lines and primary cultures from several tissues and species, including rainbow trout coelomic fluid and milt. In Vitro Cell Dev Biol Anim. 51: 361–371.

Vo NTK, Moore L, Spiteri KW, Hanner R, Wilkie M, DeWitte-Orr SJ. 2018. Assessing off-target cytotoxicity of the field lampricide 3-trifluoromethyl-4-nitrophenol using novel lake sturgeon cell lines. Ecotoxicol Environ Saf. 162:536–545.

Vo NTK, Seymour CB, Mothersill CE. 2017a. Dose rate effects of low-LET ionizing radiation on fish cells. Radiat Environ Biophys. 56: 433–441.

Vo NTK, Seymour CB, Mothersill CE. 2019a. The common field lampricide 3-trifluoromethyl-4-nitrophenol is a potential radiosensitizer in fish cells. Environ Res. 170:383–388.

Vo NTK, Seymour CB, Mothersill CE. 2019b. Radiobiological characteristics of descendant progeny of fish and amphibian cells that survive the initial ionizing radiation dose. Environ Res. 169:494–500.

Vo NTK, Sokeechand BSH, Seymour CB, Mothersill CE. 2017b. Characterizing responses to gamma radiation by a highly clonogenic fish brain endothelial cell line. Environ Res. 156:297–305.

Whicker FW, Schultz V. 1982. Radioecology: nuclear energy and the environment. Boca Raton (FL): CRC Press.

Wolf K, Quimby MC. 1962. Established eurythermic line of fish cells in vitro. Science. 135:1065–1066.

Xing JG, El-Sweisi W, Lee LE, Collodi P, Seymour C, Mothersill C, Bols NC. 2009. Development of a zebrafish spleen cell line, ZSSJ, and its growth arrest by gamma irradiation and capacity to act as feeder cells. In Vitro Cell Dev Biol Anim. 45:163–174.

Yew FH, Chang LM. 1984. DNA replication and repair in Tilapia cells. I. The effect of ultraviolet radiation. J Cell Sci. 72:213–226.

Zeng F, Yu X, Sherry JP, Dixon B, Duncker BP, Bols NC. 2016. The p53 inhibitor, pifithrin-α, disrupts microtubule organization, arrests growth, and induces polyploidy in the rainbow trout gill cell line, RTgill-W1. Comp Biochem Physiol C Toxicol Pharmacol. 179:1–10.

Environmental radiobiology of amphibians – knowledge gaps to be filled using cell lines

Nguyen T. K. Vo

ABSTRACT

Amphibians are facing an unprecedented level of population declines worldwide. The causes run the gamut from habitat loss and succumbing to opportunistic pathogen infections to vulnerability to toxic pollutants and ultraviolet (UV)-B radiation exposure. Anthropogenic activities including Chernobyl and Fukushima nuclear disasters and radioactive waste leakage into the environment raise the background radiation levels. Their immediate and chronic effects on amphibian populations are still being studied. However, the literature on environmental radiation effects on amphibian health still requires a lot more work. Laboratory and field works need to be conducted hand in hand in order to make informative and conclusive analyses to distinguish bad from good and harm from risk or to argue for or against the linear no-threshold model in radioprotection programs. Amphibian cell lines can help seek answers to important questions pertaining environmental radiobiology and amphibian health wherever they can suitably and effectively. The purpose of this work is to show that amphibian cell lines can 'rescue' important knowledge gaps in the literature, especially in the low-dose radiation mechanisms. Presently, there are 142 amphibian cell lines developed from six urodelans and 17 anurans. Amphibian cell lines can help expand and enrich the limited literature on environmental radiation effects on amphibians. They can be used to study mechanisms of radiation actions and discover reliable biomarkers for low-dose exposure. They can be used in environmental radiation monitoring and radioprotection programs. They can be used to determine the effects of co-exposure of IR and other stressors in the environment on amphibian health. They represent an ethical choice for amphibian conservation efforts in the current global amphibian declines. Lessons learned from cellular data can be useful guides to gain a better picture of effects occurring at the amphibian population and ecosystem levels.

1. Introduction

Permanent cell lines are important biotechnological novelties in basic research and clinical applications. Since their initial developments, they have been essential instruments for making advancements in biology and medicine as well as for pushing scientific boundaries and making seemingly impossible questions possible to pursue. One historic use of animal/human cell lines is to produce viruses, notably for studying infectious diseases and making vaccines. Nowadays, efforts to develop new cell lines remain continued to meet new challenges at times when existing ones are not available, suitable, or adequate.

In radiation biology, cell lines are indispensable in establishing underlying concepts. Some examples are linear-quadratic and multi-target cell survival models, oxygen effects, fractionated radiation, and genomic instability. Cell lines are also valuable as experimental tools in helping address fundamental questions such as 'how do cells survive from ionizing radiation?', 'what is an effective therapeutic dose to eradicate a specific tumor?', 'how low is a radiation dose to induce carcinogenesis?', and 'is background-radiation exposure essential for life?'. Human cell lines have empowered our understanding about radiation effects on humans and fundamental mechanisms involved. By contrast, efforts in using cell lines from anamniotes to study environmental ionizing radiation (eIR) effects on aquatic biota are considerably little.

Amphibians are an evolutionarily important vertebrate class. Typically known for their biphasic life, most amphibians are completely aquatic early on in life and after metamorphosis develop anatomical adaptations fitting for thriving on land. This unique lifestyle in both aquatic and terrestrial habitats makes amphibians often chosen as bioindicators of ecosystem health. As of 2020, 8,216 amphibian species have been described and constitute 11.24% of all vertebrates (ICUN 2020). Amphibians are also important intermediate species in the hierarchy of the biosphere's food webs. As predators, they primarily prey on smaller invertebrates, some of which, if not controlled, are nuisances for agricultural crops or carry infectious pathogens of public health concerns. Therefore, negative changes in amphibian diversity and richness have dramatic effects on the ecosystem stability as a whole.

Amphibians are facing an unprecedented level of population declines worldwide (Stuart et al. 2008). In the most recent global assessment, as high as 53% of amphibian species could be in the threatened status (including critically endangered, endangered, and vulnerable groups) (ICUN 2020). The global amphibian declines are caused by many factors. Among environmental factors that include habitat loss, infectious pathogens and chemical pollutants, ultraviolet (UV)-B radiation exposure is considered to be a significant contributor (Stuart et al. 2008; Alton and Franklin 2017). Increases in documented deformity incidences in amphibians in the wild suggest that environmentally induced genetic defects are very much likely involved and may develop synergistically along with other factors (Williams et al. 2008). Animal population shrinkages and phenotypic deformities are reminiscent of the genomic instability principles in radiation biology and radioecology. In genomic instability, alterations to the genome can be inherited and increase over generations, which can lead to early embryonic death and higher frequencies in phenotypic aberrations. Genomic instability has been previously observed in animals inhabiting heavily radioactive-contaminated Chernobyl and Fukushima areas (Hancock, Vo, Byun, et al. 2019; Hancock, Vo, Omar-Nazir, et al. 2019; Hancock et al. 2020). Incidences of genome reduction in frogs living in the Chernobyl Exclusion Zone (CEZ) have been previously reported (Vinogradov and Chubinishvili 1999).

Effects of high-dose IR on amphibian biology were studied extensively in the 1940s–1960s (Brunst 1965). This stemmed from early interests in using amphibians as experimental animal models for biomedical research and science education. Amphibians were pivotal to fundamental understanding in embryology, regeneration biology, cancer biology, immunology, and organ transplant biology, just to name a few. By contrast, studies of chronic low-dose eIR effects on amphibian health are few and sporadic. Earlier studies focused on targeted effects (TEs) of radiation damage, which are traditionally DNA-centric, and took no notice of non-targeted effects (NTEs), the idea of which was unfamiliar and unpopular at the time. NTEs occur when observed effects are due to factors other than specific DNA aberrations induced by direct IR hits (Mothersill and Seymour 2014; Mothersill et al. 2017). NTE mechanisms are fueled by intercellular communication and epigenetic mechanisms (Mothersill and Seymour 2014). Stress-induced NTEs can be good or bad; when it is bad, it poses serious health concerns (Mothersill and Seymour 2014). Low-dose response is dictated primarily by NTEs and is especially prevalent in eIR exposures where exposure levels are usually low (Mothersill et al. 2017). NTE mechanisms are complex and are still being studied both in controlled laboratory settings and natural environments. Therefore, though seemingly practical, questions about the immediate, legacy, and inheritable effects of chronic low-dose IR exposure on amphibian health cannot be answered simply by the classical DNA-centric paradigm. At the current stage of amphibian literature, dose alone cannot be used as the determinant of health risks as we simply do not have sensitive enough and reliable enough biomarkers for detecting and reporting low-dose risks. As amphibians live both in water and on land, they are susceptible to radionuclides prevalent in both aquatic and terrestrial habitats. To standardize radiation risk assessment for many amphibian species with different biology and diverse ways of life, the International Commission of Radiological Protection (ICRP)'s environmental radiation protection framework establishes that members of the family Ranidae are representatives of a Reference Frog (ICRP 2008).

Presently, efforts in taking advantage of existing amphibian cell lines to study eIR effects on amphibian biology remain infinitesimal. Amphibian cell lines can be an essential accelerator to help address many unknowns in the literature about amphibian responses to eIR.

This review will discuss the following topics:

1. eIR effects on amphibians
2. A general background on amphibian cell lines
3. Knowledge gaps and how amphibian cell lines can help improve them

2. eIR effects on amphibians

Natural exposure to background radiation is ubiquitous in the environment for all lifeforms. The natural radiation background is the result of several exposure sources: outer space, terrestrial, airborne, and diet. Radon gas is naturally produced by decay reactions of uranium in soil and rock and predominates airborne exposure. Radon gas generally contributes to a significant part of natural radiation background. There is a wide range of natural radioactivities in different biotopes worldwide (depending on, for example, the composition and concentrations of radionuclides present and the geo-architecture, latitude, and altitude of the landscape). However, natural background radioactitivies are still significantly lower than the highest radioactivities recorded to date in the CEZ (Møller and Mousseau 2013). For amphibians, background absorbed dose rates are 0.13–0.48 µGy/h (Beresford et al. 2008; Hosseini et al. 2010).

The topic of eIR effects can be divided into two overarching subtopics: (1) effects of background or below-background eIR and (2) effects of above-background eIR.

The role of background radiation on life and health has always been a topic of scientific curiosity. After all, IR precedes biology. Does IR play a role in creating biology and shaping the evolution of life? Questions such as 'Is radiation essential for life?' and 'is background radiation is beneficial for maintaining good health?' are not readily easy to test. Access to an experimental platform where radiation is reduced to below background levels is the most important limiting factor. Deep-underground facilities where above-surface radiation is essentially removed are suitable for this purpose. Biology research is conducted in nine deep-underground laboratories in Europe, North America, South America, and Asia (Liu et al. 2018). To the best of my knowledge, radiation research on amphibians has not been done in any of these laboratories.

For above-background eIR, assessing its effects would depend on the question 'in what context?', which is one of the major discussion topics in the 5th International Union of Radioecologists (IUR) workshop in Manchester, UK (Mothersill et al. 2020). If radiation backgrounds vary in 3 folds of magnitude 'naturally' around the world (Møller and Mousseau 2013) and life flourishes in habitats with higher radiation backgrounds, it is most likely that natural selection has taken place and thriving lifeforms adapted to those selective pressures. Solar UV radiation exposures naturally occur in particular biomes, so in this purview, it is considered to fall in the natural background radiation exposure. eIR levels causing health concerns are typically associated with anthropogenic activities (either regulated or not) that raise the regional background exposure level. Nuclear accidents (such as Chernobyl, Fukushima, Mayak, and Three Mile Island), industrially related radioactive wastewater and nuclear militarization activities are among those constituting anthropogenically-related eIR concerns.

Amphibians, in general, are more tolerant of ionizing radiation than are mammals (Whicker and Schultz 1982). LD50 values for acute low-LET radiation for amphibians are generally about several folds greater than that (~4 Gy) for humans (ICRP 2008). Radiosensitivities vary with developmental stages, with the embryo stage being the most radiosensitive and the limb bud stage being the most radioresistant (ICRP 2008). Urodelans (salamanders) are more radiosensitive than anurans (frogs and toads) (Conger and Clinton 1973). Still, amphibians are considered to be relatively radiosensitive among non-mammalian organisms.

However, eIR concerns never associate with such high doses at acute exposures. Chronic low-dose/dose-rate exposures are more prevalent in the environment. But data are seriously insufficient for amphibians, especially in dose rates of less than 10 mGy/d (ICRP 2008). At present, 1–10 mGy/d or 40–400 μGy/h is set as the derived consideration reference level (DCRL) for the Reference Frog by ICRP (2008), which defines DCRL to be the exposure range that may lead to harmful effects.

Laboratory data from Stark and colleagues using tadpoles of American bullfrog (*Lithobates catesbeiana*), Eastern spadefoot (*Scaphiopus holbrooki*), and Southern toad (*Bufo terrestris*) showed that the 0.13 mGy/d dose rate (gamma radiation) had equally severe effects on larval survival to metamorphosis as did the dose rate 222 mGy/d (Stark 2006). Interestingly, dose rates ranging from 0.13 mGy/d to 222 mGy/d had no effect on hatching success of frog embryos (Stark 2006).

In Japan, Fuma et al. examined Tohoku hynobiid salamanders, starting at the embryo stage, in chronic continuous exposures to low-dose gamma radiation in two laboratory studies. In the first study (Fuma et al. 2014), salamanders were irradiated for 450 days, dose rates less than 490 μGy/h had no effects on growth, survival, phenotypic abnormalities, and histological microstructures. In the second study, salamanders were irradiated for 1954 days. They showed that 33 μGy/h had a hormetic effect on growth but no effect on sexual maturation whereas 150 and 510 μGy/h had negative effects on both. All three dose rate exposures for a duration of 1954 days did not lead to any tumor development (Fuma et al. 2019). Their data suggest that the environmental radiation contamination associated with the Fukushima Daiichi nuclear power plant (NPP) accident alone would be less likely to adversely endanger the community structure and health of this species since the highest recorded absorbed dose rate in the Fukushima Exclusion Zone (FEZ) was 50 μGy/h.

Nuclear disasters present us the opportunity to study realistic eIR effects on free-living amphibians in particular and wildlife animals in general. Induction of cytogenetic aberrations in frogs was persistent many years after the Chernobyl accident despite the decrease in radioactivity in subsequent years (Yeliseeva et al. 1995). In particular, the reported dose rates at the time were 0.01–0.2 mSv/d (Yeliseeva et al. 1995); however, the dose unit Sv should not be used for non-human biota as, by convention, Sv is a unit of doses specifically to humans. Chernobyl frogs also had lower reproductive success (ICRP 2008). Among several radionuclides contaminating the surroundings after the Chernobyl disaster, Sr-90 (beta emitter) levels were particularly a lot higher in amphibians than in mammals (Gaschak et al. 2009; Beresford et al. 2020), highlighting Sr-90 as a radio-contaminant of health concerns for both internal and external exposures for amphibians. Still, given with what we know so far, environmental exposure to Sr-90 may not elevate external exposures drastically. Research activities into transgenerational/chronic effects of Chernobyl radiation on the community structures and physiological performances of surviving amphibians are being conducted in Europe. Given the evolutionary importance of amphibians in aquatic and terrestrial habitat occupancy, such work will yield evolutionary and ecological insights into radioactive contamination effects on all vertebrates.

For frogs in Mayak after the Kyshtym disaster, decreased fertility, smaller eggs, higher embryonic death, faster development, more phenotypic abnormalities were documented in the contaminated area compared to the control areas (ICRP 2008).

Studies focusing on amphibian health effects caused by the most recent major nuclear disaster from the Fukushima Daiichi NPP meltdown have been being conducted. Fukushima radiocesium concentrations were higher in tadpoles than in adults, emphasizing the radiation stress burden for tadpoles at the critical stage of amphibian development (Ishii et al. 2020). Germ cells of 5 frog species exposed to maximum of 64 μGy/h (air dose rate) in living proximity to the Fukushima Prefecture showed no notable morphological or histological differences (Matsushima et al. 2015). Also, no correlation between blood and tissue carotinoid concentrations and varied dose rates was found in Japanese tree frogs (Giraudeau et al. 2018). Most recently, Gombeau et al. (2020) observed increased levels of genomic DNA hypermethylation and mitochondrial DNA lesions in Japanese tree frogs (*Dryophytes japonicus*) living in contaminated sites.

Industrial nuclear energy plants produce radioactive wastewater. Accidental leaks or spills of radioactive waste occur from time to time (CNSC 2009; USNRC 2016). They certainly elevate the background radiation levels and may impose additional unintended radiation stress on affected local biological communities (CNSC 2009; USNRC 2016). These accidents can release radioactive materials into the air and onto the land and water, which are all geographical dimensions oxygenating the amphibian life. Let's take tritium as an example. Tritium is a common radionuclide found in the wastewater released by NPPs. In the past, the highest tritium leak/spill concentration occurred at the Pickering Nuclear facility in 2006 and was 128,800,000 Bq/L (CNSC 2009). That could yield 10 mGy radiation dose per day (Vo et al. 2019a), which is the upper threshold of the DCRL for the Reference Frog outlined by the ICRP (2008) to have potential deleterious effects on amphibians. Studies on how accidental leaks or spills of radioactive wastewater into the environment affect free-living organisms are very limited. Previous studies by Stuart and colleagues from the Chalk River Laboratories (Ontario, Canada) showed that frogs (*Lithobates pipiens*, *Lithobates septentrionalis*, and *Lithobates catesbeianus*) in environments with a higher-than-background radiation level that included increased tritium activities were more resistant to radiation damage (Audette-Stuart et al. 2011; Audette-Stuart and Yankovich 2011). Using a sophisticated approach combining both field and in vitro experimentations, Stuart and colleagues demonstrated that direct acute/chronic low-dose irradiation exerted protective effects leading to radioadaptive response (RAR) in frogs (Audette-Stuart et al. 2011; Audette-Stuart and Yankovich 2011). Their research also revealed RAR development in bystander frogs that were not exposed to above-background tritium levels when kept with frogs that had previously been (Audette-Stuart and Yankovich 2011). Irradiated frogs 'signal-communicated' with bystander frogs and the bystander signals induced bystander frogs to become more radio-resistant. Stuart and colleagues' frog data provide direct evidence that challenges the linear no-threshold (LNT) model in radiation protection, which outlines that radiation, no matter how low, is harmful.

At the moment, eIR effects on amphibians are studied primarily by radioecologists, whose foci are to document how radionuclides mobilize in the biosphere (including development, release, uptake, bioaccumulation, biotransformation, and inter-species transfer) and to describe their biologic effects on the ecosystem. However, an ecosystem approach to address various complex questions about eIR is more meaningful and effective when there is an active dialogue and collaboration between radioecologists and radiobiologists (who traditionally investigate mechanisms of radiation action on living organisms at the molecular, cellular, and physiological levels primarily under controlled laboratory settings) (Mothersill et al. 2020). This is one of the major emphases coming out of the 5th International Union of Radioecologists workshop held in Manchester, UK in 2019. Making use of in vitro tools such as cell cultures or cell lines used by radiobiologists is a technical and ethical advantage to expand the current literature on eIR effects on amphibians in the heated climate of worldwide amphibian declines.

3. Amphibian cell lines

3.1. Current status

Based on the Cellosaurus (of Expasy: Swiss Bioinformatics Resource Portal), National Center for Biotechnology Information, Web of Science, and Google Scholar databases, at least 142 amphibian cell lines have been developed to date (Table 1). They represent six urodelans and 17 anurans. Among these 23 representative amphibians, five are ranid frogs (genus Lithobates) of the Reference Frog classification, from which 41 of the cell lines were originated. The genetic animal model African clawed frog (*Xenopus laevis*) was the source of 40 of the cell lines (the largest number for a single amphibian species). Only five amphibian cell lines (A6, ICR 134, ICR-2A, XLK-WG, and FT) have been deposited into the American Type Culture Collection (ATCC).

Wolf and Quimby (1964) established the first amphibian cell line named FT, which is a fibroblastoid line from the bullfrog tongue. Most recently, Vo and colleagues have developed 11 new cell lines from American toad, green frog, wood frog, and African clawed frog species (Vo, Everson, et al. 2019; Vo, Guerreiro, et al. 2019; Vo, Moore, et al. 2019; Bui-Marinos et al. 2020).

Only seven cell lines have been used to study some aspects of radiation biology such as cell survival, chromosome aberrations, DNA repair mechanisms, radiosensitivity, and the biology of the tumor suppressor protein p53, which is central to DNA damage response (DDR) following radiation exposure (Table 2).

3.2. Advantages

Amphibian cell lines offer a number of advantages as experimental biological materials:

First, cell lines have the genetic makeup representative of the species they were derived from. They express cellular and biochemical properties characteristic of specific tissues and organs. They are useful in studying gene expressions, protein functions, and biochemical pathways. If certain genes are defective, they can be genetically engineered to re-express the wild-type genes. Thus, genetic manipulation can be done more easily with cultured cells versus whole animals.

Second, amphibian cell cultures, in general, are robust. Most basal medium formulations conventionally designed for mammallian cell cultures work for amphibians. The original basal media are diluted to 50–70% to suit the typical amphibian osmolarity (Rafferty 1969; Curtis et al. 2013). But amphibian cells can tolerate a wide range of medium tonicity (Vo, Moore, et al. 2019; Vo NTK, personal observations). Many amphibian cell lines adapt well in bicarbonate-free media, relieving the requirement of an external CO_2 gas supply. They are not terribly sensitive to temperature shifts in both range and kinetics. Growing amphibian cells can

Table 1. List of the amphibian cell lines that are described in the literature and public domains.

Reference	Cell line	Species[a]	Tissue origin (developmental stage)	Morphology
Cellosaurus* (as of July 2020)	91.1.F1	*Xenopus tropicalis*	Hindlimbs (tadpole)	Epithelioid
	A6	*Xenopus laevis*	Kidney (adult)	Epithelioid
	A6-2F3	*Xenopus laevis*	Kidney (adult)	Epithelioid
	A6-KR	*Xenopus laevis*	Kidney (adult)	Epithelioid
	A8	*Xenopus laevis*	Liver (adult)	Epithelioid
	ACTK1	*Xenopus laevis*	unknown	Unknown
	B3.2	*Xenopus laevis*	Kidney (unknown)	Unknown
	B3.2(tTA)1	*Xenopus laevis*	Kidney (unknown)	Unknown
	BA 68.1	*Anaxyrus americanus*	Unknown (adult)	Epithelioid
	Bor II	*Bombina orientalis*	Embryo	Fibroblastoid
	CPN4B	*Cynops pyrrhogaster*	unknown	unknown
	DRP 153	*Lithobates pipiens*	Embryos	Fibroblastoid
	DRP 36	*Lithobates pipiens*	Embryos	Fibroblastoid
	FT	*Lithobates catesbeianus*	Tongue (adult)	Fibroblastoid
	FUHEN	*Dryophytes japonicus*	Bone marrow (adult)	Thrombocytoid
	GSKC	*Andrias davidianus*	Kidney (tadpole)	Epithelioid
	GSSC	*Andrias davidianus*	Spleen (tadpole)	Epithelioid
	GSTC	*Andrias davidianus*	Thymus (tadpole)	Epithelioid
	HNS1A	*Hynobius nigrescens*	Unknown	Unknown
	HNS2	*Hynobius nigrescens*	Unknown	Unknown
	HNS3	*Hynobius nigrescens*	Unknown	Unknown
	HNS4	*Hynobius nigrescens*	Unknown	Unknown
	HNS5	*Hynobius nigrescens*	Unknown	Unknown
	HO(6)	*Lithobates catesbeianus*	Unknown	Unknown
	HTUD1	*Hynobius tokyoensis*	Unknown	Unknown
	ICR B20	*Lithobates pipiens*	Embryos	Fibroblastoid
	ICR-134 or RPH 67.134	*Lithobates pipiens*	Embryos	Fibroblastoid
	ICR-2A or RPH 68.2A	*Lithobates pipiens*	Embryos	Fibroblastoid
	LAH1	*Bufo japonicus formosus* (Originally thought to be of *Pelophylax nigromaculatus*)	Dorsal skin (froglet)	Epithelioid
	LAH2	*Pelophylax porosus brevipodus* (Originally thought to be of *Pelophylax nigromaculatus*)	Dorsal skin (froglet)	Mixed cultures
	LAH3	*Pelophylax porosus brevipodus*	Dorsal skin (froglet)	Epithelioid
	PNKT-3	*Lithobates pipiens*	Kidney (tadpole)	Epithelioid
	PNKT-4	*Lithobates pipiens*	Kidney (tadpole)	Epithelioid
	PNKT-4B	*Lithobates pipiens*	Kidney (tadpole)	Epithelioid
	PRKT	*Lithobates pipiens*	Kidney (tadpole)	Epithelioid
	RPD 67.7	*Lithobates pipiens*	Embryos	Fibroblastoid
	RPH 67.194	*Lithobates pipiens*	Embryos	Mixed cultures
	RPH 67.205	*Lithobates pipiens*	Embryos	Epithelioid
	RPKID	*Lithobates pipiens*	Kidney (tadpole)	Unknown
	S3-1	*Xenopus laevis*	Embryo	Unknown
	Speedy	*Xenopus tropicalis*	Hindlimbs (tadpole)	Fibroblastoid
	TB-6c	*Rhinella marina*	Bladder (adult)	Epithelioid
	TBM-54	*Rhinella marina*	Bladder (adult)	Epithelioid
	WMPa	*Lithobates pipiens*	Kidney (embryo)	Epithelioid
	XB694	*Xenopus borealis*	Embryo	unknown
	XF	*Xenopus laevis*	Unknown	Fibroblastoid
	XL-177	*Xenopus laevis*	tadpole	Epithelioid
	Xl1	*Xenopus laevis*	Bone (adult)	Fibroblastoid (Chondrocyte-like)
	XL110	*Xenopus laevis*	Liver (unknown)	Epithelioid
	XL2	*Xenopus laevis*	Tadpole	Epithelioid
	XLK-WG	*Xenopus laevis*	Kidney (adult)	Epithelioid
	XLK2	*Xenopus laevis*	Kidney (unknown)	Unknown
	XLT-15	*Xenopus laevis*	Tail (tadpole)	Myoblastoid
	XR1	*Xenopus laevis*	Retina (embryo)	Glialoid
	XR8	*Xenopus laevis*	Retina (embryo)	Unknown
	XR9	*Xenopus laevis*	Retina (embryo)	Unknown
	XRpe1	*Xenopus laevis*	Retina (embryo)	Glialoid
	XRpe2	*Xenopus laevis*	Retina (embryo)	Glialoid
	XRpe4	*Xenopus laevis*	Retina (embryo)	Unknown
	XRpe5	*Xenopus laevis*	Retina (embryo)	Unknown
	XTC-2	*Xenopus laevis*	Whole tadpole (with intestines, eyes, skin, and tail removed)	Epithelioid
	XTC-GTX-11	*Xenopus laevis*	Whole tadpole (with intestines, eyes, skin, and tail removed)	Epithelioid
	XTH-2	*Xenopus laevis*	Heart (tadpole)	Endothelioid
	XTY	*Xenopus laevis*	Subcutaneous tissues (adult)	Fibroblastoid

(*continued*)

Table 1. Continued.

Reference	Cell line	Species[a]	Tissue origin (developmental stage)	Morphology
Vo, Moore, et al. 2019	BufoTad	*Anaxyrus americanus*	Whole tadpole	Endothelioid
Vo, Everson, et al. 2019	WoodTad-rpe	*Lithobates sylvaticus*	Retina (tadpole)	Epithelioid
	WoodTad-HE1	*Lithobates sylvaticus*	Cephalic region (tadpole)	Epithelioid
	WoodTad-bone	*Lithobates sylvaticus*	Hind leg bone (tadpole)	Fibroblastoid
	GreenTad-HF1	*Lithobates clamitans*	Cephalic region (tadpole)	Fibroblastoid
	GreenTad-HF2	*Lithobates clamitans*	Cephalic region (tadpole)	Fibroblastoid
	GreenTad-HF3	*Lithobates clamitans*	Cephalic region (tadpole)	Fibroblastoid
	GreenTad-HE4	*Lithobates clamitans*	Cephalic region (tadpole)	Epithelioid
	GreenTad-gill	*Lithobates clamitans*	Gill (tadpole)	Epithelioid
Bui-Marinos et al. 2020	Xela VS2	*Xenopus laevis*	Ventral skin (adult)	Epithelioid
	Xela DS2	*Xenopus laevis*	Dorsal skin (adult)	Epithelioid
Griggs and Bender 1972	A8W243	*Xenopus laevis*	Liver (adult)	Epithelioid
van Zeeland et al. 1980	'Fibroblasts from Xenopus laevis'	*Xenopus laevis*	Unknown	Fibroblastoid
Du Pasquier and Robert 1992; Robert and Cohen 1998	B3B7	*Xenopus laevis*	Thymus	T lymphocytoid
	15/0	*Xenopus laevis*	Thymus	T lymphocytoid
	15/40	*Xenopus laevis*	Thymus	T lymphocytoid
	ff-2	*Xenopus laevis*	Thymus	T lymphocytoid
	ff-2.64	*Xenopus laevis*	Thymus	T lymphocytoid
	f/k	*Xenopus laevis*	Thymus	T lymphocytoid
	B3D3	*Xenopus laevis*	Thymus	T lymphocytoid
	MAR1	*Xenopus laevis*	Thymus	T lymphocytoid
Rafferty 1965	WERPa	*Lithobates palustris*	Whole embryo	Epithelioid
	WERPi	*Lithobates pipens*	Whole embryo	Epithelioid
	WERS	*Lithobates sylvaticus*	Whole embryo	Epithelioid
	KERS	*Lithobates sylvaticus*	Kidney (embryo)	Epithelioid
	LC-1	unknown	Renal carcinoma	Epithelioid
Roy et al. 2000; Villiard et al. 2007	AL1 or ACT1	*Ambystoma mexicanum*	Limb (adult)	Fibroblastoid
Ferretti and Brockes 1988	B1H1	*Notophthalmus viridescens*	Limb (adult)	unknown
	A1	*Notophthalmus viridescens*	Limb (adult)	Fibroblastoid
Corcoran and Ferretti 1999	TH4B	*Notophthalmus viridescens*	Limb (adult)	Unknown
Reese et al. 1976	TV-1	*Notophthalmus viridescens*	Iris	Epithelioid
Rafferty 1969	A-1	*Lithobates sylvaticus*	Kidney (large embryo)	Epithelioid
	A-2	*Lithobates pipiens*	Whole hatching embryo	Epithelioid
	A-3	*Lithobates pipiens*	Lucke tumor	Epithelioid
	A-4	*Pseudacris triseriata*	Whole hatching embryo	Mixed
	A-5	*Lithobates clamitans*	Whole hatching embryo	Fibroblastoid
	A-7	*Xenopus laevis*	Kidney (adult)	Epithelioid
	A-11	*Pseudacris crucifer* (formerly *Hyla crucifer*)	Whole hatching embryo	Mixed
	A-19	*Lithobates clamitans*	Whole hatching embryo	Fibroblastoid
	A-20	*Lithobates clamitans*	Whole hatching embryo	Fibroblastoid
	A-21	*Lithobates sylvaticus*	Whole hatching embryo	Epithelioid
	A-22	*Lithobates sylvaticus*	Whole hatching embryo	Epithelioid
	A-23F	*Lithobates sylvaticus*	Whole hatching embryo	Epithelioid
	A-23R	*Lithobates sylvaticus*	Whole hatching embryo	Epithelioid
	A-24	*Lithobates sylvaticus*	Tumor (adult)	Epithelioid
Park et al. 2000	Bok-2	*Bombina orientalis*	Embryos	Professional phagocytes
Freed et al. 1969	RPH 67.132	*Lithobates pipiens*	Embryos	Unknown
	31 additional cell lines	*Lithobates pipiens*	Embryos	Unknown

[a]*Xenopus tropicalis* (Western clawed frog); *Xenopus laevis* (African clawed frog); *Anaxyrus americanus* (American toad); *Bombina orientalis* (Oriental fire-bellied toad); *Cynops pyrrhogaster* (Japanese fire-bellied newt); *Lithobates pipiens* (Northern leopard frog); *Lithobates catesbeianus* (American bullfrog); *Dryophytes japonicus* (Japanese tree frog); *Andrias davidianus* (Chinese giant salamander); *Hynobius nigrescens* (Japanese black salamander); *Hynobius tokyoensis* (Tokyo salamander); *Bufo japonicus formosus (Eastern-Japanese common toad); Pelophylax nigromaculatus* (Black-spotted frog); *Pelophylax porosus brevipodus* (Nagoya Daruma pond frog); *Rhinella marina* (Cane toad); *Xenopus borealis* (Kenyan clawed frog); *Lithobates sylvaticus* (Wood frog); *Lithobates clamitans* (Green frog); *Lithobates* palustris (Pickerel frog); *Notophthalmus viridescens* (Eastern newt); *Pseudacris triseriata* (Western chorus frog); *Pseudacris crucifer* (formerly *Hyla crucifer*; spring peeper); *Ambystoma mexicanum* (Axolotl).

*https://web.expasy.org/cgi-bin/cellosaurus/search?input=%22Amphibian%20cell%20line%22

Table 2. List of amphibian cell lines used to study radiation biology.

Species	Cell line	References	Radiation type, exposure type	Evaluation endpoint(s)
African clawed frog (*Xenopus laevis*)	A8W243	Griggs and Bender 1972	Ultraviolet, acute Gamma, acute	Chromosomal aberrations
	Fibroblast from Xenopus leavis	van Zeeland et al. 1980	Ultraviolet, acute	Clonogenic cell survival, chromosomal aberrations
	A6	Bensaad et al. 2001; Vo et al. 2019b	Gamma, acute	p53 protein expression, clonogenic cell survival
	B3B7	Johnson et al. 2013	Ultraviolet, acute	p53 and Mdm2 protein expressions
Northern leopard frog (*Lithobates pipiens*)	ICR-2A	Freed et al. 1979	Ultraviolet, acute	Clonogenic cell survival, DNA repair
Eastern newt (*Notophthalmus viridescens*)	A1	Yun et al. 2015	Ultraviolet, acute	Senescence
Axolotl (*Ambystoma mexicanum*)	AL1 or ACT-1	Villiard et al. 2007	Ultraviolet, acute	p53 biology

rely on commercially available sera as a source of growth supplements. But some can grow in serum-free media (Park et al. 2000). Collectively, these make amphibian cell lines easy to work within the laboratory and for fieldwork when accessibility to specific resources is limited.

Third, they are unlimited cell sources. The availability of such is always advantageous, especially in the face of global amphibian population declines. Moreover, amphibian cells can be transported to and studied in laboratories in geographic locations where the amphibian species are not found or even allowed to have.

Fourth, the existing amphibian cell lines represent different life stages and diverse tissue systems, which is useful for studying developmental stage-specific and tissue-specific responses.

Fifth, many amphibian cell lines were derived from normal tissues and many retain their karyotypic stability for many passages (Rafferty 1969). This is contrary to mammalian cell lines (Rafferty 1969). To the best of my knowledge, most, if not all, amphibian cell lines were immortalized spontaneously, free of genetic manipulation or viral transformation. Many retain normal functions and differentiated characteristics as in vivo (Park et al. 2000; Bensaad et al. 2001; Villiard et al. 2007; Yun et al. 2015; Vo, Everson, et al. 2019; Bui-Marinos et al. 2020). Thus, compared to mammalian counterparts, amphibian cell lines are perhaps more representative of normal cells.

Sixth, the availability of amphibian cell lines from tumors and normal tissues of the same origin (Table 1) presents a unique position to study differences between transformed and non-transformed cell biology for the amphibian system.

Seventh, the clonogenic survival assay is a quantitative tool that assesses phenotypic consequences of cellular exposure to IR (Hall and Giaccia 2006). Several irradiated cell fates can be evaluated in one single assay and in principle yield the information about the overall reproductive integrity of cells following IR exposure. The assay requires cells to be clonable in tissue culture vessels – a property which cell lines are excellent at whereas primary cells are not.

Last but not the least, as amphibian populations are alarmingly declining, it is a more ethical approach in conservation biology to seek experimentations that do not threaten decreases in amphibian numbers. Non-lethal approaches such as small toe clippings are already used by ecologists to study amphibians in the field. The robustness of amphibian cell culture systems allows the collection and expansion of viable healthy cells from tissue biopsies (such as toe clips) that would potentially give rise to cell lines for further studies.

4. Knowledge gaps and how amphibian cell lines can help address them

There are knowledge gaps in the current literature on eIF effects on amphibians. I will discuss what the important knowledge gaps are and how amphibian cell lines can help seek answers for the existing unknowns. The optimistic view of this review is that amphibian cell lines can help 'rescue' the important knowledge gaps in the literature, especially at the present time when regional and global amphibian conservation is paramount and any effort in this forefront is welcome to pursue, particularly in practice.

4.1. Knowledge gap 1: lack of full understanding of how DNA damage sensing operates in amphibians

To understand how eIR affects the health of many amphibian species, we ought to know how amphibians sense radiation damages and damage signaling at the cellular level. As the current literature stands, we are very much behind on this important fundamental understanding.

The primordial understanding of the biological effects of IR was founded on how IR targets and damages nucleic acids, especially deoxyribonucleic acids (DNA), which is the hereditary genetic material of biological life. DDR evolves with several biochemical programs that cope with and resolve nuclear stress in the best interest of protecting genomic integrity. But how do amphibian cells sense DNA damages to alarm cells to repair them?

Development of DNA lesions is the basis of DNA damage sensing. There are different types of DNA lesions, which activate different DNA damage signaling and repair mechanisms. Common DNA lesions induced by IR are base/nucleotide damage, DNA-biomolecule crosslinks, single-strand breaks (SSBs), and double-strand breaks (DSBs). DNA breaks are serious DNA aberrations, with DSBs being the most lethal DNA lesion type (Hall and Giaccia 2006). As the building blocks of DNA are the same in organisms and radiation energy absorption in DNA is the same in principle, recognition of radiation-induced DNA lesions in amphibians appears to operate in an evolutionarily conserved fashion as seen in other animals and humans.

DNA lesions initiate DNA damage signaling, which results in the induction of several biochemical mechanisms. Details about the involved biochemical pathways from the mammalian system can be found in Hall and Giaccia's (2006) Radiobiology book, which is a go-to reference text for many radiobiologists, and other review papers in DDR mechanisms. In brief, ataxia telangiectasia mutated (ATM) and ATM and RAD3-related (ATR) are protein kinases that are activated by DNA stand break sensing and subsequently phosphorylate a number of target proteins including the tumor suppressor protein p53 and cell-cycle checkpoint kinases Chk1 and Chk2. Activated p53, Chk1, and Chk2 induce a delay in cell cycle progress and this is thought to allow time for DNA damages to be repaired prior to mitosis completion into daughter cells with an unharmed genome. Activated ATM and ATR mediate gene expressions transcriptionally and post-transcriptionally to enhance DNA repairs. On the other hand, activated p53 participates in DDR in both transcriptionally dependent and independent ways. Transcriptionally, activated p53 translocates from the cytoplasm to the nucleus and activates the expression of genes critical in cell cycle regulation, DNA repairs, and cell death programs. When DNA damages are too severe to repair, apoptosis or cellular senescence program will be induced. In a transcriptionally-independent mechanism, activated p53 interacts with mitochondria and induces the apoptotic signaling cascade following the leakage of cytochrome c from the mitochondria into the cytosol.

A number of significant differences in the biology and regulation of the biochemical players in the DNA damage signaling have been previously noted. In contrast to the human ATM protein, which is phosphorylated and increases in expression upon irradiation and finally is cleaved by caspases, none of these characteristics occur with the Xenopus ATM (Robertson et al. 1999). While both mammalian Chk1 and Chk2 phosphorylate Cdc25A, which regulates cell cycle progression important during DNA repair, only Chk1, but not Chk2, phosphorylates Cdc25A in Xenopus (Uto et al. 2004). Chromatin modification can regulate DDR. The histone variant H2AX at DSB locations is one of the phosphorylation targets of ATM/ATR and, following phosphorylation, becomes γ-H2AX. The induction of γ-H2AX formation in the cell nuclei is one of the most conserved hallmarks of radiation exposure in eukaryotes. Interestingly, in Xenopus, there exists an H2AX isoform that is constitutively phosphorylated in the absence of DSBs or radiation exposure (Shechter et al. 2009) and its functions remain to be further explored. Furthermore, the Xenopus genome completely lacks p63 and p73, which are p53 homologues in mammals (Stiewe 2007) and whose functions can make up for the loss of p53 presence or activity. Humans are prone to tumorigenesis and the human *TP53* gene encoding p53 is the most frequently mutated gene in human cancers (Hollstein et al. 1991). p53 also dictates the degree of radiosensitivity (Lee and Bernstein 1993; Ribeiro et al. 1997). However, amphibians generally rarely develop cancers (Ruben et al. 2007). Inducing tumourigenesis in amphibians using means to successfully do so in mammalian models is indeed a lot harder as well (Ruben et al. 2007). Clearly, tumourgenesis operates differently in amphibians versus mammals. Villiard et al. (2007) have revealed that mutations in mammalian p53 variants associated with tumourigenesis are found in axolotl p53, which promotes limb regeneration in amphibians. How amino acid differences between mammalian and amphibian p53 proteins regulate radiosensitivity in amphibian species remain to be further studied.

The current knowledge is largely limited to Xenopus, thanks to the two available genome sequence databases for *X. laevis* and *X. tropicalis*. However, there are 8,216 amphibian species with various complex adaptation strategies for survival in diverse habitats. Biological adaptation requires an evolution of biochemical mechanisms even for organisms within the same animal group. It should come little surprise that different adaptation mechanisms may exist in different amphibian species (e.g., anurans versus urodelans, freeze-tolerant frogs versus freeze-intolerant frogs). Additionally, some amphibians (e.g., *Xenopus laevis*) are pseudotetraploid; as a result, multiple alleles can exist for the same gene and allele inactivation appears to take place randomly so that genes are only expressed as if they were diploid (Wertheim et al. 2013). Additionally, amphibians can possess multiple isoforms of one gene product (Koubourli et al. 2018). Some amphibian proteins, in contrast to fish and mammalian counterparts, are uniquely derived from different genes that may or may not contain introns at all (Gan et al. 2017). How various differences may have an impact on DDR programs in different amphibians remain to be studied. Elucidating what they are is important and available cell lines from different amphibian species can accelerate what we can learn toward this aim.

4.2. Knowledge gap 2: lack of understanding about radiation-induced genetic mutations and epigenetic changes for amphibians

A notable absence in the amphibian literature is information about radiation-induced genetic mutations and epigenetic footprints induced by eIR (ICRP 2008).

IR is a known mutagen in the environment. There are radiation-responsive mutation hotspots in the genome. Knowing what the mutations hotspots are and for what radiation exposure type is useful as they can serve as biomarkers for environmental monitoring and diagnostics on amphibian health.

IR can also induce epigenetic changes in the genome. Epigenetic changes include DNA methylation, histone modification, and RNA silencing. These changes can be inherited and lead to epigenetic memory and transgenerational effects. eIR-induced transgenerational effects have been documented in the laboratory and in fieldwork in animals and humans (Weinberg et al. 2001; Barber et al. 2002; Hurem et al. 2018; Kamstra et al. 2018; Hancock, Vo, Byun, et al. 2019;

Hancock, Vo, Omar-Nazir, et al. 2019; Hancock et al. 2020). Discovering what the epigenetic signatures are will gain a profound insight into how eIR induces transgenerational effects in amphibian ecology and conservation.

Utilizing amphibian cell lines to figure out what mutation hotspot and epigenetic signature maps look like at the cellular level is an advantage that seems hard to ignore. Not only do we have cell lines from different amphibian species but we also have cell lines from different tissues and developmental stages. Therefore, we can learn valuable information about different mutations hotspots and epigenetic signatures in different species, tissues, and developmental stages caused by eIR.

4.3. Knowledge gap 3: lack of understanding of radioadaptive response mechanisms

RAR is a biological protective response in which low-dose radiation turns on defense mechanisms and proactively alleviate cells or organisms from radiation damages (Cohen et al. 2019). RAR was demonstrated first in frog cells by Chao and Rosenstein (1984), who documented that frog fibroblasts failed to produce UV-induced chromosomal aberrations if they were pre-exposed to gamma radiation. In vivo, frogs living in the natural or artificially set-up environment containing very low radioactivity levels also acquire RAR traits (Audette-Stuart et al. 2011; Audette-Stuart and Yankovich 2011). Cells isolated from these animals developed significantly less genomic aberrations than cells from control animals (Audette-Stuart et al. 2011; Audette-Stuart and Yankovich 2011). However, nothing is known about the molecular mechanisms at play. Among many players, p53 and survivin proteins are known to be involved in modulating RAR (Murley et al. 2017). Historically speaking, mouse and human cell lines were used extensively to elucidate RAR mechanisms in mammals. Amphibian cell lines can be used in a similar manner to tease out biochemical mechanisms governing RAR in amphibians. This will further enhance our understanding of RAR evolution in biological life.

4.4. Knowledge gap 4: lack of understanding of radiation-induced delayed lethal mutations

eIR can induce genomic instability which may not show in irradiated parents or parental cells but manifest in offspring and progeny cells that survive radiation. Delayed lethal mutations (DLMs) are one of the consequences of genomic instability (Mothersill and Seymour 1997). It is when radiation-induced mutations may not cause immediate lethal effects and are passed on to progenies in which more mutations accumulate over time to the point of inducing progeny death. The concept was demonstrated first in cultured mammalian cells (Seymour et al. 1986) and then in cultured fish cells (Shi et al. 2018). The most recent demonstration of possible DLMs in the environment was the observation of increased frequencies of embryonic death in butterflies exposed to Fukushima radiation (Hancock, Vo, Omar-Nazir, et al. 2019). No evidence of DLM development in amphibians has been shown. Most recently, Vo et al. (2019b) developed a pilot study to investigate whether DLMs could develop in cultured *X. laevis* normal kidney epithelial cells. No DLMs were induced in radiation survivor progenies at up to 43 population doublings. Interestingly, radiation survivor progeny cells became more radiosensitive (Vo et al. 2019b). If this observation manifests in wildlife, it would be concerning because offspring of amphibians initially exposed to Fukushima and Chernobyl radiation would become more sensitive to radiation exposures, which could totally contribute to the current global amphibian declines. Using amphibian cell lines to further deduce the mutation rate frequencies typically occurring in amphibian cells following the initial irradiation could yield insights into making prediction models into possible biological dystopian scenarios for perished amphibian communities and, consequently, the ecosystem as a whole.

4.5. Knowledge gap 5: lack of understanding of radiation-induced bystander effects mechanisms

Radiation-induced bystander effects (RIBEs) are one of the NTEs inducible by eIR (Mothersill and Seymour 2014; Mothersill et al. 2017). RIBEs occur when irradiated cells/organisms induce biological effects in non-irradiated counterparts via direct or indirect means. When RIBEs are adverse, they render non-irradiated cells functionally impaired or phenotypically defective. When RIBEs are beneficial, they cause non-irradiated cells to become more radioresistant. To date, Audette-Stuart and Yankovich (2011) are the only scientists who have previously studied RIBEs in bullfrog tadpoles. These authors showed that irradiated non-bystander tadpoles became more radioresistant after being kept with previously-irradiated tadpoles. Adverse RIBEs do occur in another aquatic animal group, teleosts (Mothersill et al. 2006). Clearly, intercellular/interorganismal communication modifies radiation responses and should be taken into account in evaluating and establishing environmental radioprotection measures.

The occurrence of adverse RIBEs can have serious consequences on the health status of targeted cells in the body or targeted co-inhabitants in the localized biotope. Cellular studies suggest that RIBE signal production is p53-independent whereas RIBE signal response is strongly p53-dependent (Mothersill et al. 2011). RIBEs are biochemically mediated through gap junctions between adjacent cells (Azzam et al. 2001) or by the exchange of secretable soluble factors via the extracellular medium (Lyng et al. 2002). Soluble RIBE signals can be packaged in extracellular vesicles containing genetic and non-genetic regulatory biomolecules that can induce cellular aberrations in non-irradiated cells (Le et al. 2017). Environmentally-relevant low-dose alpha irradiation of fish induces RIBEs in non-irradiated bystander fish via the shared water medium both the fish groups had swum in (Smith et al. 2013). Serotonin is a potent bystander signaling inducer as blocking serotonin uptake into cells in vitro and in an aquatic animal model in vivo abolishes the adverse RIBE (Saroya et al. 2009; Curtis et al. 2020a). However, serotonin-mediated radiosensitivity and bystander responsiveness vary with different cell types (Curtis et al. 2018, 2020a, 2020b) and assumingly with different species, which remains to be experimentally demonstrated. Increasing evidence strongly suggests that microRNAs and microRNA-containing exosomes mediate RIBEs (Xu et al. 2014, 2015), highlighting the involvement of epigenetic regulation in the RIBE phenomenon and cellular microRNAs or cell-free microRNAs (as in extracellular vesicles) as potential biomarkers for NTE detection in environmental radiobiology. These RIBE mechanisms were elucidated using fish and mammalian models and cell lines. They remain to be demonstrated for the amphibian system and amphibian cell lines would be the advantageous forefront to study the mechanisms.

4.6. Knowledge gap 6: insufficient data regarding multiple-stressor effects

Global amphibian declines are believed to be the result of multiple environmental stressors from both natural and anthropogenic sources. Amphibian cell lines are particularly useful instruments to tease out fieldwork results whose interpretations are complicated by multiple stressors naturally found in the wild. Free-living animals in the wild show higher radiosensitivity than animals kept and well cared for in the laboratory (Garnier-Laplace et al. 2013). Studies using live animals in controlled experiments in the laboratory are important and effective in determining the biological effects of IR and at least one more known factor. But these experiments can take long to get results. On the other hand, animal cell lines can be definitely equally effective biological reporters for effects of IR and other multiple stressors (e.g., Vo et al. 2019a). In combination with -omics biotechnologies, amphibian cell lines can yield a lot of important data including the discovery of tissue- and species-specific biomarkers and will enable an informative analysis of what fundamentally takes place at the cellular level in response to co-exposure of IR and multiple stressors.

Permissibility to opportunistic pathogens is one of the known contributors to global amphibian declines. Infectious pathogens constitute a biotic stressor and the effects of such infections in combination with radiation stress can have serious impact on amphibian health. Two known infectious pathogens that can be destructive to amphibian populations are ranaviruses (Vo, Guerreiro, et al. 2019) and the chytrid fungus (Lips 2016). Co-exposure of cells to IR and these two pathogens can be easily studied in amphibian cell cultures. In addition to infections and DNA damage endpoints, immunological endpoints such as inflammatory responses, antiviral/antifungal functions, cytokine expression profiles, and antigen presentation mechanisms can also be examined. These are useful information and gives us a bigger picture of cellular responses to both IR and pathogen infections, that could have profound implications at the population and ecosystem levels.

4.7. Knowledge gap 7: insufficient data regarding radiation dose-response survival curves for amphibian cells

Many fundamental concepts and underlying theories in radiation biology that we know today are partly thanks to many radiation dose-response survival curves generated over the past decades with animal and human cells. Some of those concepts and theories made based on radiation dose-response survival curves are already described in Section 1. Only a few survival curves were generated from one L. pipiens cell line and a couple of Xenopus cell lines (Griggs and Bender 1972; Freed et al. 1979; Rosenstein and Setlow 1980; van Zeeland et al. 1980) almost four decades ago. The construction of more cell survival curves using other cell lines from other tissues and species may shed light on new radiobiological properties, especially pertaining to low-dose response region where certainties are not always straight-forwardly defined.

5. Concluding remarks

The current literature on eIR effects on amphibian health is limited and is one that amphibian cell lines can help expand and enrich. Various radiation mechanisms unknown to amphibian biology can be studied with amphibian cell lines. Biomarkers for exposure to low-dose radiation can be discovered with amphibian cell lines. Environmental radiation monitoring frameworks can be developed using amphibian cell lines that are sensitive radio-responders. Additionally, amphibian cell lines would be not only a technical advantage but also an ethical choice for amphibian conservation efforts. This is especially important in this uncertain time with many unknowns facing the global amphibian declines. If mammalian cell lines have gifted us a vast library of knowledge on IR effects for the benefit of improving human health and medicine, amphibian cell lines can certainly do the same for the service of protecting amphibian health and population stability in a radioactive world and a potentially radioactive future.

Disclosure statement

No potential conflict of interest was reported by the author(s).

ORCID

Nguyen T. K. Vo http://orcid.org/0000-0002-4709-3078

References

Alton LA, Franklin CE. 2017. Drivers of amphibian declines: effects of ultraviolet radiation and interactions with other environmental factors. Clim Change Responses. 4:6.

Audette-Stuart M, Kim SB, McMullin D, Festarini A, Yankovich TL, Carr J, Mulpuru S. 2011. Adaptive response in frogs chronically exposed to low doses of ionizing radiation in the environment. J Environ Radioact. 102(6):566–573.

Audette-Stuart M, Yankovich T. 2011. Bystander effects in bullfrog tadpoles. Radioprotection. 46(6):S497–S502.

Azzam EI, de Toledo SM, Little JB. 2001. Direct evidence for the participation of gap junction-mediated intercellular communication in the transmission of damage signals from alpha-particle irradiated to nonirradiated cells . Proc Natl Acad Sci U S A. 98(2):473–478.

Barber R, Plumb MA, Boulton E, Roux I, Dubrova YE. 2002. Elevated mutation rates in the germ line of first- and second-generation offspring of irradiated male mice. Proc Natl Acad Sci U S A. 99(10): 6877–6882.

Bensaad K, Rouillard D, Soussi T. 2001. Regulation of the cell cycle by p53 after DNA damage in an amphibian cell line. Oncogene. 20(29): 3766–3775.

Beresford NA, Barnett CL, Gashchak S, Maksimenko A, Guliaichenko E, Wood MD, Izquierdo M. 2020. Radionuclide transfer to wildlife at a 'Reference site' in the Chernobyl Exclusion Zone and resultant radiation exposures. J Environ Radioact. 211:105661

Beresford NA, Barnett CL, Jones DG, Wood MD, Appleton JD, Breward N, Copplestone D. 2008. Background exposure rates of terrestrial wildlife in England and Wales. J Environ Radioact. 99(9): 1430–1439.

Brunst VV. 1965. Effects of ionizing radiation on the development of amphibians. Q Rev Biol. 40:1–67.

Bui-Marinos MP, Varga JFA, Vo NTK, Bols NC, Katzenback BA. 2020. Xela DS2 and Xela VS2: two novel skin epithelial-like cell lines from adult African clawed frog (*Xenopus laevis*) and their response to an extracellular viral dsRNA analogue. Dev Comp Immunol. 112: 103759

Canadian Nuclear Safety Commission (CNSC). 2009. Tritium releases and dose consequences in Canada in 2006. [accessed 2020 Sep 22]. https://nuclearsafety.gc.ca/pubs_catalogue/uploads/CNSC_Release_and_Dose_eng_rev2.pdf.

Chao CC, Rosenstein BS. 1984. Inhibition of the UV induction of sister-chromatid exchanges in ICR-2A frog cells by pretreatment with gamma-rays. Mutat Res. 139(1):35–39.

Cohen J, Vo NTK, Seymour CB, Mothersill CE. 2019. Parallel comparison of pre-conditioning and post-conditioning effects in human cancers and keratinocytes upon acute gamma irradiation. Int J Radiat Biol. 95(2):170–178.

Conger AD, Clinton JH. 1973. Nuclear volumes, DNA contents, and radiosensitivity in whole-body-irradiated amphibians. Radiat Res. 54(1):69–101.

Corcoran JP, Ferretti P. 1999. RA regulation of keratin expression and myogenesis suggests different ways of regenerating muscle in adult amphibian limbs. J Cell Sci. 112(Pt 9):1385–1394.

Curtis TM, Collins AM, Gerlach BD, Brennan LM, Widder MW, van der Schalie WH, Vo NTK, Bols NC. 2013. Suitability of invertebrate and vertebrate cells in a portable impedance-based toxicity sensor: temperature mediated impacts on long-term survival. Toxicol In Vitro. 27(7):2061–2066.

Curtis JJ, Seymour CB, Mothersill CE. 2018. Cell line-specific direct irradiation and bystander responses are influenced by fetal bovine serum serotonin concentrations. Radiat Res. 190(3):262–270.

Curtis JJ, Vo NTK, Seymour CB, Mothersill CE. 2020a. 5-HT_{2A} and 5-HT_3 receptors contribute to the exacerbation of targeted and non-targeted effects of ionizing radiation-induced cell death in human colon carcinoma cells. Int J Radiat Biol. 96(4):482–490.

Curtis JJ, Vo NTK, Seymour CB, Mothersill CE. 2020b. Serotonin and 5-HT_3 receptors sensitize human skin cells to direct irradiation cell death but not to soluble radiation-induced bystander signals. Environ Res. 180:108807.

Du Pasquier L, Robert J. 1992. In vitro growth of thymic tumor cell lines from Xenopus. Dev Immunol. 2(4):295–307.

Ferretti P, Brockes JP. 1988. Culture of newt cells from different tissues and their expression of a regeneration-associated antigen. J Exp Zool. 247(1):77–91.

Freed JJ, Hoess RH, Angelosanto FA, Massey HC. Jr. 1979. Survival and DNA repair in ultraviolet-irradiated haploid and diploid cultured frog cells. Mutat Res. 62(2):325–339.

Freed JJ, Mezger-Freed L, Schatz S. 1969. Characteristics of cell lines from haploid and diploid Rana pipiens embryos. In: Mizell M, editor. Biology of amphibian tumors. New York: Springer-Verlag New York Inc. p. 101–111.

Fuma S, Soeda H, Ihara S, Matsui K, Kawaguchi I, Ishikawa T, Kubota Y, Watanabe Y, Aono T. 2019. Effects of chronic γ-irradiation on growth and sexual maturation of the Tohoku hynobiid salamander, *Hynobius lichenatus*. J Environ Radioact. 196:98–103.

Fuma S, Une Y, Ihara S, Matsui K, Kudo T, Tokiwa T, Kubota Y, Soeda H, Ishikawa T, Doi K, et al. 2014. Effects of chronic γ-irradiation on growth and survival of the Tohoku hynobiid salamander, *Hynobius lichenatus*. J Environ Radioact. 135:84–92.

Gan Z, Chen SN, Huang B, Hou J, Nie P. 2017. Intronless and intron-containing type I IFN genes coexist in amphibian Xenopus tropicalis: Insights into the origin and evolution of type I IFNs in vertebrates. Dev Comp Immunol. 67:166–176.

Garnier-Laplace J, Geras'kin S, Della-Vedova C, Beaugelin-Seiller K, Hinton TG, Real A, Oudalova A. 2013. Are radiosensitivity data derived from natural field conditions consistent with data from controlled exposures? A case study of Chernobyl wildlife chronically exposed to low dose rates. J Environ Radioact. 121:12–21.

Gaschak SP, Maklyuk YA, Maksimenko AM, Bondarkov MD. 2009. Radioecology of amphibians in Chernobyl zone. Probl Chernobyl Exclusion Zone. 9:76–86.

Giraudeau M, Bonzom JM, Ducatez S, Beaugelin-Seiller K, Deviche P, Lengagne T, Cavalie I, Camilleri V, Adam-Guillermin C, McGraw KJ. 2018. Carotenoid distribution in wild Japanese tree frogs (*Hyla japonica*) exposed to ionizing radiation in Fukushima. Sci Rep. 8(1): 7438

Gombeau K, Bonzom J-M, Cavalié I, Camilleri V, Orjollet D, Dubourg N, Beaugelin-Seiller K, Bourdineaud J-P, Lengagne T, Armant O, et al. 2020. Dose-dependent genomic DNA hypermethylation and mitochondrial DNA damage in Japanese tree frogs sampled in the Fukushima Daiichi area. J Environ Radioact. 225:106429

Griggs HG, Bender MA. 1972. Ultraviolet and gamma-ray induced reproductive death and photoreactivation in a Xenopus tissue culture cell line. Photochem Photobiol. 15(6):517–526.

Hall EJ. Giaccia AJ. 2006. Radiobiology for the radiologist. 6th ed. Philadelphia: Lippincott Williams & Wilkins.

Hancock S, Vo NTK, Byun SH, Zainullin VG, Seymour CB, Mothersill CE. 2019. Effects of historic radiation dose on the frequency of sex-linked recessive lethals in Drosophila populations following the Chernobyl nuclear accident. Environ Res. 172:333–337.

Hancock S, Vo NTK, Goncharova RI, Seymour CB, Byun SH, Mothersill CE. 2020. One-decade-spanning transgenerational effects of historic radiation dose in wild populations of bank voles exposed to radioactive contamination following the Chernobyl nuclear disaster. Environ Res. 180:108816

Hancock S, Vo NTK, Omar-Nazir L, Batlle JVI, Otaki JM, Hiyama A, Byun SH, Seymour CB, Mothersill C. 2019. Transgenerational effects of historic radiation dose in pale grass blue butterflies around Fukushima following the Fukushima Dai-ichi nuclear power plant meltdown accident. Environ Res. 168:230–240.

Hollstein M, Sidransky D, Vogelstein B, Harris CC. 1991. p53 mutations in human cancers. Science. 253(5015):49–53.

Hosseini A, Beresford NA, Brown JE, Jones DG, Phaneuf M, Thørring H, Yankovich T. 2010. Background dose-rates to reference animals and plants arising from exposure to naturally occurring radionuclides in aquatic environments. J Radiol Prot. 30(2):235–264.

Hurem S, Martín LM, Lindeman L, Brede DA, Salbu B, Lyche JL, Aleström P, Kamstra JH. 2018. Parental exposure to gamma

radiation causes progressively altered transcriptomes linked to adverse effects in zebrafish offspring. Environ Pollut. 234:855–863.
International Commission of Radiological Protection (ICRP). 2008. Environmental protection: the concept and use of Reference Animals and Plants. ICRP Publication 108. Ann. ICRP. 38(:4–6.
International Union for Conservation of Nature and Natural Resources (IUCN). 2020. The IUCN Red List of Threatened Species. Version 2020-1. Table 1a "Number of species evaluated in relation to the overall number of described species, and numbers of threatened species by major groups of organisms." [accessed 2020 May 24]. https://www.iucnredlist.org/resources/summary-statistics#Summary%20Tables.
Ishii Y, Matsuzaki SS, Hayashi S. 2020. Different factors determine ^{137}Cs concentration factors of freshwater fish and aquatic organisms in lake and river ecosystems. J Environ Radioact. 213:106102
Johnson RO, Tittle TV, Sefchick MP, Zettergren LD, Ruben LN, Clothier RH, Balls M. 2013. Comparative studies of apoptosis in *Xenopus laevis* and mouse thymoma cell lines. Altern Lab Anim. 41(3):211–218.
Kamstra JH, Hurem S, Martin LM, Lindeman LC, Legler J, Oughton D, Salbu B, Brede DA, Lyche JL, Aleström P. 2018. Ionizing radiation induces transgenerational effects of DNA methylation in zebrafish. Sci Rep. 8(1):15373
Koubourli DV, Yaparla A, Popovic M, Grayfer L. 2018. Amphibian (*Xenopus laevis*) Interleukin-8 (CXCL8): a perspective on the evolutionary divergence of granulocyte chemotaxis. Front Immunol. 9: 2058
Le M, Fernandez-Palomo C, McNeill FE, Seymour CB, Rainbow AJ, Mothersill CE. 2017. Exosomes are released by bystander cells exposed to radiation-induced biophoton signals: reconciling the mechanisms mediating the bystander effect. PLoS One. 12(3): e0173685
Lee JM, Bernstein A. 1993. p53 mutations increase resistance to ionizing radiation. Proc Natl Acad Sci U S A. 90(12):5742–5746.
Lips KR. 2016. Overview of chytrid emergence and impacts on amphibians. Phil Trans R Soc B. 371(1709):20150465.
Liu J, Ma T, Liu Y, Zou J, Gao M, Zhang R, Wu J, Liu S, Xie H. 2018. History, advancements, and perspective of biological research in deep-underground laboratories: a brief review. Environ Int. 120: 207–214.
Lyng FM, Seymour CB, Mothersill C. 2002. Initiation of apoptosis in cells exposed to medium from the progeny of irradiated cells: a possible mechanism for bystander-induced genomic instability? Radiat Res. 157(4):365–370.
Matsushima N, Ihara S, Takase M, Horiguchi T. 2015. Assessment of radiocesium contamination in frogs 18 months after the Fukushima Daiichi nuclear disaster. Sci Rep. 5:9712
Møller AP, Mousseau TA. 2013. The effects of natural variation in background radioactivity on humans, animals and other organisms. Biol Rev Camb Philos Soc. 88(1):226–254.
Mothersill C, Bristow RG, Harding SM, Smith RW, Mersov A, Seymour CB. 2011. A role for p53 in the response of bystander cells to receipt of medium borne signals from irradiated cells. Int J Radiat Biol. 87(11):1120–1125.
Mothersill C, Bucking C, Smith RW, Agnihotri N, Oneill A, Kilemade M, Seymour CB. 2006. Communication of radiation-induced stress or bystander signals between fish in vivo. Environ Sci Technol. 40(21):6859–6864.
Mothersill CE, Oughton DH, Schofield PN, Abend M, Adam-Guillermin C, Ariyoshi K, Beresford NA, Bonisoli-Alquati A, Cohen J, Dubrova Y. 2020. From tangled banks to toxic bunnies; a reflection on the issues involved in developing an ecosystem approach for environmental radiation protection. Int J Radiat Biol. DOI: 10.1080/09553002.2020.1793022.
Mothersill C, Rusin A, Seymour C. 2017. Low doses and non-targeted effects in environmental radiation protection; where are we now and where should we go? Environ Res. 159:484–490.
Mothersill C, Seymour C. 1997. Lethal mutations and genomic instability. Int J Radiat Biol. 71(6):751–758.
Mothersill C, Seymour C. 2014. Implications for human and environmental health of low doses of ionising radiation. J Environ Radioact. 133:5–9.
Murley J, Miller R, Weichselbaum R, Grdina D. 2017. TP53 mutational status and ROS effect the expression of the survivin-associated radio-adaptive response. Radiat Res. 188(5):659–670.
Park SC, Lee SH, Han SS. 2000. Establishment of a phagocytic cell line from *Bombina orientalis*. Methods Cell Sci. 22(1):1–7.
Rafferty KA. Jr. 1965. The cultivation of inclusion-associated viruses from Lucké tumor frogs. Ann N Y Acad Sci. 126(1):3–21.
Rafferty KA. Jr. 1969. Mass culture of amphibian cells: methods and observations concerning stability of cell type. In: Mizell M, editor. Biology of Amphibian tumors. Berlin: Springer-Verlag. p. 52–81.
Reese DH, Yamada T, Moret R. 1976. An established cell line from the newt *Notophthalmus viridescens*. Differentiation. 6(2):75–81.
Ribeiro JC, Barnetson AR, Fisher RJ, Mameghan H, Russell PJ. 1997. Relationship between radiation response and p53 status in human bladder cancer cells. Int J Radiat Biol. 72(1):11–20.
Robert J, Cohen N. 1998. Evolution of immune surveillance and tumor immunity: studies in Xenopus. Immunol Rev. 166:231–243.
Robertson K, Hensey C, Gautier J. 1999. Isolation and characterization of Xenopus ATM (X-ATM): expression, localization, and complex formation during oogenesis and early development. Oncogene. 18(50):7070–7079.
Rosenstein BS, Setlow RB. 1980. Photoreactivation of ICR 2A frog cells after exposure to monochromatic ultraviolet radiation in the 252-313 nm range. Photochem Photobiol. 32(3):361–366.
Roy S, Gardiner DM, Bryant SV. 2000. Vaccinia as a tool for functional analysis in regenerating limbs: ectopic expression of SHH. Dev Biol. 218(2):199–205.
Ruben LN, Clothier RH, Balls M. 2007. Cancer resistance in amphibians. Altern Lab Anim. 35(5):463–470.
Saroya R, Smith R, Seymour C, Mothersill C. 2009. Injection of resperpine into zebrafish, prevents fish to fish communication of radiation-induced bystander signals: confirmation in vivo of a role for serotonin in the mechanism. Dose Response. 8(3):317–330.
Seymour CB, Mothersill C, Alper T. 1986. High yields of lethal mutations in somatic mammalian cells that survive ionizing radiation. Int J Radiat Biol Relat Stud Phys Chem Med. 50(1):167–179.
Shechter D, Chitta RK, Xiao A, Shabanowitz J, Hunt DF, Allis CD. 2009. A distinct H2A.X isoform is enriched in *Xenopus laevis* eggs and early embryos and is phosphorylated in the absence of a checkpoint. Proc Natl Acad Sci U S A. 106(3):749–754.
Shi X, Seymour C, Mothersill C. 2018. Change of cell growth and mitochondrial membrane polarization in the progeny of cells surviving low-dose high-LET irradiation from Ra-226. Environ Res. 167: 51–65.
Smith RW, Seymour CB, Mothersill CE. 2013. Short and long term bystander effect induction by fathead minnows (*Pimephales promelas*, Rafinesque, 1820) injected with environmentally relevant whole body doses of 226Ra. J Environ Radioact. 126:133–136.
Stark K. 2006. Risk from radionuclides: a frog's perspective. Accumulation of ^{137}Cs in a riparian wetland [Radiation doses, and effects on frogs and toads after low-dose rate exposure] [Doctoral thesis]. Stockholm University, Stockholm.
Stiewe T. 2007. The p53 family in differentiation and tumorigenesis. Nat Rev Cancer. 7(3):165–167.
Stuart SN, Hoffmann M, Chanson JS, Cox NA, Berridge RJ, Ramani P, Young BE, editors. 2008. Threatened Amphibians of the world. Barcelona: Lynx Edicions.
United States Nuclear Regulatory Commission (USNRC). 2016. Leaks and spills at U.S. commercial nuclear power plants. [accessed 2020 Sep 20]. https://www.nrc.gov/docs/ML1615/ML16154A745.pdf.
Uto K, Inoue D, Shimuta K, Nakajo N, Sagata N. 2004. Chk1, but not Chk2, inhibits Cdc25 phosphatases by a novel common mechanism. Embo J. 23(16):3386–3396.
van Zeeland AA, Natarajan AT, Verdegaal-Immerzeel EA, Filon AR. 1980. Photoreactivation of UV induced cell killing, chromosome aberrations, sister chromatid exchanges, mutations and pyrimidine

dimers in *Xenopus laevis* fibroblasts. Mol Gen Genet. 180(3): 495–500.

Villiard E, Brinkmann H, Moiseeva O, Mallette FA, Ferbeyre G, Roy S. 2007. Urodele p53 tolerates amino acid changes found in p53 variants linked to human cancer. BMC Evol Biol. 7:180

Vinogradov AE, Chubinishvili AT. 1999. Genome reduction in a hemiclonal frog *Rana esculenta* from radioactively contaminated areas. Genetics. 151(3):1123–1125.

Vo NTK, Everson J, Moore L, DeWitte-Orr SJ. 2019. Class A scavenger receptor expression and function in eight novel tadpole cell lines from the green frog (*Lithobates clamitans*) and the wood frog (*Lithobates sylvatica*). Cytotechnology. 71(4):757–768.

Vo NTK, Guerreiro M, Yaparla A, Grayfer L, DeWitte-Orr SJ. 2019. Class A scavenger receptors are used by Frog Virus 3 during its cellular entry. Viruses. 11(2):93.

Vo NTK, Moore L, Leis E, DeWitte-Orr SJ. 2019. Class A scavenger receptors mediate extracellular dsRNA sensing, leading to downstream antiviral gene expression in a novel American toad cell line. BufoTad. Dev. Comp. Immunol. 92:140–149.

Vo NTK, Seymour CB, Mothersill CE. 2019a. The common field lampricide 3-trifluoromethyl-4-nitrophenol is a potential radiosensitizer in fish cells. Environ Res. 170:383–388.

Vo NTK, Seymour CB, Mothersill CE. 2019b. Radiobiological characteristics of descendant progeny of fish and amphibian cells that survive the initial ionizing radiation dose. Environ Res. 169:494–500.

Weinberg HS, Korol AB, Kirzhner VM, Avivi A, Fahima T, Nevo E, Shapiro S, Rennert G, Piatak O, Stepanova EI, et al. 2001. Very high mutation rate in offspring of Chernobyl accident liquidators. Proc Biol Sci. 268(1471):1001–1005.

Wertheim B, Beukeboom LW, van de Zande L. 2013. Polyploidy in animals: effects of gene expression on sex determination, evolution and ecology. Cytogenet Genome Res. 140(2–4):256–269.

Whicker FW, Schultz V. 1982. Radioecology: nuclear energy and the environment. Boca Raton (FL): CRC Press.

Williams RN, Bos DH, Gopurenko D, Dewoody JA. 2008. Amphibian malformations and inbreeding. Biol Lett. 4(5):549–552.

Wolf K, Quimby MC. 1964. Amphibian cell culture: permanent cell line from the bullfrog (*Rana catesbeiana*). Science. 144(3626): 1578–1580.

Xu S, Ding N, Pei H, Hu W, Wei W, Zhang X, Zhou G, Wang J. 2014. MiR-21 is involved in radiation-induced bystander effects. RNA Biol. 11(9):1161–1170.

Xu S, Wang J, Ding N, Hu W, Zhang X, Wang B, Hua J, Wei W, Zhu Q. 2015. Exosome-mediated microRNA transfer plays a role in radiation-induced bystander effect. RNA Biol. 12(12):1355–1363.

Yeliseeva K, Mikhalevich L, Kartel N. 1995. Cytogenetic effects of radiation from Chernobyl nuclear accident on humans and animals in the contaminated area of Belarus. Rev Int Contam Ambient. 11(Z): 77–85.

Yun MH, Davaapil H, Brockes JP. 2015. Recurrent turnover of senescent cells during regeneration of a complex structure. Elife. 4:e05505.

Radiation-induced transgenerational effects in animals

Yuri E. Dubrova and Elena I. Sarapultseva

ABSTRACT
According to the results of recent studies, parental exposure to ionizing radiation not only leads to mutation induction in the germline of irradiated animals but also affects their non-exposed offspring. These radiation-induced transgenerational effects belong to an epigenetic phenomenon that could not be defined as a transmission of altered phenotypes from the irradiated parents to their non-exposed offspring. In this review, we present the results of laboratory studies aimed to evaluate the transgenerational effects of parental irradiation on a number of traits in the offspring of exposed parents. The results of animal studies showing compromised viability, fertility and genome stability among the non-exposed offspring of irradiated parents are presented and discussed. So far, the epigenetic phenomenon of radiation-induced transgenerational effects has been established in laboratory studies. Future work should address the important issue of manifestation of radiation-induced transgenerational effects in populations inhabiting radioactive-contaminated areas, as well as the mechanisms of transgenerational effects.

Introduction

Radiation-induced transgenerational effects belong to an epigenetic phenomenon that could not be defined as a transmission of altered phenotypes from the irradiated parents to their non-exposed offspring. The phenotypic traits epigenetically transmitted from parents to their offspring include compromised viability, fertility and genome stability. Here we present the results of recent studies in the field of radiation-induced transgenerational effects in invertebrates and mammals.

Radiation-induced transgenerational effects in invertebrates

To date, the phenomenon of radiation-induced transgenerational inheritance has been extensively analyzed by studying the progeny of irradiated *Daphnia*. According to the results of our studies, compromised survival of the directly irradiated *Daphnia magna* was observed following exposure to a wide range of doses of acute exposure to γ-rays. Thus, acute exposure to 250 and 600 Gy of γ-rays is lethal (Sarapul'tseva and Malina 2009), whereas the effects of exposure to lower doses from 10 mGy to 10 Gy are non-linear (Sarapultseva and Bychkovskaya 2010). The survival of *D. magna* exposed to 10 mGy of γ-rays does not significantly differ from that in controls, whereas it is significantly compromised in the two groups exposed to 100 and 1000 mGy (Sarapultseva and Gorski 2013). The life expectancy of irradiated animals was 20–30% lower than in the control groups. Later, we analyzed the survival of the non-exposed progeny of irradiated *D. magna* (Sarapultseva and Dubrova 2016). The results of our study are summarized in Figure 1(A). According to our data, parental exposure to 1000 and 10,000 mGy of acute γ-rays significantly compromised the F_1 viability. It should be stressed that the magnitude of transgenerational effects of parental irradiation on the F_1 viability was close to the direct effects on the F_0 parental generation. On the other hand, the survival of second-generation progeny of irradiated parents does not significantly differ from that in controls.

The transgenerational effects of parental irradiation have also been analyzed in other species of invertebrates. Thus, compromised survival and elevated incidence of morphological abnormalities were observed among the progeny of irradiated pale grass blue butterflies *Zizeeria maha* following the Fukushima accident (Hancock, Vo, Omar-Nazir, et al. 2019). The elevated genetic damage was also observed in populations of *Drosophila melanogaster* inhabiting radioactive-contaminated areas around the Chernobyl Nuclear Plant, where the later generations developed a radio-adaptive-like response (Hancock, Vo, Byun, et al. 2019). The same result was observed in the freshwater crustacean *Asellus aquaticus* 30 years since the Chernobyl accident (Fuller et al. 2019). It was also observed that exposure to external gamma radiation significantly compromised the reproduction of *Caenorhabditis elegans* as well as resulting in a delay in their growth (Lecomte-Pradines et al. 2017).

It is known from the studies on vertebrates that the most radiosensitive biological processes include oxidative

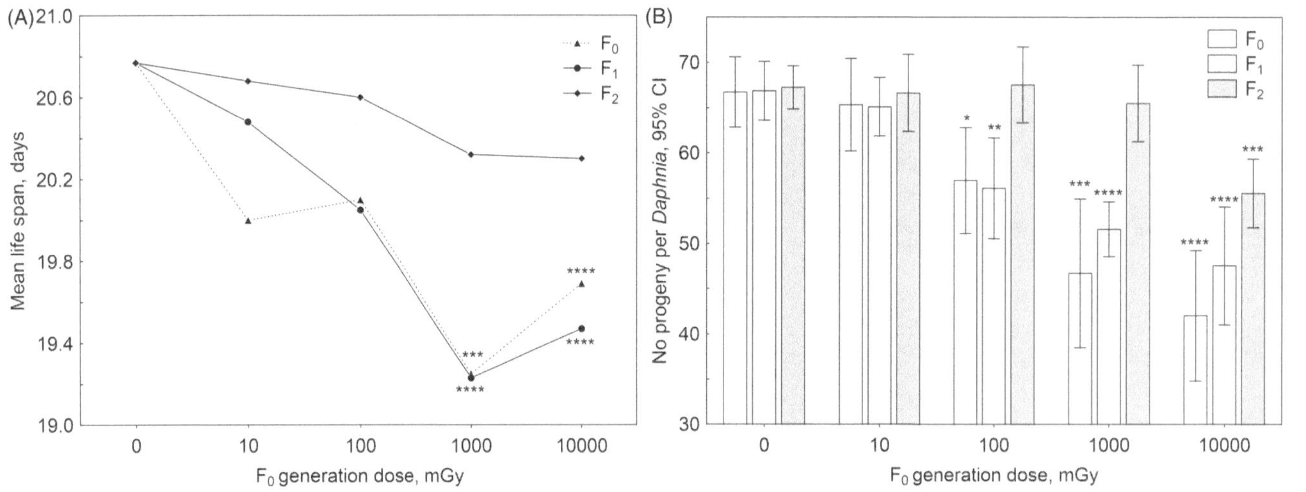

Figure 1. Radiation-induced transgenerational effects in the offspring of *Daphnia magna*. (A) The effects of parental irradiation on viability in irradiated *Daphnia* (F_0) and their non-exposed F_1 and F_2 progeny (the *p*-values were estimated using the log-rank test). (B) Fertility of irradiated *Daphnia* and their progeny (data from Sarapultseva and Dubrova 2016). On all graphs $^*p < .05$; $^{**}p < .01$; $^{***}p < .001$; $^{****}p < .0001$.

phosphorylation (Wu and Zhou 2012), free radicals damage mitochondrial and nuclear DNA, as well as cell membranes and proteins (Gaziev 2013), leading to a shortened lifespan (Moskalev et al. 2014; Van Raamsdonk 2018). The chronic external exposure to γ-rays (from 0.5 to 50 mGy h^{-1}) on zebrafish and their progeny affected reproductive success, fecundity or egg fertilization rate, together with an increase of ROS production. (Guirandy et al. 2019). It would, therefore, appear that the same processes may also manifest in invertebrates.

Over recent years, a number of techniques have been developed for the analysis of radiation cytotoxicity (Buch et al. 2012; Wu and Zhou 2012). We applied the MTT-test to *D. magna* and measured the cytotoxic effect in controls and directly exposed animals, which are commonly used in vitro to assess the viability of proliferating cells (Van Meerloo et al. 2011). The MMT-test can also be used for measuring the intensity of metabolism in cells (Bernas and Dobrucki 2002). We have shown significantly elevated cytotoxicity in *D. magna* exposed to 100 and 1000 mGy of gamma-rays (Sarapul'tseva et al. 2013). We found that the toxic effect also manifested in the F_1 progeny of irradiated *D. magna* (Sarapultseva et al. 2017). We also found that in *D. magna* the level of free radical is significantly elevated following exposure to 0.1 and 1 Gy of γ-rays (Sarapul'tseva et al. 2014). It would, therefore, appear that the abovementioned cytotoxic effects may explain the compromised viability of irradiated *D. magna*.

The observed changes in the metabolic index in *D. magna* are well consistent with the effects obtained on other species. Clonogenic survival and mitochondrial activity were reduced in vivo exposed to external gamma radiation in the three species of earthworms (Rusin et al. 2019). Parental gamma exposure at a dose rate of 53 mGy/h (cumulative dose of 31 Gy) at sensitive stages of the life cycle of zebrafish embryos (*Danio rerio)* resulted in 100% mortality in the directly exposed fish. Exposure at 8.7 mGy/h (cumulative dose of 5 Gy) induced genomic instability in the non-exposed F_1 progeny, resulting in elevated larval mortality, earlier hatching of embryos, and a high frequency of their deformations. Even one month after the parent exposure, the concentration of reactive oxygen species (ROS) remained significantly elevated but did not differ significantly from the control 1 year after the parent exposure (Hurem et al. 2017).

The compromised survival of gill cells of *Rainbow trout* together with changes in enzyme activity and mitochondrial oxidative phosphorylation was detected following low-dose exposure to ionizing radiation (Mothersill and Seymour 2006; Mothersill and Seymour 2009). When neonates were returned to non-contaminated water, the expression of mitochondrial-associated genes did not significantly differ from that in controls. However, the effects of exposure to ionizing radiation did not manifest across generations (Plaire et al. 2013; Parisot et al. 2015). Significant increases in DNA damage (double-strand DNA breaks) were observed in crustacean *Tigriopus japonicas* following exposure to 200 Gy of γ-rays (Han et al. 2014). Dose-dependent transcriptomic changes were also found in the crustacean *Paracyclopina nana* (Won and Lee 2014). An elevated level of DNA single-strand breaks in hemocytes in chronic exposure to ionizing radiation oyster *Crassostrea gigas* was observed by comet assay (Devos et al. 2015). One possible explanation for the high number of spontaneous DNA lesions can be given by the fact that the progeny cells of irradiated parents are exposed to oxidative stress. Indeed, free radicals, whose concentration increases dramatically in cells exposed to oxidative stress, cause the whole range of DNA damage. The data on this subject is controversial. The elevated level of ROS was also observed in the unstable progeny of irradiated cells (Clutton et al. 1996).

It has been established that environmental contamination mainly affects reproduction (Dallas et al. 2012). According to our results, parental acute exposure of *D. magna* at doses of 100–10,000 mGy significantly compromises the main components of total fertility – the number and size of broods (Sarapultseva and Dubrova 2016). The effect was maintained in the F_1 generation and almost leveled in F_2

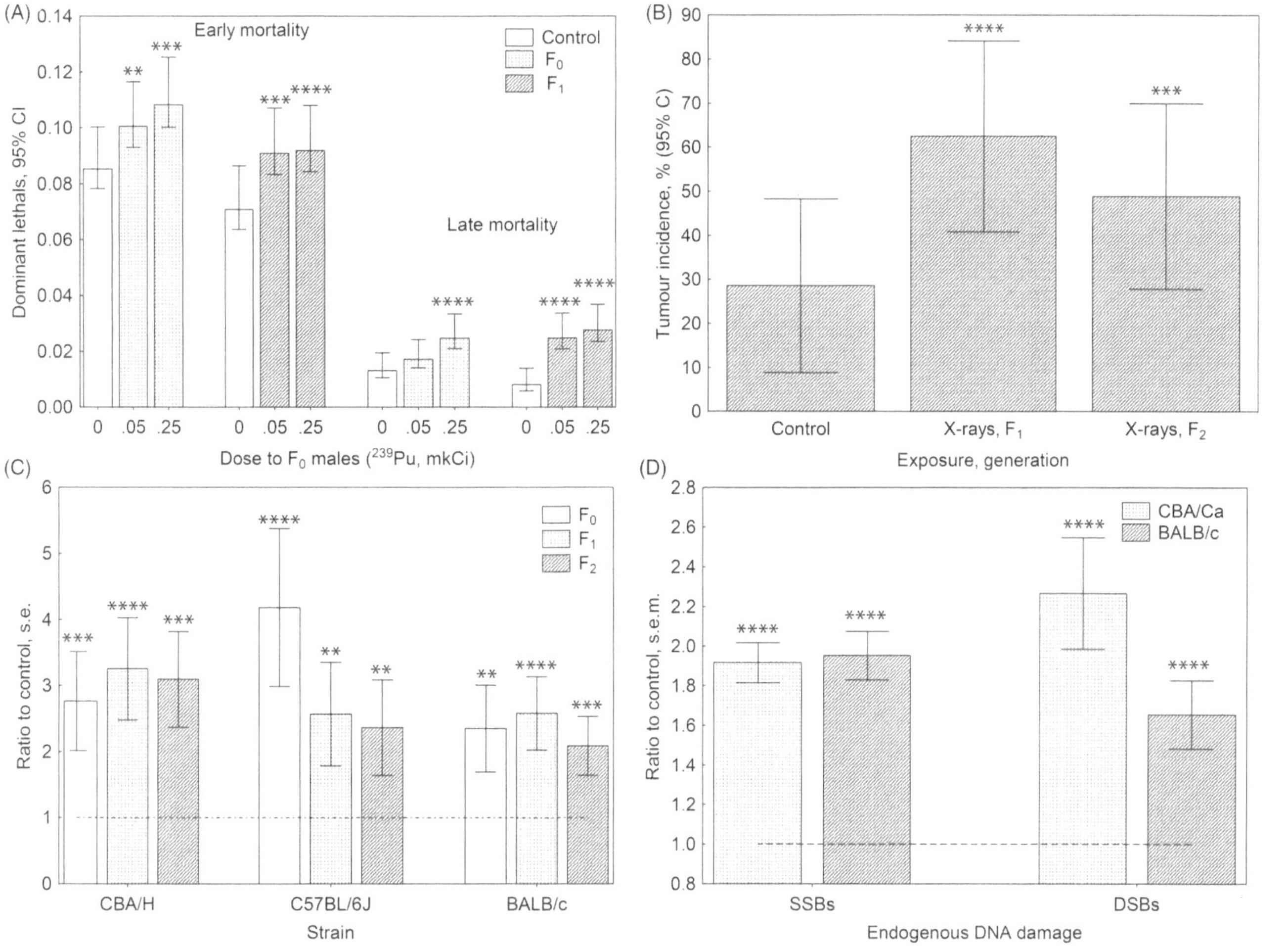

Figure 2. Radiation-induced transgenerational effects in the offspring of exposed male mice. (A) Dominant lethal mutations in the germline of irradiated males (F_0) and their F_1 offspring. The data for dominant mutations resulting in the early pre-implantation and late post-implantation losses are shown separately (data from Luning et al. 1976). (B) Promotion of skin tumors by 12-*O*-tetradecanoylphorbol-13-acetate (TPA) in the offspring of male mice acutely exposed to 4.2 Gy of X-rays (data from Vorobtsova et al. 1993). (C) ESTR mutation rates in the germline of irradiated male mice (F_0) and their non-exposed F_1 and F_2 offspring (data from Barber et al. 2002). (D) Endogenous DNA damage (single-strand breaks, SSBs and double-strand breaks, DSBs) in somatic tissues of the F_1 offspring of irradiated male mice are shown (data from Barber et al. 2006). The *p*-values for significant difference compared to control were estimated using Fisher's exact test (A–C) and Student's *t*-test (D).

and remains compromised following parental high-dose exposure to 10,000 mGy (Figure 1(B)). The decrease in the number of broods in F_0 is attributed to the cytotoxic effect of radiation on the germ cells. In our later study, we analyzed the combined effect of food supply and parental irradiation on the F_1/F_2 viability and fertility (Sarapultseva et al. 2019). We have shown that the independent effects of food supply and irradiation on the viability and reproductive success in *D. magna*. Our data are in line with the results of studies on female childhood cancer survivors, where exposure to radiotherapy leads to the depletion of the number of primordial oocytes (Wallace et al. 2005; Fuller et al. 2015). It has also been shown that high-dose acute γ-irradiation significantly affects the number of eggs produced by marine copepods *P. nana* (Won and Lee 2014), as well as their mass (Alonzo et al. 2008; Zeman et al. 2008) and the time when the neonates are washed out (Gilbin et al. 2008). Given that we analyzed the parthenogenic strain of *D. magna*, we irradiated primordial diploid eggs (Sarapultseva and Dubrova 2016), the results of the abovementioned studies provide a plausible explanation for the compromised fertility of irradiated exposed *D. magna*.

The described transgenerational effects can only be explained by epigenetic mechanisms. The main argument for this is the very high incidence of mortality among the progeny of irradiated parents, which cannot be attributable to Mendelian segregation of radiation-induced mutations. It has been suggested that changes in alterations in DNA methylation may explain the manifestation of transgenerational effects in invertebrates. Thus, profound alterations in the pattern of DNA methylation were recently observed in *D. magna* and mice following low-dose exposure to γ-rays (Asselman et al. 2015; Wang et al. 2014). It is clear that the contribution of DNA methylation to the phenomenon of transgenerational effects requires a thorough analysis.

Radiation-induced transgenerational effects in mammals

In 1976 Luning and coauthors published the results of the mouse study on the effects of paternal exposure on the alpha-particle emitter plutonium-239 on the incidence of dominant lethal mutations in the germline of directly exposed F_0 male mice and their non-exposed first-generation offspring (Luning et al. 1976). Figure 2(A) summarizes the results of this study, according to which paternal irradiation not only increased mutation rate in the germline of directly exposed F_0 male mice but also resulted in destabilization of their non-exposed F_1 offspring. The magnitude of transgenerational increase in mutation rate observed in the F_1 germline was close to that in the directly exposed F_0 male mice. Later impaired cell proliferation was observed in early embryos conceived by the F_1 offspring of irradiated male mice (Wiley et al. 1997). Another evidence for transgenerational effects of parental exposure to ionizing radiation was obtained in a number of studies aimed to analyze the incidence of cancer among the offspring of irradiated male mice exposed to recognized carcinogens (Vorobtsova et al. 1993; Lord et al. 1998; Hoyes et al. 2001). According to the results of these studies, the F_1 incidence of cancer is significantly elevated (Figure 2(B)).

In the later studies, the transgenerational effects of parental exposure to ionizing radiation were established by mutation scoring at Expanded Tandem Repeat Loci (ESTRs). These loci consist of homogeneous array of short repeats and are characterized by a very high spontaneous mutation rated in the germline and somatic tissues (Bois et al. 1998). According to our data, the mechanism of ESTR spontaneous mutation in the mouse germline and somatic tissues are close to those at microsatellite loci (Hardwick at al. 2009). In our first study, we showed a significantly elevated ESTR mutation rate in the germline of F_1 offspring of CBA/H male mice exposed to 0.5 Gy of fission neutrons (Dubrova et al. 2000). Later we analyzed ESTR mutation rates in the germline of non-exposed first- and second-generation offspring of male mice from three different inbred strains acutely exposed to X-rays of fission neutrons (Barber et al. 2002). According to our results, ESTR mutation rate remains significantly elevated in the germline of first- and second-generation offspring of irradiated males from all strains (Figure 2(C)). The manifestation of radiation-induced transgenerational instability at ESTR loci was later confirmed by a Japanese group (Niwa and Kominami 2001).

Using single-molecule PCR we later established the frequency of ESTR mutations in somatic tissues of the F_1 offspring of irradiated male mice and showed that the manifestation of transgenerational instability is not restricted to the F_1 germline but is also observed in their somatic tissues (Barber et al. 2006; Hatch et al. 2007; Barber et al. 2009; Abouzeid Ali et al. 2012; Mughal et al. 2012). The analysis of mutation frequencies in the F_1 mouse somatic tissues also revealed an important feature of transgenerational instability. It turned out that the frequency of somatic mutation was equally elevated in both alleles derived from both the irradiated fathers and the unexposed mothers (Shiraishi at al. 2002; Barber et al. 2006), thus implying genome-wide transgenerational destabilization.

According to the results of our two mouse studies, maternal irradiation does not affect F_1 stability (Barber et al. 2009; Abouzeid Ali et al. 2012), which remains unexplained.

The transgenerational effects of paternal exposure to ionizing radiation were also observed by studying other genetic endpoints. It was shown that paternal irradiation significantly increases the frequency of chromosome aberrations in their first-generation offspring (Lord et al. 1998; Vorobtsova 2000; Kropácová et al. 2002; Slovinska et al. 2004). Elevated frequency of *lac*I mutations was also detected in the F_1 offspring of irradiated male mice (Luke et al. 1997), as well as at protein-coding genes (Carls and Schiestl 1999; Shiraishi et al. 2002; Barber et al. 2006). Taking together, these observations provide a plausible explanation for an increased incidence of cancer among the offspring of irradiated male mice. As the frequency of chromosomal aberrations and mutations at protein-coding genes is elevated in the offspring of irradiated males, the destabilization of somatic genomes may promote tumor progression.

The analysis of radiation-induced transgenerational instability in mammals has shown that this phenomenon cannot be explained by conventional Mendelian genetics. As the destabilization of the genome was detected among the majority of F_1/F_2 offspring of irradiated male mice, it cannot be attributed to the segregation of mutations induced by ionizing radiation in the germline of exposed F_0 males (Barber et al. 2002). It would, therefore, appear that epigenetic mechanisms underly the phenomenon of transgenerational instability. The question is – what triggers the appearance of the yet unknown transgenerational signal in the germline of exposed male mice? To answer this question, we analyzed the F_1 offspring on male mice exposed to exposure to the alkylating agent ethylnitrosourea (ENU) or three commonly-used anticancer drugs – cyclophosphamide, mitomycin C and procarbazine (Dubrova et al. 2008; Glen and Dubrova 2012). We observed significantly elevated frequency of ESTR mutation in the germline and somatic tissues of the first-generation offspring of treated males. The results of our studies, therefore, suggest that paternal exposure to a number of mutagens can destabilize the genomes of their offspring.

Another pertinent question is – why the stability of the F_1 genome is compromised? According to the results of our study, the amount of endogenous DNA damage (single- and double-stranded DNA breaks) in the F_1 offspring of irradiated male mice is significantly elevated (Barber et al. 2006). Figure 2(D) summarized the results of our study. It would, therefore, appear that destabilization of the F_1 genome may be related to some yet unknown changes affecting the pattern of gene expression, affecting DNA replication, apoptosis or cell cycle progression. According to our recent data, paternal irradiation induces significant alterations in the expression of circadian genes in F_1 offspring of irradiated male mice, including the key components of the circadian clock *Arntl* and *Npas2* which, in turn, could potentially lead to the destabilization of their genomes (Gomes et al. 2015).

Given that the majority of epigenetic marks in mammals, such as DNA methylation, are erased after fertilization (Luo et al. 2018), it would appear that changes in DNA methylation in the germline of irradiated males may not contribute to the phenomenon of transgenerational inheritance. According to the results of recent studies, the mechanisms of epigenetic inheritance can be attributed to the non-coding

RNAs (Gapp et al. 2014; Grandjean et al. 2015; Benito et al. 2018). It was also shown that in mammals non-coding RNAs can be transmitted to maturing sperm by small extracellular vesicle epididymosomes (Sharma et al. 2018). It has been shown that miRNA-containing extracellular vesicles are present in the blood of irradiated mice (Szatmári et al. 2019). It would, therefore, appear that extracellular vesicles can be trafficked from blood to sperm. Indeed, according to the results of a recent study, the spectrum miRNA in sperm of irradiated mice significantly differs from that in control males (Paris et al. 2015). Considering the results of the abovementioned studies, future work aimed to analyze the role of miRNAs in sperm of irradiated males as an epigenetic signal capable of destabilizing the F_1 genome is clearly warranted.

Conclusions

The data reviewed here describe the transgenerational effects of parental exposure to ionizing radiation on the phenotypes of their offspring. There are a number of gaps in our knowledge. The results of studies on invertebrates have provided compelling evidence for the effects of parental exposure on the fitness of their progeny, mainly compromised viability and fertility. On the other hand, the transgenerational effects of parental irradiation on the fitness of their offspring in mammals still remain poorly understood. The analysis of transgenerational effects of parental irradiation in mammals has revealed a significant destabilization of the F_1/F_2 genome. In our opinion, this important issue requires further studies in invertebrates.

In this review, we have presented the results of laboratory studies aimed to evaluate the transgenerational effects of parental irradiation on a number of traits in the offspring of exposed parents. The key question is whether transgenerational effects can be observed in populations inhabiting radioactive-contaminated areas. Given that the organisms inhabiting radioactive-contaminated areas are continuously exposed to ionizing radiation, the analysis of transgenerational effects is complicated. The *in situ* mutational studies of the offspring of irradiated may not distinguish between the direct effects of exposure (i.e. mutation induction) and transgenerational effects. There is another issue. In the abovementioned laboratory studies, the parents were exposed to relatively high doses of acute irradiation from external sources, whereas the organisms inhabiting radioactive-contaminated areas are exposed to low doses of chronic irradiation. We have analyzed the effects of chronic low-LET γ-irradiation on the manifestation of transgenerational instability in mice (Mughal et al. 2012). According to the results of our study, chronic paternal exposure to 1 Gy of γ-rays does not destabilize the F_1 genomes, whereas acute exposure to the same dose of γ-rays significantly affects mutation rate in the F_1 germline and somatic tissues. However, it remains unknown whether paternal low-dose or chronic exposure to high-LET sources, such as α-particles and fission neutrons, can significantly compromise the offspring of irradiated parents. Future studies should address this important issue.

Disclosure statement

No potential conflict of interest was reported by the author(s).

Funding

YED work, in part, was supported by the Nuclear Community Charity Fund (NCCF) through funds received by The Armed Forces Covenant Fund Trust under the Aged Veterans Fund Grant [AVF16]. SEI work, in part, was supported by the MEPhI Academic Excellence Project (No. 02.a03.21.0005).

ORCID

Yuri E. Dubrova http://orcid.org/0000-0001-5281-7539

References

Abouzeid Ali HE, Barber RC, Dubrova YE. 2012. The effects of maternal irradiation during adulthood on mutation induction and transgenerational instability in mice. Mutat Res. 732(1–2):21–25.

Alonzo F, Gilbin R, Zeman FA, Garnier-Laplace J. 2008. Increased effects of internal alpha irradiation in *Daphnia magna* after chronic exposure over three successive generations. Aquat Toxicol. 87(3): 146–156.

Asselman J, De Coninck DI, Vandegehuchte MB, Jansen M, Decaestecker E, De Meester L, Vanden Bussche J, Vanhaecke L, Janssen CR, De Schamphelaere KA. 2015. Global cytosine methylation in *Daphnia magna* depends on genotype, environment, and their interaction. Environ Toxicol Chem. 34(5):1056–1061.

Barber RC, Hardwick RJ, Shanks ME, Glen CD, Mughal SK, Voutounou M, Dubrova YE. 2009. The effects of *in utero* irradiation on mutation induction and transgenerational instability in mice. Mutat Res. 664(1–2):6–12.

Barber RC, Hickenbotham P, Hatch T, Kelly D, Topchiy N, Almeida G, Jones GGD, Johnson GE, Parry JM, Rothkamm K, et al. 2006. Radiation-induced transgenerational alterations in genome stability and DNA damage. Oncogene. 25(56):7336–7342.

Barber RC, Plumb MA, Boulton E, Roux I, Dubrova YE. 2002. Elevated mutation rates in the germ line of first- and second-generation offspring of irradiated male mice . Proc Natl Acad Sci USA. 99(10):6877–6882.

Benito E, Kerimoglu C, Ramachandran D, Pena-Centeno T, Jain G, Stilling RM, Islam MR, Capece V, Zhou Q, Edbauer D, et al. 2018. RNA-dependent intergenerational inheritance of enhanced synaptic plasticity after environmental enrichment. Cell Rep. 23(2):546–554.

Bernas T, Dobrucki J. 2002. Mitochondrial and nonmitochondrial reduction of MTT: interaction of MTT with TMRE, JC-1, and NAO mitochondrial fluorescent probes. Cytometry. 47(4):236–242.

Bois P, Williamson J, Brown J, Dubrova YE, Jeffreys AJ. 1998. A novel unstable mouse VNTR family expanded from SINE B1 elements. Genomics. 49(1):122–128.

Buch K, Peters T, Nawroth T, Sänger M, Schmidberger H, Langguth P. 2012. Determination of cell survival after irradiation via clonogenic assay versus multiple MTT Assay-a comparative study. Radiat Oncol. 7:1.

Carls N, Schiestl RH. 1999. Effect of ionizing radiation on transgenerational appearance of p(un) reversions in mice. Carcinogenesis. 20(12):2351–2354.

Clutton SM, Townsend KM, Walker C, Ansell JD, Wright EG. 1996. Radiation-induced genomic instability and persisting oxidative stress in primary bone marrow cultures. Carcinogenesis. 17(8):1633–1639.

Dallas LJ, Keith-Roach M, Lyons BP, Jha AN. 2012. Assessing the impact of ionizing radiation on aquatic invertebrates: a critical review. Radiat Res. 177(5):693–716.

Devos A, Dallas LJ, Voiseux C, Lecomte-Pradines C, Jha AN, Fiévet B. 2015. Assessment of growth, genotoxic responses and expression of stress related genes in the Pacific oyster *Crassostrea gigas* following chronic exposure to ionizing radiation. Mar Pollut Bull. 95(2): 688–698.

Dubrova YE, Hickenbotham P, Glen CD, Monger K, Wong H-P, Barber RC. 2008. Paternal exposure to ethylnitrosourea results in transgenerational genomic instability in mice. Environ Mol Mutagen. 49(4):308–311.

Dubrova YE, Plumb M, Gutierrez B, Boulton E, Jeffreys AJ. 2000. Transgenerational mutation by radiation. Nature. 405(6782):37.

Fuller N, Ford AT, Lerebours A, Gudkov DI, Nagorskaya LL, Smith JT. 2019. Chronic radiation exposure at Chernobyl shows no effect on genetic diversity in the freshwater crustacean, *Asellus aquaticus* thirty years on. Ecol Evol. 9(18):10135–10144.

Fuller N, Lerebours A, Smith JT, Ford AT. 2015. The biological effects of ionising radiation on Crustaceans: a review. Aquat Toxicol. 167: 55–67.

Gapp K, Jawaid A, Sarkies P, Bohacek J, Pelczar P, Prados J, Farinelli L, Miska E, Mansuy IM. 2014. Implication of sperm RNAs in transgenerational inheritance of the effects of early trauma in mice. Nat Neurosci. 17(5):667–669.

Gaziev AI. 2013. [Pathways for maintenance of mitochondrial DNA integrity and mitochondrial functions in cells exposed to ionizing radiation]. Radiats Biol Radioecol. 53(2):117–136. In Russian.

Gilbin R, Alonzo F, Garnier-Laplace J. 2008. Effects of chronic external gamma irradiation on growth and reproductive success of *Daphnia magna*. J Environ Radioact. 99(1):134–145.

Glen CD, Dubrova YE. 2012. Exposure to anticancer drugs can result in transgenerational genomic instability in mice. Proc Natl Acad Sci USA. 109(8):2984–2988.

Gomes A, Barber RC, Dubrova YE. 2015. Paternal irradiation perturbs the expression of circadian genes in offspring. Mutat Res. 775:33–37.

Grandjean V, Fourré S, De Abreu DA, Derieppe MA, Remy JJ, Rassoulzadegan M. 2015. RNA-mediated paternal heredity of diet-induced obesity and metabolic disorders. Sci Rep. 5:18193.

Guirandy N, Gagnaire B, Frelon S, Munch T, Dubourg N, Camilleri V, Cavalié I, Floriani M, Arcanjo C, Murat EI, et al. 2019. Adverse effects induced by chronic gamma irradiation in progeny of adult fish not affecting parental reproductive performance. Environ Toxicol Chem. 38(11):2556–2567.

Han J, Won EJ, Lee BY, Hwang UK, Kim IC, Yim JH, Leung KM, Lee YS, Lee JS. 2014. Gamma rays induce DNA damage and oxidative stress associated with impaired growth and reproduction in the copepod *Tigriopus japonicus*. Aquat Toxicol. 152:264–272.

Hancock S, Vo NTK, Byun SH, Zainullin VG, Seymour CB, Mothersill C. 2019. Effects of historic radiation dose on the frequency of sex-linked recessive lethals in *Drosophila* populations following the Chernobyl nuclear accident. Environ Res. 172:333–337.

Hancock S, Vo NTK, Omar-Nazir L, Batlle JVI, Otaki JM, Hiyama A, Byun SH, Seymour CB, Mothersill C. 2019. Transgenerational effects of historic radiation dose in pale grass blue butterflies around Fukushima following the Fukushima Dai-ichi Nuclear Power Plant meltdown accident. Environ Res. 168:230–240.

Hardwick RJ, Tretyakov MV, Dubrova YE. 2009. Age-related accumulation of mutations supports a replication-dependent mechanism of spontaneous mutation at tandem repeat DNA loci in mice. Mol Biol Evol. 26(11):2647–2654.

Hatch T, Derijck A, Black PD, van der Heijden GW, De Boer P, Dubrova YE. 2007. Maternal effects of the *scid* mutation on radiation-induced transgenerational instability in mice. Oncogene. 26(32):4720–4724.

Hoyes KP, Lord BI, McCann C, Hendry JH, Morris ID. 2001. Transgenerational effects of preconception paternal contamination with ^{55}Fe. Radiat Res. 156(5):488–494.

Hurem S, Gomes T, Brede DA, Lindbo Hansen E, Mutoloki S, Fernandez C, Mothersill C, Salbu B, Kassaye YA, Olsen AK, et al. 2017. Parental gamma irradiation induces reprotoxic effects accompanied by genomic instability in zebrafish (*Danio rerio*) embryos. Environ Res. 159:564–578.

Kropácová K, Slovinská L, Misúrová E. 2002. Cytogenetic changes in the liver of progeny of irradiated male rats. J Radiat Res. 43(2): 125–133.

Lecomte-Pradines C, Hertel-Aas T, Coutris C, Gilbin R, Oughton D, Alonzo F. 2017. A dynamic energy-based model to analyze sublethal effects of chronic gamma irradiation in the nematode *Caenorhabditis elegans*. J Toxicol Environ Health Part A. 80(16–18): 830–844.

Lord BI, Woolford LB, Wang L, Stones VA, McDonald D, Lorimore SA, Papworth D, Wright EG, Scott D. 1998. Tumour induction by methyl-nitroso-urea following preconceptional paternal contamination with plutonium-239. Br J Cancer. 78(3):301–311.

Luke GA, Riches AC, Bryant PE. 1997. Genomic instability in haematopoietic cells of F_1 generation mice of irradiated male parents. Mutagenesis. 12(3):147–152.

Luning KG, Frolen H, Nilsson A. 1976. Genetic effects of ^{239}Pu salt injections in male mice. Mutat Res. 34(3):539–542.

Luo C, Hajkova P, Ecker JR. 2018. Dynamic DNA methylation: in the right place at the right time. Science. 361(6409):1336–1340.

Moskalev AA, Aliper AM, Smit-McBride Z, Buzdin A, Zhavoronkov A. 2014. Genetics and epigenetics of aging and longevity. Cell Cycle. 13(7):1063–1077.

Mothersill C, Seymour CB. 2006. Radiation-induced bystander effects and the DNA paradigm: an "out of field" perspective. Mutat Res. 597(1-2):5–10.

Mothersill C, Seymour C. 2009. Communication of ionising radiation signals-a tale of two fish. Int J Radiat Biol. 85(11):909–919.

Mughal SK, Myazin AE, Zhavoronkov LP, Rubanovich AV, Dubrova YE. 2012. The dose and dose-rate effects of paternal irradiation on transgenerational instability in mice: a radiotherapy connection. PLoS One. 7(7):e41300.

Niwa O, Kominami R. 2001. Untargeted mutation of the maternally derived mouse hypervariable minisatellite allele in F_1 mice born to irradiated spermatozoa. Proc Natl Acad Sci USA. 98(4):1705–1710.

Paris L, Giardullo P, Leonardi S, Tanno B, Meschini R, Cordelli E, Benassi B, Longobardi MG, Izzotti A, Pulliero A, et al. 2015. Transgenerational inheritance of enhanced susceptibility to radiation-induced medulloblastoma in newborn $Ptch1^{+/-}$ mice after paternal irradiation. Oncotarget. 6(34):36098–36112.

Parisot F, Bourdineaud JP, Plaire D, Adam-Guillermin C, Alonzo F. 2015. DNA alterations and effects on growth and reproduction in *Daphnia magna* during chronic exposure to gamma radiation over three successive generations. Aquat Toxicol. 163:27–36.

Plaire D, Bourdineaud JP, Alonzo A, Camilleri V, Garcia-Sanchez L, Adam-Guillermin C, Alonzo F. 2013. Transmission of DNA damage and increasing reprotoxic effects over two generations of *Daphnia magna* exposed to uranium. Comp Biochem Physiol C Toxicol Pharmacol. 158(4):231–243.

Rusin A, Lapied E, Le M, Seymour C, Oughton D, Haanes H, Mothersill C. 2019. Effect of gamma radiation on the production of bystander signals from three earthworm species irradiated in vivo. Environ Res. 168:211–221.

Sarapultseva EI, Bychkovskaya IB. 2010. Peculiar low radiation effects as a risk factor. Assessment of organism viability in the model experiments with *Daphnia magna*. Int J Low Rad. 7(1):1–9.

Sarapultseva EI, Dubrova YE. 2016. The long-term effects of acute exposure to ionising radiation on survival and fertility in *Daphnia magna*. Environ Res. 150:138–143.

Sarapultseva EI, Gorski A. 2013. Low-dose γ-irradiation affects the survival of exposed *Daphnia* and their offspring. Dose Response. 11: 460–468.

Sarapul'tseva EI, Malina II. 2009. The change of *Daphnia magna* viability after low doses of gamma-irradiation. Radiats Biol Radioecol. 49: 82–84.

Sarapul'tseva EI, Melekhova OP, Kossova GV, Igolkina IV, Alekseeva NV. 2014. Free-radical reactions in *Daphnia* after low doses of γ-irradiation in vivo. Radiats Biol Radioecol. 54(3):305–308.

Sarapul'tseva EI, Riabchenko NI, Igolkina IV. 2013. Methyltetrazolium bromide (MTT) was applied for bioassay of environmental radioactivity in vivo. Radiats Biol Radioecol. 53:634–638.

Sarapultseva EI, Uskalova D, Savina N, Ustenko K. 2017. Medical-biological aspects of radiation effects in *Daphnia magna*. J Phys: Conf Ser. 784:012052.

Sarapultseva EI, Ustenko K, Dubrova YE. 2019. The combined effects of acute irradiation and food supply on survival and fertility in Daphnia magna. J Environ Radioact. 199–200:75–83.

Sharma U, Sun F, Conine CC, Reichholf B, Kukreja S, Herzog VA, Ameres SL, Rando OJ. 2018. Small RNAs are trafficked from the epididymis to developing mammalian sperm. Dev Cell. 46(4): 481–494.

Shiraishi K, Shimura T, Taga M, Uematsu N, Gondo Y, Ohtaki M, Kominami R, Niwa O. 2002. Persistent induction of somatic reversions of the pink-eyed unstable mutation in F-1 mice born to fathers irradiated at the spermatozoa stage. Radiat Res. 157(6):661–667.

Slovinska L, Elbertova A, Misurova E. 2004. Transmission of genome damage from irradiated male rats to their progeny. Mutat Res. 559: 29–37.

Szatmári T, Persa E, Kis E, Benedek A, Hargitai R, Sáfrány G, Lumniczky K. 2019. Extracellular vesicles mediate low dose ionizing radiation-induced immune and inflammatory responses in the blood. Int J Radiat Biol. 95(1):12–22.

Van Meerloo J, Kaspers GJL, Cloos J. 2011. Cell sensitivity assays: the MTT assay. Meth Mol Biol. 731:237–245.

Van Raamsdonk JM. 2018. Mechanisms underlying longevity: a genetic switch model of aging. Exp Gerontol. 107:136–139.

Vorobtsova IE, Aliyakparova LM, Anisimov VN. 1993. Promotion of skin tumors by 12-O-tetradecanoylphorbol-13-acetate in two generations of descendants of male mice exposed to X-ray irradiation. Mutat Res. 287(2):207–216.

Vorobtsova IE. 2000. Irradiation of male rats increases the chromosomal sensitivity of progeny to genotoxic agents. Mutagenesis. 15(1): 33–38.

Wallace WH, Anderson RA, Irvine DS. 2005. Fertility preservation for young patients with cancer: who is at risk and what can be offered. Lancet Oncol. 6:209–218.

Wang J, Zhang Y, Xu K, Mao X, Xue L, Liu X, Yu H, Chen L, Chu X. 2014. Genome-wide screen of DNA methylation changes induced by low dose X-ray radiation in mice. PLoS One. 9(3):e90804.

Wiley LM, Baulch JE, Raabe OG, Straume T. 1997. Impaired cell proliferation in mice that persists across at least two generations after paternal irradiation. Radiat Res. 148(2):145–151.

Won EJ, Lee JS. 2014. Gamma radiation induces growth retardation, impaired egg production, and oxidative stress in the marine copepod *Paracyclopina nana*. Aquat Toxicol. 150:17–26.

Wu Y, Zhou Q. 2012. Dose- and time-related changes in aerobic metabolism, chorionic disruption, and oxidative stress in embryonic medaka (*Oryzias latipes*): underlying mechanisms for silver nanoparticle developmental toxicity. Aquat Toxicol. 124–125:238–246.

Zeman FA, Gilbin R, Alonzo F, Lecomte-Pradines C, Garnier-Laplace J, Aliaume C. 2008. Effects of waterborne uranium on survival, growth, reproduction and physiological processes of the freshwater cladoceran *Daphnia magna*. Aquat Toxicol. 86(3):370–378.

OPEN ACCESS

A review of the impact on the ecosystem after ionizing irradiation: wildlife population

Georgetta Cannon and Juliann G. Kiang

ABSTRACT

Purpose: On 26 April 1986, reactor 4 at the Chernobyl power plant underwent a catastrophic failure leading to core explosions and open-air fires. On 11 March 2011, a combination of earthquake and tsunami led to a similar disaster at the Fukushima Daiichi power plant. In both cases, radioactive isotopes were released and contaminated the air, soil and water in a substantial area around the power plants. Humans were evacuated from the immediate regions but the wildlife stayed and continued to be affected by the ongoing high radiation exposure initially and later decayed amounts of fallout dusts with time. In this review, we will examine the significant effects of the increased radiation on vegetation, insects, fish, birds and mammals.

Conclusions: The initial intense radiation in these areas has gradually begun to decrease but still remains high. Adaptation to radiation is evident and the ecosystems have dynamically changed from the periods immediately after the accidents to the present day. Understanding the molecular mechanisms that allow the adaptation and recovery of wildlife to chronic radiation challenges would aid in future attempts at ecosystem remediation in the wake of such incidents.

Introduction

Twenty-one years later after the Chernobyl power plant explosion, various isotopes of plutonium, strontium-90, americium-241, and cesium-137 were still detected at high levels causing adverse biological effects across the nearby areas (Voitsekhovych et al. 2007). Wildlife continued to be exposed to substantial radiation doses after humans were evacuated from these areas. The half-life of cesium-137 is approximately 30 years and it decays by β emission to a metastable isomer of barium-137. The half-life of barium-137 isomer is 2 minutes. Subsequently, the metastable isomer emits γ radiation and becomes ground state barium (Baum et al. 2002). Food or water contaminated with cesium-137 that are ingested lead to internal β and γ radiation doses in addition to external radiation doses. The half-life of cesium-134 is about 2 years. Cesium-134 emits β particles. The half-life of strontium-90 is approximately 29 years. Strontium-90 emits pure β radiation. Most of the plutonium isotopes emit α particles, which are ionizing and harmful, but have a short penetration distance. The half-life of plutonium-241 is approximately 14 years. It emits β radiation to become americium-241. The half-life of americium-241 is 432 years, and it emits α particles to become neptunium-237, with a by-product of γ emissions (Baum et al. 2002). This is the composition of radiation released and retained in the soil, water and air across the Chernobyl landscape. In addition to external radiation exposure, ingestion of contaminated food and water by wildlife occurred from the beginning of the disaster and continues to the present.

In contrast to the Chernobyl disaster, for Fukushima, a relatively smaller area of land and water was contaminated with cesium-134 and cesium-137 with radioactive noble gases released to air. The cesium isotopes are expected to be responsible for 98% of the next 30-year-cumulative radiation dose in the surrounding area of Fukushima (Imanaka et al. 2015).

Many scientific studies were begun at once after the Chernobyl disaster (Copplestone et al. 2008; UNSCEAR 2011) and mathematical models used for predicting environmental impacts based on the data were adopted to predict how the Fukushima disaster would affect the ecosystem (Aliyu et al. 2015; Bréchignac et al. 2016). Unfortunately, studies on the impact of chronic radiation to the physiology of the remaining organisms in the environment are limited. Changes in the food chain are surely inevitable, but relatively few studies have been focused on the ecosystem as a whole exposed to chronic medium to low levels of radiation. Many laboratory experimental animal models, such as mice, rats, dogs, pigs, and nonhuman primates, indicate that ionizing radiation is a poly-traumatic hit resulting in multi-organ injury to individuals (Kiang and Olabisi 2019). Presumably ionizing radiation produces similar results in wild animals and plants, and over time to the ecological network as a whole particularly when the hits keep on coming for an extended time.

In this brief review, we summarize observations with insects, plants, birds, fish, and mammals chronically exposed to radiation released in contaminated soil, water and air. Observations of adaptation to radiation is included. Molecular mechanisms underlying responses of the wildlife to radiation as well as identification of biomarkers for evaluating changes in the ecosystem will be discussed and recommended.

Effects of radiation on insects

In 2009, around 23 years after the disaster, the insect abundance within the Chernobyl zone during 2006–2008 versus insects from other forested locations around Chernobyl, Ukraine and Belarus were investigated (Moller and Mousseau 2009). The study observed 298 bumble-bees with 72.6% of them being *Bombus terrestris*. The abundance of bumble-bees decreased in the radiation-contaminated areas in a radiation dose-dependent fashion. The reductions also depended on environmental elements such as year, temperature, wind and habitat where bumble-bees were collected. Three hundred eighty-nine butterflies with 36.6% of them being *Aporis Crataegi* were collected. As with the bumble-bees, the butterfly population was significantly reduced in the radiation-contaminated area depending on the severity of radiation contamination. The time of the day, wind and type of habitat introduced variance into the population reductions. Three hundred and five grasshoppers were collected around Chernobyl, Ukraine and Belarus. The grasshopper population was significantly more diminished if radiation levels were higher. Dragonflies were investigated as well. One hundred and five dragonflies were collected. The population significantly declined based on the severity of radiation doses. The results were varied and correlated with time of day, date, habitat, temperature, and cloud cover. Finally, seven hundred seventy-five spider webs were observed. The spider population appeared to decrease depending on radiation severity, year, temperature and wind. At this time, most radiation around Chernobyl was in the topmost soil, where most insects lived (Shestopalov 1996). For example, butterfly eggs, larvae and pupae developed in the radiation-contaminated soil layer. Pollinator abundance was reduced, thereby, affecting plant fecundity and seed set. This impacted the insect habitat (Proctor et al. 1996). Spiders were one of the key players in predation (Wise 1993) and the attenuation of their abundance led to increases in abundance of other insects (Snyder and Wise 2001). Shortly after the Chernobyl disaster, studies performed in the 30 Km impact zone have demonstrated that 30-Gy doses did not directly affect adult animals in the soil and litter, but impacted their eggs and juveniles. After two to two and half years, the significant differences between insect populations in the contaminated areas and uncontaminated areas were not found any more (Krivolutzkii and Pokarzhevskii 1992). This may be because insects from uncontaminated areas migrated into the contaminated areas, leading to the recovery of the soil animal population (Krivolutzkii and Pokarzhevskii 1992). These results were different from the 2009 results which show the effect of long-term chronic radiation (Moller and Mousseau 2009).

Comparing the observations among bumble-bees, butterflies, grasshopper, dragonflies and spider webs, Table 1 indicates the relationship between insect abundance and environmental elements. All five species of insects are sensitive to radioactive contamination. All except spiders (as indicated by web production) were also sensitive to other changes in their habitats. The habitat temperature varied along with cloud cover, wind, time of day, date, and year. Most interestingly, the population of grasshoppers was affected by cloud cover and date but not wind and year, whereas populations of bumble-bees and spider webs were influenced by all these factors. The results indicate a holistic approach is necessary to understand environmental impacts. It should be noted that different sites which had the same radiation levels might have differing habitat (i.e. ground cover), soil composition, or water level/access. Since the collections took place over 3 years, temperature, cloud cover, wind, time of day, actual date and year were also considered and tested as potential confounding variables.

In general, adult insects are believed to be resistant to radiation. These insects were significantly affected by radiation possibly because of the ingestion of a combination of plutonium, strontium-90, americium-241, and/or cesium-137 leading to internal contamination with α, β and γ combined radiation in these areas (Voitsekhovych et al. 2007). The eggs and larva forms are also more sensitive to radiation in general and some grow in close contact with the contaminated ground (Bonisoli-Alquati et al. 2018).

Besides the impact of environmental elements on the insect population, persistent genetic damage was detected. Hancock and colleagues (Hancock et al. 2019) investigated the relationship between historic radiation doses and the genetic damage found in populations of Drosophila

Table 1. Factors affecting variability of insect population responses to different doses of irradiation as reported by Moller and Mousseau (2009).

	Location*	Habitat	Soil	Water	Temperature	Cloud Cover	Wind	Time of Day	Date	Year
Bumble-bees	Yes	Yes	No	No	Yes	No	Yes	No	No	Yes
Butterflies	Yes	Yes	Yes	No	No	No	Yes	Yes	No	No
Grasshoppers	Yes	Yes	No	No	Yes	Yes	No	Yes	Yes	No
Dragonflies	Yes	Yes	No	Yes	No	No	No	Yes	No	No
Spider Webs	Yes	No	No	No	Yes	No	Yes	No	No	Yes

*Different sites (there were a total of 700) that had the same radiation levels might have differing habitat (i.e. ground cover), soil composition, or water level/access. Since the collections took place over 3 years, temperature, cloud cover, wind, time of day, actual date and year were also considered and tested as potential confounding variables.

(*Drosophila melanogaster*) around the Chernobyl Nuclear power Plant. For earlier generations of Drosophila living in the radioactive contaminated locations, the sex-linked recessive lethal (SLRL) rates appeared to correlate with the dose in a linear, non-threshold relationship. The later descendent generations exhibited a radio-adaptive-like response with a plateau in SLRL frequencies. This adaptation is thought to be important for future Drosophila survival through competition and selection, or evolution.

Shortly after the Fukushima disaster, pale grass blue butterflies (*Pseudozizeeria maha*), whose populations were overwintering as larvae at time of the accident were examined. This species lives concurrently with humans as well as other wildlife and is multivoltine, taking about one month per generation. The first-voltine adults, the F_1 generation, and the F_2 generation were collected to study their wing size, half-closing time, and morphological malformations including wings, abdomen, legs, eyes, palpi, and antennae (Hiyama et al. 2012). Some of first-voltine adults in the Fukushima area, collected in May 2011, manifested relatively mild abnormalities. However, the F_1 offspring from the first-voltine females exhibited more severe abnormalities, which continued to be seen in the F_2 offspring. Adult butterflies collected in September 2011 showed more severe abnormalities than those collected in May 2011. Most importantly, similar abnormalities were experimentally confirmed when butterflies from noncontaminated areas were exposed to artificial external and internal radiation at a low dose. The authors concluded that the physiological and genetic damage observed in the butterflies resulted from exposure to radionuclides from the Fukushima Nuclear Power Plant. This species was also monitored from 2011–2013 (Taira et al. 2015). The distribution of abnormality and mortality rates continued to be radiation dose-dependent. Abnormality and mortality rates reached a peak level primarily in the fall of 2011 and returned to normal levels afterwards. Early radiation exposure had a higher impact than later radiation exposure. Radiation effects were accumulated trans-generationally over a specific period. The population regained normality relatively quickly after about 15 generations within 3 years. They concluded that even low doses induced morphological abnormalities and death for some butterflies of this species, whereas other individuals were not affected, at least morphologically. This variable sensitivity implied the possibility of radiation resistance or adaptive evolution. However, the data observed from 298 haploid genomes of *Drosophila melanogaster* collected in 2012 showed no chromosomal inversions, and 2006 genomes collected in 2013 contained only 2 chromosomal inversions. The authors claimed no convincing data to suggest that the chronic radiation released in the Fukushima disaster caused genetic damage (Itoh et al. 2018). This result in Fukushima was different from that found in Chernobyl where chronic radiation increased the SLRL rate in Drosophila (Hancock et al. 2019). The chronic dose rates in Fukushima were much lower than the ones in Chernobyl which may explain why no genetic damage was found in this case. However, molecular level changes in Chernobyl's and Fukushima's wildlife are not yet fully understood. Further studies should be explored.

Effects on plant life

Seed set, seeding consumption, seed dispersal and maintenance of plant communities in the ecosystem are impacted by co-living animals. Moller et al. (Moller et al. 2012) studied fruit trees such as rowan (*Sorbus aucuparia*), pear (*Pyrus communis*), and apple (*Malus domestica*), and bushes such as European cranberry bush (*Viburnum opulus*), twisting-wood (*Viburnum lantana*), and wild rose (*Rosa rugose*) along with butterflies and bumble-bees in the area of Chernobyl. In the areas with high radiation contamination, they found insect-pollinated fruit plants produced fewer fruits and this correlated with the local reduction in pollinating insects and with the generally smaller size of fruit trees. Lower fruit abundance led to lower levels of fruit-eating birds thus further limiting seed dispersal. These observations suggest that the abundance of frugivores, pollinating insect populations, and fruit production were directly associated with radiation levels. In other words, the number of fruit-bearing trees was low when the level of radiation was high, while the fruit sets were low when pollinating insect populations was low due to high radiation levels. The results taken together suggest that the close interplay among many species and between animals and plants is strongly affected by chronic radiation. Direct killing of the fruit trees by high radiation resulting in the overgrowth of other plants years later should be also be considered.

During the period of 1986–1991, after the Chernobyl disaster, pine trees (*Pinus silvestris*) within a 10-km radius zone from the Chernobyl Power Plant were significantly hurt (Arkhipov et al. 1994). Radiation doses at 10–60 Gy, 1–10 Gy and 0.1–1 Gy resulted in high, medium and low damage to pine trees, respectively. Radiation doses more than 60 Gy led to a massive fatality without recovery or regeneration of pine trees since 1987. The radiation-exposed pine trees dried-up and/or burned down. Those dying trees were infested by pathogenic insects. Pine trees exposed to radiation doses less than 0.1 Gy did not display any visible injury. Pine trees appeared to be capable of repairing the medium and low degrees of injury (Arkhipov et al. 1994), suggesting their regeneration ability remains.

It was thought that radiation would cause more intense stress to plant growth at certain stages. Boratynaki and colleagues (Boratyński et al. 2016) reported that they used 660 seeds collected from 33 wild carrots (*Daucus carota*) from various regions near Chernobyl (that significantly varied in radioactive contamination) to conduct a common garden experiment in an uncontaminated greenhouse. They found that the higher the level of radiation received by the maternal plants was, the longer the timing and the slower the rates of seed germination were. Radiation exposure also prolonged leaf production. Again, the higher the radiation dose was, the longer leaf production took. The results suggest that radiation slows down the cell proliferation capacity and that is consistent with what we have observed in recovery of mouse skin wound after irradiation (Kiang et al. 2010).

Effects on birds

When the lens of the eye becomes opaque, the eye's ability to see is reduced. The lens opacity is known as a cataract and can occur when eyes are exposed to radiation that can be either non-ionizing or ionizing. We would expect cataracts to make free-living animals less fit in a radiation-contaminated environment. Cataracts in free living birds were increased in a dose-dependent manner with the increasing level of background radiation, but cataract development showed no threshold of radiation dose. Birds with opacities in one lens had a greater probability of developing opacities in the lens of the other eye as well, resulting in even poorer vision. The incidence of lens opacities, typical of cataracts, in more than 1100 free-living birds in the Chernobyl region was investigated to elucidate their relation to the background radiation (Moller and Mousseau 2007). The incidence of cataracts was increased when increasing levels of background radiation was detected. The data were analyzed based both on a dichotomous score and on continuous scores of intensities of cataracts. The odds ratio per unit change in the regressor was 0.722 (95% CI 0.648, 0.804). It was lower than odds ratios from investigations of radiation cataracts in humans. The small odds ratio in birds may be driven by increased fatality in birds with cataracts in both eyes, because cataracts impede a bird's mobility to hunt food, to stay away from predators, to fly safely and even to mate and raise offspring. No increases in incidence of cataracts with increasing age were found, suggesting that yearlings and older birds were similarly affected by radiation-induced cataracts. The authors found a robust inverse correlation between bird abundance and background radiation, suggesting that radiation lowers bird abundance via not only an increase in the probability of cataracts in bird populations, but also via effects on disease susceptibility, food abundance and interactions with other species. As part of the ecosystem, for example, if fruit-eating birds like warblers and thrushes suffered cataracts, which would decrease their ability to find and eat fruits to disperse seeds, there would be soon be fewer fruit-bearing trees available in areas leading to a lower fruit abundance, even after controlling the directly detrimental effects of radiation on fruit trees. This could cause remaining birds to starve to death in an adverse cycle (Moller and Mousseau 2007). Therefore, fruit-eating birds, seed dispersal, fruit trees, fruit abundance, and pathogen invasion are affected not only individually, but also collectively by radiation leading to adverse outcomes that synergize to amplify the detrimental impacts on the ecosystem.

There are no reports of radiation induced cataract occurrence in insects mentioned above or fish discussed below. However, radiation-induced cataracts are not unique to birds. Humans and other mammals are known to develop radiation-induced cataracts (Laskowski et al. 2020). More research in this area with other animals including pigs, cows and nonhuman primates should be considered.

Studies on birds at Fukushima have been more limited but have been shown that both bird species and abundances decrease abruptly since the disaster (Moller et al. 2015), suggesting the low-dose radiation induces a similar outcome as the high-dose radiation does, although the underlying mechanisms for the low-dose radiation and the high-dose radiation may be different. More studies in this regard should be explored.

Effects on fish

Fish are radiation sensitive. At Chernobyl, fish in the freshwater environment have been highly exposed to radiation since the disaster happened (Lerebours et al. 2018). Little is known about the biological effects of chronic radiation exposure at low level on fish. Lerebours and colleagues collected 142 roach (*Rulilus ruitilus*) and 162 perch (*Perca fluviatilis*) in total from seven lakes in Belarus and Ukraine. Specific activities of americium-241, plutonium-238 − 239 − 240, strontium-90, and cesium-137, index conditions, size and distribution of oocytes, as well as biological and environmental confounding factors in these two fish species were investigated. The results indicated that the general physiological and reproductive health of roach and perch were not affected. Nonetheless, roach appeared to be less effected by radiation than perch. In most radiation-contaminated lakes, numbers of perch exhibited radiation-delayed maturation of gonads and the presence of several undeveloped phenotypes not seen in roach. Perch are a carnivorous fish and as such accumulated more cesium-137 than their prey fish, roach which may account for the differences. The differences may also be due to variations in epigenesis, genetic composition, signal transduction and energy production. The metabolomics approach would probably be able to shed light on why roach are less sensitive to radiation than perch.

The biological effects of the radiological Fukushima disaster on fish are little known, because it is difficult to separate them from the effects of the preceding earthquake followed by tsunami and because most models imply that radioactivity doses seen there are too low to induce biological effects. However, the prohibitions on fishing in the area would have been anticipated to lead to increased populations of target fish, which has not yet happened (Kodama et al. 2018). With all the uncontrolled variables, it is difficult to evaluate how low-dose radiation might chronically impact the ecosystem. With this in mind, studies with low-dose radiation in the laboratory setting should be encouraged.

Effects on mammals

Pomerantseva and colleagues (Pomerantseva et al. 1997) caught house mice from 1986 to 1994 in radionuclides-polluted regions after the Chernobyl disaster. The dose rates of γ radiation on the soil surface ranged from 0.0002 to 2 mGy/h. Using the air-drying method of Evans et al. (1964)

for meiotic preparations from mammalian testes, they found reciprocal translocation occurred in these house mice. The frequency of reciprocal translocation in mouse spermatocytes was increased when dose rates were increased. Embryonic lethality was higher in the progeny of male mice caught in 1987 in the most contaminated area compared to data obtained in the less contaminated area. These changes were reduced with time after the disaster. In another study (Shevchenko et al. 1992), among 74 mice caught around the Chernobyl nuclear power station, four mice showed reciprocal translocation heterozygotes but increased embryonic lethality and an increased number of abnormal sperm heads were found. These results suggest that radiation severely harms the reproductive system but that the effects decrease over time perhaps due to the die-off of the most affected families and adaptation of the less affected ones. Similar field studies should be continued to verify whether adaptation appears. The competition and the selection that are fundamental principles of evolution require time. Studies with subjects collected from Chernobyl may now show evidence evolution/adaptation.

Radiation-induced DNA double-strand breaks (DSB) can be detected by measuring phosphorylated H2AX (i.e. γ-H2AX) at the DNA break sites, which is a well-established biomarker. γ-H2AX is so sensitive to radiation and can be induced even with radiation exposure at 1 mGy (Rothkamm and Lobrich 2003). Cows abandoned at Fukushima were examined for γ-H2AX foci in their peripheral circulating lymphocytes. Nakamura and colleagues (Nakamura et al. 2017) collected blood samples from cows ($N = 70$) from every location of the ex-evacuated area in Fukushima and blood samples from cows at radiation-uncontaminated cities ($N = 8$ per region) as a control group. Using immunohistochemistry, they found that the number of γ-H2AX foci in peripheral circulating lymphocytes of cows from the area contaminated with radiation was twofold higher compared with the data from the control group. The levels of DNA damage slightly decreased over the 700-day sample collection period. The extent of damage did not appear to be associated with the distance from the accident site, whereas an age-dependent accumulation of DNA damage was apparent. The results suggest that age and time are confounding factors for measuring DNA damage. Notably, such low-dose radiation is sufficient to cause a significantly persistent DNA damage even 700 days later. Whether this persistent DNA damage would result in teratogenic effects should be explored.

When human residents at areas near the Fukushima Daiichi Nuclear Power plant were evacuated, Japanese macaques (*Macaca fuscata*) residing there were also evacuated. Japanese macaques ($N = 72$) were captured between May 2013 and December 2014 within a 40-kilometer radius of the power plant. The nonexposed (control) group consisted of 23 macaques. They were divided into the immature group ($\leq$1-year old) and the mature group ($\geq$5 years old). The median internal dose-rate was 7.6 μGy/day (ranging from 1.8 to 219 μGy/day) and the external dose rate was 13.9 μGy/day (ranging from 6.7 to 35.1 μGy/day). White blood cells (WBCs), red blood cells (RBCs), hemoglobin (HGB), hematocrits (HTC) and platelets (PLTs) were measured in their blood samples. Bone marrow samples were used for histopathological analysis. The white blood cell and platelet counts appeared to be low when the internal dose rate in mature macaques was high. Likewise, the myeloid cell, megakaryocyte, and hematopoietic cell counts in the bone marrow samples were low, but adipocyte counts in the bone marrow were high when internal dose-rate in femoral bone marrow of mature macaques was high (Urushihara et al. 2018). The observations are in agreement with the data obtained in mice (Kiang et al. 2010; Kiang and Olabisi 2019), minipigs (Kiang and Smith, unpublished data) and nonhuman primates (Wong et al. 2020). These results suggest that both acute radiation exposure and chronic radiation exposure lead to hematopoietic depletion, which is a typical radiation syndrome and fatal if no proper care was provided. Acute radiation syndrome (ARS) is well characterized (Kiang and Olabisi 2019), but delayed acute radiation effect (DEARE) and long-term radiation impact are not well understood in wildlife. More future effort should be focused on the latter two areas.

Adaptation to radiation

Radiation intensity in both Chernobyl and Fukushima has decreased with time. Under these conditions, organisms may evolve ways of surviving the radiation stress, also called adapting. Bonisoli-Alquati and colleagues (Bonisoli-Alquati et al. 2018) captured grasshoppers (*Chorthippus albomarginatus*) in the Chernobyl exclusion zone and raised their offspring in a common garden. They found that offspring that matured quicker than average grasshoppers contained more DNA damage. Faster growth rates cause elevated oxidative stress and subsequent injury, but the grasshoppers may gain time to reach adulthood and to reproduce before the radiation released from the environment kills them. The result suggests the presence of an adaption by maintaining the population at the expense of more DNA damage to the individual.

Similar results may be found in other species. For example, pale grass blue butterflies (*Zizeeria maha*) that survived in Fukushima at the low radiation dose may be examples of adaptation (Taira et al. 2015). Also, the frequency of sex-linked recessive lethal genes in *Drosophila melanogaster* was mitigated in later descent generations compared to the data observed with earlier generations (Hancock et al. 2019). These examples reinforce the idea that radiation adaptation can occur. Table 2 lists more studies describing data from insects, plants and mammals in support of adaptation.

It is possible that resistance to acute radiation and to chronic radiation may involve different underlying mechanisms. Studying the chronic radio-resistance phenomenon is complex in the field where environmental elements such as wind, cloud cover, temperature, seasonal changes, interspecies competition, and habitat (Table 1) are not controllable. On the other hand, it is almost impossible to design experiments with consideration of all parameters mentioned above

Table 2. Brief summary of recent papers that addressed adaptation to ionizing radiation. Papers were varied as to dose levels (and how they were measured), sample sizes, experimental designs and significance levels.

Species	Design	Measured	Adapt.	Area	Reference
Plants					
Evening primrose/*Cenothera biennis*	Field samples	DNA repair	Y	Chern	Boubriak et al. 2008
Thale cess/*Arabidopsis thaliana*	Transplant	Antioxidants	N	Chern	Morozova et al. 2020
	Transplant	MR, Antioxidants, Methylation	Y	Chern	Kovalchuk et al. 2004
Wild carrot/*Daucus carota*	Transplant	Germination	N	Chern	Boratynski et al. 2016
Soybean/*Glycine max*	Transplant	DNA repair, Methylation	Y	Chern	Georgieva et al. 2017
	Field samples	Proteomics comparison	Y	Chern	Danchenko et al. 2009
	Field samples	Proteomics comparison	Y	Chern	Klubicova et al. 2012
Flax/*Linum usitatissimum*	Field samples	Proteomics comparison	N	Chern	Klubicova et al. 2010
White birch/*Betula verrucosa*	Field samples	DNA repair	Y	Chern	Boubriak et al. 2008
Scots pine/*Pinus sylvstris*	Field samples	Antioxidants	Y	Chern	Volkova et al. 2017
	Field samples/2° expo.	Aberrant cells/root meristems	N	Chern	Geras'kin et al. 2011
	Field samples	AFLP-PCR	Y	Chern	Kuchma and Finkeldey 2011
	Field samples	Methylation	Y	Chern	Kovalchuk et al. 2003
Insects					
Fruit flies/*Drosophila melanogaster*	Field samples	SLRL frequencies	Y	Chern	Hancock et al. 2019
Grasshopper/*Chorthippus*	Transplant	Development	Y	Chern	Bonisoli-Alquati et al. 2018
albomargintus	Transplant	Wing size/shape, development	N	Chern	Beasley et al. 2012
Butterfly/*Zizeeria maha*	Transplant/2° expo.	forewing size over generations	Y	Fuku	Hiyama et al. 2015
Birds					
multiple	Field samples	DNA damage, antioxidants	Y		Galvan et al. 2014
Mammals					
Voles; Vole-moles	Field samples	Oxidative stress/Antiox.	Y	EURT	Rasina et al. 2017
	Field samples	Micronuclei/Immunological Indices	Y	EURT	Grigorkina and Olenev 2009
Bank vole/*Myodes glareolus*	FFS/2° expo.	Antiox./ resistance to DNA damage	Y	Chern	Mustonen et al. 2018
Field mice/*Apodemus speciosus*	Field samples	Spermatogenesis	Y	Fuku	Takino et al. 2017

Adapt: adaptation; Antiox: antioxident; Chern: Chernobyl; EURT: East ural radioactive trace; Expo.: exposure; FFS: Fibroblasts from field samples; Fuku: Fukushima; Immunol: immunological; MR: Metabolic Rate; N: No; SLRL: Sex-link recessive lethal; Y: Yes.

under control in the laboratory. Both all encompassing (but messy) field studies and controlled (but limited) laboratory studies are necessary. A system biology approach is needed (Shuryak 2020).

Molecular mechanisms

Almost no underlying molecular mechanisms for radiation-induced biological changes are defined in wildlife. Offspring of grasshoppers (*Chorthippus albomarginatus*) captured in the Chernobyl exclusion zone and raised in a common garden manifested DNA damage (Bonisoli-Alquati et al. 2018). γ-H2AX foci in lymphocytes, as a biomarker for DNA damage, was two times higher in cattle in the Fukushima area captured in 2011 compared to the data obtained in uncontaminated areas. The DNA damage decreased after 2012 and 2013 (Nakamura et al. 2017). The investigation of the hematopoiesis of acute radiation syndrome in Japanese nonhuman primates exposed to radiation in Fukushima is more or less touching the edge of mechanistic elucidation (Urushihara et al. 2018). More studies of this type performed using the state-of-art technology such as epigenomics, genomics, metabolomics, and proteomics should be explored.

In experimental laboratory animal models where radiation doses, radiation dose rates, temperature, air pressure, and water and food availability are controlled, acute radiation exposure results in hematopoietic acute radiation syndrome, gastrointestinal acute radiation syndrome, cutaneous radiation syndrome, cardiovascular syndrome and neurological syndrome depending on the dose of radiation. Changes in DNA damage, microRNA-34a, cytokine/chemokine storms in circulation and tissues, inflammasome formation, inducible nitric oxide synthase (Kiang et al. 2010; Williams et al. 2010), AKT activation, MAPK activation (Wang et al. 2015; Kiang and Olabisi 2019), ATP production, apoptosis (Kiang et al. 2019, 2020), autophagy, and necroptosis (Hei et al. 2011; Kiang and Olabisi 2019) are well characterized and targeted for drug development. Chronic radiation exposure has been less studied, the definition of chronic radiation in the laboratory is temporally limited and the mechanisms studied are also limited (Fuller et al. 2019; Murat El Houdigui et al. 2019; Ojima et al. 2019; Overbey et al. 2019; Koval et al. 2020; Shuryak 2020). A problem developing in one species can have an effect on another species (i.e. interspecies interaction and competition such as deficiency of food chain), possibly struggling species, and that is also difficult to address in the laboratory. Although it is evident that some organisms in their natural environments are more sensitive to radiation than in the laboratory (Garnier-Laplace et al. 2013), investigating the mechanisms underlying that sensitivity is very difficult to do in the wild and nearly impossible to fully model in the laboratory.

Biomarkers for ecosystems post-irradiation

As far as elucidation of molecular mechanisms, there is also very little investigation into dose biomarkers for wildlife post-irradiation. Biomarkers for experimental animal models in laboratories have been identified for each organ/radiation syndrome and are well established. Table 3 lists biomarkers, in particular those well suited for human triage after a mass casualty event such as a radiation accident or a nuclear

Table 3. Effects of radiation doses, radiation dose rates and genders on blood cells, cytokines, thrombopoietin, erythropoietin and saliva amylase in blood.

Blood	Rad Dose	Rad Dose Rate	Gender
WBCs	YES	NO	NO
RBCs	YES	NO	NO
Platelets	YES	YES	NO
Flt-3 ligand	YES	NO	NO
G-CSF	YES	NO	NO
IL-18	YES	NO	NO
TPO	YES	NO	NO
EPO	YES	NO	NO
SAA	YES	NO	NO

WBC: white blood cell; RBC: red blood cell; Flt-3 ligand: Fms-related tyrosine kinase 3 ligand; G-CSF: granulocyte-colony stimulating factor; IL-18: interleukin-18; TPO: thrombopoietin; EPO: erythropoietin; SAA: saliva amylase.

detonation. Changes in white blood cells (WBCs), red blood cells (RBCs), platelets, Fms-related tyrosine kinase 3 ligand (Flt-3 ligand), granulocyte-colony stimulating factor (G-CSF), interlukin-18 (IL-18), thrombopoietin (TPO), erythropoietin (EPO) and saliva amylase (SAA) all vary in a radiation dose-dependent manner (Hegge and King 2017; Kiang et al. 2018; Ossetrova et al. 2018a, 2018b). All of them are independent of gender. Except platelets, these are independent of dose rates. Generally, circulating Flt-3 ligand and citrulline/IL-18 concentrations are the biomarkers for hematopoietic syndrome (Kiang and Olabisi 2019) and gastrointestinal syndrome (Bujold et al. 2016; Kiang et al. 2020) respectively. CD31 in skin injury (Doctrow et al. 2013) and ATP production in brain tissues (Kiang et al. 2019) were reported as good biomarkers. It is highly possible that there are similar biomarkers for wildlife species as well. Synergistic effects of combined radiation injury have been documented in survivors from atomic bombing and laboratory animals (Kiang et al. 2010; Kiang and Olabisi 2019). Confounding factors (gender, age, wound, burn, hemorrhage, bone fracture, insecticide, fertilizer, etc.) must be considered when studying biomarkers for the ecosystem after irradiation with time as a key variable. The long-term effects of radiation on changes in the ecosystem will have a profound effect on humans. The knowledge of how to prevent or mitigate radiation or other negative impacts on the ecosystem is likely to be key to human survival.

Challenges and perspective

There are a vast number of questions about how radiation alters the ecosystem. New techniques and tools available for laboratory experiments are also available for detecting changes in wildlife samples collected from radiation contaminated areas. Major challenges for such work include: (i) gene maps for wildlife are not available, (ii) antibodies against proteins responsible for signal transduction identified in mammals are not available or tested/proved for wildlife, and (iii) systems biology approaches utilized in laboratory experiments are not well developed for studying wildlife. Additionally, if wildlife adaptation is indeed occurring, the collection of samples before adaptation sets in becomes very critical. Collective effort from this research community is imperative to overcome these challenging issues. Ultimately, further funding is required to advance our understanding at the systemic, organismic, cellular, and molecular levels.

Conclusions

It has been 34 years since the Chernobyl Power Plant accident. The impact of radiation on wildlife is evident, but only few studies have been performed on long-term effects. More observations in a greater variety of species are needed. Nonetheless, it is delightful to detect the presence of adaptation. The wildlife population and abundance are in recovery. However, it would be useful to design experimental models that can be more closely to mimic the natural environment, to aid in elucidating the molecular mechanisms of radiation responses and to establish biomarkers of ecosystems impacted by radiation.

Disclosure statement

The authors report no financial conflicts of interest. The manuscript was cleared for publication by the Armed Forces Radiobiology Research Institute, the Uniformed Services University of the Health Sciences. The views, opinions, and findings contained in this report are those of the authors and do not reflect official policy or positions of the Armed Forces Radiobiology Research Institute, the Uniformed Services University of the Health Sciences, the Department of Defense, or the United States government.

Funding

This review has been funded in part by Armed Forces Radiobiology Research Institute (AFRRI) intramural program RBB34363 and RAB310934.

References

Aliyu AS, Ramli AT, Garba NN, Saleh MA, Gabdo HT, Liman MS. 2015. Fukushima nuclear accident: preliminary assessment of the risks to non-human biota. Radiat Prot Dosimetry. 163(2):238–250.

Arkhipov NP, Kuchma ND, Askbrant S, Pasternak PS, Musica VV. 1994. Acute and long-term effects of irradiation on pine (Pinus silvestris) strands post-Chernobyl. Sci Total Environ. 157(1-3): 383–386.

Baum ED, Knox HD, Miller TR. 2002. Nuclides and isotopes: chart of the nuclides. 16th ed. Schenectady, New York: Knolls Atomic Power Laboratory, Inc.

Beasley DE, Bonisoli-Alquati A, Welch SM, Moller AP, Mousseau TA. 2012. Effects of parental radiation exposure on developmental instability in grasshoppers. J Evol Biol. 25(6):1149–1162.

Bonisoli-Alquati A, Ostermiller S, Beasley AE, Welch SM, Moller AP, Mousseau TA. 2018. Faster development covaries with higher DNA damage in grasshoppers (Chorthippus albomarginatus) from chernobyl. Physiol Biochem Zool. 91(2):776–787.

Boratyński Z, Arias JM, Garcia C, Mappes T, Mousseau TA, Møller AP, Pajares AJM, Piwczyński M, Tukalenko E. 2016. Ionizing radiation from Chernobyl affects development of wild carrot plants. Sci Rep. 6:39282.

Boubriak II, Grodzinsky DM, Polischuk VP, Naumenko VD, Gushcha NP, Micheev AN, McCready SJ, Osborne DJ. 2008. Adaptation and impairment of DNA repair function in pollen of Betula verrucosa and seeds of Oenothera biennis from differently radionuclide-contaminated sites of Chernobyl. Ann Bot. 101(2):267–276.

Bréchignac F, Oughton D, Mays C, Barnthouse L, Beasley JC, Bonisoli-Alquati A, Bradshaw C, Brown J, Dray S, Geras'kin S, et al. 2016. Addressing ecological effects of radiation on populations and ecosystems to improve protection of the environment against radiation: Agreed statements from a Consensus Symposium. J Environ Radioact. 158–159:21–29.

Bujold K, Hauer-Jensen M, Donini O, Rumage A, Hartman D, Hendrickson HP, Stamatopoulos J, Naraghi H, Pouliot M, Ascah A, et al. 2016. Citrulline as a biomarker for gastrointestinal-acute radiation syndrome: species differences and experimental condition effects. Radiat Res. 186(1):71–78.

Copplestone D, Hingston J, Real A. 2008. The development and purpose of the FREDERICA radiation effects database. J Environ Radioact. 99(9):1456–1463.

Danchenko M, Skultety L, Rashydov NM, Berezhna VV, Mátel L, Salaj T, Pret'ová A, Hajduch M, 2009. Proteomic analysis of mature soybean seeds from the Chernobyl area suggests plant adaptation to the contaminated environment. J Proteome Res. 8(6):2915–2922.

Doctrow SR, Lopez A, Schock AM, Duncan NE, Jourdan MM, Olasz EB, Moulder JE, Fish BL, Mader M, Lazar J, et al. 2013. A synthetic superoxide dismutase/catalase mimetic EUK-207 mitigates radiation dermatitis and promotes wound healing in irradiated rat skin. J Invest Dermatol. 133(4):1088–1096.

Evans EP, Breckon G, Ford CE. 1964. An air-drying method for meiotic preparations from mammalian testes. Cytogenetics. 3:289–294.

Fuller N, Smith JT, Ford AT. 2019. Impacts of ionising radiation on sperm quality, DNA integrity and post-fertilisation development in marine and freshwater crustaceans. Ecotoxicol Environ Saf. 186: 109764.

Garnier-Laplace J, Geras'kin S, Della-Vedova C, Beaugelin-Seiller K, Hinton TG, Real A, Oudalova A. 2013. Are radiosensitivity data derived from natural field conditions consistent with data from controlled exposures? A case study of Chernobyl wildlife chronically exposed to low dose rates. J Environ Radioact. 121:12–21.

Galván I, Bonisoli-Alquati A, Jenkinson S, Ghanem G, Wakamatsu K, Mousseau TA, Møller AP. 2014. Chronic exposure to low-dose radiation at Chernobyl favours adaptation to oxidative stress in birds. Funct Ecol. 28(6):1387–1403.

Georgieva M, Rashydov NM, Hajduch M. 2017. DNA damage, repair monitoring and epigenetic DNA methylation changes in seedlings of Chernobyl soybeans. DNA Repair (Amst). 50:14–21.

Geras'kin S, Oudalova A, Dikareva N, Spiridonov S, Hinton T, Chernonog E, Garnier-Laplace J. 2011. Effects of radioactive contamination on Scots pines in the remote period after the Chernobyl accident. Ecotoxicology. 20(6):1195–1208.

Grigorkina E, Olenev G. 2009. Radioadaptation of rodents in the zone of local radioactive contamination (Kyshtim Accident, Russia): 50 Years on. Radioprotection. 44(5):129–134.

Hancock S, Vo NTK, Byun SH, Zainullin VG, Seymour CB, Mothersill C. 2019. Effects of historic radiation dose on the frequency of sex-linked recessive lethals in Drosophila populations following the Chernobyl nuclear accident. Environ Res. 172:333–337.

Hegge SR, King GL. 2017. Radiation dose-rate effects on select biomarkers in a mouse total-body irradiation model. Radiat Appl. 2(3): 158–163.

Hei TK, Zhou H, Chai Y, Ponnaiya B, Ivanov VN. 2011. Radiation induced non-targeted response: mechanism and potential clinical implications. Curr Mol Pharmacol. 4(2):96–105.

Hiyama A, Nohara C, Kinjo S, Taira W, Gima S, Tanahara A, Otaki JM. 2012. The biological impacts of the Fukushima nuclear accident on the pale grass blue butterfly. Sci Rep. 2:570.

Hiyama A, Taira W, Nohara C, Iwasaki M, Kinjo S, Iwata M, Otaki JM. 2015. Spatiotemporal abnormality dynamics of the pale grass blue butterfly: three years of monitoring (2011-2013) after the Fukushima nuclear accident. BMC Evol Biol. 15:15

Imanaka T, Hayashi G, Endo S. 2015. Comparison of the accident process, radioactivity release and ground contamination between Chernobyl and Fukushima-1. J Radiat Res. 56(Suppl 1):i56–61.

Itoh M, Kajihara R, Kato Y, Takano-Shimizu T, Inoue Y. 2018. Frequencies of chromosomal inversions in Drosophila melanogaster in Fukushima after the nuclear power plant accident. PLoS One. 13(2):e0192096.

Kiang JG, Jiao W, Cary LH, Mog SR, Elliott TB, Pellmar TC, Ledney GD. 2010. Wound trauma increases radiation-induced mortality by activation of iNOS pathway and elevation of cytokine concentrations and bacterial infection [Research Support, N.I.H., Extramural]. Radiat Res. 173(3):319–332.

Kiang JG, Olabisi AO. 2019. Radiation: a poly-traumatic hit leading to multi-organ injury. Cell Biosci. 9:25.

Kiang JG, Smith JT, Anderson MN, Umali MV, Ho C, Zhai M, Lin B, Jiang S. 2019. A novel therapy, using Ghrelin with pegylated G-CSF, inhibits brain hemorrhage from ionizing radiation or combined radiation injury [Research]. PPIJ. 7(3):133–145.

Kiang JG, Smith JT, Cannon G, Anderson MN, Ho C, Zhai M, Cui W, Xiao M. 2020. Ghrelin, a novel therapy, corrects cytokine and NF-κB-AKT-MAPK network and mitigates intestinal injury induced by combined radiation and skin-wound trauma. Cell Biosci. 10:63.

Kiang JG, Smith JT, Hegge SR, Ossetrova NI. 2018. Circulating cytokine/chemokine concentrations respond to ionizing radiation doses but not radiation dose rates: granulocyte-colony stimulating factor and interleukin-18. Radiat Res. 189(6):634–643.

Klubicova K, Danchenko M, Skultety L, Berezhna VV, Uvackova L, Rashydov NM, Hajduch M. 2012. Soybeans grown in the Chernobyl area produce fertile seeds that have increased heavy metal resistance and modified carbon metabolism. PLoS One. 7(10):e48169

Klubicova K, Danchenko M, Skultety L, Miernyk JA, Rashydov NM, Berezhna VV, Pret'ova A, Hajduch M. 2010. Proteomics analysis of flax grown in Chernobyl area suggests limited effect of contaminated environment on seed proteome. Environ Sci Technol. 44(18): 6940–6946.

Kodama K, Aramaki T, Horiguchi T. 2018. Current status of the megabenthic community in coastal Fukushima Prefecture, Japan, in the wake of the Great East Japan Earthquake. Mar Environ Res. 140: 358–374.

Koval L, Proshkina E, Shaposhnikov M, Moskalev A. 2020. The role of DNA repair genes in radiation-induced adaptive response in Drosophila melanogaster is differential and conditional. Biogerontology. 21(1):45–56.

Kovalchuk I, Abramov V, Pogribny I, Kovalchuk O. 2004. Molecular aspects of plant adaptation to life in the Chernobyl zone. Plant Physiol. 135(1):357–363.

Kovalchuk O, Burke P, Arkhipov A, Kuchma N, James SJ, Kovalchuk I, Pogribny I. 2003. Genome hypermethylation in Pinus silvestris of Chernobyl–a mechanism for radiation adaptation? Mutat Res. 529(1-2):13–20.

Kuchma O, Finkeldey R. 2011. Evidence for selection in response to radiation exposure: Pinus sylvestris in the Chernobyl exclusion zone. Environ Pollut. 159(6):1606–1612.

Krivolutzkii DA, Pokarzhevskii AD. 1992. Effects of radioactive fallout on soil animal populations in the 30 km zone of the Chernobyl atomic power station. Sci Total Environ. 112(1):69–77.

Laskowski L, Williams D, Seymour C, Mothersill C. 2020. Environmental and industrial developments in radiation cataractogenesis. Int J Radiat Biol. 1–33. DOI:10.1080/09553002.2020.1767820

Lerebours A, Gudkov D, Nagorskaya L, Kaglyan A, Rizewski V, Leshchenko A, Bailey EH, Bakir A, Ovsyanikova S, Laptev G, et al. 2018. Impact of environmental radiation on the health and reproductive status of fish from Chernobyl. Environ Sci Technol. 52(16):9442–9450.

Moller AP, Barnier F, Mousseau TA. 2012. Ecosystems effects 25 years after Chernobyl: pollinators, fruit set and recruitment. Oecologia. 170(4):1155–1165.

Moller AP, Mousseau TA. 2007. Birds prefer to breed in sites with low radioactivity in Chernobyl. Proc Biol Sci. 274(1616):1443–1448.

Moller AP, Mousseau TA. 2009. Reduced abundance of insects and spiders linked to radiation at Chernobyl 20 years after the accident. Biol Lett. 5(3):356–359.

Moller AP, Nishiumi I, Mousseau TA. 2015. Cumulative effects of radioactivity from Fukushimaon the abundance and biodiversity of birds. J Ornithol. 156(S1):297–S305.

Morozova V, Kashparova E, Levchuk S, Bishchuk Y, Kashparov V. 2020. The progeny of Chernobyl Arabidopsis thaliana plants does not exhibit changes in morphometric parameters and cellular antioxidant defence system of shoots. J Environ Radioact. 211:106076

Mustonen V, Kesaniemi J, Lavrinienko A, Tukalenko E, Mappes T, Watts PC, Jurvansuu J. 2018. Fibroblasts from bank voles inhabiting Chernobyl have increased resistance against oxidative and DNA stresses. BMC Cell Biol. 19(1):17

Murat El Houdigui S, Adam-Guillermin C, Loro G, Arcanjo C, Frelon S, Floriani M, Dubourg N, Baudelet E, Audebert S, Camoin L, et al. 2019. A systems biology approach reveals neuronal and muscle developmental defects after chronic exposure to ionising radiation in zebrafish. Sci Rep. 9(1):20241.

Nakamura AJ, Suzuki M, Redon CE, Kuwahara Y, Yamashiro H, Abe Y, Takahashi S, Fukuda T, Isogai E, Bonner WM, et al. 2017. The causal relationship between DNA damage induction in bovine lymphocytes and the fukushima nuclear power plant accident. Radiat Res. 187(5):630–636.

Ojima M, Hirouchi T, Etani R, Ariyoshi K, Fujishima Y, Kai M. 2019. Dose-rate-dependent PU.1 inactivation to develop acute myeloid leukemia in mice through persistent stem cell proliferation after acute or chronic gamma irradiation. Radiat Res. 192(6):612–620.

Ossetrova NI, Stanton P, Krasnopolsky K, Ismail M, Doreswamy A, Hieber KP. 2018a. Biomarkers for radiation biodosimetry and injury assessment after mixed-field (neutron and gamma) radiation in the mouse total-body irradiation model. Health Phys. 115(6):727–742.

Ossetrova NI, Stanton P, Krasnopolsky K, Ismail M, Doreswamy A, Hieber KP. 2018b. Comparison of biodosimetry biomakers for radiation dose and injury assessment after mixed-field (neutron and gamma) and pure gamma radiation in the mouse total body irradiation model. Health Phys. 115(6):743–759.

Overbey EG, Paul AM, da Silveira WA, Tahimic CGT, Reinsch SS, Szewczyk N, Stanbouly S, Wang C, Galazka JM, Mao XW. 2019. Mice exposed to combined chronic low-dose irradiation and modeled microgravity develop long-term neurological sequelae. Int J Mol Sci. 20(17):4094.

Pomerantseva MD, Ramaiya LK, Chekhovich AV. 1997. Genetic disorders in house mouse germ cells after the Chernobyl catastrophe. Mutat Res. 381(1):97–103.

Proctor M, Yeo PF, Lack AJ. 1996. Natural history of pollination. Portland, OR: Timber Press.

Rothkamm K, Lobrich M. 2003. Evidence for a lack of DNA double-strand break repair in human cells exposed to very low x-ray doses. Proc Natl Acad Sci USA. 100(9):5057–5062.

Rasina LN, Grigorkina EB, Orekhova NA. 2017. Adaptation of small mammals to radiocontaminated environment: Oxidative metabolism and radioresistance. Dokl Biochem Biophys. 476(1):340–343.

Shevchenko VA, Pomerantseva MD, Ramaiya LK, Chekhovich AV, Testov BV. 1992. Genetic disorders in mice exposed to radiation in the vicinity of the Chernobyl nuclear power station. Sci Total Environ. 112(1):45–56.

Shuryak I. 2020. Review of resistance to chronic ionizing radiation exposure under environmental conditions in multicellular organisms. J Environ Radioact. 212:106128.

Shestopalov VM. 1996. Atlas of Chernobyl exclusion zone. Ukrainian Academy of Science, Kiev, Ukraine.

Snyder WE, Wise DH. 2001. Contrasting trophic cascades generated by a community of generalist predators. Ecology. 82(6):1571–1583

Taira W, Hiyama A, Nohara C, Sakauchi K, Otaki JM. 2015. Ingestional and transgenerational effects of the Fukushima nuclear accident on the pale grass blue butterfly. J Radiat Res. 56(Suppl 1): i2–18.

Takino S, Yamashiro H, Sugano Y, Fujishima Y, Nakata A, Kasai K, Hayashi G, Urushihara Y, Suzuki M, Shinoda H, et al. 2017. Analysis of the Effect of Chronic and Low-Dose Radiation Exposure on Spermatogenic Cells of Male Large Japanese Field Mice (Apodemus speciosus) after the Fukushima Daiichi Nuclear Power Plant Accident. Radiat Res. 187(2):161–168.

UNSCEAR 2011. Sources and effects of ionizing radiation: UNSCEAR 2008 report to the General Assembly. New York, NY: United Nations. p. 45–219.

Urushihara Y, Suzuki T, Shimizu Y, Ohtaki M, Kuwahara Y, Suzuki M, Uno T, Fujita S, Saito A, Yamashiro H, et al. 2018. Haematological analysis of Japanese macaques (Macaca fuscata) in the area affected by the Fukushima Daiichi Nuclear Power Plant accident. Sci Rep. 8(1):16748.

Voitsekhovych OV, Kanivets V, Onishi Y. 2007. The Chernobyl accident and its aquatic impacts on the surrounding area. In: Onishi Y, Voitsekhovych OV, Zheleznyak V, editors. Chernobyl – What have we learned? Dordrecht, the Netherlands: Springer. p. 9–47.

Volkova PY, Geras'kin SA, Kazakova EA. 2017. Radiation exposure in the remote period after the Chernobyl accident caused oxidative stress and genetic effects in Scots pine populations. Sci Rep. 7:43009

Wang Z, Yang WL, Jacob A, Aziz M, Wang P. 2015. Human ghrelin mitigates intestinal injury and mortality after whole body irradiation in rats. PLoS One. 10(2):e0118213

Williams JP, Brown SL, Georges GE, Hauer-Jensen M, Hill RP, Huser AK, Kirsch DG, Macvittie TJ, Mason KA, Medhora MM, et al. 2010. Animal models for medical countermeasures to radiation exposure. Radiat Res. 173(4):557–578.

Wise DH. 1993. Spiders in ecological webs. New York, NY: Cambridge University Press.

Wong K, Chang PY, Fielden M, Downey AM, Bunin D, Bakke J, Gahagen J, Iyer L, Doshi S, Wierzbicki W, et al. 2020. Pharmacodynamics of romiplostim alone and in combination with pegfilgrastim on acute radiation-induced thrombocytopenia and neutropenia in non-human primates. Int J Radiat Biol. 96(1): 155–166.

Impact of nonionizing electromagnetic radiation on male infertility: an assessment of the mechanism and consequences

Rohit Gautam, Eepsita Priyadarshini, JayPrakash Nirala, and Paulraj Rajamani

ABSTRACT

Purpose: Environment and lifestyle factors are being attributed toward increased instances of male infertility. Rapid technological advancement, results in emission of electromagnetic radiations of different frequency which impacts human both biologically as well as genetically. Devices like cell phone, power line and monitors emit electromagnetic radiation and are a major source of the exposure. Numerous studies describe the detrimental consequence of radiation on physiological parameters of male reproductive system including sperm parameters (morphology, motility, and viability), metabolism and genomic instability. While the thermal and nonthermal interaction of nonionizing radiations with biological tissues can't be ruled out, most studies emphasize the generation of reactive oxygen species. Oxidative stress alters redox equilibrium and disrupts morphology and normal functioning of sperms along with declination of total anti-oxidant capacity.

Conclusion: In this paper, we describe a detailed literature review with the intent of analyzing the impact of electromagnetic radiation and understand the consequence on male reproductive system. The underlying mechanism suggesting ROS generation and pathway of action has also been discussed. Additionally, the safety measures while using electronic gadgets and mobile phones has also been presented.

1. Introduction

The World Health Organization (WHO) defines infertility as the failure to attain pregnancy even after a year of regular unprotected sexual intercourse (WHO 2016). According to a study, 7% of world population suffer from infertility, almost half of which are instances of male infertility (Barati et al. 2020). As per WHO, primary infertility in India occurs in the range of 3.9 – 16.8% and affects about 15% of reproductive aged individuals (Zahid 2016). Besides physiological factors, various environmental factors contribute toward infertility. Electromagnetic radiations, alcohol and smoking are some of the exogenous determinates responsible for declining fertility (Tremellen 2008; Pacey 2010). Human are exposed to both ionizing and nonionizing radiation. Among this, nonionizing radiation is major concern as the sources of these radiations are mobile phone, laptop, Wi-Fi and Base Transvier station (BTS) high power electric lines. Exogenous factors such as alcohol consumption and smoking affect the quality of sperms by altering morphology and motility (Guthauser et al. 2014). A strong correlation has been established between reactive oxygen species (ROS) generation and male infertility, suggesting that almost 30–80% cases of male infertility are because of variations in ROS concentration. While significantly lower amount of ROS is necessary for sperm capacitation, hyperactivation and acro-mediated damage (Bui et al. 2018). Therefore, it is essential that normal ROS level is maintained by cells.

Continued exposure to electromagnetic radiations (EMR) stimulates chromosomal damage, genomic instability and also result in cancer development (Martin et al. 1986; Baverstock 2000). Electronic devices emitting electromagnetic radiation (EMR) have become an essential part of everyday life. Mobile phones, smart watches, laptops, cell phone towers and Wi-Fi are some sources that emit EMR. Since, there has been an increasing usage of such EMR devices; the instance of exposure of humans to these EMR is increasing. Number of investigations have focused on establishing the impact of radiations on different human organs amongst which male testis with the germinal epithelium is found to be most sensitive to EMR (Xu et al. 2008). Considerable changes in PKC (Protein Kinase C) and ODC (Ornithine decarboxylase) activity, structural alterations and declination in antioxidants level in brain, liver and testis of animals exposed to radiations has been reported (Paulraj et al. 2004; D'Silva et al. 2017). Figure 1 shows a schematic illustration of the various sources of EMR, their underlying mechanism of action and consequence on male reproductive system.

Radio frequency (RF) radiation comes under nonionizing radiations. The drastic increase in cell phone and Wi-Fi instigated researchers to analyze the impact of the radiations

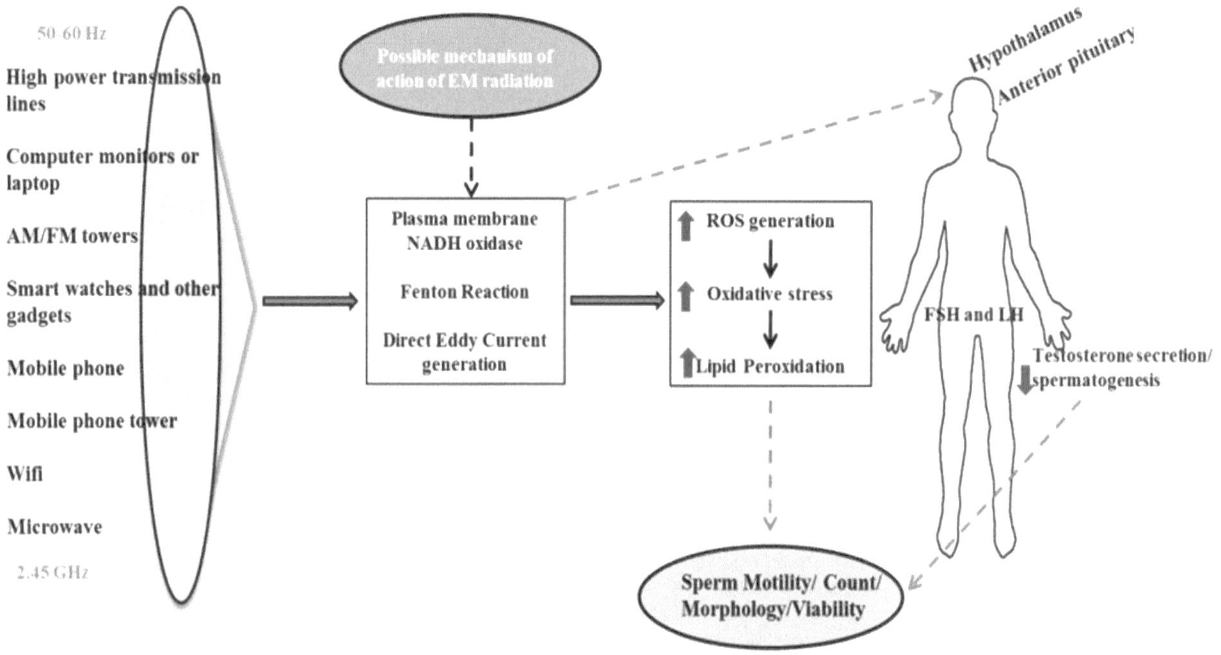

Figure 1. Diagrammatic illustration suggesting the consequence of EMR exposure on male reproductive system.

obtained results, International Agency for Research on Cancer has categorized radio-frequency emission as possibly carcinogenic to humans and has been grouped in 2B category (Baan et al. 2011). Previous studies, reported that the exposure to electromagnetic radiation may increase the risk of cancer. Exposure to radiation in *in vitro* models showed correlation with increased incidence of glioma. Patients with brain tumor were analyzed with respect to interphone study components. Patients from Australia, Canada, France, Israel and New Zealand were included in the study and 553 patients with glioma, 676 patients with meningioma along with 1911 control individuals were included in the study and correlated on basis of age, sex, region and location of tumor. They inferred that long-term usage of mobile phone with high RF radiation exposure have an increased risk of glioma tumor and lesser incidence of meningioma (Cardis et al. 2011; Miller et al. 2018; Gupta et al. 2020). Human glioma cancer is found to be associated with radiofrequency exposure (Yang et al. 2017) In this review, we have therefore assessed the impact of radiation on male fertility with respect to variations brought about in sperm morphology and other biological effects. Furthermore, we have summarized the possible mechanism of interaction of EMR with male reproductive organs after a careful and critical analysis of earlier reported studies.

2. Sources of radio-frequency radiation

Commonly used sources of radiations include radiofrequency (100 kHz to 300 GHz), AM radio transmission (540 to 1600 kHz), FM radio transmission (76 to 108 MHz), mobile phones (800 MHz to 3 GHz), microwaves (2.45 GHz) and Wi-Fi (2.4 GHz) (Kesari et al. 2018). In contrast to non-ionizing radiations, ionizing radiations are more deleterious exhibiting harmful effects via both thermal and non-thermal means. X rays, γ-rays and α-particles emitted from radioactive sources are classified as ionizing radiations. On exposure to these radiation, free radicals are generated which lead to oxidative stress-related damage to DNA, lipids and proteins (Reisz et al. 2014). Short-term as well as long-term exposure to these radiation affects both animals and humans. While in short-term studies (few minutes to hours) sperms are exposed to radiations under *in vitro* set-up, long-term exposing animal models to radiations for weeks to months (Singh et al. 2018). We have discussed in brief some of the widely used sources of radiofrequency radiations.

2.1. High power transmission lines

The widespread usage of electricity has resulted in an increase in exposure to magnetic fields (MF) emitted from household appliances or power lines (50–60Hz). Studies confirms that exposure to 50 Hz MF (0.4 mT) for varied time-periods increases mitochondrial permeability in exposed human amniotic epithelial cells. Yet another study in which neuro blastoma and glioma cell lines were exposed to 50 Hz MFs for 24 h induced increase in micronuclei with simultaneous enhancement in mitochondrial and cytosolic superoxide concentrations. Interestingly, the results suggest that threshold to induce biological response for 50 Hz MFs et at 10 mT or less. Likewise, the threshold for biological effects of ELF MFs is 10 mT or less and cytosolic superoxide, was unexpectedly large at 10 mT (Kesari et al. 2016). Similar data were reported in a case study summarizing sixteen research reports, published from 2000 to 2007 which signify the risk of human breast cancer cell line on exposure to extremely low frequency electromagnetic field (ELF-EMFs) and was performed at magnetic field exposures of 0.2-1 μT, 10 μT, 50 μT, 100 μT (Zhao et al. 2014).

2.2. Laptop, PC monitors and television sets

The modern era has witnessed an increase in the use of laptop devices and other electronic gadgets among the young age group. These gadgets emit electromagnetic of low frequency of up to 300 Hz. Such low-frequency radiation has deleterious impact of reproductive system and reported to induce considerable damage in sperm morphology and viability. In previous studies reported that, Wistar rat exposed to a maximum magnetic field 1.15 μT which is emitted by laptop computers for a period 7 hrs a day for one week showed a dose dependent decrease in sperm count and motility (Mortazavi et al. 2010; McGill and Agarwal 2014; Brodić and Amelio 2015). Exposure to radiations from computer monitors induces oxidative stress in the corneal tissues with increase in malondialdehyde (MDA), catalase, glutathione peroxidase activity (Balci et al. 2009). Additionally, studies showed after exposure a decrease in sperm count, motility (Mortazavi et al. 2010) and the frequency of aging was enhanced in exposed set of rats compared to control sets (Falone et al. 2008). In yet another study exposure to RF-EMR emitted from laptop computers display unit for a period of 2,4,6,8 hours respectively for 4 weeks signified considerable decrease in sperm count, motility and viability. As well as decline in concentration of testosterone and follicle stimulating hormone level were observed (Stephen et al. 2019).

Constant use of laptop also has thermal effects as people place it over their laps while working. This results in an increase in scrotal temperature. It was observed that in men using a lap pad ad sitting with legs closed the scrotal temperature increased by 1° C in 15 minutes (right hemiscrotum) and 13 minutes (left hemiscrotum) respectively (Sheynkin et al. 2011). As testis are temperature sensitive organ and spermatogenesis is temperature sensitive process, it is found that with change in male scrotum temperature (1–2 °C) may cause scrotal hypothermia and leads to oxidative stress in testis, cell apoptosis and affect sperm DNA integrity (Shiraishi et al. 2010).

2.3. Mobile phone, AM, FM and mobile towers

With the advent of 3 G and 4 G along with enhancement in cell phone is leading to increase in number of mobile phone towers installation. Countries like the USA, Australia, Japan and Canada follows ICNIRP guidelines according to which power density limits for cell towers are 4.5, 9.0, 10.5 W/m^2 for 900, 1800 and 2100 MHz frequency respectively. In India revised standards limits for power density are 0.45, 0.9, 1.05 W/m^2 for 900, 1800 and 2100 MHz frequency respectively, which are now 10 times more stringent than many other counties. Besides this many countries have their own specified radiation standard according to their environmental and physiological factors (Bhawan and Marg 2014). The radiations emitted from these mobile towers is quite harmful for the biological system. Inhabitants living within 50 m radius of mobile towers complain of fatigue, nausea, dizziness and muscle pain compared to individuals living beyond 50 m. The Power density of the RF radiation measured from the studied mobile towers was 0.021 W/m^2 (Pachuau and Pachuau 2014). Headache, memory changes, dizziness, tremors, depression, and sleep cycle disturbance are some of the neuropsychiatric conditions being reported in individuals exposed to radiations from mobile towers (Abdel-Rassoul et al. 2007). In yet another study, symptoms of depression, memory status, insomnia, and hair loss were observed on exposure to 0.0122 W/m^2 radiation power density in humans (Gulati et al. 2016).

Mobile phone has become an integral part of daily life of human beings. As mobile phone is placed in pockets, exposure to the electromagnetic radiation emitted from these devices may affect reproductive system. A study focusing on analyzing the effect of 4 G mobile phone radiation (2.104 GHz frequency) and 3.0 W/kg SAR with respect to time period and distance, signified alteration in male reproductive parameters. Exposure of 18 hrs per day, at a distance of 3 cm for 28 days showed alterations in spermatogenesis with significant decrease in sperm cell, spermatogonia and leydig cells which effect the fertility (Oh et al. 2018). In an *in vitro* study, exposure of spermatozoa to 900 MHz radiation for one hour with a SAR value of 2.0 W/Kg affects sperm morphometry and binding to hemizona which decreases sperm fertilization capacity (Falzone et al. 2011). Similar results were also observed in another study performed on 124 semen samples, exposed to 850 MHz (SAR =1.46 W/kg) for one hour. Results showed a significant decrease in sperm motility, acrosin activity and sperm DNA fragmentation (Zalata et al. 2015).

2.4. Wi-Fi

Wi-Fi (Wi-Fi or WLAN) is a wireless network being used in developed as well as developing countries. Literature suggests that exposure to Wi-Fi radiations can also have harmful consequences on the biological system (Doré and Chignol 2012; Ding et al. 2018). In one such experimental study, human semen sample from 29 donors were incubated underneath an active Wi-Fi–connected laptop (2.4 GHz) for 4 h at 25 °C. Progressive decrease in sperm motility was found along with DNA fragmentation in the exposed set of samples (Pall 2018b). On the basis of the results, they hypothesized that working on Wi-Fi connected laptop kept on lap may result in decreased male fertility because of the close exposure of testis with the radiations (Avendano et al. 2012). Similarly, male Wistar albino rats exposed to 2.437 GHz (24 h/day for 20 weeks) led to enhancement in serum 8-hydroxy-2′-deoxyguanosine in the testes suggesting damage to DNA on exposure (Atasoy et al. 2013b).

Thus, the results obtained from such exposure raises concerns on the safety of RF exposure in the reproductive age groups affecting both fertility and germ cell integration (Atasoy et al. 2013a).

3. Biophysical parameters and its impact on infertility

As described in the previous section, EM radiation exerts harmful effect and deleteriously affects male fertility. In our recent study we postulated that 3 G cell phone radiations alter sperm tail morphology and decrease sperm count in male Wistar rats (Gautam et al. 2019). However, the extent of harmful effect of radiations on the biological system is determined by various biophysical parameters. Biophysical parameter refers to the physical and biological factors that determine the sensitivity of EMR radiations and generally measures the absorption rate of radiation. Hence, the extent of effect of radiations is determined by calculating the Specific Absorption Rate (SAR) and power density.

SAR refers to the rate of energy absorbed per unit area and is measured in terms of W/kg. Equation (1) shows the formula for calculating SAR value (Kesari et al. 2013).

$$\mathrm{SAR} = \sigma\, \mathrm{E}^2/\rho \quad (1)$$

where, σ = liquid conductivity of tissues, E = Electric Field and ρ = density of tissues

Federal Communications Commission (FCC) limits the SAR value of cell phones at 1.6 W/kg, in order to ensure safe exposure and has accordingly set guidelines for mobile phone manufactures (FCC 2019). As per the compliance, only if the SAR value for any cell phone is less or equal to the stated value then it's considered as safe phone.

Power density is another important parameter which represents the emitted power per unit area. It is articulated in watts per square meter (W/m^2) or miliwatt per centimeter square (mW/cm^2). Power density is given by expression depicted in Equation (2).

$$\text{Power density } (\mathrm{P_d}) = \mathrm{E}^2/1200\Pi \mathrm{mW/cm^2} \quad (2)$$

$$\mathrm{P_d} = \mathrm{E}^2/377\ \mathrm{W/m^2}$$

(E represents electric field strength) (Biolnitiative Working Group CSaDC 2012).

Power level measurement can be done for near field or far-field points. R value is $>2\,D^2/\lambda$ than it will consider as far field exposure. Where R is distance from mobile tower, λ wavelength of transmitting frequency and D is diameter of the antenna.

The maximum power for individual macro cellular base station transmitter is 20 watts. In India Antennas on mobile phone towers transmit in the frequency range of 869–890 MHz for Code-division multiple access (CDMA), 935-960 MHz for Global System for Mobile Communications (GSM) 900, 1805–1880 MHz for GSM1800 and 2110–2170 MHz for 3 G. Wi-Fi frequency range is 2.4 GHz and 4 G Long-Term Evolution (LTE) is 2.99 GHz.

Apart from this, the potential harmful effect of EMR depends on other biophysical parameters such as penetration rate which in turn depends on various parameters such as intensity, power density, length of the biological system, dielectric properties like conductivity, permittivity, shape and geometry and object orientation. The amount of energy absorbed by biological system is a function of emitted frequency and composition of the tissue under exposure.

4. Mechanisms of EMR exposure on biological system

Depending on the nature of emitted radiations, EMR exposure can be continuous and pulsed of which, the latter is more harmful to the biological system. Radiations emitted from cell phone, laptops or mobile towers is basically a coupling of both electrical and magnetic field. The velocity and impact of this is different in the environment and after it penetrates the cell interior. Literature suggests thermal and non-thermal effects as the major mediators via which EMR interact with living system, while the electric and magnetic field are detrimental in determining the consequence (Kesari et al. 2013). The mode of interaction of EMR has been described in detail in the below sections.

4.1. Thermal effect

The absorbed EM radiation when converted in the form of heat represents thermal effect. The extent of heat generation depends on SAR and power density of emitted EMradiation. Biological system is affected by thermal effect when the heat generated exceeds $100\,mW/cm^{2.}$ (Habash 2007). Previous studies showed using laptop and exposure to low frequency EMR can increase the temperature in testis or near to (Sheynkin et al. 2005) And rise in scrotal temperature effects show adverse effects in sperm production and maturation and in adverse condition it may lead to DNA damage and apoptosis of germ cell (Durairajanayagam et al. 2015). Microwave operates by rotating the polar molecules assisted by generation of electromagnetic field. Leakage of radiation from microwave ovens can be a cause of exposure to radiations, causing heating of living tissues, the degree of which depends on the dielectric constant of the material. Literature signifies a variation in rectal temperature before and after exposure to radiation. While blood has the capability of disposing the excessive heat generated, sensitive organs as eye cornea and testis do not possess any temperature regulation mechanism. Hence depending on the degree of heat generated it will have a variable effect on the organs (Dasdag et al. 1999).

4.2. Non-thermal effect

Both electrical and magnetic field of EM radiation exhibit non-thermal effect on biological system. Magnetic field developed by alternating currents affects human body via its high permeability and strength of magnetic field (Fields 1997). Generally, electric field has poor penetration capability as skin layer prevents it but it act by producing eddy currents. EMR exposure alters the membrane permeability and make it porous when the electric field voltage exceeds the threshold value (Hamada et al. 2011).

Besides thermal and non-thermal effect, Fenton, Calcium ion concentration and oxidative stress generation have also been explained for exerting harmful effect by EMR.

Fenton method

In yet other studies, Fenton method has been suggested as a means of EM radiation effect. Exposure to radiation enhances generation of free hydroxyl radicals from hydrogen peroxide in cell (Bandyopadhyay et al. 1999).

Fenton reaction can be summarized in three different steps and explained in form of chemical equations as follows:

i. The interaction of Fe $^{2+}$ salt with hydrogen peroxide results in the generation of free hydroxyl ions ($^{\bullet}OH$).

$$Fe^{2+} + H_2O_2 \rightarrow Fe^{3+} + OH^{\bullet} + OH^{-}$$

ii. Any trace iron (Fe^{3+}) present further reacts with hydrogen peroxide leading to the formation of hydrogen ion and superoxides, given by the following formula

$$Fe^{3+}H_2O_2 \rightarrow Fe^{2+} + O_2^{-} + H^{+}$$

iii. Thereafter, hydrogen peroxide interacts with the superoxide ion leading to the formation of $^{\bullet}OH$

$$O_2^{-} + H_2O_2 \rightarrow OH^{-} + {}^{\bullet}OH + O_2$$

Calcium ion concentration

Seminiferous tubules and leydig cells of rat testis have PKC enzyme complex that modulate ion conductance (Nikula et al. 1987). PKC complex facilitates via calcium-dependent phosphorylation of membrane and ion exchange proteins. This enzyme complex along with cAMP and calcium ion plays a pivotal role in regulating sperm motility (Kimura et al. 1984) and hence variation in calcium ion concentration affects sperm motility (Kubitz et al. 2004; Crocenzi et al. 2008). There exist several evidences which suggest that non-thermal effects by EMR leads to enhanced calcium ion concentration that results in downstream activation of VGCC (Voltage-gated calcium channels) (Pall 2013, 2015). Cultured cells when exposed to pulsed EMR showed considerable rise in calcium ion followed by generation of high concentration of nitric oxide within a span of 5 seconds of exposure. This study validity the calcium/calmodulin-dependent nitric oxide synthesis facilitated by plasma membrane depolarization. They thus suggested that EM radiations act directly on VGCC voltage gated sodium channels via high calcium influx (Pilla 2012). The involvement of VGCC channel over other voltage gated ion channels is further supported that the fact that resting cells possess 10,000-fold concentration gradient and million-fold electrochemical gradient of calcium that results in high influx of calcium when the VGCC channel is activated. Moreover, calcium has various cell regulatory functions, the activation of which can have high physiological consequences as in the elevation of calcium causes major cellular damage (Pall 2018a). Thus, the concentration of calcium has an integral role on functioning of reproductive system, the activation of VGCC channel and high calcium concentration has also been associated with polyspermy and can also prevent fertilization if triggered earlier.

Oxidative stress and ROS generation:

Levels of ROS and nitrogen must be regulated at appropriate levels for normal physiological functions of cells (Wagner et al. 2018). Radiation exposure stimulates oxidative stress that results in generation of ROS leading to tissue damage and apoptosis. Desai et al. (2009) suggested that radiations emanating from cell phone have effect through cell plasma membrane. They further stated that ROS generation by radiations is enhanced by altering NADH oxidase activity of cell membrane and postulated a detailed mechanism underlying the interaction of radiation with biological system by involving free radicals. ROS are generated by means of electron transfer reaction wherein ground state oxygen is converted to singlet oxygen, $O_2^{\bullet -}$, $OH^{\bullet}$, and H_2O_2. An increase in these ROS level causes oxidative stress. Thus induction of oxidative stress is explained by an enhancement in free radicals such as $O_2^{\bullet -}$, $OH^{\bullet}$, and H_2O_2, NO or a decrease in concentration of superoxide dismutase (SOD), glutathione peroxidases (GPs) and catalase, the antioxidant defensive enzymes. Sperm capacitation is a necessary process to obtain fertilizing facility and ROS such as H_2O_2, superoxide, nitrogen oxide (NO), and superoxide anion radical are contributed in this procedure (Wagner et al. 2018). Sperm cells possess their own antioxidant defense mechanism. These may be antioxidant enzymes or nonenzyme, those consumed through supplements or food (Sies 1993). Important antioxidant enzymes which reduce oxidative stress includes Catalase, Superoxide dismutase (SOD) and glutathione peroxidase (GPX). High levels SOD were found in sertoli cells during spermatogenesis and sperm maturation as it scavenges superoxide anion and convert it into H_2O_2 and O_2 (Bauché et al. 1994). CAT is also found in developing sperm and it converts H_2O_2 into H_2O and O_2. GPX is found in sertoli cells and has a role in protecting plasma membrane of sperm from lipid peroxidation (Tavilani et al. 2008). Earlier study reported that male mice, exposed to 900 MHz with a SAR value of 2.2 W/kg for 12 hrs per day for 1, 3 or 5 weeks showed a significant increase in a two-fold ROS generation in the sperm mitochondria (Houston et al. 2019). Based on these theories radiations does affect the reproductive system and has been supported by a series of experiments, details of which have been described under the section titled "EMR and male infertility".

5. EMR and male infertility

Exposure to radiofrequency EM radiation emitted from mobile phone, laptops and other devices adversely affects male fertility. Epidemiological studies suggest that people using such devices have decreased fertility (Al-Quzwini et al. 2016). A study conducted on semen samples of 371 individuals exposed to mobile phone radiations, showed decrease in sperm motility (Fejes et al. 2005). *In vitro* studies on human sperm also showed correlation between exposure to such radiation and changes in reproductive parameters (Agarwal et al. 2009; Gorpinchenko et al. 2014b). Likewise Wi Fi users are increasing as it is being used by laptop and mobile user as well (Jaffar et al. 2019). Jonwal et al. 2018

Table 1. EMF exposure on reproductive system (in vivo and in vitro studies).

	Exposure parameters				
Organism/ Species	Frequency	SAR/Power density	Duration	Observations	Reference
Wistar rat, n = 8	3G (1915 MHz)	0.26 W/kg	45 days (2 hrs/day)	Upsurge in ROS, lipid peroxidation level Declination in sperm count, changes in sperm tail morphology	Gautam et al. (2019)
Sprague-Dawley rats, n = 20	2.104 GHz	3.0 W/kg.	28 days/6-18 hrs/day	Exposure for longer duration resulted in significant decrease of spermatogenesis	Oh et al. (2018)
Sprague-Dawley rats n = 20	900 MHz	1.22 W/kg	30 days 4hrs/day	Considerable increases in testicular proteins associated with carcinogenicity and reproductive damage.	Sepehrimanesh et al. (2017)
Wistar Albino rats n = 8	890-915 MHz	0.96 W/kg	2 – 12 hrs/day 7 weeks	caused morphologic and histological changes by the affecting germinal epithelium tissue negatively	Cetkin et al. (2017)
Human spermatozoa, n = 26	850 MHz	1.46 W/kg	60 minutes	Exposure showed negative impact on sperm motility, sperm acrosin activity, sperm DNA fragmentation.	Zalata et al. (2015)
Sprague Dawley rats, n =8	900 MHz	0.025 W/kg	1 hrs/day for 30 days	Deleterious effect on the morphology of testis and significant changes in biochemical markers.	Odacı & Özyılmaz, (2015)
Wistar rats, n = 6	2.45 GHz	0.14 W/kg	2 hrs/day for 45 days	LDH activity, XO, ROS, protein carbonyl, DNA damage, apoptosis, testosterone.MDA level altered considerably	Meena et al. (2014)
Albino rats n = 6	3G (1910.5 MHz)	0.28 W/kg to .0226 W/kg	2 hrs/day for 60 days	Declination in sperm count, enhancement in lipid peroxidation, reduced seminiferous tubules and testicular weight. DNA damage	Kumar et al. (2014)
Adult Wistar rats, n = 7	900 MHz-GSM Band	1.60 mW/cm^2	14-21 days	Reduction in viability and motility of sperms. Decreases total sperm antioxidant activity	Ghanbari et al. (2013)
Sertoli cells (Wistar rat)	S-band microwave radiation	100 mW/ cm^2	4 min	Cytokines produced by microwave-radiated Sertoli cells may disrupt spermatogenesis.	Wu et al. (2012)
Drosophila melanogaster	GSM 900 and 1800	0.89 W/kg.	6min/ day for 6 days	Induces cell death and hence decreases reproductive capacity	Panagopoulos et al. (2010)
Adult Wistar rat, n= 6	2.45GHz	0.11W/kg	2 hrs/ day for 35 days	Reduced sperm count cell apoptosis, decrease in GPx and SOD activities. Enhanced catalase activity	Kesari and Behari (2010)
Human Spermatozoa	900 MHz	2.0 W/Kg	1 hour	Studied showed declination in efficiency of sperm to bind with hemizona.	Falzone et al. (2011)
Human spermatozoa	1.8 GHz	0.4–27.5 W/kg	…..	Reduction in motility and viability of sperm, increases ROS generation and DNA fragmentation	De Iuliis et al. (2009)

reported that male swiss albino mice exposed to 2.45 GHz microwave radiation for 2 hrs/day (30 days) cause alterations testis histopathology, significant increase in ROS level and lipid per-oxidation. Decrease in SOD, GPx and testosterone level was also observed. Similarly, in study combinational exposure to 3 G + WiFi EM radiation showed considerable decrease in sperm motility and velocity (Kamali et al. 2017)

Testis where the development and maturation of sperms occurs is quite sensitive to radiation and leads to genotoxicity. Depth of penetration, exposure time, duration and water content of the organ are affect the level of sensitivity (Lee et al. 2010; Ulubay et al. 2015). Since, sperm cellular membrane is rich in polyunsaturated fatty acids (PUFA) which are susceptible to ROS damage and leads to lipid peroxidation. Peroxidase activity causes alteration in membrane integrity and decreases sperm motility (De Iuliis et al. 2009). Several studies have highlighted the negative effect of EMR on male testis and associated fertility (Agarwal et al. 2011; Kesari et al. 2018).

Our group has well established the undesirable consequence of radiation on Wistar rat model. 60 days (2 h/day) exposure of male albino rats to EMR with SAR value 0.28 W/Kg brought about significant declination in sperm number, seminiferous tubule diameter, testicular weight along with significant increase in lipid peroxidation and DNA damage (Kumar et al. 2014).

Quite recently, we established that ROS generated in rats was the key reason behind sperm membrane integrity, declination in viability and alteration in sperm mitochondrial activity when exposed to 3 G radiation (1915 MHz) for 2 hrs/day for 45 days with a SAR value of 0.26 W/kg (Gautam et al. 2019). Similarly, exposing rats to microwave radiations (2.45 GHz) for 2 hrs/day for 45 days with power density (0.21 mW/cm^2) and SAR value 0.14 W/Kg was found to affect lactate dehydrogenase, xanthine oxidase, ROS and protein carbonyl content. Along with this DNA damage and cellular apoptosis was induced (Meena et al. 2014). In yet another study, exposure of rats to 900 MHz radiation at 2 hrs/day for 45 days and SAR of 0.9 W/kg resulted a decrease in testosterone and rise in caspase-3 when exposed to radiofrequency. Alterations in sperm head and mid-piece was evident (Kesari and Behari 2012). Testosterone is the principal male hormone involved in reproduction. Variation in its concentration has deleterious consequence on reproductive and general health. A study states that exposure to radiations from cell phone for 60 min/day up to 3 months leads to dramatic decline in serum testosterone (Meo et al. 2010). In similar studies, exposure of mouse Leydig (TM3) cell with 1950 MHz frequency for a period of 24 hrs showed decrease in testosterone synthesis, cell proliferation and cell cycle disturbance (Fejes et al. 2005).

Exposure of adult male rats to EMR (900 MHz, 1 hour/30 days) showed considerable changes in sperm parameters. Histopathological analysis demonstrated generation of vacuolar spaces in seminiferous tubules membrane along with decrease in seminiferous tubule diameter, germinal epithelium thickness, high apoptotic index and declination in antioxidant enzymes (Odacı and Özyılmaz 2015). Koppers et al. (2008) suggested that mitochondria are the first site of free radical generation and that radiations induces NADH oxidase activity and lead to ROS production.

To understand the molecular mechanism various epigenetics studies have also been performed to evaluate the effect of EM radiation on DNA methylation, histone modification etc. (Monk 1995; Consales et al. 2018). Exposure to low dose ionizing radiation (LDIR) suggested oxidative stress induced epigenetic modifications which affects pregnant women and fetal development (Tharmalingam et al. 2017). Liu et al. 2015 observed epigenetic changes, in GC-2 cell line, a mouse spermatocyte, exposed to 50 Hz for 72 hrs (5 min on and 10 min off) and different magnetic field. study results showed higher global methylation after exposure to 3 mT magnetic field as compared to control. They also observed that some genes with aberrant DNA methylation and expression after exposure to 50 Hz EMF.

Besides these, studies also demonstrate the harmful effect of radiations on humans leading to reduction in sperm motility, viability along with increased ROS generation (Agarwal et al. 2009). In one such study, men exposed to combination of EMR (1800 MHz) and wi-fi (2.45 GHz) were divided into 3 groups. While the 1st group was exposed for 30 min duration, 2nd and 3rd groups were exposed to 31-120 min and >120 mins respectively. The results showed that increase in exposure time results in enhancement of ROS and increased expression of 8-OHdG leading to DNA fragmentation. Besides this, decrease in total antioxidant capacity, glutathione and superoxide dismutase were also observed. Sperm count, vitality, and motility was also found to decrease significantly (Ding, Ping, et al. 2018). In another study, sperm samples were collected from 32 healthy individuals and exposed to 900/1800 MHz for 5 hours. Analysis of sperm parameters signified progressive decrease in sperm motility and increase in DNA fragmentation formation on exposure (Gorpinchenko et al. 2014a). Table 1 summarizes some of studies performed for determination of the effect of EMR on *in vivo* and *in vitro* system.

It is well known that, ROS generated in mammalian spermatozoa has a vital functioning in regulating the process of capacitation, without which fertilization is not possible (de Lamirande and Cagnon 1993; Agarwal and Sekhon 2010). Mild oxidative stress assists in the ability of sperm to bind to zona pellucida layer of egg cells. Epididymal maturation is also facilitated by ROS as it assists in condensing sperm chromatin (Pfeifer et al. 2001). Sperm plasma membrane is rich in unsaturated fatty acids that provides structural fluidity required for participation in membrane fusion (Aitken and Curry 2011). Hence, disturbances in the equilibrium between ROS and antioxidant system can affect spermatozoa and fertilization process as well. Thus, uninhibited ROS production may damage reproductive tissues and destabilize normal cell regulation thereby adversely affecting male fertility (Darbandi et al. 2018). The presence of restricted cytoplasmic antioxidant capacity and highly sensitive ROS targets makes sperm highly susceptible to oxidative stress. Decosahexaenoic and arachidonic acids are the major PUFA constituents of sperm which generally contain six and four double bonds (Koppers et al. 2010; Aitken et al. 2012). The abundance of these double bonded fatty acid moieties along with the absence of an endogenous antioxidant protection ability, makes sperm highly prone to radiation-induced damage. EMR radiations thus bring about a series of biochemical change that increase sperm vulnerability to damage by high concentration of ROS ultimately leading to alterations in sperm morphology, sperm motility, count, genomic instability, mitochondrial damage and micronuclei formation.

On the contrary, there are few studies that suggest no significant change due to EM radiation. In one such study, males exposed to mobile phone on talking mode for 120 to 240 minutes did not show any considerable variations in sperm parameter. The study was conducted on 571 individuals and sperm count, progressive motility, viability, strict sperm morphology and HOS (Hypo osmolarity swelling) test results were analyzed (Feijo et al. 2011). Some studies on *in vivo* animal models as male Wistar rats exposed to 900 and 1800 MHz frequency, 2 hours daily for 90 days did not show any negative changes in sperm motility, seminiferous tubule structure, caspases activity in testis and other sperm parameters. Further, caspase 3 levels were unaltered and no alterations in histopathology of seminiferous tubules was found. They thus suggested that low SAR does not have significant effect on puberty of rat testicles (Nisbet et al. 2012; Tumkaya et al. 2016). The extent of harmful effect of mobile phone radiation is still debatable and requires large sample size, long term studies and exposure to varied levels of frequency to determine the impact.

Conclusion

EM radiations are emitted from commonly used electronic gadgets (laptop, mobile phone), Wi-Fi as well as from cell phone towers. These radiations may affect normal functioning of various biological systems and the male reproductive system in particular. However, the extent of negative effect of radiations on male fertility varies depending on exposure time, dielectric properties, dosimetry, emitted power density. Radiation induces generation of ROS, NADH oxidase and variation in calcium ion concentration are responsible for changes in reproductive parameters (sperm count, viability, motility, morphology). But there are few studies which suggest that exposure to radiation has no or minimal effect on reproductive system. The topic is thus debatable and requires long term animal and epidemiological studies. Additionally, assessment and determination of molecular mechanisms involved in these may provide in-depth analysis.

Acknowledgments

The authors are thankful to Indian Counsel of Medical Research (ICMR) for project funding. EP is grateful to SERB (Scientific and Engineering Research Board) for fellowship under NPDF scheme of DST-SERB.

Disclosure statement

The authors declare that they have no conflict of interest.

ORCID

Paulraj Rajamani http://orcid.org/0000-0003-4922-9028

References

Abdel-Rassoul G, El-Fateh OA, Salem MA, Michael A, Farahat F, El-Batanouny M, Salem E. 2007. Neurobehavioral effects among inhabitants around mobile phone base stations. Neurotoxicology. 28(2): 434–440.

Agarwal A, Desai NR, Makker K, Varghese A, Mouradi R, Sabanegh E, Sharma R. 2009. Effects of radiofrequency electromagnetic waves (RF-EMW) from cellular phones on human ejaculated semen: an in vitro pilot study. Fertil Steril. 92(4):1318–1325.

Agarwal A, Sekhon LH. 2010. The role of antioxidant therapy in the treatment of male infertility. Hum Fertil (Camb)). 13(4):217–225.

Agarwal A, Singh A, Hamada A, Kesari K. 2011. Cell phones and male infertility: a review of recent innovations in technology and consequences. Int Braz J Urol. 37(4):432–454.

Aitken RJ, Curry BJ. 2011. Redox regulation of human sperm function: from the physiological control of sperm capacitation to the etiology of infertility and DNA damage in the germ line. Antioxid Redox Signal. 14(3):367–381.

Aitken RJ, Gibb Z, Mitchell LA, Lambourne SR, Connaughton HS, De Iuliis GN. 2012. Sperm motility is lost in vitro as a consequence of mitochondrial free radical production and the generation of electrophilic aldehydes but can be significantly rescued by the presence of nucleophilic thiols. Biol Reprod. 87(5):110–110111.

Al-Quzwini OF, Al-Taee HA, Al-Shaikh S. 2016. Male fertility and its association with occupational and mobile phone towers hazards: an analytic study. Middle East Fertil Soc J. 21(4):236–240.

Atasoy HI, Gunal MY, Atasoy P, Elgun S, Bugdayci G. 2013a. Immunohistopathologic demonstration of deleterious effects on growing rat testes of radiofrequency waves emitted from conventional Wi-Fi devices. J Pediatr Urol. 9(2):223–229.

Atasoy HI, Gunal MY, Atasoy P, Elgun S, Bugdayci G. 2013b. Immunohistopathologic demonstration of deleterious effects on growing rat testes of radiofrequency waves emitted from conventional Wi-Fi devices. J Pediatr Urol. 9(2):223–229.

Avendano C, Mata A, Sarmiento CAS, Doncel GF. 2012. Use of laptop computers connected to internet through Wi-Fi decreases human sperm motility and increases sperm DNA fragmentation. Fertil Steril. 97(1):39–45.

Baan R, Grosse Y, Lauby-Secretan B, El Ghissassi F, Bouvard V, Benbrahim-Tallaa L, Guha N, Islami F, Galichet L, Straif K, WHO International Agency for Research on Cancer Monograph Working Group 2011. Carcinogenicity of radiofrequency electromagnetic fields. Lancet Oncol. 12(7):624–626.

Balci M, Namuslu M, Devrim E, Durak İ. 2009. Effects of computer monitor-emitted radiation on oxidant/antioxidant balance in cornea and lens from rats. Mol Vis. 15:2521–2525.

Bandyopadhyay U, Das D, Banerjee RK. 1999. Reactive oxygen species: oxidative damage and pathogenesis. Curr Sci. 77:658–666.

Barati E, Nikzad H, Karimian M. 2020. Oxidative stress and male infertility: current knowledge of pathophysiology and role of antioxidant therapy in disease management. Cell Mol Life Sci. 77(1):21–93.

Bauché F, Fouchard M-H, Jégou BJFl. 1994. Antioxidant system in rat testicular cells. FEBS Lett. 349(3):392–396.

Baverstock K. 2000. Radiation-induced genomic instability: a paradigm-breaking phenomenon and its relevance to environmentally induced cancer. Mutat Res. 454(1-2):89–109.

Bhawan MD, Marg J. 2014. Effects of electromagnetic field radiation from mobile towers and handsets. Telecom Regulatory Authority of India. https://www.trai.gov.in/sites/default/files/EMF_Information_Paper_30.07.2014.pdf

BioInitiative Working Group CSaDC. 2012. BioInitiative report: A rationale for a biologically-based public exposure standard for electromagnetic radiation. Chapter 18, p. 1–37. https://bioinitiative.org

Brodić D, Amelio A. 2015. Classification of the extremely low frequency magnetic field radiation measurement from the laptop computers. Meas Sci Rev. 15(4):202–209.

Bui A, Sharma R, Henkel R, Agarwal A. 2018. Reactive oxygen species impact on sperm DNA and its role in male infertility. Andrologia. 50(8):e13012

Cardis E, Armstrong B, Bowman J, Giles G, Hours M, Krewski D, McBride M, Parent M-E, Sadetzki S, Woodward AJO, et al. 2011. Risk of brain tumours in relation to estimated RF dose from mobile phones: results from five Interphone countries. Occup Environ Med. 68(9):631–640.

Cetkin MURA, Kızılkan N, Demirel C, Bozdağ Z, Erkılıç S, Erbağcı H. 2017. Quantitative changes in testicular structure and function in rat exposed to mobile phone radiation. Andrologia. 49(10):e12761.

Consales C, Cirotti C, Filomeni G, Panatta M, Butera A, Merla C, Lopresto V, Pinto R, Marino C, Benassi BJMn. 2018. Fifty-hertz magnetic field affects the epigenetic modulation of the miR-34b/c in neuronal cells. Mol Neurobiol. 55(7):5698–5714.

Crocenzi FA, Sánchez Pozzi EJ, Ruiz ML, Zucchetti AE, Roma MG, Mottino AD, Vore M. 2008. Ca(2+)-dependent protein kinase C isoforms are critical to estradiol 17beta-D-glucuronide-induced cholestasis in the rat. Hepatology. 48(6):1885–1895.

D'Silva MH, Swer RT, Anbalagan J, Rajesh B. dr. 2017. Effect of radiofrequency radiation emitted from 2G and 3G cell phone on developing liver of chick Embryo – A Comparative Study. J Clin Diagn Res. 11(7):AC05–AC09.

Darbandi M, Darbandi S, Agarwal A, Sengupta P, Durairajanayagam D, Henkel R, Sadeghi MR. 2018. Reactive oxygen species and male reproductive hormones. Reprod Biol Endocrinol. 16(1):1–14.

Dasdag S, Ketani M, Akdag Z, Ersay A, Sari I, Demirtas Ö, Celik M. 1999. Whole-body microwave exposure emitted by cellular phones and testicular function of rats. Urol Res. 27(3):219–223.

De Iuliis GN, Newey RJ, King BV, Aitken RJ. 2009. Mobile phone radiation induces reactive oxygen species production and DNA damage in human spermatozoa in vitro. PLoS One. 4(7):e6446.

de Lamirande E, Cagnon C. 1993. Human sperm hyperactivation and capacitation as parts of an oxidative process. Free Radical Biol Med. 14(2):157–166.

Desai NR, Kesari KK, Agarwal A. 2009. Pathophysiology of cell phone radiation: oxidative stress and carcinogenesis with focus on male reproductive system. Reprod Biol Endocrinol. 7(1):114.

Ding S-S, Ping S, Hong T. 2018. Association between daily exposure to electromagnetic radiation from 4G smartphone and 2.45-GHz wi-fi and oxidative damage to semen of males attending a genetics clinic: a primary study. Int J Clin Exp Med. 11(3):2821–2830.

Ding SS, Sun P, Zhang Z, Liu X, Tian H, Huo YW, Wang LR, Han Y, Xing JP. 2018. Moderate dose of trolox preventing the deleterious effects of Wi-Fi radiation on spermatozoa in vitro through reduction of oxidative stress damage. Chinese Med J. 131(4):402.

Doré JF, Chignol MC. 2012. Laptop computers with Wi-Fi decrease human sperm motility and increase sperm DNA fragmentation. Fertil Steril. 97(4):e12.

Durairajanayagam D, Agarwal A, Ong C. 2015. Causes, effects and molecular mechanisms of testicular heat stress. Reprod Biomed Online. 30(1):14–27.

Falone S, Mirabilio A, Carbone MC, Zimmitti V, Di Loreto S, Mariggiò MA, Mancinelli R, Di Ilio C, Amicarelli F. 2008. Chronic exposure to 50Hz magnetic fields causes a significant weakening of antioxidant defence systems in aged rat brain. Int J Biochem Cell Biol. 40(12):2762–2770.

Falzone N, Huyser C, Becker P, Leszczynski D, Franken D. 2011. The effect of pulsed 900-MHz GSM mobile phone radiation on the acrosome reaction. Head Morphometry and Zona Binding of Human Spermatozoa. 34(1):20–26.

FCC 2019. Specific Absorption Rate (SAR) for cellular telephones. [Updated 2019 August 22] https://www.fcc.gov/general/specific-absorption-rate-sar-cellular-telephones

Feijo C, Junior SV, Esteves S. 2011. P-044 Lack of evidence that radiofrequency electromagnetic waves (RF-EMW) emitted by cellular phones impact semen parameters of Brazilian men. Human Reproduction. 26(suppl_1):i139–i140.

Fejes I, Závaczki Z, Szöllosi J, Koloszár S, Daru J, Kovács L, Pál A. 2005. Is there a relationship between cell phone use and semen quality? Arch Androl. 51(5):385–393.

Fields RE. 1997. Evaluating compliance with FCC guidelines for human exposure to radiofrequency electromagnetic fields. OET Bulletin. 65(10):1–84.

Gautam R, Singh KV, Nirala J, Murmu NN, Meena R, Rajamani PJA. 2019. Oxidative stress-mediated alterations on sperm parameters in male Wistar rats exposed to 3G mobile phone radiation. Andrologia. 51(3):e13201.

Ghanbari M, Mortazavi SB, Khavanin A, Khazaei M. 2013. The effects of cell phone waves (900 MHz-GSM band) on sperm parameters and total antioxidant capacity in rats. Int J Fertil Steril. 7(1): 21.

Gorpinchenko I, Nikitin O, Banyra O, Shulyak A. 2014a. The influence of direct mobile phone radiation on sperm quality. Cent European J Urol. 67(1):65–71.

Gorpinchenko I, Nikitin O, Banyra O, Shulyak A. 2014b. The influence of direct mobile phone radiation on sperm quality. Cent European J Urol. 67(1):65–71.

Gulati S, Yadav A, Kumar N, Aggarwal NK, Kumar R, Gupta R. 2016. Effect of GSTM1 and GSTT1 polymorphisms on genetic damage in humans populations exposed to radiation from mobile towers. Arch Environ Contamin Toxicol. 70(3):615–625.

Gupta S, Sharma RS, Singh R. 2020. Non-ionizing radiation as possible carcinogen. Int J Environ Health Res. 1–25. DOI:10.1080/09603123.2020.1806212

Guthauser B, Boitrelle F, Plat A, Thiercelin N, Vialard F. 2014. Chronic excessive alcohol consumption and male fertility: a case report on reversible azoospermia and a literature review. Alcohol Alcohol. 49(1):42–44.

Habash R. 2007. Bioeffects and therapeutic applications of electromagnetic energy. Boca Raton: CRC press.

Hamada AJ, Singh A, Agarwal A. 2011. Cell phones and their impact on male fertility: fact or fiction. Open Reprod Sci J. 5(4):125–137.

Houston BJ, Nixon B, McEwan KE, Martin JH, King BV, Aitken RJ, De Iuliis G. 2019. Whole-body exposures to radiofrequency-electromagnetic energy can cause DNA damage in mouse spermatozoa via an oxidative mechanism. 9(1):1–14.

Jaffar FHF, Osman K, Ismail NH, Chin K-Y, Ibrahim S. 2019. Adverse effects of Wi-Fi radiation on male reproductive system: a systematic review. Tohoku J Exp Med. 248(3):169–179.

Jonwal C, Sisodia R, Saxena VK, Kesari K. 2018. Effect of 2.45 GHz microwave radiation on the fertility pattern in male mice. biophysics. 37(4):453–460.

Kamali K, Atarod M, Sarhadi S, Nikbakht J, Emami M, Maghsoudi R, Salimi H, Fallahpour B, Kamali N, Momtazan A, et al. 2017. Effects of electromagnetic waves emitted from 3G + wi-fi modems on human semen analysis. Urologia. 84(4):209–214.

Kesari KK, Agarwal A, Henkel R. 2018. Radiations and male fertility. Reprod Biol Endocrinol. 16(1):118.

Kesari KK, Behari J. 2010. Effect of microwave at 2.45 GHz radiations on reproductive system of male rats. Toxicol Environ Chem. 92(6): 1135–1147.

Kesari KK, Behari J. 2012. Evidence for mobile phone radiation exposure effects on reproductive pattern of male rats: role of ROS. Electromag Biol Med. 31(3):213–222.

Kesari KK, Juutilainen J, Luukkonen J, Naarala J. 2016. Induction of micronuclei and superoxide production in neuroblastoma and glioma cell lines exposed to weak 50 Hz magnetic fields. J R Soc Interface. 13(114):20150995.

Kesari KK, Kumar S, Nirala J, Siddiqui MH, Behari JJCb. 2013. Biophysical evaluation of radiofrequency electromagnetic field effects on male reproductive pattern. Cell Biochem Biophys. 65(2):85–96.

Kimura K, Katoh N, Sakurada K, Kubo S. 1984. Phospholipid-sensitive Ca2+-dependent protein kinase system in testis: localization and endogenous substrates. Endocrinology. 115(6):2391–2399.

Koppers AJ, De Iuliis GN, Finnie JM, McLaughlin EA, Aitken RJ. 2008. Significance of mitochondrial reactive oxygen species in the generation of oxidative stress in spermatozoa. J Clin Endocrinol Metab. 93(8):3199–3207.

Koppers AJ, Garg ML, Aitken RJ. 2010. Stimulation of mitochondrial reactive oxygen species production by unesterified, unsaturated fatty acids in defective human spermatozoa. Free Radical Biol Med. 48(1): 112–119.

Kubitz R, Saha N, Kühlkamp T, Dutta S, Vom Dahl S, Wettstein M, Häussinger D. 2004. Ca2+-dependent protein kinase C isoforms induce cholestasis in rat liver. J Biol Chem. 279(11):10323–10330.

Kumar S, Nirala JP, Behari J, Paulraj R. 2014. Effect of electromagnetic irradiation produced by 3G mobile phone on male rat reproductive system in a simulated scenario. Indian J Exp Biol. 52(9):890–897.

Lee HJ, Pack JK, Kim TH, Kim N, Choi SY, Lee JS, Kim SH, Lee YS. 2010. The lack of histological changes of CDMA cellular phone-based radio frequency on rat testis . Bioelectromagnetics. 31(7): 528–534.

Liu Y, Liu W-b, Liu K-j, Ao L, Zhong JL, Cao J, Liu J-y. 2015. Effect of 50 Hz extremely low-frequency electromagnetic fields on the DNA methylation and DNA methyltransferases in mouse spermatocyte-derived cell line GC-2. Biomed Res Int. 2015:237183.

Martin RH, Hildebrand K, Yamamoto J, Rademaker A, Barnes M, Douglas G, Arthur K, Ringrose T, Brown IS. 1986. An increased frequency of human sperm chromosomal abnormalities after radiotherapy. Mutat Res Lett. 174(3):219–225.

McGill JJ, Agarwal A. 2014. The impact of cell phone, laptop computer, and microwave oven usage on male fertility. Male Infertility. New York (NY): Springer; p. 161–177.

Meena R, Kumari K, Kumar J, Rajamani P, Verma H, Kesari KK. 2014. Therapeutic approaches of melatonin in microwave radiations-induced oxidative stress-mediated toxicity on male fertility pattern of Wistar rats. Electromagn Biol Med. 33(2):81–91.

Meo SA, Al-Drees AM, Husain S, Khan MM, Imran MB. 2010. Effects of mobile phone radiation on serum testosterone in Wistar albino rats. Saudi Med J. 31(8):869–873.

Miller AB, Morgan LL, Udasin I, Davis D. 2018. Cancer epidemiology update, following the 2011 IARC evaluation of radiofrequency electromagnetic fields (Monograph 102). Environ Res. 167:673–683.

Monk MJDg. 1995. Epigenetic programming of differential gene expression in development and evolution. Dev Genet. 17(3): 188–197.

Mortazavi SMJ, Tavassoli A, Ranjbari F, Moammaiee P. 2010. Effects of laptop computers' electromagnetic field on sperm quality. Infertility. 11(4):251–258.

Nikula H, Naor Z, Parvinen M, Huhtaniemi I. 1987. Distribution and activation of protein kinase C in the rat testis tissue. Mol Cell Endocrinol. 49(1):39–49.

Nisbet HO, Nisbet C, Akar A, Cevik M, Karayigit MO. 2012. Effects of exposure to electromagnetic field (1.8/0.9 GHz) on testicular function and structure in growing rats. Res Vet Sci. 93(2):1001–1005.

Odacı E, Özyılmaz C. 2015. Exposure to a 900 MHz electromagnetic field for 1 hour a day over 30 days does change the histopathology and biochemistry of the rat testis. Int J Radiat Biol. 91(7):547–554.

Oh JJ, Byun S-S, Lee SE, Choe G, Hong S. 2018. Effect of electromagnetic waves from mobile phones on spermatogenesis in the era of 4G-LTE. BioMed Res Int. 2018:1–8.

Pacey A. 2010. Environmental and lifestyle factors associated with sperm DNA damage. Hum Fertil (Camb)). 13(4):189–193.

Pachuau L, Pachuau Z, Department of Physics, Pachhunga University College, Aizawl -796001, India 2014. Study of Cell Tower Radiation and its Health Hazards on human body. IOSRJAP. 6(1):01–06. e-ISSN.2278-4861.

Panagopoulos DJ, Chavdoula ED, Margaritis LH. 2010. Bioeffects of mobile telephony radiation in relation to its intensity or distance from the antenna. Int J Radiat Biol. 86(5):345–357.

Pall ML. 2013. Electromagnetic fields act via activation of voltage-gated calcium channels to produce beneficial or adverse effects. J Cell Mol Med. 17(8):958–965.

Pall ML. 2015. Scientific evidence contradicts findings and assumptions of Canadian safety panel 6: microwaves act through voltage-gated calcium channel activation to induce biological impacts at non-thermal levels, supporting a paradigm shift for microwave/lower frequency electromagnetic field action. Rev Environ Health. 30(2): 99–116.

Pall ML. 2018a. How cancer can be caused by microwave frequency electromagnetic field (EMF) exposures: EMF activation of voltage-gated calcium channels (VGCCs) can cause cancer including tumor promotion, tissue invasion, and metastasis via 15 mechanisms. Mobile Communications and Public Health. Boca Raton (FL) CRC press; p. 165–186.

Pall ML. 2018b. Wi-Fi is an important threat to human health. Environ Res. 164:405–416.

Paulraj R, Behari J, Mutagenesis MMo. 2004. Radio frequency radiation effects on protein kinase C activity in rats' brain. Mutat Res. 545(1-2):127–130.

Pfeifer H, Conrad M, Roethlein D, Kyriakopoulos A, Brielmeier M, Bornkamm GW, Behne D. 2001. Identification of a specific sperm nuclei selenoenzyme necessary for protamine thiol cross-linking during sperm maturation. Faseb J. 15(7):1236–1238.

Pilla AA. 2012. Electromagnetic fields instantaneously modulate nitric oxide signaling in challenged biological systems. Biochem Biophys Res Commun. 426(3):330–333.

Reisz JA, Bansal N, Qian J, Zhao W, Furdui CMJA, Signaling r. 2014. Effects of ionizing radiation on biological molecules-mechanisms of damage and emerging methods of detection. Antioxid Redox Signal. 21(2):260–292.

Sepehrimanesh M, Kazemipour N, Saeb M, Nazifi S, Davis DL. 2017. Proteomic analysis of continuous 900-MHz radiofrequency electromagnetic field exposure in testicular tissue: a rat model of human cell phone exposure. Environ Sci Pollut Res. 24(15):13666–13673.

Sheynkin Y, Jung M, Yoo P, Schulsinger D, Komaroff EJHr. 2005. Increase in scrotal temperature in laptop computer users. Fertil Steril. 20(2):452–455.

Sheynkin Y, Welliver R, Winer A, Hajimirzaee F, Ahn H, Lee KJF, sterility. 2011. Protection from scrotal hyperthermia in laptop computer users. Fertil Steril. 95(2):647–651.

Shiraishi K, Takihara H, Matsuyama H. 2010. Elevated scrotal temperature, but not varicocele grade, reflects testicular oxidative stress-mediated apoptosis. World J Urol. 28(3):359–364.

Sies H. 1993. Strategies of antioxidant defense. Eur J Biochem. 215(2): 213–219.

Singh R, Nath R, Mathur AK, Sharma RS. 2018. Effect of radiofrequency radiation on reproductive health. Indian J Med Res. 148(Suppl):S92–S99.

Stephen SA, Kelechi LN, Adeola AA. 2019. The effect of electromagnetic radiation (EMR) from laptop on reproductive hormones, sperm quality and prostate specifec anticen of male albino rats (rattus norvegicus). Ecol Safety Balance Use Resour. 1(19):43–52.

Tavilani H, Goodarzi MT, Doosti M, Vaisi-Raygani A, Hassanzadeh T, Salimi S, Joshaghani H. 2008. Relationship between seminal antioxidant enzymes and the phospholipid and fatty acid composition of spermatozoa. Reproductive BioMedicine Online. 16(5):649–656.

Tharmalingam S, Sreetharan S, Kulesza AV, Boreham DR, Tai TJRr. 2017. Low-dose ionizing radiation exposure, oxidative stress and epigenetic programing of health and disease. Radiat Res. 188(4.2): 525–538.

Tremellen K. 2008. Oxidative stress and male infertility-a clinical perspective. Hum Reprod Update. 14(3):243–258.

Tumkaya L, Kalkan Y, Bas O, Yilmaz A. 2016. Mobile phone radiation during pubertal development has no effect on testicular histology in rats. Toxicol Ind Health. 32(2):328–336.

Ulubay M, Yahyazadeh A, Deniz ÖG, Kıvrak EG, Altunkaynak BZ, Erdem G, Kaplan S. 2015. Effects of prenatal 900 MHz electromagnetic field exposures on the histology of rat kidney. Int J Radiat Biol. 91(1):35–41.

Wagner H, Cheng JW, Ko EY. 2018. Role of reactive oxygen species in male infertility: An updated review of literature. Arab J Urol. 16(1): 35–43.

WHO 2016. Multiple definitions of infertility. [updated 21 October; accessed]. https://www.who.int/reproductivehealth/topics/infertility/multiple-definitions/en/.

Xu G, Intano GW, McCarrey JR, Walter RB, McMahan CA, Walter CA. 2008. Recovery of a low mutant frequency after ionizing radiation-induced mutagenesis during spermatogenesis. Mut Res/Genetic Toxicol Environ Mutagenesis. 654(2):150–157.

Yang M, Guo W, Yang C, Tang J, Huang Q, Feng S, Jiang A, Xu X, Jiang GJPo. 2017. Mobile phone use and glioma risk: A systematic review and meta-analysis. PLoS One. 12(5):e0175136

Wu H, Wang D, Shu Z, Zhou H, Zuo H, Wang S, Li Y, Xu X, Li N, Peng R. 2012. Cytokines produced by microwave-radiated Sertoli cells interfere with spermatogenesis in rat testis. Andrologia. 44: 590–599.

Zahid 2016. Infertility. National Health Portal; [accessed 5 August]. https://www.nhp.gov.in/disease/reproductive-system/infertility.

Zalata A, El-Samanoudy AZ, Shaalan D, El-Baiomy Y, T. Mostafa 2015. In vitro effect of cell phone radiation on motility DNA Fragmentation and Clusterin Gene Expression in Human Sperm. Int J Fertil Steril. 9(1):129–36.

Zhao G, Lin X, Zhou M, Zhao J. 2014. Relationship between exposure to extremely low-frequency electromagnetic fields and breast cancer risk: a meta-analysis. Eur J Gynaecol Oncol. 35(3):264–269.

Environmental and industrial developments in radiation cataractogenesis

Lukasz Laskowski, David Williams, Colin Seymour, and Carmel Mothersill

ABSTRACT

Purpose: This review discusses recent developments in our understanding of biological and physiological mechanisms underlying radiation cataractogenesis. The areas discussed include effects of low-dose exposures to the lens including potential relevance of non-targeted effects, the development of new personal-protective equipment (PPE) and standards in clinical and nuclear settings motivated by the updated ICRP recommendations to mitigate exposures to the lens of the eye. The review also looks at evidence from the field linking cataracts in birds and mammals to low dose exposures.

Conclusions: The review suggests that there is evidence that cataractogenesis is not a tissue reaction (deterministic effect) but rather is a low dose effect which shows a saturable dose response relationship similar to that seen for non-targeted effects in general. The review concludes that new research is needed to determine the dose response relationship in environmental studies where field data are contradictory and lab studies confined to rodent models for human exposure studies

Background to biological and physiological mechanisms

The biology of the lens is, in general, well understood. Textbooks such as Hall and Giaccia (2019) provides a useful introduction to radiation cataractogenesis by looking at perturbations caused by ionizing radiation to the regular biological processes. However, more and more complex pathways are being discovered all the time which while outside the scope of a textbook, maybe important given the ICRP report 118 (2012), which recommended, for consideration by the radiation protection community, a new threshold value for the absorbed dose to the eye lens at 0.5 Gy for acute or protracted exposures. The also recommended consideration of a proposal that doses to the lens for workers need to be reduced from 150mSv/year to 20mSv/year averaged over 5 years. This was chiefly in response to new lower dose epidemiological information indicating cataract formation at <1 Gy and was a precautionary response to new evidence that cataract induction due to radiation might not be purely a tissue reaction (deterministic effect) as had previously been thought (Rafnsson et al. 2005; Worgul et al. 2007). Due to lack of evidence the potential effects of dose rate or lifetime accumulated dose are not considered in the new limits which mainly apply for nuclear workers and the medical profession.

An extensive review was completed by Ainsbury et al. (2016). This looked at the most recent developments in radiation-induced cataract biological research. The main focus of Ainsbury's review was new developments in the field especially in relation to biological mechanisms. A significant finding emerged following their review of two independent studies (Ruotolo et al. 2003; Hamada and Fujimichi 2015). These studies compared expression of genotypes in human subjects with and without cataracts. Both of the studies found expression of some genotypes was suppressed or even absent, however, the genotypes affected differed in the two studies and there were no common pathways. This could suggest there are multiple pathways that could result in cataract formation as a result of radiation exposure and that genetic background may be very important. The most significant and most heavily researched pathway in radiobiology is the DNA damage response (DDR). It was proposed by Ainsbury et al. (2016) that damaged DNA, specifically double strand breaks, when repaired incorrectly result in errors in cell division that cause opacities. This is because the process by which lens precursor epithelial cells differentiate into fiber cells is highly orchestrated and involves a large number of proteins. Lens epithelial cells (LEC) have DNA, and the hypothesis around DNA damage is that changes occur before the translation of the LEC into full anuclear lens fiber cells. The same review cited a study by Markiewicz et al. (2015) which showed that DNA damage could be seen in the lens epithelial cell line FHL124 exposed to doses as low as 280 mGy, which is under half of the current ICRP recommended threshold. Large numbers of in-vivo animal studies have been performed due to the logistic simplicity in comparison to human studies. Markiewicz et al. (2015) also investigated DNA damage in mouse lens epithelium and found double strand breaks in the lens epithelium following doses as low as 20 and 100 mGy suggesting a greater level of radiosensitivity manifesting as DNA damage than was previously thought.

With regard to the lens however changes caused by ionizing radiation to proteins may be more important in cataractogenesis than DNA damage. This is because the vast majority of the lens is comprised of protein in lens fibers, cells devoid of nuclear material although as noted above, the relevant DNA damage is likely to occur before the lens fibers become anuclear. While there is DNA in the mature LEC, the vast majority 90%+ of the lens is protein. There is no DNA in the bulk of the lens but only in the cells on the very outside of the lens and at its equator - giving the view widely held by lens biologists that cataract is a post-translational effect of photons on crystallin proteins, binding them together through photo-oxidation of thiol groups of cysteine amino acids which aggregates these lens proteins, resulting in cataract. The changes which occasion cataract, involving protein aggregation, are for the most part post-translational modifications to the lens crystallin proteins where thiol (-SH) groups are photo-oxidatively joined to form disulfide bridges. This leads to protein aggregation and cataract formation in the protein-rich lens nucleus where no DNA is to be found. These changes occur with sunlight, particularly in the ultraviolet end of the spectrum (Truscott 2005; Michael and Bron 2011) and are particularly marked with higher energy ionizing radiation.

DNA damage may have some part to play in the generation of cataract following irradiation in childhood (Wilde and Sjöstrand 1997) but the formation of cataract photon irradiation, be this by ultraviolet radiation (Taylor et al. 1988), visible light (Zigman 1983) or gamma irradiation (Lipman et al. 1988) but the fact that antioxidants reduce the effect of gamma radiation on lens opacification in experimental models strongly suggests that it is photo-oxidation of lens proteins that is the pathogenic pathway in cataract formation after lens irradiation be it by visible, ultraviolet or ionizing irradiation (Kocer et al. 2007). The vast majority of cataract formation is from photon damage to the proteins that are already there - new lens formation is very slow in an adult lens. Kleiman and Spector (1993) reported that there was no association between ag, e and DNA damage while cataract clearly worsens with age. In half of the lenses with cataract there were higher numbers of cells with greater DNA breaks than in control eyes but not in the other half. While oxidative stress damages proteins the free radicals produced can also affect the DNA to be found in lens epithelial cells and the replicating lens fiber cells at the lens equator. Oxidative stress was found to alter gene expression in the lens epithelial cells of mice (Kubo et al. 2008; Liu et al. 2008; Cohen et al. 2009). Affected mice were found to develop cataracts more quickly than unirradiated controls although how much of this effect is caused by DNA damage and how much by protein aggregation is not clear. This mechanism is important as oxidative stress is associated with both bystander signaling and with delayed effects such as lethal mutations or genomic instability (Clutton et al. 1996; Lyng et al. 2001; Maguire et al. 2007; Hurem et al. 2017). While the relevance of these non-targeted effects (NTE) in radiation protection is highly controversial (Mothersill et al. 2017; Mothersill and Seymour 2018; Cucinotta et al, 2019; Shuryak 2019) they dominate the dose response after low dose radiation (LDR) being triggered by doses as low as 5mGy and are triggered to the same extent by high or low dose rates (Schettino et al. 2005; Liu et al. 2006; Gow et al. 2008; Tomita and Maeda, 2015; Meade et al. 2016) meaning they could be a relevant mechanism in cataractogenesis. NTE are of particular interest in the light of recent evidence that UVA photons are released by irradiated epithelial cells and that these are associated with exosome release and appear to vary depending on the p53 status of the irradiated cells (Le, Fernandez-Palomo, et al. 2017; Le, Mothersill, et al. 2017). The subcellular target of the UVA photons is the mitochondrial electron transport chain where complex 1 is completely supressed by exposure to the photons leading to oxidative stress and loss of ATP energy production (Le et al. 2018). While these experiments have not yet been done on lens epithelial cells, any activity resulting in UV photon release is of potential interest in cataractogenesis. Given the importance of p53 in orchestrating the response to NTE signals (Le et al 2017; Mothersill et al. 2011) this could stimulate a damage signaling pathway induced by p53. In the normal situation, activation of p53 would activate a G1 cell cycle block the cell cycle and possibly result in apoptosis rather than outright facilitate opacification but following exposure to x-rays, lens epithelial cells were unexpectedly found to begin to divide at a rate faster than usual, perhaps to replace the lens fiber cells that have been lost (Bai et al. 2003). While radiosensitivity differs at different stages of the cell cycle, the speed of the cycle and its fidelity is regulated by processes during the G1 and G2 phases which prevent damaged cells from dividing (Blakely et al. 1989; Short et al. 2003; Hall and Giaccia 2019, p. 60). A faster cell cycle could mean these safety mechanisms may not be happening leading to errors. Greater radiosensitivity might be expected to occur because of the relatively longer time spent in radiosensitive phases of the cycle (Schöllnberger, Scott, et al. 2001; Schöllnberger, Mebust, et al. 2001; Ahmed et al. 2018). Normally, cell differentiation is a very controlled process but because of the faster rate of cell division noted above (Bai et al. 2003), it is possible that the compromised feedback mechanism can result in more opacities within the lens. It is important also to consider the population kinetics of cells as opposed to individual cell cycle kinetics particularly if NTE are important since these clearly operate at the level of the population and progeny rather than at the level of the individual cell. The proportion of cells in the population that are dividing at any one time and the probability of reinitiating need to be considered. Extended G2 caused by damage repair also needs to be considered although it is not known if the doses used in the studies cited above are sufficient at any one time to delay G2 in enough cells to cause a significant effect.

There is also evidence that shortened telomeres associated with both aging and LDR (Frenzel et al. 2017) may cause division errors in lens epithelial cells (Babizhayev and Yegorov 2010; Babizhayev et al. 2011). It is unknown, however, whether this is simply because both cataract incidence and telomere shortening correlate with aging or because there is some link between the two. Other secondary mechanisms that could result in cataractogenesis include changes in crystallins following radiation exposure (Khan et al. 2015; Barnard et al. 2018). These are proteins that make up the majority of the protein content in the lens cells.

Most of the mechanisms discussed above provide evidence of correlation of the mechanism with radiation cataractogenesis but there is little direct evidence of causation. To address this, more research needs to be done on the mechanisms of cataractogenesis by radiation, monitoring the pathways and mechanisms discussed in relation to dose, dose rate and radiation quality.

Low-dose exposures and cataract incidence

After the latest update the ICRP and NCRP still consider cataracts to be a tissue reaction or deterministic effect just with a lower threshold of 0.5 Gy (Hall and Giaccia 2019). Traditionally, a 'tissue reaction' is one where cell damage will not occur at doses below the threshold. There is evidence from some studies, however, that suggests that this may not be the case for cataracts. In addition, as seen in the previous section, there is evidence suggesting that 500 mGy, may be too high a threshold. However it is necessary to be cautious when suggesting that evidence from mechanistic studies at low doses that look at hypothesized steps in cataract formation, is meaningful when considering actual cataract. There are certainly some epidemiological studies with evidence at doses < 500 mGy but not many. The mechanistic and animal studies do indicate the threshold should be lower, but as yet the evidence is not very strong

Studies performed by Little et al. (2018, 2020) regarding occupational exposure of a cohort of US Radiological Technologists found some statistically significant effects in workers with exposures below the threshold. The statistical accuracy of this completed study is largely due to the high sample size: 67,246. According to their study, individuals which had received between 50 and 99.9 mGy of equivalent dose to the eye had a hazard ratio for cataract history that fell within the range (1.04–1.27). This measurement indicates the relative probability (with $p < .001$) that individuals receiving this dose would be diagnosed with cataracts in comparison to the group that received less than 10 mGy. It should be noted that this dose was received over the career of the individuals and that a saturated dose response relationship was observed with no increase in incidence with increasing dose under 100mGy (Figure 1).

This measurement gets as high as (1.17–1.50) for individuals that received between 200 and 499.9 mGy of dose. Both of these ranges are below the new 'threshold value for deterministic effects', yet they show statistically significant effects. Based on these findings, Little et al. claimed, 'Based on our results, a threshold of 500 mGy is inconsistent with the pattern of excess risk… ' (Little et al. 2018). In

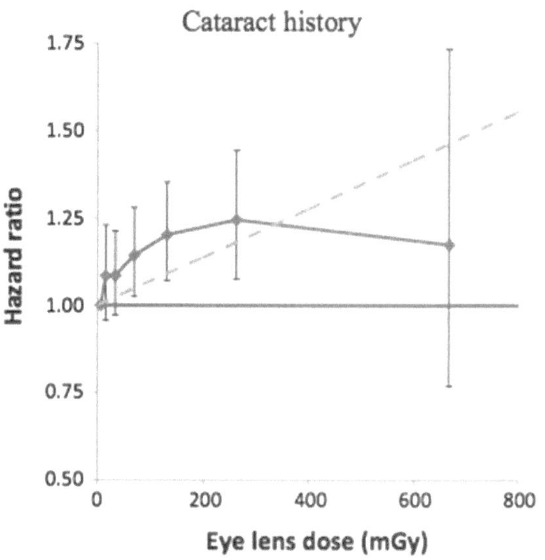

Figure 1. taken from Little et al: Adjusted Hazard Ratios plotted against Cumulative Eye Lens Dose. Note that the "Adjusted" Hazard Ratio, has taken into account other factors which contribute to cataracts risk. The factors considered in this were diabetes, BMI as well as smoking status. The extensive error on the final value is due to the lower sample size (Little et al. 2018).

addition, since these results were seen at cumulative doses below the threshold, it further indicates that cataract formation has characteristics of a stochastic effects in that there are effects at low doses. However, it differs from classic stochastic effects in the fact that cataracts can result from a sufficient cumulative dose over a lifetime. This suggests that the concept of historic dose described by our group in a series of papers focusing on exposure in contaminated environments may be useful to consider (Mothersill et al. 2017). Historic dose represented by HGy, takes into consideration the impact of heritable genomic instability induced at the time of initial exposure, which increases the probability of mutations and epigenetic effects occurring – this would increase the 'speed' at which the threshold life time cumulative *effect* would be seen but would not manifest as an increased lifetime *dose*. The paper by Uwineza et al. 2019, is perhaps relevant here. They proposed that radiation increases the 'cataractogenic load', which means that cataracts that would have appeared later anyway manifest earlier. The cell cycle discussion above may also be relevant for the HGy concept because it highlights the importance of dose rate and duration of exposure to radiation in relation to accumulation of populations of cells in G2. G2/M block is linked to both low dose hypersensitivity, genomic instability and bystander signaling (Löbrich and Jeggo 2007; Tauchi et al. 2009; Krueger et al. 2010; Fernandez-Palomo et al. 2016) at the population level. It would be interesting to look at whether the extent of the HGy burden is related to disproportionately increased G2 delay following acute exposure compared with chronic exposure over the long-term.

Some of these issues are currently being modeled by the MODARIA group as part of an IAEA sponsored project (https://wwwns.iaea.org/projects/modaria/modaria2.asp?s=8&l=129).

To summarize the main point of this part of the discussion, an interaction between radiation and a dividing epithelial cell may result in an opacity. Enough of these opacities will result in cataracts. Since there is no natural mechanism for the removal of opacities, they will accumulate. This means that repeated exposures can have an accumulating effect on cataract development and indicates further deviation from the ICRP tissue reaction classification (Little et al. 2018, 2020).

Another study with different conclusions was completed by Karatasakis et al. (2018) using interventional cardiologists and catheterization lab staff as subjects. This study did not have a sample size nearly as large as the one done by Little et al. (2018), 99 exposed individuals and 18 controls, but it still had some interesting findings. According to this study, 'The prevalence of overall cortical and posterior subcapsular lens changes (including subclinical findings) was higher in exposed participants compared with controls (47% vs. 17%, $p = .015$)'. Pre-cataract lens changes were found in individuals with cumulative doses as low as 0.5 Gy. However, this study found no statistically significant correlation between cumulative exposure and the prevalence of cataracts. This does not necessarily disagree with the findings of the previous study as it may simply be a result of the low sample size (Karatasakis et al. 2018). It could also be due to the LDR characteristically 'flat' dose response referred to earlier.

Asides from studies on clinical workers, there was a notable study done on industrial radiography workers. Industrial radiography is one of the most common nondestructive methods of checking materials for flaws and for checking the integration of welded joints. In a Chinese study, 1401 industrial radiographers and 1878 unexposed workers were investigated over a period of 12 years. It was found that the radiographers were significantly more likely to be diagnosed with cataracts. More so, the study did not find a statistically significant hazard ratio for the development of nuclear cataracts (Lian et al. 2015). This suggests that the difference in cataract incidence in the two groups is likely due to radiation exposure since nuclear cataracts are not associated with radiation exposure (Hall and Giaccia 2019). All other forms of cataracts investigated all saw hazard ratios greater than 1 with 95% confidence. The radiographers were very evenly distributed in terms of the cumulative dose ranges with 350 in the dose ranges 0–48.4 mSv, 48.4–66.1 mSv, and 66.1–88.6 mSv; the last 351 were had doses ranging from 88.6–146.1 mSv. There was no statistically significant difference found in the hazard ratio determined by dose range however, they are all still greater than 1 with statistical significance. This again agrees with the flat dose response seen in the low dose range. On the topic of the cumulative doses investigated, the average cumulative dose for the radiographers investigated over 12 years was 38.8 mSv. If this was received in even increments over the 12 years, this is significantly below the ICRP recommendations and further suggests that cumulative dose should be used to assess cataracts risk (Lian et al. 2015).

The above studies suggest that the ICRP's classification of cataracts in Publication 118 is incorrect and that effects at low doses follow the classic pattern of LDR response. They further suggest that historic dose may be a useful concept to include in future analysis of available cohorts. In addition to the modified threshold, ICRP 118 recommends that the equivalent dose to the lens of the eye should not exceed a 20 mSv average over 5 years with a single year exceeding 50 mSv; a similar recommendation was made by the ICRP regarding effective dose to protect against stochastic effects. However, based on the data from the studies provided, an argument can be made that this limit is still too high. It is particularly important to emphasize again that studies have found that cataract prevalence correlates with *accumulated* dose (Lian et al. 2015; Little et al. 2018). Assuming occupational exposure for 45 years (20 years old to 65 years old) at the maximum permissible dose, 20 mSv per year, 900 mSv of lifetime dose would be accumulated. Effects were seen at cumulative doses much less than this. Should the findings of Little et al. (2018) be verified by additional studies, a decision will need to be made by the ICRP on whether or not some statistically significant cataract risk will be allowed based on their recommendations.

Industry reaction to the updated ICRP recommendations

Since the ICRP's new recommendations were released, research has been done on effective methods to measure eye dose as well as radiological personal protective equipment (RPPE). A significant portion of this research has been done in the clinical setting with some investigation in other industries. ICRP and other international bodies such as IAEA (International Atomic Energy Agency) have also produced a large number of publications (available on their websites) targeting specific groups of workers such as pregnant women. These include mention of the risk of cataract for example in the woman or the fetus but these are outside the scope of this review.

One of the areas of greatest concern in the clinical setting is the performance of interventional radiology procedures. This is because operators must treat the patient while the patient is being imaged, a rarity in majority of radiological imaging and radiation therapy procedures. For this reason, RPPE and shielding in clinical settings have been an area of intense research after the ICRP recommendations. Today, lead glasses are used clinical occupational settings to reduce dose to the eye. Properly fitted lead eye-wear was found to reduce eye equivalent dose by factors as high as 2.1 in a clinical setting. In experimental settings, it was found that this factor was between 7.9 and 10.0. The difference was most likely due to non-optimized room geometry and the training of the staff (van Rooijen et al. 2014). In the study regarding radiography cohorts cited previously, the effects of different PPE were investigated as well. Although lead eyeglasses, on average, showed smaller hazard ratios for individuals that wore them for a longer duration during work the trend could not be replicated to give a similar confidence interval. The use of a shielded enclosure however, did show statistically significant evidence of reducing the hazard ratio for cataract development (Lian et al. 2015). As far as shielding goes for radiological and interventional cardiology staff, multiple studies have shown that the use of ceiling suspended shielding can reduce eye dose by a factor of 5–8 (Vanhavere et al. 2011). It was found that using ceiling-suspended shielding as well as an under-couch x-ray unit (as opposed to an over-couch x-ray unit) was the optimized configuration for a IR procedures with respect to radiation protection.

In addition to this equipment layout, it was also recommended that the primary operator and the nurses, that must remain close to the patient, all be given lead glasses since the protection they receive from the ceiling hanging shield is limited (O'Connor et al. 2013). This research and future optimizations are very significant in the clinical setting since physicians are expected to exceed the newly recommended annual dose limit set by the ICRP based on current practices in the hospitals where these studies were completed (Vanhavere et al. 2011). In the future, it is possible that radioprotectors may be used, particularly for work where eye exposure will be prevalent. An experiment was completed on rats using Nigella sativa oil (NSO) and thymoquinone (TQ). All experimental groups were given a single 5 Gy of total cranial irradiation. The control group developed cataracts in 80% of the subject population where the subjects that were given NSO and TQ had developed it in 20% and 50% respectively. NSO and TQ both have protective effects which reduce the oxidative stress that occurred in the eye as a result of irradiation. The focus is to develop a compound using these to prevent cataract development after radiotherapy. Asides from this, there is motivation to extend this into occupational protection applications (Taysi et al. 2015).

With the changes to the ICRP recommendations, developing an effective method to measure dose to the lens of the eye, has been a priority to both the clinical and nuclear occupational settings. Various methods have been developed over a multitude of studies. Some groups have sought various formulae and systems of equations to estimate the dose to the eye, but these methods were, in general, found to be too inaccurate and irreproducible for industry standard (Khan et al. 2017). Currently, $H_p(3)$ is the best measurement for eye lens dose. The EYE-D is an example of a dosimeter capable of measuring $H_p(3)$. It has a TLD pellet attached to an adjustable headband. Its sensitivity to doses as low as 10 μSv as well as a satisfactory photon energy response makes it satisfy the requirements for IR application (O'Connor et al. 2013). However, dosimeters that measure $H_p(3)$, including the EYE-D, are very expensive and calibrating standard dosimeters to measure $H_p(3)$ is impossible. Therefore, $H_p(0.07)$, which measures local skin dose and partial whole body, is used to estimate the dose to the eye. Based on a study completed by Scalchi et al., it was found that the predicted dose using this method tended to be higher than the true dose (Scalchi et al. 2018). In an industry setting, a conservative measure such as this one is preferred over a measurement that has a possibility of being lower than what is expected. All-in-all, methods for lens dosimetry have been developed to standards that meet industry requirements.

Compliance with dosimeter-use protocols was found to be an issue in several studies. Khan et al. found that physicians did not routinely use dosimeters even when made available. Similar findings were echoed in O'Connor et al. study. Matsubara et al. went as far to say that there is an, 'urgent need to improve training in radiation protection and to promote the use of radiation protection tools to reduce the lens exposure' (Matsubara et al. 2017). Unfortunately, the studies reviewed for this paper have all suggested this. A study of 83 participants at a level 1 trauma centire and children's hospital in Zurich found that only 54% demonstrated regular compliance with the radiation protection procedures. Complying with procedures consisted of the usage of dosimetry, an apron, and a thyroid shield (Jentzsch et al. 2015). An alarming general lack of knowledge surrounding radiation protection has also been observed in some professionals. A study looking at 780 Italian radiographers found that only 12.1% of them attended radiation protection courses on a regular basis. 7% of the sample believed that mammograms did not use ionizing radiation and roughly half were not able to differentiate between deterministic and stochastic effects. With the now more strict recommendations made by the ICRP, it has become increasingly more important for clinical workers who are being exposed to be aware of the potential risks of radiation.

This significant research has put the clinical setting and other places of work where radiation exposure occurs regularly into an excellent position to fall in line with the ICRPs new requirements. An interesting area requiring radiological protection is space travel. Based on a 2002 study, astronauts and cosmonauts are slightly to significantly more likely to have opacity in the lens due to exposure to cosmic rays (Rastegar et al. 2002). As the technolgy behind space travel and exploration improves, the field of radiation protection, including that of the lens, will need to keep up due to the elevated levels of radiation in space (National Aeronautics and Space Administration 2018).

Radiation cataractogenesis in animals

Radiation cataractogenesis in laboratory rodents is well studied as referred to earlier, as they are models for humans, particularly those used to derive dose-response relations in humans. However, radiation cataractogenesis in other species as well as animal field studies, are important to consider when trying to develop environmental radiation protection guidelines and when establishing benchmarks.

One controversial study looked at the prevalence of cataracts in voles (Myodes glareolus), in Chernobyl (Lehmann et al. 2016). A significant proportion of these voles, 57 out of 80, were found to have cataracts. There was no statistically significant difference between the proportions of female and male voles that had cataracts. However, the correlation between accumulated dose and cataract severity was only found to be statistically significant in females and not in males. Cataracts were scored for severity on a scale from 0 to 4; each eye was scored individually (before being summed) on a scale from 0 to 2, with 2 being most severe. When it came to female voles, a negative correlation was found between cataract severity and the number of offspring the females would bear. It was suggested that this was due to the oxidative stress that came as a result of radiation exposure (Lehmann et al. 2016). Although this study may make it seem like there is a clear connection between low doses of radiation and cataracts, there is doubt whether the exposure is truly the causative factor. The voles in the study above were killed by the experimenters and examined when dead. Other research groups have not found cataracts in rodents trapped around Chernobyl at higher levels than in similar species trapped in the UK (Williams 2019). These workers suggested that the lens opacity being observed by Lehman et al. are changes in lens transparency occurring after death rather than a one caused by cataracts, given that the rodents in Lehmann et al. report were examined some time after death having been stored at −20 degrees centigrade and then thawed (Lehmann et al. 2016). These manipulations will necessarily have caused lens opacifications quite apart from any effects caused by radiation (Mastromonaco et al. 2018).

Either way the R^2 values do not indicate strong correlations and the data are prone to error due to the covariance of and vastness of parameters that cannot be controlled in the wild. To track all parameters in field studies is, in most cases, not feasible and this is one of the biggest drawbacks of field research.

A similar study was completed with 1111 birds from Chernobyl as well (Mousseau and Møller 2013). This study had a much smaller proportion of subjects with cataracts, 391 of 1111 (35.2%) and the illustrations of purported lens opacities in the publication were of variable quality. Measurements of radiation in the field were completed by the group completing the study and then cross validated with already reported measurements by the Ukraine Ministry of Emergencies. Note that the measurements of background radiation was made after the examination of the birds to prevent bias. Although ground measurements were used to determine the radiation exposure of the birds, the researchers defended this by stating that the majority of dose came from eating contaminated food on the ground. In spite of this, an accurate radiation dose received would be difficult to determine due to the dynamics of bird flight. The odds ratio per unit change was determined as 0.722 (0.648, 0.804). The article suggested a relationship between radiation and development of cataracts in the bird population but it should be noted that the plot referenced had an R^2 of only 0.08. An R^2 was not provided for the plot of mean cataract score and background radiation. The study's analysis section ends with a path analysis done by Mousseau and Møller where it is claimed that radiation accounts for 14% of the variation in cataracts incidence and severity.

The studies of both the birds and mammals have been disputed in large part due to concerns over dose measurements but they illustrate the difficulty of doing good field studies particularly in contaminated areas in foreign countries. Such studies are essential however due to the published data suggesting that in the field species may be much more sensitive to radiation than when measured in the laboratory. (Garnier-Laplace et al. 2013). The discrepancy is thought to be due to the multiple stresses in the field that reduce the fitness of species such as parasitism and disease, predation and starvation. But these 'confounders' are the reality and a huge challenge in the environment is isolating the component of harm, which can be attributed to the radiation stress. The other major challenge is determining dose accurately – there are almost as many confounders here as there are in the measurement of attributable effect! These are discussed in another paper by our group (Mothersill et al. 2017).

When seeking answers to the issues posed above, it is likely that cross validation of laboratory and field data is a valuable way forward. A notable study was completed by Dalke et al. (2018). looking at the effect of an acute-low dose exposure on mice over their lifetime, with a special focus on cataractogenesis. The mice were divided into four groups: one control, one group exposed to 0.063 Gy, one exposed to 0.125 Gy and one exposed to 0.5 Gy. An increase in cataract incidence was seen after 2 years in mice that were exposed to 0.5 Gy. The increase is smaller than what has been reported by the previous studies (Worgul et al. 2002, 2005; Hall et al. 2006; Kleiman et al. 2007). The increase in lens opacity over 24 months after a single acute exposure to 0.5 Gy was 1% where previous reports have observed cataract formation after a dose of this magnitude in four weeks. Earlier papers observed acute doses at an early life stage of just 0.33 Gy result in cataract formation in 20 months. A possible reason for this was the difference of cataract measuring methodologies. This study used Scheimpflug imaging to diagnose cataracts by directly measuring lens density and opacity while earlier studies relied on subjective scoring as discussed in Mastromonaco et al. (2018). The subjective scoring technique uses Merriam-and-Focht scoring, a four-point system which was developed for classifying cataracts caused by exposure to high-dose radiation (40–50 Gy). Vision-impairing cataracts have a score of 2 or more. The Scheimpflug system, unlike Merriam-and-Focht, allows for objective quantification of the opacities much easier because the images are free of flash reflections. This method has a near perfect correllation with the classification system used in clinical opthamology today, LOCS III. Studies (cited within the article by Dalke et al) have also found Scheimpflug imaging to be a better indicator of visual acuity since the lens density was found to be directly related to visual impairment. This new methoology has opened a door in this field of study and needs to be applied in field studies to resolve the controversies. Another possible reason for the lack of opacities in the Dalke study is the low dose rate (0.063 Gy/min; ^{60}Co) or the genetic background (F1 hybrids of male C3HeB/FeJ and C57BL/6J females). Genetic background as noted in the introduction is very important in determining low dose radiation exposure outcomes (e.g. see Mothersill et al. 1999, 2005). However another paper appearing in 2018 also looking at cataract formation in mice

(C57Bl6) and provides a more nuanced explanation. Barnard et al. (2018) found an inverse dose rate effect when looking at 53BP1 foci (a marker for DNA damage) in mouse LEC exposed to a range of dose rates of gamma radiation. However they did not find an inverse dose rate effect when looking at lymphocytes from the same animals. They suggest that DNA damage may be but one of many factors leading to radiation induced cataracts. This view is shared by Hamada (2017) who reviewed the dose rate effect literature for ocular lens and concluded that there was some evidence for an inverse dose rate effect after low dose exposure. This highlights the importance of considering targets other than DNA when studying very low doses and dose rates of any type of radiation.

Conclusion

Radiation cataractogenesis is arguably the most discussed field right now in health physics due to the new ICRP recommendations. In Canada, the equivalent dose limit for the lens of the eye is still 150 mSv (Government of Canada 2019). This is despite the fact that there is statistically significant risk associated with this dose, and it is only taking into account the dose after one year. As seen in nearly every cited study, radiation cataractogenesis is by no means a closed book or even one that is close to being finished. All the fields described in this paper require more research to either pursue applications or answer current questions. The field of biological mechanisms requires more research for several reasons, but an immediate application would be for radioprotectors to prevent cataracts in both radiation therapy patients and nuclear energy workers. In addition, research on cataract incidence rates in workers with occupational exposure should be continued in order to reduce the error in current studies. As far as field studies go, although the publications do make some conclusions, it is difficult to develop quantitative models based on the existing data. A recommendation suggested in this review is to have laboratory studies where subsets of field parameters are simulated. This will allow for more controlled measurements. There are drawbacks, however, with cost and limitations on species that can be accurately measured this way.

This review aimed to focus on less studied areas of cataract research which nevertheless are important. These include the controversies in the non-human biota studies and the limitations of laboratory animal studies when trying to assess cataract risk in human or wild populations exposed over long periods of time to multiple agents including radiation. Another aim was to integrate the non-targeted effects well documented in radiobiology into the discussion on cataract radiation protection. A further aim was to give a Canadian perspective to the general discussion.

Disclosure statement

No potential conflict of interest was reported by the author(s).

References

Ahmed EA, Rosemann M, Scherthan H. 2018. NHEJ contributes to the fast repair of radiation-induced DNA double-strand breaks at late Prophase I telomeres. Health Phys. 115(1):102–107.

Ainsbury EA, Barnard S, Bright S, Dalke C, Jarrin M, Kunze S, Tanner R, Dynlacht JR, Quinlan RA, Graw J, et al. 2016. Ionizing radiation induced cataracts: recent biological and mechanistic developments and perspectives for future research. Mutat Res. 770:238–261.

Babizhayev MA, Vishnyakova KS, Yegorov YE. 2011. Telomere-dependent senescent phenotype of lens epithelial cells as a biological marker of aging and cataractogenesis: the role of oxidative stress intensity and specific mechanism of phospholipid hydroperoxide toxicity in lens and aqueous. Fundam Clin Pharmacol. 25(2): 139–162.

Babizhayev MA, Yegorov YE. 2010. Telomere attrition in lens epithelial cells - a target for N-acetylcarnosine therapy. Front Biosci (Landmark Ed). 15:934–956.

Bai F, Xi JH, Wawrousek EF, Fleming TP, Andley UP. 2003. Hyperproliferation and p53 status of lens epithelial cells derived from alphaB-crystallin knockout mice. J Biol Chem. 278(38): 36876–36886.

Barnard SGR, Moquet J, Lloyd S, Ellender M, Ainsbury EA, Quinlan RA. 2018. Dotting the eyes: mouse strain dependency of the lens epithelium to low dose radiation-induced DNA damage. Int J Radiat Biol. 94(12):1116–1124.

Blakely E, Chang P, Lommel L, Bjornstad K, Dixon M, Tobias C, Kumar K, Blakely WF. 1989. Cell-cycle radiation response: role of intracellular factors. Advances in Space Research : The Official Journal of the Committee on Space Research (COSPAR). Adv Space Res. 9(10):177–186.

Clutton SM, Townsend KM, Walker C, Ansell JD, Wright EG. 1996. Radiation-induced genomic instability and persisting oxidative stress in primary bone marrow cultures. Carcinogenesis. 17(8):1633–1639.

Cohen B, Addadi Y, Sapoznik S, Meir G, Kalchenko V, Harmelin A, Ben-Dor S, Neeman M. 2009. Transcriptional regulation of vascular endothelial growth factor C by oxidative and thermal stress is mediated by lens epithelium-derived growth factor/p75. Neoplasia. 11(9): 921–933.

Cucinotta FA, Cacao E, Kim MY, Saganti PB. 2019. Non-targeted effects lead to a paridigm shift in risk assessment for a mission to the earth's moon or martian moon phobos. Radiat Prot Dosimetry. 183(1–2):213–218.

Dalke C, Neff F, Bains SK, Bright S, Lord D, Reitmeir P, Rößler U, Samaga D, Unger K, Braselmann H, et al. 2018. Lifetime study in mice after acute low-dose ionizing radiation: a multifactorial study with special focus on cataract risk. Radiat Environ Biophys. 57(2): 99–113.

Fernandez-Palomo C, Seymour C, Mothersill C. 2016. Inter-relationship between low-dose hyper-radiosensitivity and radiation-induced bystander effects in the human t98g glioma and the epithelial HaCaT cell line. Radiat Res. 185(2):124–133.

Frenzel M, Ricoul M, Benadjaoud MA, Bellamy M, Lenain A, Haddy N, Diallo I, Mateus C, de Vathaire F, Sabatier L. 2017. Retrospective cohort study and biobanking of patients treated for hemangioma in childhood – telomeres as biomarker of aging and radiation exposure. Int J Radiat Biol. 93(10):1040–1053.

Garnier-Laplace J, Geras'kin S, Della-Vedova C, Beaugelin-Seiller K, Hinton TG, Real A, Oudalova A. 2013. Are radiosensitivity data derived from natural field conditions consistent with data from controlled exposures? A case study of Chernobyl wildlife chronically exposed to low dose rates. J Environ Radioact. 121:12–21.

Government of Canada. 2019. Radiation Protection Regulations. Available from Justice Laws; [accessed 2020 May 14]. https://laws-lois.justice.gc.ca/PDF/SOR-2000-203.pdf.

Gow MD, Seymour CB, Byun SH, Mothersill CE. 2008. Effect of dose rate on the radiation-induced bystander response. Phys Med Biol. 53(1):119–132.

Hall EJ, Giaccia AJ. 2019. Radiobiology for the radiologist, 8th ed. Philadelphia: Wolter Kluwers.

Hall EJ, Worgul BV, Smilenov L, Elliston CD, Brenner DJ. 2006. The relative biological effectiveness of densely ionizing heavy-ion radiation for inducing ocular cataracts in wild type versus mice heterozygous for the ATM gene. Radiat Environ Biophys. 45(2):99–104.

Hamada N. 2017. Ionizing radiation sensitivity of the ocular lens and its dose rate dependence. Int J Radiat Biol. 93(10):1024–1034.

Hamada N, Fujimichi Y. 2015. Role of carcinogenesis related mechanisms in cataractogenesis and its implications for ionizing radiation cataractogenesis. Cancer Lett. 368(2):262–274.

Hurem S, Gomes T, Brede DA, Lindbo Hansen E, Mutoloki S, Fernandez C, Mothersill C, Salbu B, Kassaye YA, Olsen AK, et al. 2017. Parental gamma irradiation induces reprotoxic effects accompanied by genomic instability in zebrafish (Danio rerio) embryos. Environ Res. 159:564–578.

Jentzsch T, Pietsch CM, Stigler B, Ramseier LE, Seifert B, Werner C. 2015. The compliance with and knowledge about radiation protection in operating room personnel: a cross-sectional study with a questionnaire. Arch Orthop Trauma Surg. 135(9):1233–1240.

Karatasakis A, Brilakis HS, Danek BA, Karacsonyi J, Martinez-Parachini JR, Nguyen-Trong P-KJ, Alame AJ, Roesle MK, Rangan BV, Rosenfield K, et al. 2018. Radiation-associated lens changes in the cardiac catheterization laboratory: results from the IC-CATARACT (CATaracts Attributed to RAdiation in the CaTh lab) study . Catheter Cardiovasc Interv. 91(4):647–654.

Khan D, Lacasse M, Khan R, Murphy K. 2017. Radiation cataractogenesis: the progression of our understanding and its clinical consequences. J Vasc Interv Radiol. 28(3):412–419.

Khan SY, Hackett SF, Lee MC, Pourmand N, Talbot CC, Jr, Riazuddin SA. 2015. Transcriptome profiling of developing murine lens through RNA sequencing. Invest Ophthalmol Vis Sci. 56(8): 4919–4926.

Kleiman NJ, David J, Elliston CD, Hopkins KM, Smilenov LB, Brenner DJ, Worgul BV, Hall EJ, Lieberman HB. 2007. Mrad9 and Atm haploinsufficiency enhance spontaneous and X-ray-induced cataractogenesis in mice. Radiat Res. 168(5):567–573.

Kleiman NJ, Spector A. 1993. DNA single strand breaks in human lens epithelial cells from patients with cataract. Curr Eye Res. 12(5): 423–431.

Kocer I, Taysi S, Ertekin MV, Karslioglu I, Gepdiremen A, Sezen O, Serifoglu K. 2007. The effect of L-carnitine in the prevention of ionizing radiation-induced cataracts: a rat model. Graefes Arch Clin Exp Ophthalmol. 245(4):588–594.

Krueger SA, Wilson GD, Piasentin E, Joiner MC, Marples B. 2010. The effects of G2-phase enrichment and checkpoint abrogation on low-dose hyper-radiosensitivity. Int J Radiat Oncol Biol Phys. 77(5): 1509–1517.

Kubo E, Fatma N, Akagi Y, Beier DR, Singh SP, Singh DP. 2008. TAT-mediated PRDX6 protein transduction protects against eye lens epithelial cell death and delays lens opacity. Am J Physiol Cell Physiol. 294(3):C842–55.

Le M, Fernandez-Palomo C, McNeill FE, Seymour CB, Rainbow AJ, Mothersill CE. 2017. Exosomes are released by bystander cells exposed to radiation-induced biophoton signals: reconciling the mechanisms mediating the bystander effect. PLoS One. 12(3): e0173685

Le M, Mothersill CE, Seymour CB, Rainbow AJ, McNeill FE. 2017. An observed effect of p53 status on the bystander response to radiation-induced cellular photon emission. Radiat Res. 187(2):169–185.

Le M, McNeill FE, Seymour CB, Rusin A, Diamond K, Rainbow AJ, Murphy J, Mothersill CE. 2018. Modulation of oxidative phosphorylation (OXPHOS) by radiation- induced biophotons. Environ Res. 163:80–87.

Lehmann P, Boratyński Z, Mappes T, Mousseau TA, Møller AP. 2016. Fitness costs of increased cataract frequency and cumulative radiation dose in natural mammalian populations from Chernobyl. Sci Rep. 6:19974.

Lian Y, Xiao J, Ji X, Guan S, Ge H, Li F, Ning L, Liu J. 2015. Protracted low-dose radiation exposure and cataract in a cohort of Chinese industry radiographers. Occup Environ Med. 72(9):640–647.

Lipman RM, Tripathi BJ, Tripathi RC. 1988. Cataracts induced by microwave and ionizing radiation. Surv Ophthalmol. 33(3):200–210.

Little MP, Cahoon EK, Kitahara CM, Simon SL, Hamada N, Linet MS. 2020. Occupational radiation exposure and excess additive risk of cataract incidence in a cohort of US radiologic technologists. Occup Environ Med. 77(1):1–8.

Little MP, Kitahara CM, Cahoon EK, Bernier M, Velazquez-Kronen R, Doody MM, Borrego D, Miller JS, Alexander BH, Simon SL, et al. 2018. Occupational radiation exposure and risk of cataract incidence in a cohort of US radiologic technologists. Eur J Epidemiol. 33(12): 1179–1191.

Liu WB, Li Y, Zhang L, Chen HG, Sun S, Liu JP, Liu Y, Li DW. 2008. Differential expression of the catalytic subunits for PP-1 and PP-2A and the regulatory subunits for PP-2A in mouse eye. Mol Vis. 14: 762–773.

Liu Z, Mothersill CE, McNeill FE, Lyng FM, Byun SH, Seymour CB, Prestwich WV. 2006. A dose threshold for a medium transfer bystander effect for a human skin cell line. Radiat Res. 166(1 Pt 1): 19–23.

Löbrich M, Jeggo PA. 2007. The impact of a negligent G2/M checkpoint on genomic instability and cancer induction. Nat Rev Cancer. 7(11):861–869.

Lyng FM, Seymour CB, Mothersill C. 2001. Oxidative stress in cells exposed to low levels of ionizing radiation. Biochem Soc Trans. 29(Pt 2):350–353.

Maguire P, Mothersill C, McClean B, Seymour C, Lyng FM. 2007. Modulation of radiation responses by pre-exposure to irradiated cell conditioned medium. Radiat Res. 167(4):485–492.

Markiewicz E, Barnard S, Haines J, Coster M, van Geel O, Wu W, Richards S, Ainsbury E, Rothkamm K, Bouffler S, et al. 2015. Nonlinear ionizing radiation-induced changes in eye lens cell proliferation, cyclin D1 expression and lens shape. Open Biol. 5(4). DOI: 10.1098/rsob.150011

Mastromonaco C, Balazsi M, Zoroquiain P, Esposito E, Coblentz J, Logan P, Burnier MN. 2018. Removing subjective post-mortem grading from posterior capsular opacification: a new automated detector opacification software, ADOS. Curr Eye Res. 43(11): 1362–1368.

Matsubara K, Lertsuwunseri V, Srimahachota S, Krisanachinda A, Khambhiphant B, Tulvatana W, Sudchai W, Rehani M. 2017. Eye lens dosimetry and the study on radiation cataract in interventional cardiologists. Phys Med. 44:232–235.

Meade AD, Howe O, Unterreiner V, Sockalingum GD, Byrne HJ, Lyng FM. 2016. Vibrational spectroscopy in sensing radiobiological effects: analyses of targeted and non-targeted effects in human keratinocytes. Faraday Discuss. 187:213–234.

Michael R, Bron AJ. 2011. The ageing lens and cataract: a model of normal and pathological ageing. Phil Trans R Soc B. 366(1568): 1278–1292.

Mothersill C, Rusin A, Seymour C. 2017. Low doses and non-targeted effects in environmental radiation protection; where are we now and where should we go? Environ Res. 159:484–490.

Mothersill C, Seymour C. 2018. Old data-new concepts: integrating "indirect effects" into radiation protection. Health Phys. 115(1): 170–178.

Mothersill CE, O'Malley KJ, Murphy DM, Seymour CB, Lorimore SA, Wright EG. 1999. Identification and characterization of three subtypes of radiation response in normal human urothelial cultures exposed to ionizing radiation. Carcinogenesis. 20(12):2273–2278.

Mothersill C, Bristow RG, Harding SM, Smith RW, Mersov A, Seymour CB. 2011. A role for p53 in the response of bystander cells to receipt of medium borne signals from irradiated cells. Int J Radiat Biol. 87(11):1120–1125.

Mothersill C, Lyng F, Seymour C, Maguire P, Lorimore S, Wright E. 2005. Genetic factors influencing bystander signaling in murine bladder epithelium after low-dose irradiation in vivo. Radiat Res. 163(4):391–399.

Mousseau TA, Møller AP. 2013. Elevated frequency of cataracts in birds from chernobyl. PLoS One. 8(7):e66939.

National Aeronautics and Space Administration. 2018. Why Space Radiation Matters. Available from NASA; [accessed 2020 May 14]. https://www.nasa.gov/analogs/nsrl/why-space-radiation-matters.

O'Connor U, Gallagher A, Malone L, O'Reilly G. 2013. Occupational radiation dose to eyes from endoscopic retrograde cholangiopancreatography procedures in light of the revised eye lens dose limit from the International Commission on Radiological Protection. Br J Radiol. 86(1022):20120289.

Rafnsson V, Olafsdottir E, Hrafnkelsson J, Sasaki H, Arnarsson A, Jonasson F. 2005. Cosmic radiation increases the risk of nuclear cataract in airline pilots: a population-based case-control study. Arch Ophthalmol. (Chicago, IL: 1960). 123(8):1102–1105

Rastegar ZN, Eckart P, Manfred M. 2002. Radiation-induced cataract in astronauts and cosmonauts. Graefes Arch Clin Exp Ophthalmol. 240(7):543–547.

Ruotolo R, Grassi F, Percudani R, Rivetti C, Martorana D, Maraini G, Ottonello S. 2003. Gene expression profiling in human age-related nuclear cataract. Mol Vis. 9:538–548.

Short SC, Woodcock M, Marples B, Joiner MC. 2003. Effects of cell cycle phase on low-dose hyper-radiosensitivity. Int J Radiat Biol. 79(2):99–105.

Scalchi P, Magagna J, Montenegro GM, Beghetto M, Boi L, Dall'Acqua J, Francescon P, Savastano S. 2018. Eye lens dose monitoring in interventional radiology. Physica Medica. 52:83–84.

Schettino G, Folkard M, Michael BD, Prise KM. 2005. Low-dose binary behavior of bystander cell killing after microbeam irradiation of a single cell with focused c(k) x rays. Radiat Res. 163(3):332–336.

Schöllnberger H, Mebust MR, Crawford-Brown DJ, Eckl PM, Hofmann W. 2001. Significance of cell-cycle delay, multiple initiation pathways, misrepair and replication errors in a model of radiobiological effects. Int J Radiat Biol. 77(4):519–527.

Schöllnberger H, Scott BR, Hanson TE. 2001. Application of Bayesian inference to characterize risks associated with low doses of low-LET radiation. Bull Math Biol. 63(5):865–883.

Shuryak I. 2019. Enhancing low-dose risk assessment using mechanistic mathematical models of radiation effects. J Radiol Prot. 39(4): S1–S13.

Tauchi H, Waku H, Matsumoto E, Yara S, Okumura S, Iwata Y, Komatsu K, Furusawa Y, Eguchi-Kasai K, Tachibana A. 2009. Two major factors involved in the reverse dose-rate effect for somatic mutation induction are the cell cycle position and LET value. J Radiat Res. 50(5):441–448.

Taylor HR, West SK, Rosenthal FS, Muñoz B, Newland HS, Abbey H, Emmett EA. 1988. Effect of ultraviolet radiation on cataract formation. N Engl J Med. 319(22):1429–1433.

Taysi S, Abdulrahman ZK, Okumus S, Demir E, Demir T, Akan M, Saricicek E, Saricicek V, Aksoy A, Tarakcioglu M. 2015. The radioprotective effect of Nigella sativa on nitrosative stress in lens tissue in radiation-induced cataract in rat. Cutan Ocul Toxicol. 34(2): 101–106.

Tomita M, Maeda M. 2015. Mechanisms and biological importance of photon-induced bystander responses: do they have an impact on low-dose radiation responses. J Radiat Res. 56(2):205–219. Mar.

Truscott RJ. 2005. Age-related nuclear cataract-oxidation is the key. Exp Eye Res. 80(5):709–725.

Uwineza A, Kalligeraki AA, Hamada N, Jarrin M, Quinlan RA. 2019. Cataractogenic load – A concept to study the contribution of ionizing radiation to accelerated aging in the eye lens. Mutation research. 779:68–81.

van Rooijen B, de Haan MW, Das M, Arnoldussen C, de Graaf R, Backes WH, van Zwam WH, Jeukens C. 2014. Efficacy of Radiation Safety Glasses in Interventional Radiology. Cardiovasc Intervent Radiol. 37(5):1149–1155.

Vanhavere F, Carinou E, Domienik J, Donadille L, Ginjaume M, Gualdrini G, Koukorava C, Krim S, Nikodemova D, Ruiz-Lopez N, et al. 2011. Measurements of eye lens doses in interventional radiology and cardiology: final results of the ORAMED project. Radiation Measurements. 46(11):1243–1247.

Wilde G, Sjöstrand J. 1997. A clinical study of radiation cataract formation in adult life following gamma irradiation of the lens in early childhood. Br J Ophthalmol. 81(4):261–266.

Williams D. (2019, Apr 1). Do rodents in radioactive areas around Chernobyl Ukraine have high prevalence of cataracts? In BSAVA Congress Proceedings (pp. 458–458). BSAVA Library.

Worgul BV, Smilenov L, Brenner DJ, Junk A, Zhou W, Hall EJ. 2002. Atm heterozygous mice are more sensitive to radiation-induced cataracts than are their wild-type counterparts. Proc Natl Acad Sci USA. 99(15):9836–9839.

Worgul BV, Smilenov L, Brenner DJ, Vazquez M, Hall EJ. 2005. Mice heterozygous for the ATM gene are more sensitive to both X-ray and heavy ion exposure than are wildtypes. Adv Space Res. 35(2): 254–259.

Worgul BV, Smilenov L, Brenner DJ, Vazquez M, Hall EJ. 2005. Mice heterozygous for the ATM gene are more sensitive to both X-ray and heavy ion exposure than are wildtypes. Adv Space Res. 35(2): 254–259.

Worgul BV, Kundiyev YI, Sergiyenko NM, Chumak VV, Vitte PM, Medvedovsky C, Bakhanova EV, Junk AK, Kyrychenko OY, Musijachenko NV. et al. 2007. Cataracts among Chernobyl clean-up workers: implications regarding permissible eye exposures. Radiation Research. Radiat Res. 167(2):233–243.

Zigman S. 1983. The role of sunlight in human cataract formation. Surv Ophthalmol. 27(5):317–325.

Bio-acoustic signaling; exploring the potential of sound as a mediator of low-dose radiation and stress responses in the environment

Bruno F. E. Matarèse, Jigar Lad, Colin Seymour, Paul N. Schofield, and Carmel Mothersill

ABSTRACT

Objectives: This commentary reviews and evaluates the role of sound signals as part of the infosome of cells and organisms. Emission and receipt of sound has recently been identified as a potentially important universal signaling mechanism invoked when organisms are stressed. Recent evidence from plants, animals and microbes suggests that it could be a stimulus for specific or general molecular cellular stress responses in different contexts, and for triggering population level responses. This paper reviews the current status of the field with particular reference to the potential role of sound signaling as an immediate/early bystander effector (RIBE) during radiation-induced stress.

Conclusions: While the chemical effectors involved in intercellular and inter-organismal signaling have been the subject of intense study in the field of Chemical Ecology, less appears to be known about physical signals in general and sound signals in particular. From this review we conclude that these signals are ubiquitous in each kingdom and behave very like physical bystander signals leading to regulation of metabolic pathways and gene expression patterns involved in adaptation, synchronization of population responses, and repair or defence against damage. We propose the hypothesis that acoustic energy released on interaction of biota with electromagnetic radiation may represent a signal released by irradiated cells leading to, or complementing, or interacting with, other responses, such as endosome release, responsible for signal relay within the unirradiated individuals in the targeted population.

Introduction

A variety of mechanisms have evolved throughout the animal, microbe and plant kingdoms by which communication can be mediated. At the level of whole organisms the development of special sensing mechanisms, smell, taste, touch, sight and hearing are all involved in inter-organismal communication as well as in sensing the outside world, and have evolved to mediate the interaction of organisms from bacteria to humans with each other and their environment, including individuals of their own and different species (Stevens 2013).

Of particular interest is the evolution of signaling between individuals in response to threat or distress which may be due to predators or physiological stress. Such communication largely involves the interpretation of the environment and relay to other individuals. It can be intra- or interspecific (Goodale et al. 2010) but is generally advantageous to a species, its kin, or local guilds. For example, chemical cues from urine affect behavior both within and between species, (Apfelbach et al. 2015), and volatile compounds such as jasmonic acid (Farmer and Ryan 1990; Heil and Karban 2010; Karban et al. 2010) are widely used by plants to communicate threats both between and within species. Such communication may involve multiple mediators, singly or in ensemble. For example, a recent study elegantly describes the visual and olfactory modes of communication between fruit flies as in the acute response of fruit flies to the presence of predatory wasps (Kacsoh et al. 2018). Other types of cues are widely used and for example sound is extensively used in animal species and, as in the case of the fruit fly example above may be species discriminatory and information rich. The examples of acoustic communication between cetaceans, bats and higher mammals are well known, but the generation of and response to sonic vibratory signals of plants and prokaryotes is only recently becoming understood, with some remarkable findings such as the ability of plants to respond metabolically and with changes in gene expression to specific sound stimuli (Md. Emran Khan et al. 2014; Ghosh et al. 2019), frequency-specific growth-enhancement in, for example, *Phaseolus spp.* (Collins and Foreman 2001) and defensive action initiated by plants whose leaves are being chewed by caterpillars (Appel and Cocroft 2014).

At the level of cells and tissues, Intercellular communication relies on the release, detection and transduction of physical or molecular signals which can range from electromagnetic (EM) radiation of various wavelengths (Le, McNeill, et al. 2015; Le et al. 2018), small molecules such as hydroxyl radicals (Schieber and Chandel 2014), peptide growth factors and extracellular vesicles (Margolis and Sadovsky 2019) to complex extracellular matrix interactions (Hastings et al. 2019). While such cues are essential for morphogenesis and tissue homeostasis, we are particularly interested here in the cues regulating cell proliferation and death in response to extrinsic and intrinsic damage. Cell death or senescence are classically induced in response to radiation damage (Sia et al. 2020). This may be through p53-associated apoptosis or other mechanisms which are p53 independent, or related to necrosis, such as necroptosis, seen in some systems (Kuwahara et al. 2018). Evidence has recently been accumulating from non-radiation studies that suggests there are significant community effects following apoptosis induction (Eroglu and Derry 2016). Apoptotic cells release what are often described as 'find me' and 'eat me' signals for macrophage clearance, but it is becoming apparent that neighboring cells can be induced to undergo apoptosis in a non-autonomous fashion (Pérez-Garijo and Steller 2015). In some cases this due to effects mediated by macrophages (Lemke 2019) but in other cases by direct effects of factors released by dying cells such as TNF-related compounds (Pérez-Garijo et al. 2013) or the mechanical effects of crowding (Eisenhoffer et al. 2012). This form of assisted suicide can have profound non-cell-autonomous effects on surrounding tissues by affecting the cell division, cell fate and tissue remodeling, involving mechanical forces exerted on the cytoskeleton as well as cell replacement and proliferation (Gudipaty et al. 2018). It has also become clear recently that elements of metastatic behavior can be transmitted from cell to cell via extracellular vesicles (Zomer et al. 2015).

Developments in our understanding of the effects of low-dose radiation on cells and whole organisms in recent years have raised some puzzling questions concerning the evident transmission of cellular phenotypes from one cell or organism to other non-irradiated organisms or cells, the radiation induced bystander effect (RIBE) (Mothersill and Seymour 2012). To date various mediators have been suggested, amongst them exosomes (Al-Mayah et al. 2012; Jella et al. 2014; Le, McNeill, et al. 2017; Du et al. 2020) and UV photons (Le, Fernandez-Palomo, et al. 2015; Mothersill et al. 2019). However, these mechanisms still fail to explain all the reported phenomena (Mothersill et al. 2006; Mothersill and Seymour 2009; Mothersill et al. 2012). While our accumulated knowledge of communication between animals, plants and cells at multiple levels implicates the likely interaction of multiple systems, we still have no idea what other forms of communication might be responsible for the transfer of the 'irradiated phenotype'.

The observation that the 'irradiated state' can be passed from one whole organism to another raises the question of how such a transfer might occur between individuals in the same or different species in a contaminated environment and how such transfer of information might affect the response of the ecosystem to contamination as well as how the genetics of the response mechanism might be subject to selective pressures over time, as we see happening with the response to anthropogenic sound in many species (Harding et al. 2019).

In this commentary we present an integrated view of sound as a means of stress communication and to highlight areas of biology where the contribution of acoustic mechanisms is still unknown or understanding in its infancy. We explore the possibility that passively-generated acoustic signals may act as direct or indirect mediators of transfer of the 'irradiated state' at multiple levels of organization, to stimulate discussion of what is emerging as a novel and potentially important mode of communication between cells and organisms.

Sound and electromagnetic waves as communication mechanisms

Sound waves are a longitudinal mechanical harmonic oscillation of vibrational energy or pressure waves inside a molecular lattice medium (Rayleigh 1877). The density of this medium fluctuates by compressions and rarefactions (migrates into low and high pressure regions) in such a way that a mechanical sensor, such as the ear, can detect them. They exert a physical pressure at the wavefront, which along with amplitude and frequency constitute the characteristics of acoustic waves.

The properties of electromagnetic waves differ fundamentally from acoustic mechanical waves due to the fact that they do not require a medium through which to travel. They are transverse oscillations of electric fields and magnetic fields, traveling with the speed of light in vacuum and electromagnetic waves and sound waves differ fundamentally in their spatial shape and mode of propagation. Sound and EM waves represent complementary mechanisms for information transfer between entities that are acoustically and electromagnetically active (e.g. Preisig 2007).

While this description of acoustic phenomena may be familiar, the related vibrational concept of the phonon, which is relevant here to the generation of acoustic force from the ionizing radiation induced interaction with matter, is not. A phonon is a discrete unit or quantum of vibrational mechanical energy, therefore having both wave and particle duality, which arises from oscillation of a collection of atoms or molecules. It may be considered as a Nambu-Goldstone Boson with broken symmetry consequent on translation across a long range structure (Leutwyler 1997) but may be considered naively as a 'particle of heat'.

Atoms behave as if they are connected by tiny springs which establish the shape of molecules and components of tissue structure. The atoms of the lattice vibrate with their own thermal energy and in response to applied forces/energy. The generation of a packet of mechanical waves can be defined today as 'bio-phonon' in the context of biological systems, and which can travel and carry heat and sound through biomaterials with a discrete energy and momentum. The more dense the biological medium the faster bio-phonons travel through it; the more dense a medium the faster sound waves travel through it. A recent report suggests that phonons may traverse a vacuum at the nanoscale through quantum fluctuation, but the implications for the role of acoustic signals discussed here are unclear (Fong et al. 2019). We return to the interaction between electromagnetic and acoustic waves below.

Communication through acoustic signals requires an actuator process and a receiver, an environment permissive to the propagation of an acoustic signal, and some kind of discrimination/filtration and transcription among signals by receiver. Such signals can travel long distances in a relatively short period of time (Bradbury and Vehrencamp 2011), are independent of time of day and availability of light, are rapidly transmitted at the cellular level, as they do not rely on the speed of diffusion of small molecules like paracrine growth factor and are not precluded by lack of lines of sight. Both energy and information can be transmitted due to the capacity of acoustic signals to undergo modulation of frequency, amplitude and intensity. They can modulate direction depending on temporal and spatial coherence of mode of generation and propagation, through transmittance, reflection and refraction dependent upon the dielectric and sound wave absorption coefficient and attenuation coefficient properties of each molecular constituent of the environmental milieu.

The limitation of acoustic communication is directly influenced by the composition of environment. The environment is a heterogeneous medium composed of different matter with specific acoustic properties (Lorén et al. 2009). Sound waves represent a physical energy field successfully transferred from one point to another via diverse communication mechanisms exploited by a biological entity as part of adaptation to a specific complex environment. Such environments, composed of heterogeneous and irregularly distributed media, make inter-organism or intercellular biological communicative system very complex (Gillooly and Ophir 2010).

Interactions with environmental media

Sound waves are intrinsically short wavelength and have a frequency-dependent attenuation with increased scattering of higher frequencies (D'Astous and Foster 1986). This is medium-dependent and high frequencies of sound are absorbed/attenuated rapidly in dry air compared to those of lower frequencies (Sharan Verma 1950; Sakai et al. 1990; Bond et al. 1992). However, in humid air, water droplets massively reduce attenuation/absorbance at low frequency but increase attenuation and thermal loss at high frequency as the speed of sound propagation $\sim$340 m/sec at sea level in air and five times more in far more dense medium such as water (1480 m/s) and can reach several km/s in metals (Ludwig 1950; Shyu and Gaspari 1969; Del Grosso and Mader 1972). Electromagnetic waves by comparison travel a million times faster than sound waves ($\sim 3 \times 10^8$ m/sec). In contrast to sound waves that travel faster in a denser environment, light waves travel slower in a denser environment (Bludman and Ruderman 1968). These characteristics determine their differential utility in different environments and for different purposes.

Sound and electromagnetic waves are different forms of energy and can be interchanged/transduced from one to another in order to effect an efficient way of communication through different media. This exchange of forms of energy can be defined today as biological electromagnetic acoustic transducers (bio-EMATs) for biological systems. Both, separately or combined, are potentially important as modes of information transmission, and reward investigation in order to understand the mechanism of biological communication and environmental stress (Gagliano et al. 2012b). While electromagnetic waves such as light photons are successful carriers of information for long distance communication in air, in this review we mainly focus on sound waves to reflect their efficient transmission in water or electrolyte solutions such as sea water or tissue fluid (which are poor transmitters of EM waves) and the fact that biological systems are mainly composed of water and dense biomaterial (Woodard and White 1986).

Acoustic signals interact with the environment in which they are generated and, depending on its complexity and the physical relationship between signal, source and receipt offer a rich mode of communication at all ranges of scale from single cell to single cell, to the macroscopic level of whole organisms. There is considerable literature on the acoustics of metazoan vocalization and other forms of biophonic communication in the environment. While aquatic environments offer an ideal medium for low frequency biophony the relationship between distance, frequency and energy is much more complex in a terrestrial macroenvironment and for high frequency acoustics signals. What has not so far been considered is the acoustic equivalent of paracrine communication, acting over short distances between cells in close physical proximity, such as those in tissues, where signal attenuation will be minimal and it is possible to consider phonon mediated transmission through contiguous cellular structures, such as the membrane or gap junctions.

Electromagnetic induction of acoustic signals (EMA)

There is an intrinsic relationship between electromagnetic forces (mediated by photons) and acoustic vibration (mediated by phonons), experimentally established in the 19th century (Bell 1881). Photons are responsible for all the interactions between atoms that produce the compression or shear forces that generate sound waves. It is not therefore surprising to see this interaction reflected in the ability of ionizing radiation induce an acoustic signal passing through water (Sulak et al. 1979), and in 1983 with the demonstration that synchrotron X-ray photons could generate an acoustic response from tissues (Kim and Sachse 1983) later giving rise to the strategy for probing the deep structure of materials, such as tissues, with X-ray-induced acoustic emissions (Bowen et al. 1991).

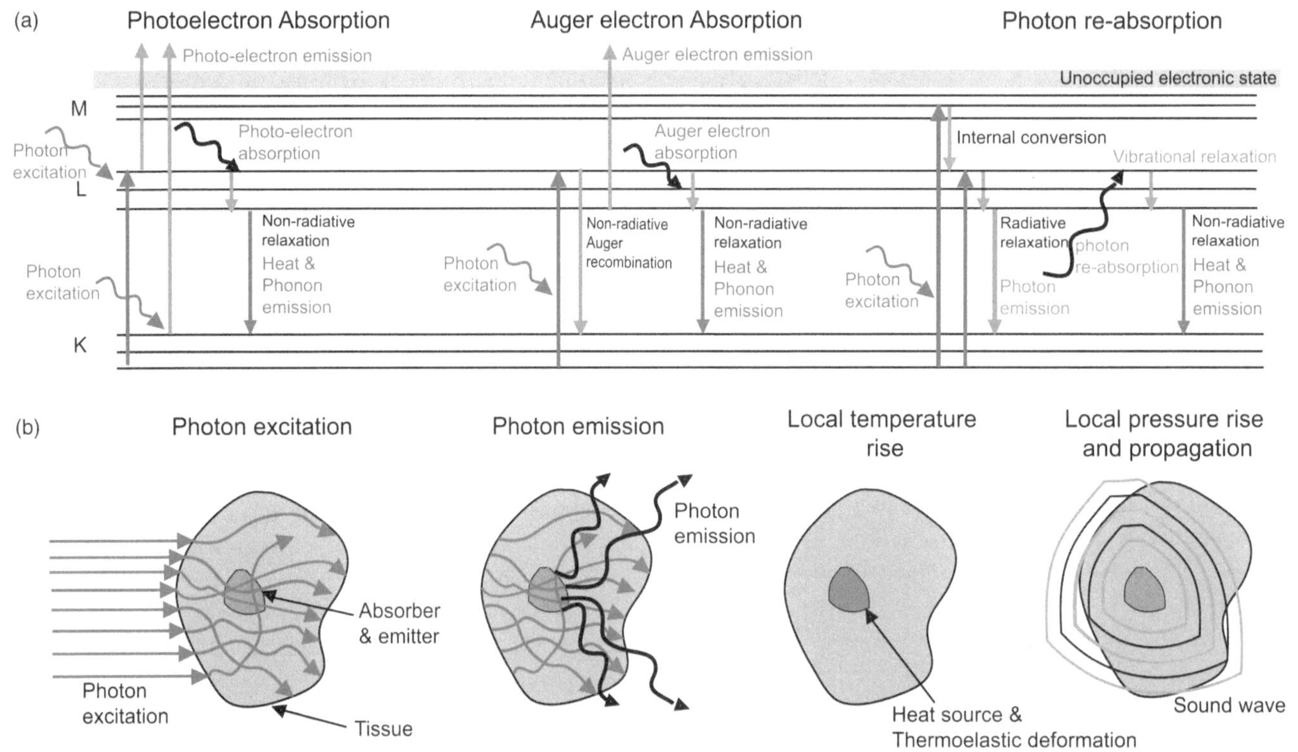

Figure 1. Mechanisms by which electromagnetic radiation induce the emission of sound signals and phonons. (a) Diagram of Photoelectron absorption, Auger electron absorption and Photon re-absorption and (b) Schematic of a biological tissue receiving EM photon excitation, the photo-luminescence from a biological absorber/emitter, the local temperature rise and the sound wave generation associated from thermo-elastic deformation.

The mechanism by which ionizing and non-ionizing electromagnetic radiation induces an acoustic wave from matter originates through the interaction of photons and electrons (see Figure 1). On exposure to ionized photons, inner-shell electrons of the biological molecules are excited, generating photoelectrons, Auger electrons, or electromagnetic radiation (e.g. photoluminescence) - which decay producing cascades of secondary electrons as these decay processes transfer kinetic energy to surrounding atoms to reach thermal equilibrium. These processes generate thermal energy into the biological structure as fluorescence reabsorption, Auger electron absorption, and photoelectron absorption (Garcia et al. 1988; Nie et al. 2008).

Ionizing photons themselves may also interact with outer shell orbital electrons and undergo Compton scattering, ejecting the orbital electron with kinetic energy net of the incident and scattered photons. Compton processes play a key role in ionized energy deposition into the material structures (Tong et al. 1995).

The resulting electron-phonon interactions in the surrounding atoms created by these processes lead to a localized increase of temperature in the irradiated matter and resulting a transient thermoelastic expansion of the biological structure generating pressure waves which constitute X-ray induced acoustic signals (Garcia et al. 1988). These are directionless and propagated in three dimensions and form, for example, the basis of X-ray acoustic imaging investigated for the recent past decades. (Bowen et al. 1991; Hickling et al. 2014). Exploitation of this process is under development for particularly dosimetry and radiation monitoring during radiotherapy (Liangzhong et al. 2013; Hickling et al. 2016). While resolution of whole tissue imaging by this method is poor, work is now being undertaken to examine the possibility for use in X-ray acoustic computed tomography (XACT) for medical diagnosis (Liangzhong et al. 2013; Xiang et al. 2013, 2014; Samant et al. 2020) allowing low-dose, real-time, three dimensional imaging requiring only single site access by an electro-acoustic probe.

The signal amplitude of EMA is directly proportional to electromagnetic absorber density and the deposited dose of energy. This thermal elastic wave expansion process is widely exploited in the optical frequency range to induce photoacoustic (PA) signals generated from the visible absorbers (e.g. myoglobin, hemoglobin, melanin, cytochrome, and DNA/RNA) or near-infrared absorber (e.g. glucose, lipid Cao et al. 2018; Chi et al. 2013; Christison and MacKenzie 1993; Danielli et al. 2015; Da-Kang et al. 2020; Longo et al. 2017). The EMA signal requires only a short pulse of light excitation for efficient signal generation. Furthermore, an EMA wave propagates in 3 dimensions regardless of the angle or the geometry of the electromagnetic excitation beam. From these principles we can adduce that the interaction of electromagnetic radiation of any kind (e.g. RF radiation, thermal radiation, or optical radiation), and that of energy high enough to be ionizing, will induce an acoustic signal on interaction with matter, of a greater or lesser energy and wavelength depending on the linear energy transferred, the molecular nature of the target and the medium through which the acoustic wave is propagated, together with temperature and the acoustic impedance. All organisms are therefore expected to elicit a bio-acoustic signal of some type on irradiation as a consequence of their physicochemical makeup.

Effects of ionizing radiation on non-biotic sound emission

Before discussing the sound generation and response of biota to ionizing radiation it is worth noting that sources of ionizing radiation otherwise contained, such as in nuclear power plants or in radiaoactive waste repositories are expected to emit an acoustic signal. Radioactive heat sources produce a unique spatial and temporal thermoelastic stress in which an acoustic wave expands and will affect surrounding elastic media (such as soil, sand, rocks and minerals) generating acoustic pressure waves (APW) which may interact with the surrounding biota. Such radiation induced sonic emissions are in widespread use for noninvasive structural testing (Tang et al. 2018) and may be a way of remotely monitoring radioactivity and subsequent heat generation within containment structures.

Acoustic emissions may have pleiotropic effects on affected cells through several mechanisms, one of which is the generation of reactive oxygen species which accompanies thermoelastic expansion. The generation of ROS in this way is well characterized and here is evidence for free radicals produced in aqueous and nonaqueous solutions induced by ultrasound (Riesz et al. 1985; Carmichael et al. 1986; Christman et al. 1987). Recent experiments link the generation of intracellular ROS by low intensity ultrasound to cell killing in a number of cell lines (e.g. Xia et al. 2020).

Acoustic signals in the production of and response to stress in the environment

Acoustic signals in the environment may be divided into two main categories. Firstly, biogenic signals can be generated as active signals or passively through activities such as movement and feeding, *biophony*, including those present in the ambient environment such as those resulting from human activity, *anthropophony*. Secondly from the natural abiotic environment, *geophony*. These comprise the components of the acoustic ecology of an environment (Pijanowski et al. 2011). Active or passive biophony may either inadvertently elicit a stress response or be calculated to do so as a form of information transfer.

Metazoan animals

The phenomena of active vocalisations or mechanically generated signals, such as tail slapping or foot thumping is out of scope for this commentary. However, we do note the detrimental and stressing effect of anthropogenic sound on non-human ecosystems. We do not know yet how such acoustic contamination affects multiscale systems below those mediated through a cognitive response.

Anthropophonic contamination effects have become increasingly prominent in the 20th and 21st centuries and result in numerous consequences, some of which are not well appreciated. For example, the overfishing of trout primarily known to cause population reduction, environmental pollution (caused by fishing boats), and has overall effects on the food chain. However, elevated fishing activities have also led to rising, unwarranted levels, of underwater noise. In the presence of boat noise, tuna schools were observed to be less coordinated, resulting in fish straying off by either going to the surface or the bottom of the ocean floor to avoid the disruption (Slabbekoorn et al. 2010b). The effects on whale populations are also well known since anthropogenic noise interferes with the highly complex 'songs' of whales which can travel trans-oceanic distances (Erbe et al. 2019). Studies performed on captive fish, which were subject to anthropogenic noise, had increased levels of heart rate, cortisol secretion and muscle metabolism. Other effects include potential lower egg viability and reduced larval growth rates (Slabbekoorn et al. 2010b). Over the years, since the trout overfishing events, authorities have enforced stricter laws for fishermen and fishing companies to abide by, in order to protect all fish species. One of which includes, placing a ban on fishing during mating seasons. The increased underwater noise levels in mating areas can skew the information that lies within a males' mating call. When combined with potential harmful effects on reproduction, the presence of fishing boats during mating seasons can greatly affect the ability of these fish to procreate. Similar consequences have also been observed in other species (i.e. birds and frogs) that are subject to alternate forms of anthropogenic noise (Fay and Popper 2000; Kaiser and Hammers 2009; Slabbekoorn et al. 2010a, Sun and Narins 2005).

Effects at a cellular level

Molecular electromechanical coupling is the basis of sound perception in a wide range of metazoan species. The coupling between mechanical stimulus generated by acoustic waves and cellular response provides a paradigm for the receipt of a passively or actively generated acoustic signal and its effects on cellular metabolism. For example when outer hair cells (OHC) of guinea pigs when subject to ultrasonic waves, Fridberger et al. (1998), found that overstimulation of the outer hair cells led to increased concentrations of calcium influx (Fridberger and Ulfendahl, 1996; Fridberger et al. 1998). Sustained levels of calcium ions can be detrimental to the targeted cell and the surrounding cells, as this can lead to apoptosis (Sen 1992; Duchen 2000). It is interesting to note that calcium influx is the first observable event in cells receiving a radiation-induced bystander signal (Lyng et al. 2000) and has been observed to mediate cell death in response to a variety of stressors including low doses of ionizing radiation (Stevenson et al. 1987). The proposed mechanism of conversion from mechanical to chemical energy is that the rising calcium levels activate proteinases that rearrange the actin cytoskeleton. Subsequent to this, rounding of the cell or cytoplasmic blebs occur (Fridberger et al. 1998; Olwell et al. 2005). Similar to gas bubbles from cavitation, cytoplasmic blebs are bulges of the plasma membrane which arise from changes in the cytoplasmic pressure due to contractions produced by actin, ultimately leading to apoptosis (Fridberger et al. 1998).

Another consequence of higher calcium levels, leading to apoptosis, is the activation of endonucleases which cleave DNA, producing multiple fragments (Fridberger et al. 1998; Szydlowska and Tymianski 2010). Overstimulation can lead to degeneration of the outer hair cells, due to loss of tip links connecting neighboring stereocilia (Fridberger et al. 1998; White et al. 2020). With outer hair cells relying on calcium for electromechanical transduction, the associated ion channels serve as its primary source. However, with acoustical damages, calcium cannot enter through these ion channels and instead enters through its basal membrane (Fridberger et al. 1998; Hudspeth 1989). With a 50-fold increase in endolymph calcium concentrations, this huge influx as result of overstimulation, can severely reduce functioning of outer hair cells and lead to overall hearing loss (Fridberger and Ulfendahl, 1996; Fridberger et al. 1998). Much of the calcium post-treatment were found to coalesce in areas of dense mitochondria concentrations, again paralleling the response to bystander signaling (Hei et al. 2008), in the nuclei of cellular debris and in the cytoplasm of the hair cells. Mitochondria also play a major role in the death of outer hair cells subject to overstimulation, via activation of mitochondria-mediated cell death pathways (Vicente-Torres and Schacht 2006). It begins with the activation of calcineurin, a calcium-dependent phosphatase, leading to the activation of Bcl-2-associated death promoter (BAD) and its translocation to the mitochondria (Vicente-Torres and Schacht 2006). As is known, *Bcl-2* and *Bax*, are apoptotic-regulatory genes that when activated (through dephosphorylation), increase the permeability of the mitochondrial membrane releasing cytochrome C and endonuclease G, both of which are proapoptotic compounds (Vicente-Torres and Schacht 2006). It was observed that overstimulation causes mitochondria to localize; however, certain studies also suggest that this may occur near the nucleus (Vicente-Torres and Schacht 2006). These compounds would then be able to easily bind to the DNA inducing cell death. Similar studies performed by Ashush et al. (2000), on human myeloid leukemia cells (HL-60, K562, U397 and M1/2) were testing for the pathways involved with apoptotic induction when subject to therapeutic ultrasound of 750 kHz (Ashush et al. 2000). The results were similar to those cells subjected to γ-irradiation (Mothersill et al. 1999, 2001), suggesting a potential role for ultrasound waves in facilitating two of the most common types of low-dose radiobiological responses: the bystander effect and the radioadaptive response (Bonner 2003).

As well as pathological calcium influx the generation of reactive oxygen species by acoustic energy has also been implicated in cell death (Feng et al. 2010). It has been recently established that at least in the OHC of the vertebrate ear acoustic trauma gives rise to redox imbalance, alteration of distribution of NADPH within the cell, lipid peroxidation and changes in membrane structure accompanied by ROS generation (Maulucci et al. 2014) with the implication that the resulting mitochondrial stress might engage apoptotic signaling in exposed cells (Raimundo et al. 2012). Such cellular generated ROS are in addition to the ROS known to be generated in aqueous media by acoustic energy itself (Riesz et al. 1985; Christman et al. 1987).

Bio-acoustic signaling in vascular plants and microbes

While the reception and transmission of sound signals are usually associated with specialized sound sensing organs in metazoan animals, various studies have shown the existence of acoustic emissions and responses in plants, invertebrates and bacteria.

Vascular plants

There is considerable evidence for bio-acoustic emissions from vascular plants, and increasing indications that plants are responsive to external acoustic stimuli (Gagliano et al. 2012a; Khait et al. 2019). In general these are associated with stress responses and there are several mechanisms postulated for their generation. Much less well developed are models for the sensing of acoustic signals, and evidence for these and potential signal transduction mechanisms is discussed below. It is important to stress that such mechanisms are not 'voluntary' emissions by the plant but a consequence of the physicochemical and anatomical properties of their constitution.

Drought stress and predation constitute two of the major physiological stressors of vascular plants. Under drought stress some plants produce measurable bio-acoustic emissions (De Roo et al. 2016). The mechanisms for generating these signals are not fully understood but may involve the effects of decreasing hydrostatic pressure in xylem, leading to the production of ultrasonic sound emissions variously measured as >20 kHz (Tyree and Dixon 1983) and from 10 to 300 kHz (Laschimke et al. 2006). With rapidly decreasing pressure in the xylem, collapse of bubbles caused by cavitation has been suggested as one mechanism for the generation of sound, but an alternative hypothesis derived to explain the 'violent acoustic activity' detected in *Ulmus sp.* in response to drought stress, is release of energy from the xylem-adherent bubble system that normally contributes to water flow (Laschimke et al. 2006; Zweifel and Zeugin 2008; Gagliano et al. 2012a, 2012b; Gagliano 2013). Respiration and metabolic growth activity of the cambium is another method suggested to be involved (Zweifel and Zeugin 2008). The cambium is the portion between the xylem and phloem where cells are rapidly dividing and is responsible for secondary growth of stems and roots (Zweifel and Zeugin 2008; Schöner et al. 2016). At nighttime when the plant is subject to drought stress, the cambium has increased turgor pressure due to increased respiration. This increased pressure causes greater levels of carbon dioxide to enter the xylem, resulting in more gas bubbles and subsequent acoustic emissions. In the absence of drought stress and consequent xylem cavitation, young corn roots are able to produce clicking sounds under water – the reason for retaining or developing this mechanism is unknown (Schöner et al. 2016). It is apparent that a variety of plant species have developed mechanisms for sound production.

Acoustic responsiveness

Our understanding of the mechanisms for sensing and transducing acoustic signals is in its infancy and currently there are no mechanisms elucidated (ten Cate 2013). However there is increasing evidence for a physiological response to external signals (Collins and Foreman 2001; Khait et al. 2019; Veits et al. 2019), and some molecular correlates which implies that there must be a mechanism at work.

There is evidence that acoustic signaling may modify growth and maturation of plants and fruits. Jung et al produced a review of this field in 2018 (Jung et al. 2018), showing that in a wide range of plants, physical triggers could modulate physiological traits such as ripening and could influence downstream signaling pathways. They concluded that sound wave treatment could be a new way to improve fitness and protect against unfavorable conditions. The 'organ' responding to sound has not been identified in plants but may be a systemic response to vibrational waves through liquids in the plant as discussed above. A table in Jung et al. (2018) lists the results of studies on thirteen different plants and fruits, where effects range from yield changes and delayed ripening in tomatoes (Hou et al. 2009; Hassanien et al. 2014; Kim et al. 2018) to effects on photosynthesis (Kwon et al. 2012; Hassanien et al. 2014). Kim et al. (2018) later linked the tomato delayed ripening effects to regulation of both coding and non-coding RNAs and transcription factor genes. An earlier study performed by Kim et al. (2015), found that tomatoes that were exposed to 1 kHz sound waves for 6 hours displayed delayed onset of ripening due to reduced ethylene levels (hormone responsible for early development and senescence in fruits), and reduced expression of the associated genes. These genes include the ethylene biosynthesis-related genes (*LeACS2, LeACS4 and Le ACO1*) and the ethylene-inducible genes (*E4* and *E8*). Reduced expression was not only observed in genes responsible for ethylene production but also in those regulating the on-set of ripening: *RIN, LeHB-1, CNR, NOR* and *TAGL1*. The potential adaptive effects and ecological relevance of phytoacoustics is an emerging field which could be harnessed to manipulate agriculturally relevant features such as synchronized or staggered ripening of fruits (Khait et al. 2019).

Additional molecular evidence is emerging that plants have stress or signal-activated genes that enable them to process sound. Jeong and coworkers (Jeong et al. 2008), found that transgenic rice plants up-regulated the expression of the *ald* gene, when subject to frequencies of 125 and 250 Hz. Similarly, when an *ald* promoter was administered with a GUS reporter, GUS expressions also increased at 250 Hz. This suggests that the *ald* gene is associated with (transgenic rice plants in this case) processing of specific acoustic frequencies for gene regulation instead of light, to which it is also responsive. Other mechanisms include phytohormone production and germination and growth, through which plants are able to respond to sound waves (Schöner et al. 2016). Remarkably, recent studies link an adaptive response in *Arabidopsis* to acoustic signals (Ghosh et al. 2019; Bhandawat et al. 2020), the former suggesting some form of learning response, thus linking together signal perception, transduction, changes in gene expression and a physiological response with an advantageous adaptation to stress.

The details of a plant acoustic perception apparatus are as yet unknown but it has been suggested they may similar to the way outer hair cells function in mammals, by altering membrane potentials of subcellular structures. Altering cell membrane and cell wall potentials have been shown to produce acoustic waves from kHz to THz range (Gagliano et al. 2012b), myosin is also believed to be involved as it can produce mechanical vibrations within cells by sliding against actin filaments (Gagliano 2013). These vibrations can then propagate through cytoplasm and create a vibrational cascade with surrounding cells causing cytoplasmic streaming. A process known as coherent excitation, where multiple cells work collectively could also produce sound signals in frequencies between 150 and 200 kHz (Gagliano 2013). We discuss hypotheses for potential mechanisms below. At the range of frequencies most often described it is likely that any plant to plant signal transmission will occur at a very short distance given air movement and acoustic attenuation but in a natural context it is not possible to rule out transmission underground or at the interface of substrate and air via soil water and mass where signals might be expected to travel much further.

Microbes

Bacteria have highly developed acoustic communication. Matsuhashi et al. (Matsuhashi et al. 1998), found that certain bacterial cells, such as *Bacillus subtilis* and *B. carboniphilus*, process sound waves as a form of growth-regulation when the two populations are in close proximity (Matsuhashi et al. 1998). *B. carboniphilus* under high-temperature and potassium-chloride stress becomes extremely sensitive to growth-promoting acoustic signals (Matsuhashi et al. 1998) leading to a regulation of growth rate which is highly controlled by environmental factors. The acoustic signals from Bacillus subtilis were able to cross petri dishes, even when there was a 2 mm iron barrier, leading to the production of more spores and greater overall growth (Matsuhashi et al. 1998). *B. subtilis* were found to emit sound waves of frequencies between 8 and 43 kHz, while *B. carboniphilus* were seen to form colonies when subjected to frequencies between 6 and 38 kHz. The overlapping frequency ranges suggests communicative purposes for the acoustic emissions. Similarly, increase in colony counts was also observed in E. coli, when subject to frequencies of 1, 5, and 15 kHz (Lee et al. 2009).

As previously mentioned adverse effects of cell membrane thinning, localized heat production and free radical formation, along with bacterial inactivation and deagglomeration of bacterial clusters, all arise from acoustical cavitation (Joyce et al. 2003). Shear forces from ultrasound waves produce a pressure gradient, subsequently, creating low-pressure gas bubbles (Joyce et al. 2003). When the surrounding pressure becomes too high, these gas bubbles burst, sending a shockwave damaging the bacterial membrane and cell wall (Joyce et al. 2003). Consequently, giving rise to either a positive (increased resistance) or negative (cell death) signaling cascade, as a result of calcium ion leakage.

Although ultrasound frequencies are known for such adverse effects as cell membrane thinning, localized heating and free radical production (Joyce et al. 2003; Lee et al. 2009), they are also responsible for growth-regulatory signaling as observed with *Staphylococcus epidermis* and *Pseudomonas aeruginosa* (Pitt and Ross 2003). The mechanisms involved in bacterial acoustic signaling are not clear but are believed to involve vibration of bacterial membranes, while perception of sound is thought to be mediated through activating ion channels (Matsuhashi et al. 1998). The production and transduction of these acoustic signals, in bacteria, could provide insight into its role at the cellular level.

Puzzles and knowledge gaps in the field of acoustic biology

Before discussing the puzzles in the field of radiation-induced bystander effects and the potential of acoustic biology to inform radiation biology, we should draw attention to knowledge gaps and questions in the field of acoustic biology itself; these include:

1. 'Listening' as a biological event.
   While much is known about acoustic receptors in organisms, little is currently known about the cellular processes involved in acoustic signal reception outside specialized sensory organs. There is however some limited evidence of the activation of specific mechanotransduction receptors which we discuss below. Any specificity is likely to lie in such macromolecular signaling mechanisms as opposed to chemical signaling through generation of ROS, although in principle both might occur separately or together (discussed below).
2. How might the cell or tissue distinguish an acoustic 'alarm' from an 'adapt' signal?
   The answer to this is unknown at present. It is likely that the cell does not distinguish an alarm from an adapt response. Alarm is probably the default response but if the environment is favorable, the cell or its progeny (or the population of organisms) may adapt. Those outcomes will probably depend more on what else is happening in the cell/population, signal cross-talk, and the responsive state of the cells stimulated.
3. How might an acoustic signal be sensed by non-sensory cells?
   We understand the nature and physics of acoustic waves and therefore there are existing paradigms on which to base hypotheses. The physical displacement of molecules as part of an acoustic wave produces a pressure and that pressure in principle will act to reversibly deform any elastic structure with which they interact. It is now very well established that low-intensity ultrasound causes cytoskeletal structural and gene expression changes in cells (Louw et al. 2013; Miller et al. 2017; Samandari et al. 2017). Potential mechanisms for this involve changes in the structure and composition of the cell membrane and cytoskeleton (reviewed in (Kruglikov and Scherer 2019)), (Mizrahi et al. 2012; Ye et al. 2016), redistribution or modification of expression of caveolin (Ye et al. 2016), and stimulating exosome release (Yuana et al. 2017; Li et al. 2019; Zeng et al. 2019). The opening of specific mechanosensory channels and activation of receptors has been reported (Prieto et al. 2018; Liao et al. 2020), changes in membrane structure (Krasovitski et al. 2011) and possibly, in some circumstances, sonoporation (Li et al. 2018). There is an existing well-developed model for the opening of Ca^{2+} channels by acoustic stimulus in the Piezo 1 channel (Liao et al. 2019), although whether this might be feasible in vivo depends on measurement of the acoustic pressure or energy exerted at the surface of an individual cell, its geometry, acoustic modulus and possibly its substrate. Such measurements have not yet been made, but it is hoped that given the known ability of mechanosensory receptors to be opened by ultrasound, such a mechanism is attractive and worthy of investigation. Mechanosensory receptors are an unusual family of proteins, both functionally and structurally and currently and there are currently thought to be only 18 mechanosensitive ion channels in humans. However similar molecules are present in plants, bacteria and archaea, so are very widespread. It is unclear if a potential acoustic signal receptor may be one of the already characterized genes or a novel function for another functionally associated gene.
   The profound effects that these signal sensing mechanisms have on cell death, cell division (Zhou et al. 2004) and differentiation (Miller et al. 2017) suggest that acoustic signals might contribute an important component to the bystander response. The involvement of exosome release, changes in miR expression (Costa et al. 2019) and calcium influx in well characterized sound responses (Zhang et al. 2012; Takahashi et al. 2019; Lee et al. 2020) is very reminiscent of observations in the RIBE (see below), and is persuasive of a hypothesis in which local acoustic emissions may be the trigger for the RIBE, followed sequentially by other more familiar processes of intercellular communication. To what extent this might be operative at a macroscopic level in the environment is currently unknown.
4. How could we identify and monitor the receptor mechanisms?
   Identification of a receptor protein or structure would be greatly assisted by identification of a response spectrum to radiation-induced acoustic emission, and in principle analysis of cellular response using knockout cells or model organisms. Human genetic studies on radiation sensitivity or possibly therapeutic response might also yield evidence for involvement of mechanoreceptor involvement.

5. Could we block, inhibit or boost this reception machinery? This is possible in principle once a mechanism has been determined. One interesting avenue might be to use sonic enhancing agents such as Sonovue, a stabilized microbubble preparation used as a contrast agent in clinical US studies. Similarly membrane composition manipulation may provide some circumstantial evidence. Known mechanoreceptors do have specific inhibitors such as gadolinium (Bae et al. 2011) and ruthenium red (Syeda et al. 2015) and activators such as Yoda, but these do not affect all classes of mechanoreceptors.

Puzzles in the RIBE at whole organism and cellular level; a communication problem

While many of the mechanisms involved in RIBE have been worked out – particularly the response to signals (Hei et al. 2011; Le, Fernandez-Palomo, et al. 2017), some key puzzles remain.

- What is the 'first cause'; what happens to the energy from the ionizing radiation that is deposited in the target?
- How does the energy (dose) measured in joules per kilogram get converted to chemical energy?
- How are the contents of exosomes, that are the putative vehicles carrying information, modified by the irradiated target, and what stimulates their release?
- What is the relationship between RIBE and genomic instability (non-clonal transgenerational instability)?
- Is inter-organism RIBE the same as inter-cellular RIBE or is it a different process?
- Is acoustic signaling a potentially fundamental 'first cause' or just one of many possible signals?

Most of these questions remain unanswered at this time but it is worth discussing why we consider them to be *puzzles* and why we believe that the induction of acoustic signals is a neglected potential mechanism by which the RIBE might be initiated or propagated.

1. A perusal of any of the many textbooks of radiobiology would suggest this question was answered years ago; energy deposition caused generation of free radicles and direct or indirect damage to cellular components. DNA damage is most critical and damage to DNA is proportional to the dose of energy deposited. Most damage gets repaired after low-dose exposures (generally accepted to be <100mGy, 0.05mGy/day) but mis-repair can happen leading to the dose-related generation of mutations (e.g. Elkind et al. 1967; Hall and Giaccia 2012). The puzzles start with the discontinuity between low dose effects and high dose effects – non-linearity is seen and both 'good' and 'bad' effects can occur (although these terms are highly dependent on context). Genomic instability, hormesis and adaptive responses are examples of effects that are not accommodated by classical target theories of energy deposition (Mothersill and Seymour 2019). The missing element in this first puzzle may be a failure to consider the importance of excitation after low dose exposure. It is well known that both ionization and excitation occur during irradiation but because of the paradigm linking energetic free radicals to biological damage, excitation is dismissed as unimportant. Puzzle 2 assumes it is actually VERY important at low doses.
2. The evidence for the importance of excitation mainly comes from experiments of Le et al. (Le, McNeill, et al. 2015, Le, Mothersill, et al. 2015, 2017, Le, Fernandez-Palomo, et al. 2017, Le et al. 2018) and (Cohen et al. 2020). These document the generation of photons, assumed to result from excitation decay, during ionizing radiation exposure to beta and gamma radiations. These photons were collected in the UVA range (around 340nm) but unpublished spectral analysis shows emissions right across the EM spectrum. Photon emission has been directly and quantitatively linked to RIBE and exosomes harvested from photon exposed bystander cells can substitute for photons and induce secondary RIBE (Le, Mothersill, et al. 2017). The puzzle here is all the data were collected using a filter collecting UVA photons but can other photon energies or acoustic energies initiate biological effects?
3. The third puzzle concerns how energy is transduced in a way that modifies the cargo of exosomes to transmit altered instructions to the bystander cells (or organisms)? It is known that calcium channels are critical in this process but the biophysical processes are far from clear. If the excitation decay products open channels, perhaps acting like a nerve impulse, how do the cells discriminate between the process induced by radiation and natural processes involving calcium channels induced by normal metabolic processes? Exosomes generated by photon exposed cells induce the RIBE phenotype but exosomes harvested from control or sham exposed cells do not (Le, Fernandez-Palomo, et al. 2017).
4. Another puzzle that is critically important when trying to ascertain the role of RIBE in radiation risk assessment involves the relationship between RIBE and radiation-induced genomic instability (RIGI). In the last 40 years RIGI was actually documented before RIBE although evidence of both under other names can be found in the old literature (Mothersill and Seymour 2019). The key issue here is what drives RIGI? The manifestation of RIGI is effectively permanent and once induced cannot be 'reset'. However, it is not due to a mutation that can be identified but rather it is a state resulting in a higher tolerance for mutations of all types. Non-clonal mutations arise sporadically and unpredictably in the system in perpetuity (Mothersill et al. 2000). Some try to explain the state as resulting from a genetic mutation leading to a 'mutator phenotype' (Loeb 2016) but mutator phenotypes lead to accumulation of mutation over time while RIGI leads to sporadic non-clonal mutations occurring at a higher than normal rate in cells which have up to that point been perfectly normal. The suggestion is that the RIBE signal-generating phenotype is an epigenetic maintainer of the RIGI

phenotype. Why this phenotype would persist is unknown and how it would not be selected out in cultures or in organisms is a puzzle taxing the minds of radiobiologists concerned about evolution of radiation-induced traits in populations.

5. When inter-organism transfer of the RIBE phenotype was first seen in the radiation field in mice (Surinov et al. 2004) and in fish (Mothersill et al. 2006) it was a cause of amazement but quickly dismissed as not the same as cellular RIBE. Subsequent demonstration of inter animal, yeast and plant transfer (reviewed in (Mothersill et al. 2018)) established the phenomenon as real but it is still not clear if the mechanisms are similar or if whole organism transfer is a sub-category of the type of acoustic, light or chemical communication well documented in the ecology field. The paper that gives us the strongest evidence for a role for EM signaling other than light is one where irradiated fish were placed in an aquarium inside another larger tank in with unirradiated fish were swimming (Mothersill et al. 2012). The unirradiated fish picked up the bystander signals and showed evidence of having acquired the irradiated phenotype. We then repeated the experiment but covered the inner tank with aluminum foil to prevent light signals from reaching the unirradiated fish. The same result was seen. Calculations performed to study the attenuation of the signal could not exclude sound as a possible signal but others such as photo-voltaic reactions are also possible. This work has not been followed up.
6. Acoustic signaling is well documented in the animal and plant world so question 5 above really pertains to whether acoustic or light signals are fundamental first causes of RIBE which can manifest at multiple levels of organization from sub-cellular to organismal. Ultimately even in a whole organism, the signal has to be generated in a cell. Systems which allow the signals from multiple cells to be coordinated so as to be strong enough include quorum sensing in bacteria emitting light signals (Pena et al. 2019) and chemical sensing in slime molds where aggregation is triggered by a range of species-specific ions or peptides (Loomis 2014).

These puzzles have no immediate answers and raise many new questions which will undoubtedly keep us busy exploring the rise of bioacoustics as evidence in the years to come.

A cellular acoustic response in the bystander effect?

We do not consider that the received acoustic signal is in itself a damaging agent but a secondary relay from the cell hit by radiation and damaged. We hypothesize that the initial radiation interaction produces acoustic pressure in the surrounding medium or cellular milieu and the surrounding cells respond to that pressure. We believe it unlikely that the acoustic signal itself induces damage but rather we propose that there is some form of signal detection and transduction, either in the way of direct molecular interaction – discussed below, or through the mediation of induced reactive oxygen species (ROS). The former might not be expected to damage DNA directly, although the latter might.

There is evidence that ultrasound can generate DNA damaging effects, both as naked DNA or DNA within cells. (Milowska and Gabryelak 2007; Furusawa et al. 2012; Saliev et al. 2018). However, the acoustic energy released from a single irradiated cell is likely to be much less than that directly applied across volumes or fields of cells in these published studies. It is difficult to establish a prediction of local acoustic pressure generated by irradiation in a field of cells. We can however demonstrate that a relationship may be established between absorbed dose and generation of acoustic pressure and we can predict that radiation energy and quality, and the material absorbance spectrum, will affect the range of wavelengths emitted [Supplemental file 1].

If each randomly irradiated cell generates an acoustic signal it is very difficult to estimate the energy deposition into a neighboring cell in the surrounding field or volume without direct measurement. This will be affected by the geometry of both activated and responding cells, the relative distribution of activated cells and the attenuation effect of both cytoplasm (between cell border and nucleus) and any intervening cells. The acoustic attenuation effect of cells is of the order of in $10^3 \times$ greater in comparison to water (Pasternak et al. 2015) and so a significant modulator of signal strength. It is consequently unlikely that the effect of ultrasound signal from an irradiated cell is likely to damage DNA in the nucleus of a neighboring cell.

We consider that generation of reactive oxygen species by the acoustic wave or a direct effect on a macromolecular structure are more likely methods of signal detection and transduction. As to whether acoustic pressures at this scale might be enough to elicit an active signal; 3mPa externally applied ultrasound has been shown to elicit a biological response with Ca^{2+} influx in HUVEC cells (Hwang et al. 2014) but with no effect on cell viability, although no direct titrations are available in the literature and the authors of the study did not use stimuli lower than 3mPa. Whether elicited acoustic signals from medium to low doses of radiation fall in to this range is yet to be directly ascertained.

Conclusions

We have reviewed the evidence for the emission of acoustic signals from irradiated biota of many different species, and conclude that all biological material has the potential to emit sound on interaction with ionizing radiation. There is not so far any evidence of sustained acoustic emission outside the time period of dose delivery, which would require either extremely inefficient relaxation processes or the triggering of an unknown energy-dependent cellular mechanism for producing sound. However, in situations of chronic exposure, such as external irradiation or internal contamination such as are found in contaminated ecosystems we would expect there to be sustained sound production if doses are sufficiently high. This is complicated by issues of intense local irradiation for example in internal contamination and raises problems already discussed about the meaning of environmental dosimetry (Beaugelin-Seiller et al. 2020). Transmission distances of such sound in air are likely to be short, given the acoustic modulus of air, its movement, and the presence of physical barriers to transmission such as interspersed objects in the environment. The physical characteristics of sound however suggest that transmission through the underlying matrix or at the interface of matrix and air, especially in wet environments, could be more efficient. Persuasive evidence exists that biota of all types, at the level of cells and organisms can respond to sound, and that in some cases to biophony. This strongly suggests that the ability to perceive and respond to the sounds emitted by biota under stress may be a widespread phenomenon.

Experiments on the RIBE raise many issues about the mechanism and response to the irradiation of neighboring cells and organisms. The properties of the information-carrying entities predicted by these experiments are often contradictory and are at odds with well-established paradigms. We were struck however by the parallels with the experiments of Gagliano and Renton (Gagliano and Renton 2013), who showed that *Ocimum basilicum* can stimulate the germination of chilli seeds, even when all obvious routes of signal transmission were blocked, implying that neither chemical nor light signals were involved.

It was the intention of this commentary to stimulate discussion on the neglected role of radiation-induced acoustic emission in the response of cells and organisms to radiation and we hope that in turn this will stimulate experimentation to establish its role.

Acknowledgements

CM/CBS acknowledge the National CFIDS Foundation Inc.

Disclosure statement

No potential conflict of interest was reported by the author(s).

Funding

The present research was financially supported by the Canada Research Chairs Programme [Grant Number: 950-221284], NSERC Collaborative Research and Development Grant [Grant Number: RGPIN293153-12], the National CFIDS Foundation Inc., and Bruce Power.

ORCID

Bruno F. E. Matarèse http://orcid.org/0000-0001-5276-5636
Colin Seymour http://orcid.org/0000-0002-4424-0968
Paul N. Schofield http://orcid.org/0000-0002-5111-7263
Carmel Mothersill http://orcid.org/0000-0001-8501-3377

References

Al-Mayah AH, Irons SL, Pink RC, Carter DR, Kadhim MA. 2012. Possible role of exosomes containing RNA in mediating nontargeted effect of ionizing radiation. Radiat Res. 177(5):539–545.

Apfelbach R, Parsons MH, Soini HA, Novotny MV. 2015. Are single odorous components of a predator sufficient to elicit defensive behaviors in prey species? Front Neurosci. 9:263.

Appel HM, Cocroft RB. 2014. Plants respond to leaf vibrations caused by insect herbivore chewing. Oecologia. 175(4):1257–1266.

Ashush H, Rozenszajn LA, Blass M, Barda-Saad M, Azimov D, Radnay J, Zipori D, Rosenschein U. 2000. Apoptosis induction of human myeloid leukemic cells by ultrasound exposure. Cancer Res. 60: 1014–1020.

Bae C, Sachs F, Gottlieb PA. 2011. The mechanosensitive ion channel Piezo1 is inhibited by the peptide GsMT × 4. Biochemistry. 50(29): 6295–6300.

Beaugelin-Seiller K, Garnier-Laplace J, Beresford NA. 2020. Estimating radiological exposure of wildlife in the field. J Environ Radioact. 211:105830.

Bell AG. 1881. The production of sound by radiant energy. Science. os-2(49):242–253.

Bhandawat A, Jayaswall K, Sharma H, Roy J. 2020. Sound as a stimulus in associative learning for heat stress in Arabidopsis. Commun Integr Biol. 13(1):1–5.

Bludman SA, Ruderman MA. 1968. Possibility of the speed of sound exceeding the speed of light in ultradense matter. Phys Rev. 170(5): 1176–1184.

Bond LJ, Chiang CH, Fortunko CM. 1992. Absorption of ultrasonic waves in air at high frequencies (10–20 MHz). J Acoust Soc Am. 92(4):2006–2015.

Bonner WM. 2003. Low-dose radiation: thresholds, bystander effects, and adaptive responses. Proc Natl Acad Sci USA. 100(9):4973–4975.

Bowen T, Chen CX, Liew SC, Lutz WR, Nasoni RL. 1991. Observation of ultrasonic emission from edges of therapeutic X-ray beams. Phys Med Biol. 36(4):537–539.

Bradbury JW, Vehrencamp SL. 2011. Principles of animal communication. Second Edition. Oxford: Oxford University Press.

Cao Y, Kole A, Hui J, Zhang Y, Mai J, Alloosh M, Sturek M, Cheng J-X. 2018. Fast assessment of lipid content in arteries in vivo by intravascular photoacoustic tomography. Sci Rep. 8(1):2400.

Carmichael AJ, Mossoba MM, Riesz P, Christman CL. 1986. Free radical production in aqueous solutions exposed to simulated ultrasonic diagnostic conditions. IEEE Trans Ultrason Ferroelectr Freq Cont. 33(2):148–155.

Chi Z, Yu Shrike Z, Da-Kang Y, Younan X, Lihong VW. 2013. Label-free photoacoustic microscopy of cytochromes. J Biomed Opt. 18: 1–4.

Christison GB, MacKenzie HA. 1993. Laser photoacoustic determination of physiological glucose concentrations in human whole blood. Med Biol Eng Comput. 31(3):284–290.

Christman CL, Carmichael AJ, Mossoba MM, Riesz P. 1987. Evidence for free radicals produced in aqueous solutions by diagnostic ultrasound. Ultrasonics. 25(1):31–34.

Cohen J, Vo NTK, Chettle DR, McNeill FE, Seymour CB, Mothersill CE. 2020. Quantifying biophoton emissions from human cells directly exposed to low-dose gamma radiation. Dose Resp. 18(2): 1559325820926763.

Collins ME, Foreman JEK. 2001. The effect of sound on the growth of plants. Can Acoust. 29(2):3–8.

Costa V, Carina V, Conigliaro A, Raimondi L, De Luca A, Bellavia D, Salamanna F, Setti S, Alessandro R, Fini M, et al. 2019. miR-31-5p Is a LIPUS-mechanosensitive MicroRNA that targets HIF-1α signaling and cytoskeletal proteins. Int J Mol Sci. 20(7):1569.

D'Astous FT, Foster FS. 1986. Frequency dependence of ultrasound attenuation and backscatter in breast tissue. Ultrasound Med Biol. 12(10):795–808.

Da-Kang Y, Konstantin M, Kirk KS, Qifa Z, Lihong VW. 2020. In vivo label-free photoacoustic microscopy of cell nuclei by excitation of DNA and RNA. Opt Lett. 35:4139–4141.

Danielli A, Maslov K, Favazza CP, Xia J, Wang LV. 2015. Nonlinear photoacoustic spectroscopy of hemoglobin. Appl Phys Lett. 106(20): 203701.

De Roo L, Vergeynst LL, De Baerdemaeker NJF, Steppe K. 2016. Acoustic emissions to measure drought-induced cavitation in plants. Applied Sciences. 6(3):71.

Del Grosso VA, Mader CW. 1972. Speed of sound in pure water. J Acoust Soc Am. 52(5B):1442–1446.

Du Y, Du S, Liu L, Gan F, Jiang X, Wangrao K, Lyu P, Gong P, Yao Y. 2020. Radiation-induced bystander effect can be transmitted through exosomes using mirnas as effector molecules. Radiat Res. 194(1):89–100.

Duchen MR. 2000. Mitochondria and calcium: from cell signalling to cell death. J Physiol. 529(Pt 1):57–68.

Eisenhoffer GT, Loftus PD, Yoshigi M, Otsuna H, Chien C-B, Morcos PA, Rosenblatt J. 2012. Crowding induces live cell extrusion to maintain homeostatic cell numbers in epithelia. Nature. 484(7395): 546–549.

Elkind MM, Whitmore GF, American Institute of Biological S. 1967. The radiobiology of cultured mammalian cells [by] Mortimer M, Elkind [and] Gordon F. Whitmore New York, London: Gordon and Breach.

Erbe C, Marley SA, Schoeman RP, Smith JN, Trigg LE, Embling CB. 2019. The effects of ship noise on marine mammals—a review. Front Mar Sci. 6:344.

Eroglu M, Derry WB. 2016. Your neighbours matter – non-autonomous control of apoptosis in development and disease. Cell Death Differ. 23(7):1110–1118.

Farmer EE, Ryan CA. 1990. Interplant communication: airborne methyl jasmonate induces synthesis of proteinase inhibitors in plant leaves. Proc Natl Acad Sci USA. 87(19):7713–7716.

Fay RR, Popper AN. 2000. Evolution of hearing in vertebrates: the inner ears and processing. Hear Res. 149(1–2):1–10.

Feng Y, Tian Z, Wan M. 2010. Bioeffects of low-intensity ultrasound in vitro: apoptosis, protein profile alteration, and potential molecular mechanism. J Ultrasound Med. 29(6):963–974.

Fong KY, Li H-K, Zhao R, Yang S, Wang Y, Zhang X. 2019. Phonon heat transfer across a vacuum through quantum fluctuations. Nature. 576(7786):243–247.

Fridberger A, Flock Å, Ulfendahl M, Flock B. 1998. Acoustic overstimulation increases outer hair cell Ca^{2+} concentrations and causes dynamic contractions of the hearing organ. Proc Natl Acad Sci USA. 95(12):7127–7132.

Fridberger A, Ulfendahl M. 1996. Acute mechanical overstimulation of isolated outer hair cells causes changes in intracellular calcium levels without shape changes. Acta Otolaryngol. 116(1):17–24.

Furusawa Y, Fujiwara Y, Campbell P, Zhao Q-L, Ogawa R, Ali Hassan M, Tabuchi Y, Takasaki I, Takahashi A, Kondo T. 2012. DNA double-strand breaks induced by cavitational mechanical effects of ultrasound in cancer cell lines. PLoS One. 7(1):e29012.

Gagliano M. 2013. Green symphonies: a call for studies on acoustic communication in plants. Behav Ecol. 24(4):789–796.

Gagliano M, Mancuso S, Robert D. 2012a. Towards understanding plant bioacoustics. Trends Plant Sci. 17(6):323–325.

Gagliano M, Renton M. 2013. Love thy neighbour: facilitation through an alternative signalling modality in plants. BMC Ecol. 13:19.

Gagliano M, Renton M, Duvdevani N, Timmins M, Mancuso S. 2012b. Acoustic and magnetic communication in plants: is it possible? Plant Signal Behav. 7(10):1346–1348.

Garcia ME, Pastor GM, Bennemann KH. 1988. Theory for the photoacoustic response to X-ray absorption. Phys Rev Lett. 61(1):121–124.

Ghosh R, Choi B, Kwon YS, Bashir T, Bae D-W, Bae H. 2019. Proteomic changes in the sound vibration-treated arabidopsis thaliana facilitates defense response during *Botrytis cinerea* infection. Plant Pathol J. 35(6):609–622.

Gillooly JF, Ophir AG. 2010. The energetic basis of acoustic communication. Proc Royal Soc B: Biolog Sci. 277(1686):1325–1331.

Goodale E, Beauchamp G, Magrath RD, Nieh JC, Ruxton GD. 2010. Interspecific information transfer influences animal community structure. Trends Ecol Evol. 25(6):354–361.

Gudipaty SA, Conner CM, Rosenblatt J, Montell DJ. 2018. Unconventional ways to live and die: cell death and survival in development, homeostasis, and disease. Annu Rev Cell Dev Biol. 34: 311–332.

Hall EJ, Giaccia AJ. 2012. Radiobiology for the radiologist. Philadelphia: Wolters Kluwer Health/Lippincott Williams & Wilkins.

Harding HR, Gordon TAC, Eastcott E, Simpson SD, Radford AN. 2019. Causes and consequences of intraspecific variation in animal responses to anthropogenic noise. Behav Ecol. 30(6):1501–1511.

Hassanien RHE, Hou T-z, Li Y-f, Li B-m. 2014. Advances in effects of sound waves on plants. J Integr Agric. 13(2):335–348.

Hastings JF, Skhinas JN, Fey D, Croucher DR, Cox TR. 2019. The extracellular matrix as a key regulator of intracellular signalling networks. Br J Pharmacol. 176(1):82–92.

Hei TK, Zhou H, Chai Y, Ponnaiya B, Ivanov VN. 2011. Radiation induced non-targeted response: mechanism and potential clinical implications. Curr Mol Pharmacol. 4(2):96–105.

Hei TK, Zhou H, Ivanov VN, Hong M, Lieberman HB, Brenner DJ, Amundson SA, Geard CR. 2008. Mechanism of radiation-induced bystander effects: a unifying model. J Pharm Pharmacol. 60(8): 943–950.

Heil M, Karban R. 2010. Explaining evolution of plant communication by airborne signals. Trends Ecol Evol. 25(3):137–144.

Hickling S, Hobson M, El Naqa I. 2014. Feasibility of X-ray acoustic computed tomography as a tool for noninvasive volumetric in vivo dosimetry. Int J Radiat Oncol Biol Phys. 90(1):S843.

Hickling S, Lei H, Hobson M, Leger P, Wang X, El Naqa I. 2016. Sci-Thur AM: YIS – 02: imaging dose distributions through the detection of radiation-induced acoustic waves. Med Phys. 43(8 Part 2): 4928–4928.

Hou T, Li B, Teng G, Zhou Q, Xiao Y, Qi L. 2009. Application of acoustic frequency technology to protected vegetable production. Nongye Gongcheng Xuebao/Trans Chinese Soc Agri Eng. 25: 156–160.

Hudspeth AJ. 1989. Mechanoelectrical transduction by hair cells of the Bullfrog's Sacculus. Prog Brain Res. 80:129–135.

Hwang JY, Lim HG, Yoon CW, Lam KH, Yoon S, Lee C, Chiu CT, Kang BJ, Kim HH, Shung KK. 2014. Non-contact high-frequency ultrasound microbeam stimulation for studying mechanotransduction in human umbilical vein endothelial cells. Ultrasound Med Biol. 40(9):2172–2182.

Jella KK, Rani S, O'Driscoll L, McClean B, Byrne HJ, Lyng FM. 2014. Exosomes are involved in mediating radiation induced bystander signaling in human keratinocyte cells. Radiat Res. 181(2):138–145.

Jeong MJ, Shim CK, Lee JO, Kwon HB, Kim YH, Lee SK, Byun MO, Park SC. 2008. Plant gene responses to frequency-specific sound signals. Mol Breeding. 21(2):217–226.

Joyce E, Phull SS, Lorimer JP, Mason TJ. 2003. The development and evaluation of ultrasound for the treatment of bacterial suspensions. A study of frequency, power and sonication time on cultured Bacillus species. Ultrason Sonochem. 10(6):315–318.

Jung J, Kim S-K, Kim JY, Jeong M-J, Ryu C-M. 2018. Beyond chemical triggers: evidence for sound-evoked physiological reactions in plants. Front Plant Sci. 9:25–25.

Kacsoh BZ, Bozler J, Bosco G. 2018. Drosophila species learn dialects through communal living. PLoS Genet. 14(7):e1007430.

Kaiser K, Hammers JL. 2009. The effect of anthropogenic noise on male advertisement call rate in the neotropical Treefrog, *Dendropsophus triangulum*. Behav. 146(8):1053–1069.

Karban R, Shiojiri K, Ishizaki S. 2010. An air transfer experiment confirms the role of volatile cues in communication between plants. Am Nat. 176(3):381–384.

Khait I, Obolski U, Yovel Y, Hadany L. 2019. Sound perception in plants. Semin Cell Dev Biol. 92:134–138.

Kim JY, Lee JS, Kwon TR, Lee SI, Kim JA, Lee GM, Park SC, Jeong MJ. 2015. Sound waves delay tomato fruit ripening by negatively regulating ethylene biosynthesis and signaling genes. Postharvest Biol Technol. 110:43–50.

Kim KY, Sachse W. 1983. X-ray generated ultrasound. Appl Phys Lett. 43(12):1099–1101.

Kim S-K, Jeong M-J, Ryu C-M. 2018. How do we know that plants listen: advancements and limitations of transcriptomic profiling for the identification of sound-specific biomarkers in tomato. Plant Signal Behav. 13(12):e1547576.

Krasovitski B, Frenkel V, Shoham S, Kimmel E. 2011. Intramembrane cavitation as a unifying mechanism for ultrasound-induced bioeffects. Proc Natl Acad Sci U S A. 108(8):3258–3263.

Kruglikov IL, Scherer PE. 2019. Caveolin-1 as a pathophysiological factor and target in psoriasis. NPJ Aging Mech Dis. 5:4.

Kuwahara Y, Tomita K, Urushihara Y, Sato T, Kurimasa A, Fukumoto M. 2018. Association between radiation-induced cell death and clinically relevant radioresistance. Histochem Cell Biol. 150(6): 649–659.

Kwon YS, Jeong M-J, Cha J, Jeong S-W, Park S-C, Shin S, Chung W, Bae H, Bae D-W. 2012. Comparative proteomic analysis of plant responses to sound waves in *Arabidopsis*. J Plant Biotechnol. 39(4): 261–272.

Laschimke R, Burger M, Vallen H. 2006. Acoustic emission analysis and experiments with physical model systems reveal a peculiar nature of the xylem tension. J Plant Physiol. 163(10):996–1007.

Le M, Fernandez-Palomo C, McNeill FE, Seymour CB, Rainbow AJ, Mothersill CE. 2017. Exosomes are released by bystander cells exposed to radiation-induced biophoton signals: reconciling the mechanisms mediating the bystander effect. PLoS One. 12(3):e0173685.

Le M, McNeill FE, Seymour C, Rainbow AJ, Mothersill CE. 2015. An observed effect of ultraviolet radiation emitted from beta-irradiated HaCaT cells upon non-beta-irradiated bystander cells. Radiat Res. 183(3):279–290.

Le M, McNeill FE, Seymour CB, Rusin A, Diamond K, Rainbow AJ, Murphy J, Mothersill CE. 2018. Modulation of oxidative phosphorylation (OXPHOS) by radiation-induced biophotons. Environ Res. 163:80–87.

Le M, Mothersill CE, Seymour CB, Ahmad SB, Armstrong A, Rainbow AJ, McNeill FE. 2015. Factors affecting ultraviolet-A photon emission from β-irradiated human keratinocyte cells. Phys Med Biol. 60(16):6371–6389.

Le M, Mothersill CE, Seymour CB, Rainbow AJ, McNeill FE. 2017. An observed effect of p53 status on the bystander response to radiation-induced cellular photon emission. Radiat Res. 187(2):169–185.

Lee NS, Yoon CW, Wang Q, Moon S, Koo KM, Jung H, Chen R, Jiang L, Lu G, Fernandez A, et al. 2020. Focused ultrasound stimulates ER localized mechanosensitive PANNEXIN-1 to mediate intracellular calcium release in invasive cancer cells. Front Cell Dev Biol. 8:504.

Lee Y, Cho J, Dayou J, Phin CK. 2009. Experimental investigation on the effects of audible sound to the growth of *Escherichia coli*. Modern App Sci. 3.

Lemke G. 2019. How macrophages deal with death. Nat Rev Immunol. 19(9):539–549.

Leutwyler H. 1997. Phonons as goldstone bosons. Helv Phys Acta. 70: 275–286.

Li F, Yang C, Yuan F, Liao D, Li T, Guilak F, Zhong P. 2018. Dynamics and mechanisms of intracellular calcium waves elicited by tandem bubble-induced jetting flow. Proc Natl Acad Sci. 115(3): E353–E362.

Li X, Li X, Lin J, Sun X, Ding Q. 2019. Exosomes derived from low-intensity pulsed ultrasound-treated dendritic cells suppress tumor necrosis factor-induced endothelial inflammation. J Ultrasound Med. 38(8):2081–2091.

Liangzhong X, Bin H, Colin C, Guillem P, Yu K, Lei X. X-ray induced photoacoustic tomography. Proceedings of the ProcSPIE; 2013.

Liao D, Li F, Lu D, Zhong P. 2019. Activation of Piezo1 mechanosensitive ion channel in HEK293T cells by 30 MHz vertically deployed surface acoustic waves. Biochem Biophys Res Commun. 518(3):541–547.

Liao W-H, Hsiao M-Y, Kung Y, Liu H-L, Béra J-C, Inserra C, Chen W-S. 2020. TRPV4 promotes acoustic wave-mediated BBB opening via Ca^{2+}/PKC-δ pathway. J Adv Res. doi:10.1016/j.jare.2020.06.012

Loeb LA. 2016. Human cancers express a mutator phenotype: hypothesis, origin, and consequences. Cancer Res. 76(8):2057–2059.

Longo DL, Stefania R, Aime S, Oraevsky A. 2017. Melanin-based contrast agents for biomedical optoacoustic imaging and theranostic applications. Int J Mol Sci. 18(8):1719.

Loomis WF. 2014. Cell signaling during development of Dictyostelium. Dev Biol. 391(1):1–16.

Lorén N, Nydén M, Hermansson A-M. 2009. Determination of local diffusion properties in heterogeneous biomaterials. Adv Colloid Interface Sci. 150(1):5–15.

Louw TM, Budhiraja G, Viljoen HJ, Subramanian A. 2013. Mechanotransduction of ultrasound is frequency dependent below the cavitation threshold. Ultrasound Med Biol. 39(7):1303–1319.

Ludwig GD. 1950. The velocity of sound through tissues and the acoustic impedance of tissues. J Acoust Soc Am. 22(6):862–866.

Lyng FM, Seymour CB, Mothersill C. 2000. Production of a signal by irradiated cells which leads to a response in unirradiated cells characteristic of initiation of apoptosis. Br J Cancer. 83(9):1223–1230.

Margolis L, Sadovsky Y. 2019. The biology of extracellular vesicles: the known unknowns. PLoS Bio. 17(7):e3000363.

Matsuhashi M, Pankrushina AN, Takeuchi S, Ohshima H, Miyoi H, Endoh K, Murayama K, Watanabe H, Endo S, Tobi M, et al. 1998. Production of sound waves by bacterial cells and the response of bacterial cells to sound. J Gen Appl Microbiol. 44(1):49–55.

Maulucci G, Troiani D, Eramo SL, Paciello F, Podda MV, Paludetti G, Papi M, Maiorana A, Palmieri V, De Spirito M, et al. 2014. Time evolution of noise induced oxidation in outer hair cells: role of NAD(P)H and plasma membrane fluidity. Biochim Biophys Acta. 1840(7):2192–2202.

Md. Emran Khan C, Lt Sup Gt Lt Sup Gt Hyoun-Sub L, Lt Sup, et al. 2014. Update on the effects of sound wave on plants. Res Plant Dis. 20:1–7.

Miller AD, Chama A, Louw TM, Subramanian A, Viljoen HJ. 2017. Frequency sensitive mechanism in low-intensity ultrasound enhanced bioeffects. PLoS One. 12(8):e0181717.

Milowska K, Gabryelak T. 2007. Reactive oxygen species and DNA damage after ultrasound exposure. Biomol Eng. 24(2):263–267.

Mizrahi N, Zhou EH, Lenormand G, Krishnan R, Weihs D, Butler JP, Weitz DA, Fredberg JJ, Kimmel E. 2012. Low intensity ultrasound perturbs cytoskeleton dynamics. Soft Matter. 8(8):2438–2443.

Mothersill C, Bucking C, Smith RW, Agnihotri N, Oneill A, Kilemade M, Seymour CB. 2006. Communication of radiation-induced stress or bystander signals between fish in vivo. Environ Sci Technol. 40(21):6859–6864.

Mothersill C, Kadhim MA, O'Reilly S, Papworth D, Marsden SJ, Seymour CB, Wright EG. 2000. Dose- and time-response relationships for lethal mutations and chromosomal instability induced by ionizing radiation in an immortalized human keratinocyte cell line. Int J Radiat Biol. 76:799–806.

Mothersill C, Le M, Rusin A, Seymour C. 2019. Biophotons in radiobiology: inhibitors, communicators and reactors. Radiat Prot Dosimetry. 183(1–2):136–141.

Mothersill C, Lyng F, Mulford A, Seymour C, Cottell D, Lyons M, Austin B. 2001. Effect of low doses of ionizing radiation on cells cultured from the hematopoietic tissue of the Dublin Bay prawn, *Nephrops norvegicus*. Radiat Res. 156(3):241–250.

Mothersill C, Rusin A, Fernandez-Palomo C, Seymour C. 2018. History of bystander effects research 1905-present; what is in a name? Int J Rad Bio. 94(8):696–707.

Mothersill C, Seymour C. 2009. Communication of ionising radiation signals-a tale of two fish. Int J Rad Biol. 85(11):909–919.

Mothersill C, Seymour C. 2012. Changing paradigms in radiobiology. Mutat Res. 750(2):85–95.

Mothersill C, Seymour C. 2019. Targets, pools, shoulders, and communication – a reflection on the evolution of low-dose radiobiology. Int J Rad Biol. 95(7):851–860.

Mothersill C, Smith RW, Fazzari J, McNeill F, Prestwich W, Seymour CB. 2012. Evidence for a physical component to the radiation-induced bystander effect? Int J Rad Biol. 88(8):583–591.

Mothersill CE, O'Malley KJ, Murphy DM, Seymour CB, Lorimore SA, Wright EG. 1999. Identification and characterization of three subtypes of radiation response in normal human urothelial cultures exposed to ionizing radiation. Carcinogenesis. 20(12):2273–2278.

Nie L, Xing D, Zhou Q, Yang D, Guo H. 2008. Microwave-induced thermoacoustic scanning CT for high-contrast and noninvasive breast cancer imaging. Med Phys. 35(9):4026–4032.

Olwell PM, Cottell DC, Ni Shuilleabhain S, Maderna P, Seymour C, Mothersill C, Lyng FM. 2005. Cytoskeletal reorganization and altered phagocytotic ability in primary cultures of rainbow trout hemopoietic tissue exposed to low-level ionizing radiation. Radiat Res. 164(1):45–52.

Pasternak MM, Strohm EM, Berndl ES, Kolios MC. 2015. Properties of cells through life and death – an acoustic microscopy investigation. Cell Cycle. 14(18):2891–2898.

Pena RT, Blasco L, Ambroa A, González-Pedrajo B, Fernández-García L, López M, Bleriot I, Bou G, García-Contreras R, Wood TK, et al. 2019. Relationship between quorum sensing and secretion systems. Front Microbiol. 10:1100.

Pérez-Garijo A, Fuchs Y, Steller H. 2013. Apoptotic cells can induce non-autonomous apoptosis through the TNF pathway. Elife. 2:e01004.

Pérez-Garijo A, Steller H. 2015. Spreading the word: non-autonomous effects of apoptosis during development, regeneration and disease. Development. 142(19):3253–3262.

Pijanowski BC, Villanueva-Rivera LJ, Dumyahn SL, Farina A, Krause BL, Napoletano BM, Gage SH, Pieretti N. 2011. Soundscape ecology: the science of sound in the landscape. BioSci. 61(3):203–216.

Pitt WG, Ross SA. 2003. Ultrasound increases the rate of bacterial cell growth. Biotechnol. Prog. 19(3):1038–1044.

Preisig J. 2007. Acoustic propagation considerations for underwater acoustic communications network development. Sigmobile Mob Comput Commun Rev. 11(4):2–10.

Prieto ML, Firouzi K, Khuri-Yakub BT, Maduke M. 2018. Activation of Piezo1 but Not NaV1.2 channels by ultrasound at 43 MHz. Ultrasound Med Biol. 44(6):1217–1232.

Raimundo N, Song L, Shutt Timothy E, McKay Sharen E, Cotney J, Guan M-X, Gilliland Thomas C, Hohuan D, Santos-Sacchi J, Shadel Gerald S. 2012. Mitochondrial stress engages E2F1 apoptotic signaling to cause deafness. 148:716–726.

Rayleigh JWS. 1877. The theory of sound. London: Macmillan and Co.

Riesz P, Berdahl D, Christman CL. 1985. Free radical generation by ultrasound in aqueous and nonaqueous solutions. Environ Health Perspect. 64:233–252.

Sakai H, Minamisawa A, Takagi K. 1990. Effect of moisture content on ultrasonic velocity and attenuation in woods. Ultrasonics. 28(6): 382–385.

Saliev T, Begimbetova D, Baiskhanova D, Abetov D, Kairov U, Gilman CP, Matkarimov B, Tachibana K. 2018. Apoptotic and genotoxic effects of low-intensity ultrasound on healthy and leukemic human peripheral mononuclear blood cells. J Med Ultrason. 45(1):31–39.

Samandari M, Abrinia K, Mokhtari-Dizaji M, Tamayol A. 2017. Ultrasound induced strain cytoskeleton rearrangement: an experimental and simulation study. J Biomecha. 60:39–47.

Samant P, Trevisi L, Ji X, Xiang L. 2020. X-ray induced acoustic computed tomography. Photoacoustics. 19:100177.

Schieber M, Chandel NS. 2014. ROS function in redox signaling and oxidative stress. Curr Biol. 24(10):R453–462.

Schöner MG, Simon R, Schöner CR. 2016. Acoustic communication in plant-animal interactions. Curr Opin Plant Biol. 32:88–95.

Sen S. 1992. Programmed cell death: concept, mechanism and control. Biolog Rev. 67(3):287–319.

Sharan Verma G. 1950. Effects of humidity on ultrasonic absorption in Air at 1.46 Megacycles. J Acoust Soc Am. 22(6):861–862.

Shyu W-M, Gaspari GD. 1969. Sound velocity in metals. Phys Rev. 177(3):1041–1043.

Sia J, Szmyd R, Hau E, Gee HE. 2020. Molecular mechanisms of radiation-induced cancer cell death: a primer. Front Cell Dev Biol. 8:41.

Slabbekoorn H, Bouton N, van Opzeeland I, Coers A, ten Cate C, Popper AN. 2010a. A noisy spring: the impact of globally rising underwater sound levels on fish. Trends Ecol Evol. 25(7):419–427.

Slabbekoorn H, Bouton N, van Opzeeland I, Coers A, ten Cate C, Popper AN. 2010b. A noisy spring: the impact of globally rising underwater sound levels on fish. Trend Ecol Evol. 25(7):419–427.

Stevens M. 2013. Sensory ecology, information, and decision-making. In Sensory Ecology, Behaviour, and Evolution. Oxford: OUP.

Stevenson MA, Calderwood SK, Hahn GM. 1987. Effect of hyperthermia (45 degrees C) on calcium flux in Chinese hamster ovary HA-1 fibroblasts and its potential role in cytotoxicity and heat resistance. Cancer Res. 47(14):3712–3717.

Sulak L, Armstrong T, Baranger H, Bregman M, Levi M, Mael D, Strait J, Bowen T, Pifer AE, Polakos PA, et al. 1979. Experimental studies of the acoustic signature of proton beams traversing fluid media. Nucl Instrum Methods. 161(2):203–217.

Sun JWC, Narins PM. 2005. Anthropogenic sounds differentially affect amphibian call rate. Biol Conserv. 121(3):419–427.

Surinov BP, Isaeva VG, Dukhova NN. 2004. Postirradiation volatile secretions of mice: syngeneic and allogeneic immune and behavioral effects. Bull Exp Biol Med. 138(4):384–386.

Syeda R, Xu J, Dubin AE, Coste B, Mathur J, Huynh T, Matzen J, Lao J, Tully DC, Engels IH, et al. 2015. Chemical activation of the mechanotransduction channel Piezo1. Elife. 4:e07369.

Szydlowska K, Tymianski M. 2010. Calcium, ischemia and excitotoxicity. Cell Calcium. 47(2):122–129.

Takahashi T, Nakagawa K, Tada S, Tsukamoto A. 2019. Low-energy shock waves evoke intracellular Ca^{2+} increases independently of sonoporation. Scient Rep. 9(1):3218.

Tang S, Ramseyer C, Samant P, Xiang L. 2018. X-ray-induced acoustic computed tomography of concrete infrastructure. Appl Phys Lett. 112(6):063504.

ten Cate C. 2013. Acoustic communication in plants: do the woods really sing? Behav Ecol. 24(4):799–800.

Tong XM, Yamaoka H, Nagasawa H, Watanabe T. 1995. Heat-energy deposition in x-ray interaction with materials application to Si and Be. J Appl Phys. 78(4):2288–2297.

Tyree MT, Dixon MA. 1983. Cavitation events in *Thuja occidentalis* L.?: Utrasonic acoustic emissions from the sapwood can be measured. Plant Physiol. 72(4):1094–1099.

Veits M, Khait I, Obolski U, Zinger E, Boonman A, Goldshtein A, Saban K, Seltzer R, Ben-Dor U, Estlein P, et al. 2019. Flowers respond to pollinator sound within minutes by increasing nectar sugar concentration. Ecol Lett. 22(9):1483–1492.

Vicente-Torres MA, Schacht J. 2006. A BAD link to mitochondrial cell death in the cochlea of mice with noise-induced hearing loss. J Neurosci Res. 83(8):1564–1572.

White H, Helwany M, Peterson D. 2020. Anatomy, head and neck, ear organ of corti. Treasure Island (FL): StatPearls Publishing. https://www.ncbi.nlm.nih.gov/books/NBK538335/.

Woodard HQ, White DR. 1986. The composition of body tissues. Br J Radiol. 59(708):1209–1218.

Xia C, Zeng H, Zheng Y. 2020. Low-intensity ultrasound enhances the antitumor effects of doxorubicin on hepatocellular carcinoma cells through the ROS-miR-21-PTEN axis. Mol Med Rep. 21(3): 989–998.

Xiang L, Ahmad M, Nikoozadeh A, Pratx G, Khuri-Yakub B, Xing L. 2014. TU-A-9A-07: X-Ray Acoustic Computed Tomography (XACT): 100% sensitivity to X-ray absorption. Med Phys. 41(6 Part 26):448–448.

Xiang L, Han B, Carpenter C, Pratx G, Kuang Y, Xing L. 2013. X-ray acoustic computed tomography with pulsed x-ray beam from a medical linear accelerator. Med Phys. 40(1):010701.

Ye Q, Meng C, Shen Y, Ji J, Wang X, Zhou S, Jia L, Wang Y. 2016. Caveolin-1 mediates low-intensity ultrasound-induced apoptosis via downregulation of signal transducer and activator of transcription 3 phosphorylation in laryngeal carcinoma cells. Ultrasound Med Biol. 42(9):2253–2260.

Yuana Y, Jiang L, Lammertink BHA, Vader P, Deckers R, Bos C, Schiffelers RM, Moonen CT. 2017. Microbubbles-assisted ultrasound triggers the release of extracellular vesicles. Int J Mol Sci. 18(8):1610.

Zeng Q, Hong S, Wang X, Cheng Y, Sun J, Xia W. 2019. Regulation of exosomes secretion by low-intensity pulsed ultrasound in lung cancer cells. Exp Cell Res. 383(1):111448.

Zhang S, Cheng J, Qin YX. 2012. Mechanobiological modulation of cytoskeleton and calcium influx in osteoblastic cells by short-term focused acoustic radiation force. PLoS One. 7(6):e38343.

Zhou S, Schmelz A, Seufferlein T, Li Y, Zhao J, Bachem MG. 2004. Molecular mechanisms of low intensity pulsed ultrasound in human skin fibroblasts. J Biol Chem. 279(52):54463–54469.

Zomer A, Maynard C, Verweij FJ, Kamermans A, Schäfer R, Beerling E, Schiffelers RM, de Wit E, Berenguer J, Ellenbroek SIJ, et al. 2015. In vivo imaging reveals extracellular vesicle-mediated phenocopying of metastatic behavior. Cell. 161(5):1046–1057.

Zweifel R, Zeugin F. 2008. Ultrasonic acoustic emissions in drought-stressed trees-more than signals from cavitation? New Phytologist. 179(4):1070–1079.

Radiation induces stress and transgenerational impacts in the cricket, *Acheta domesticus*

Xiaobing Li and C. D. Rollo

ABSTRACT

Purpose: Radiation exposure of crickets during their fourth juvenile molt inflicted ionizing radiation damage and altered growth rate, adult size at sexual maturity. High levels of ionizing radiation also impacted the subsequent generation, likely via heritable epigenetic mechanisms. Using radiation as a proxy for external stress, we aim to understand the transgenerational impacts of stress on non-irradiated offspring.

Methods and materials: We assess the impacts of ionizing radiation on maturation mass and growth rate in F0 male and female house crickets (*Acheta domesticus*). We also assessed trans-generational impacts of irradiation on growth rate and maturation mass on non-irradiated offspring of irradiated parents compared to non-irradiated controls.

Results: Early-life exposure to high levels of ionizing radiation-induced lower growth rate and maturation mass compared to controls ($p < .0001$). Non-irradiated male F1 offspring of irradiated parents demonstrated significantly lower mass at maturation ($p = .0012$) and significantly faster time of maturation ($p < .0001$) compared to F1 non-irradiated controls.

Conclusion: Our results show that a single early-life exposure to ionizing radiation can alter male offspring development through accelerated maturation and reduced maturation mass.

Introduction

Ionizing radiation is ideal for applying precise stress on large number of individuals because the dose-rate can be precisely controlled across all irradiation samples. With the distance between individuals and the source determining the dose-rate for the exposure (Crofton et al. 1996). Crickets are an ideal research subject as their short lifespan and small size, allows rapid generation of large populations for controlled dose-response experiments as well as studying transgenerational impacts. Furthermore, juvenile crickets undergo multiple molts and maturation is easily identified by development of wings.

Exposure to ionizing radiation has a wide range of impacts on development and behavior of organisms. Radiation impacts on organisms can be divided into targeted and non-targeted effects. Targeted or direct effects include damage to DNA and cellular features. DNA damage can also result in loss of cell integrity and ultimately cell death. Besides direct impacts, radiation also hydrolyzes water to produce reactive oxygen species (ROS) that can attack DNA and other cellular features to induce cellular damage, and death at high levels (Desouky et al. 2015).

The linear no-threshold (LNT) model is commonly applied to estimate targeted and non-targeted impacts of ionizing radiation (Christensen et al. 2014). The LNT model estimates that radiation damage is proportional to dose and dose-rate. The LNT model predicts that even if exposure is insufficient to inflict significant direct damage, generation of free radicals can damage DNA (Christensen et al. 2014; Desouky et al. 2015).

The 'Hormetic' model of radiation stress holds that low-level radiation can have positive effects on physiology, development and behavior, even though higher doses can cause damage (Luckey 1991). Hormetic and damaging doses vary among species, but invertebrates are generally more radio-resistant compared to vertebrates (Paithankar et al. 2017; Timbadiya et al. 2018). Insect resistance to irradiation compared to vertebrates likely reflects that most adult tissues in insects are post-mitotic (proliferating mitotic cells are more vulnerable) (Bhatnagar et al. 1965).

Regardless, gonads of mature insects are mitotic and ionizing radiation can induce sterility. In the house fly, *Musca domestic L.*, exposure of pupae to high levels of ionizing radiation induces gender-specific responses (Bhatnagar et al. 1965). In *A. domesticus*, female hormetic doses are between $0.5 - 2\,\text{Gy}$ of radiation (dose-rate $= 0.25\,\text{Gy/min}$) (Shephard et al. 2018).

Previous research of stress in invertebrates looked primarily at post-fertilization prenatal stress. In the water fleas, *Daphnia pulex* and *D. longispina*, pregnant females exposed to predators cues, produced offspring expressing defensive phenotypes (i.e. longer tail spines, wider bodies, and expression of neck teeth) (Imai et al. 2009; Sperfeld et al. 2020). In the cuttlefish *Sepia officinalis*, prenatal maternal stress altered offspring behavior (O'Brien et al. 2017). In humans, offspring of parents with early-life dietary stress can express a similar obesity syndrome as their parent, even if they never experienced the stress (Franklin et al. 2010; Curley et al. 2011; Skinner 2014). Such trans-generational transfer of stress responses suggests a heritable, potentially epigenetic mechanism.

Research on trans-generational stress typically focuses on maternal stress. Working with insects that lack parental care and that were exposed to radiation stress (both parents) allows exploration of trans-generational stress response without parental care. In this case, stress associated with parents can still impact sperm or eggs, and thus, offspring.

Here, we examine trans-generational consequences of a single, early life exposure to ionizing radiation on maturation size and growth rate from parent to offspring. Specifically, we exposed groups of juvenile house crickets (*Acheta domesticus*) to a moderate dose (13.92 Gy @ 0.58 Gy/min) at 14 days post-hatch and examined the impacts on maturation mass and growth rate within and across generations.

Methods

Study animals and experimental groups

Breeding colony

Common house crickets (*Acheta domesticus*) were initially acquired from a local pet store, and a long-term breeding colony was established. Crickets were maintained in a large acrylic terrarium (93 × 64.2 × 46.6 cm) with a constant temperature at 29 °C ± 2 °C using 60 volt UV heat lamps, and a photoperiod of 12 h light/12 h dark photoperiod generated with overhead LED strip lights and florescent lamps. Egg carton shelters were provided, distilled water was available in soaked cellulose sponges, and food was *ad libitum* access to Quicks Country Range Multi-Fowl Grower Rations (17.0% crude protein, 2.5% crude fat, and 4.0% crude fiber). The enclosure was insulated with 1.5 cm thick Durofoam insulation, fans provided air circulation from top of enclosure.

Experimental groups

When most crickets have reached sexual maturity (expressed wings), the breeding colony was provided with oviposition mediums in small plastic containers (7 × 7 × 7 cm) (Organic Garden Soil, Swiss Farms Products Inc., Marysville, USA). One container of oviposition medium was left in the colony to collect eggs for 48 h and attracted numerous females for oviposition. Eggs were incubated at 29 °C ± 2 °C and they hatched in approximately 11–13 days. Several replicates of hatchlings were subsequently reared to maturity to ensure quality of experimental crickets.

Juveniles were removed from soil 24 h post-hatch to ensure individuals were of the same age. Juvenile *A domesticus* were collected on day 14 (4th instar) after hatching and randomly assigned to one of the 3 experimental treatments.

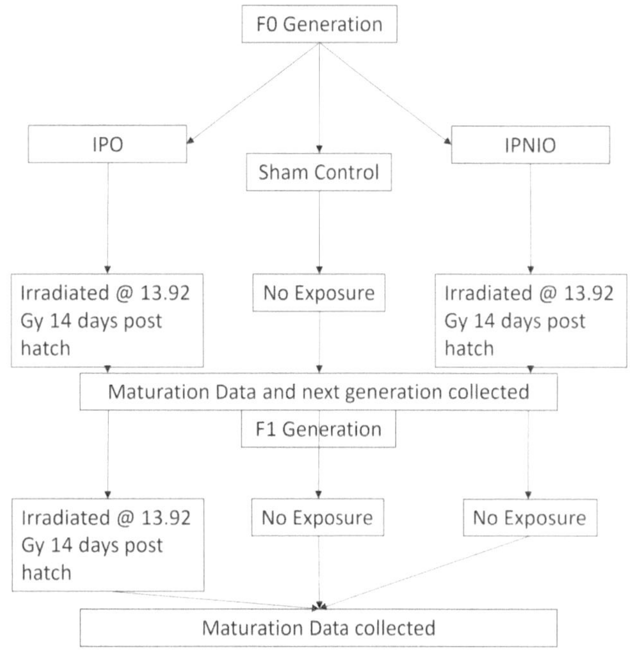

Figure 1. Experimental design of experiment. Eggs were collected in ovipositing mediums placed inside the main breeding colony and reared until hatch. 14 days post hatch crickets were randomly assigned one of three groups, sham control, irradiated parents and offspring (IPO), or Irradiated Parents and Non-Irradiated Offspring (IPNIO). All three F0 groups were kept separately for the duration of the experiment. At 14 days post hatch, all three groups were transported to the Taylor Source, but only IPO and IPNIO were exposed to radiation (13.92 Gy). F0 Maturation data and F1 eggs were then collected from the three F0 groups. 14 days after the F1 groups hatched, all three groups were transported to the Taylor Source, but only IPO was exposed to radiation (13.92 Gy). Maturation data was collected from the F1 groups.

The groups (n ≈ 200 per group) consisted of the negative sham control group (sham control), who had no exposure to radiation in both generations but were brought over to the Taylor Source and exposed to the stress of handling and travel to control for external stress factors. Irradiated Parents and Offspring (IPO) group, where both the parents and offspring were independently irradiated at 13.92 Gy at 14 days of age. And Irradiated Parents and Non-Irradiated Offspring (IPNIO) group, where only the parents were irradiated at 13.92 Gy at 14 days old, the offspring were not irradiated, but were brought the Taylor Source (Figure 1).

Irradiation

Crickets were irradiated at the Taylor Radiobiology Source (caesium-137) at McMaster University (dose rate = 0.58 Gy/min) totaling 13.92 Gy 14 days post hatch. The 13.92 Gy dose was chosen as prior testing found this dose to have negative impacts on growth.

Housing terrariums were brought to the Taylor Source in insulating Styrofoam boxes, where crickets were sorted and temporarily housed in cylindrical plastic vials (9.5 cm height x 2.5 cm width) with approximately 30–35 crickets per vial for the duration of the exposure. Crickets were irradiated one group at a time (six vials per group ~190–200 crickets). Replicates were performed to obtain a total of ~400 crickets per group. Vials were evenly spaced around another cylindrical vial to ensure even radiation exposure. Following

Table 1. Average time to sexual maturation (days) and survival to sexual maturation (n).

F0 males	Survival to maturation (*n*)	Average maturation time (days)	F1 males	Survival to maturation (*n*)	Average maturation time (days)
Sham control	173	47.71 ± 3.677	Sham control	223	47.92 ± 2.608
Irradiated parents & offspring	170	48 ± 2.955	Positive control	194	49.38 ± 3.589
Irradiated parents & non-irradiated Offspring	142	49.30 ± 4.194	Non-irradiated offspring	200	45.30 ± 3.483
F0 females	**Survival to maturation (n)**	**Average maturation time (days)**	**F1 females**	**Survival to maturation (n)**	**Average maturation time (days)**
Sham control	150	46.19 ± 3.928	Sham control	159	46.44 ± 2.592
Irradiated parents & offspring	175	47.13 ± 3.037	Positive control	151	47.41 ± 3.116
Irradiated parents & non-irradiated offspring	143	47.38 ± 4.185	Non-irradiated offspring	135	43.84 ± 3.891

Containers were monitored daily multiples times a day, all recently matured crickets were weighted immediately, and growth rate was calculated using maturation weight and time. Comparing average time to maturation, every group was maturing significantly or trending toward maturing significantly later than the sham controls, except for the IPNIO males and females, who matured significantly faster when compared to the shams.

irradiation, crickets were housed in plastic containers (30×19 x 12 cm, $n \approx 200$ per container, 2 containers per treatment group) and housed in same environment as the colony.

Maturation age and growth rate

Experimental groups were monitored daily for sexual maturity (adult molt) as indicated by expression of wings. Newly mature male and female crickets were immediately weighted with an Accuris analytical balance (readability of 0.001 g ± 0.002 g) and separated into group-dependent adult containers. Containers were monitored several times daily to ensure newly matured crickets were not missed.

Reproductive effort and next generation

After all crickets matured in experimental groups, oviposition containers were placed into the housing containers for 24 h as an oviposition medium to collect viable eggs for the next generation, this was done twice within 2 days (2 oviposition mediums per container, 4 total per treatment).

Two weeks post-hatch ~200 juvenile *A domesticus* were randomly selected from there, respectively, groups while the rest were culled. All three groups of crickets (14 days old) were taken to The Taylor Source but only F1 IPO were exposed to 13.92 Gy of radiation.

Groups were monitored daily for sexual maturity (adult molt) and were recorded as described above. Growth rate was calculated by dividing the mass at maturation in mg over days from hatch to maturation. Sample sizes and average time to maturation for all three groups across generations are presented in Table 1.

Statistical analysis

No significant difference was observed between the two containers of each treatment, and all data moving forward are combined data for an approximate sample size of 400 per group (Table 1).

Maturation weight and growth rate were analyzed using one-way ANOVA and Dunnett's multiple comparisons test to detect difference between experimental groups and control. Unpaired t tests were performed to compare differences between parents and offspring. All statistical analyses were carried out with GraphPad Prism 8.4.3.

Figures are presented as percentage differences from the sham controls. The zero-difference line signifies the average maturation mass or growth rate for the sham controls for that generation or sex. Maturation weights (mg) for all groups are presented as mean percentage differences from sham controls ± SEM. Growth rates (mg/day) were calculated as mean maturation mass (mg) over maturation time (days) and presented as mean percentage differences from sham controls ± SEM. Significances were calculated using empirical data before transforming into percentages.

Results

F_0 radiation impact on growth rate and mature mass

We examined developmental responses of crickets to damaging levels of ionizing radiation. At 13.92 Gy (0.58 Gy/min), radiation exposure reduced maturation mass (mg) and growth rates mg/day compared to non-irradiated controls.

Significant differences in maturation mass between F_0 sham, IPO, and IPNIO males were observed among F_0 males F (3, 485) = 48.43, $p < .0001$ across the three experimental groups (Figure 2). A follow-up Dunnett's multiple comparison test showed a significant decrease ($p < .0001$) in mean maturation mass of both male IPO and IPNIO compared to the sham controls. Male IPO and IPNIO crickets weighted 11.35% (334.6 mg) and 13.25% (327.5 mg) less at maturation, respectively, than male F_0 sham controls (377.5 mg).

Comparing growth rate between F_0 sham, IPO, and IPNIO males found a significant difference among the F_0 males F (3, 485) = 81.86, $p < .0001$ compared across the

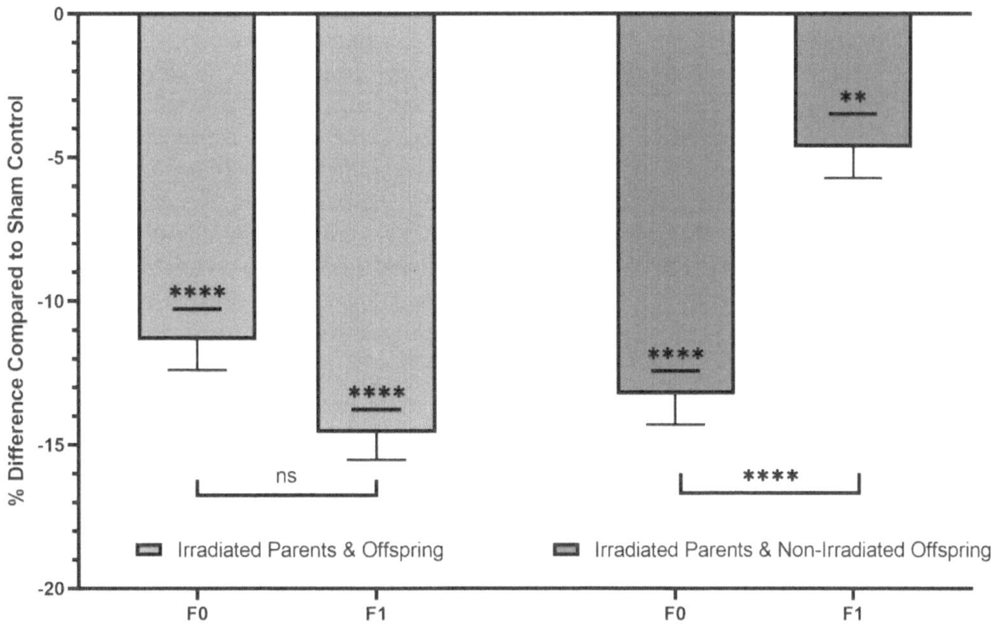

Figure 2. Effects of early life ionizing radiation on mean maturation mass (mg) ± SEM of F0 and F1 male *Acheta domesticus* Compared to Sham Controls. Percent difference in mean maturation mass of Irradiated Parents & Offspring (IPO) and Irradiated Parents & Non-Irradiated Offspring (IPNIO) compared to sham control. Sham crickets had no exposure to radiation but were exposed to same environmental stressors of being handled and traveling to the source. F0 and F1 Crickets in the IPO group were exposed to 13.92 Gy of ionizing radiation (dose rate = 0.58 Gy/min) at 14 d old. F0 crickets in IPNIO group were exposed to 13.92 Gy of radiation at 14 day old, F1 crickets were exposed to the same environmental stressors as the other groups but were not exposed to radiation. Radiation significantly decreased mean maturation mass in crickets exposed to 13.92 Gy of radiation. F1 IPNIO crickets exhibits significant difference both from the parents and the sham control, suggesting a transgenerational stress response. Dunnett's multiple comparison test indicating significant differences between IPO and IPNIO compared to sham, and significant differences within and between groups. $^{*}p < .05$, $^{**}p < .01$, $^{***}p < .001$, $^{****}p < .0001$.

three groups (Figure 4). A Dunnett's multiple comparison test detected a significant decrease ($p < .0001$) in growth rate in both male IPO and IPNIO groups independently compared to the sham control. Male IPO and IPNIO crickets grew on average 12.07% (6.968 mg/day) and 16.04% (6.653 mg/day) less than the sham controls (7.924 mg/day).

Significant differences in maturation mass between F_0 sham, IPO, and IPNIO males was observed among F_0 females F (3,468) = 35.09, $p < .0001$ among the three groups (Figure 3). Follow up Dunnett's multiple comparison test found that mean maturation mass of female IPO and IPNIO crickets were significantly lighter ($p < .0001$ for both groups, independently) than female F_0 sham crickets. Female F_0 IPO and IPNIO crickets weighed 7.773% (380 mg) and 13.57% (356.1 mg) lighter than the female F_0 sham controls (412 mg), respectively.

Comparing growth rate between F_0 sham, IPO, and IPNIO females found a significant difference among the F_0 females F (3,468) = 54.54, $p < .0001$ compared among the three groups (Figure 5). Dunnett's tests further showed that there is a significant decrease ($p < .0001$) of growth rate of female IPO and IPNIO crickets compared to the sham control. Female IPO and IPNIO crickets grew on average 9.79% (8.064 mg/day) and 15.54% (7.550 mg/day) less than the sham controls (8.939 mg/day).

Trans-generational impacts of ionizing radiation between F_0 and F_1

One-way ANOVAs demonstrated significant differences in mean maturation mass in F_1 generation among the three groups of both male (F (3,617) = 59.34, $p < .0001$) and female (F (3,445) = 23.81, $p < .0001$) crickets compared within their respective sex. T tests were performed within the IPO group where offspring were re-exposed to radiation and found no significant difference in mean body weight of both sexes when comparing between parents and offspring (Figures 2 and 3).

A Dunnett's multiple comparison test comparing among F_1 groups found mean maturation mass of both male and female IPO crickets are significantly ($p < .0001$) lighter than that of their sham counterparts (14.59% and 12.21%, respectively).

Non-Irradiated offspring maturation mass and growth rate

T tests were performed in the IPNIO group between parents and offspring where the offspring were not re-exposed to radiation. A significant difference ($p < .0001$) in male and female mean maturation size was observed when comparing between F_0 and F_1 generations (16.27% and 17.91% increase, respectively). A Dunnett's multiple comparison test between F_1 IPNIO and sham controls offspring found IPNIO males weighed significantly less ($p = .0012$) at maturation (4.658%) compared to the sham controls while there was no significant difference in mean maturation mass between female offspring.

T tests were performed between parents and offspring of the experimental groups and found no significant difference in average growth rate of F_0 and F_1 IPO males and females. T tests between F_0 and F_1 IPNIO crickets found a significant

($p < .0001$) difference in average growth rate of both males and females (26.8% and 27.1%, respectively).

Significant differences were observed in average growth rate in F_1 generation groups of both male (F (3,617) = 113.8, $p < .0001$) and female (F (3,445) = 52.65, $p < .0001$) crickets compared to their respective groups.

Dunnett's multiple comparison tests were performed and a significant difference ($p < .0001$) in average growth rate was observed in male and female F_1 IPO crickets compared to F_1 sham control crickets (16.8% and 13.64% decrease). No differences in average growth rate were observed in neither male nor female F_1 IPNIO crickets compared to F_1 sham controls.

Average time to maturation

Significantly lower mean maturation time in male and female IPNIO F_1 crickets compared to sham were observed F (3, 552) = 12.11, $p < .0001$ and F (3, 445) = 13.23, $p < .0001$, respectively (Table 1).

Discussion

High levels of radiation are linked to free radical generation, oxidative damage, and increased mortality. In insects, most of the focus is on sterilization and lethal dosages related to agricultural pest species (Paithankar et al. 2017; Timbadiya et al. 2018). Trans-generational impacts of acute damaging doses, or non-sterilization doses of radiation are poorly understood. House crickets (*Acheta domesticus*) prove to be an excellent model to study non-lethal, non-sterilization levels of ionizing radiation.

A key objective of this study was to determine whether developmental shifts caused by radiation stress were heritable. A secondary objective was to understand impacts of early-life radiation stress on development. Here we characterized responses from *A domesticus* associated with radiation stress for both irradiated parents (both parents??) and their irradiated or non-irradiated offspring. Impacts of ionizing radiation were assessed in terms of maturation mass and growth rate as key measures of development.

We found that early-life radiation stress induces heritable detrimental impacts on maturation mass without impacting survival to reproductive maturity. Developmental responses were also observed in crickets irradiated at 13.92 Gy (0.58 Gy/min). Results showed that a single early-life exposure to radiation in the first generation will reduce maturation mass of both the parents *and* their non-irradiated male offspring. These results are consistent with current models of stress inheritance (Soubry et al. 2014). We found that at reproductive maturity, crickets exposed to ionizing radiation had lower growth rate and maturation size compared to non-irradiated sham controls (Figures 2–5). These results are consistent with other radiation studies finding that early life/prenatal exposure to radiation stunted growth (Rugh et al. 1964).

Despite not being exposed to radiation, similar stress responses to their parents were observed in non-irradiated male offspring of irradiated parents. Non-irradiated male offspring of irradiated parents exhibited a similar decrease

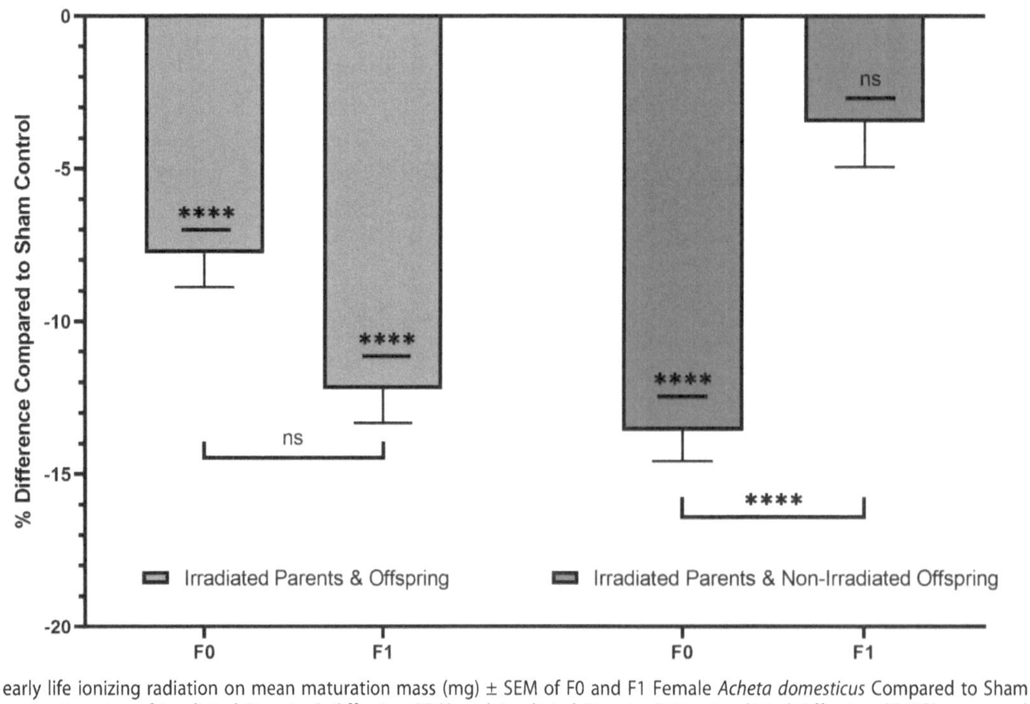

Figure 3. Effects of early life ionizing radiation on mean maturation mass (mg) ± SEM of F0 and F1 Female *Acheta domesticus* Compared to Sham Controls. Percent difference in mean maturation size of Irradiated Parents & Offspring (IPO) and Irradiated Parents & Non-Irradiated Offspring (IPNIO) compared to sham control. Sham crickets had no exposure to radiation but were exposed to same environmental stressors of being handled and traveling to the source. F0 and F1 Crickets in the IPO group were exposed to 13.92 Gy of ionizing radiation (dose rate = 0.58 Gy/min) at 14 d old. F0 crickets in IPNIO group were exposed to 13.92 Gy of radiation at 14 day old, F1 crickets were exposed to the same environmental stressors as the other groups but were not exposed to radiation. Radiation significantly decreased mean maturation mass in crickets exposed to 13.92 Gy of radiation. F1 IPNIO crickets exhibits significant difference from the parents but no significant difference from the sham control. Dunnett's multiple comparison test indicating significant differences between IPO and IPNIO compared to sham, and significant differences within and between groups. *$p < .05$, **$p < .01$, ***$p < .001$, ****$p < .0001$.

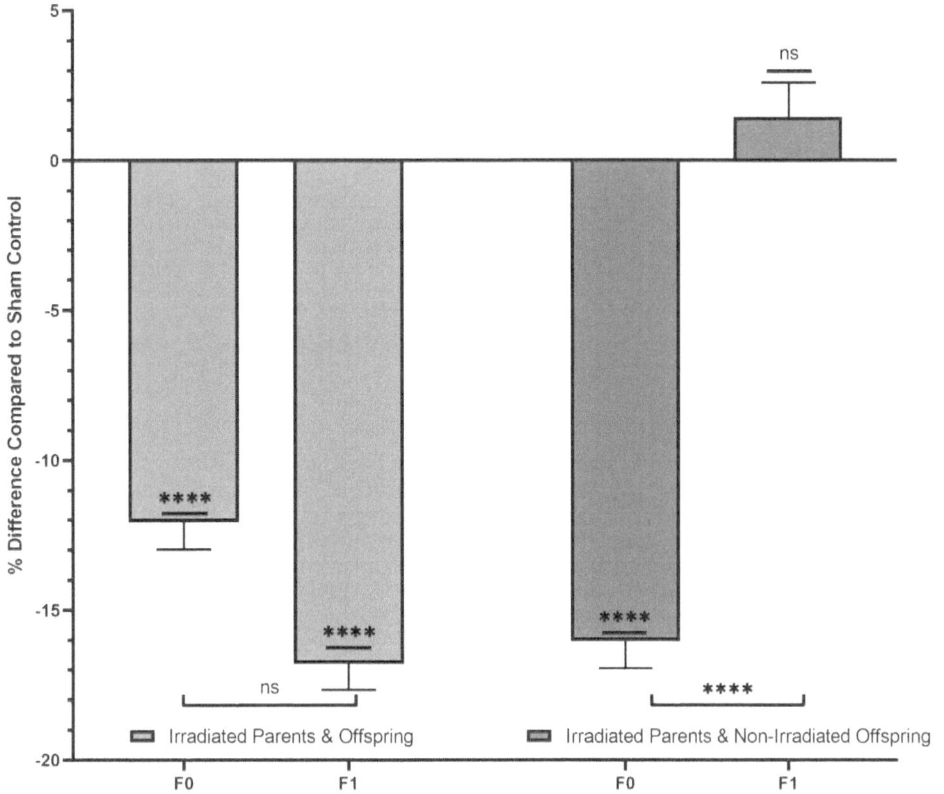

Figure 4. Effects of Early Life Ionizing Radiation on Average Growth Rate (mg/day) ± SEM of F0 and F1 Male *Acheta domesticus* Compared to Sham Controls. Percent difference in average growth rate of Irradiated Parents & Offspring (IPO) and Irradiated Parents & Non-Irradiated Offspring (IPNIO) compared to sham control. Sham crickets had no exposure to radiation but were exposed to same environmental stressors of being handled and traveling to the source. F0 and F1 Crickets in the IPO group were exposed to 13.92 Gy of ionizing radiation (dose rate = 0.58 Gy/min) at 14 d old. F0 crickets in IPNIO group were exposed to 13.92 Gy of radiation at 14 day old, F1 crickets were exposed to the same environmental stressors as the other groups but were not exposed to radiation. Radiation significantly decreased mean growth rate in crickets exposed to 13.92 Gy of radiation. F1 IPNIO crickets exhibits significant difference from the parents but no significant difference from the sham control. Dunnett's multiple comparison test indicating significant differences between IPO and IPNIO compared to sham, and significant differences within and between groups. $^{*}p < .05$, $^{**}p < .01$, $^{***}p < .001$, $^{****}p < .0001$.

in maturation mass to their irradiated parents. Male second-generation IPNIO crickets weighed significantly less than male sham control crickets ($p = .0012$, 4.65%, Figure 2) while female offspring had no significant difference in maturation weight or growth rate (Figures 3 and 5).

Our results are consistent with other studies examining offspring epigenetic profiles and health status (Pembrey et al. 2006). Paternal exposure to chemicals also causes defects in gametes and development in offspring, and this effect is more pronounced in male offspring. This has been correlated with altered DNA methylation patterns in germ lines (Anway et al 2006; Filkowski et al. 2010). In humans, radiation treatment for childhood cancers has been linked to elevated sperm DNA fragmentation in adulthood (Romerius et al. 2010). Our results are consistent with human research in suggesting a sex-linked epigenetic mechanism (Cordier 2008; Guerrero-Bosagna et al. 2012). We propose that a similar phenomenon likely occurred in our experiment, where early-life irradiation fragmented gamete DNA that extended into adulthood. This then can pass down via altered methylation patterns to offspring.

While there was a significant impact on maturation mass of nonirradiated male offspring, there was no significant difference in growth rate of these offspring. Our results instead showed a significantly faster time to maturation (at lower maturation mass) in F_1 male and female offspring compared to sham controls (Table 1). This might suggest beneficial parental shifts in epigenetic markers that cause offspring to develop faster in the presence of ionizing radiation. Faster maturation and decreased maturation mass suggest a trade-off mechanism, where organisms under stress mature faster as a defense mechanism. Alternatively, slowly growth may impact maturation.

In water fleas, offspring developing in waters with predator cues express longer tail spines and other defensive features not found in predator-free waters (Imai et al. 2009; Sperfeld et al. 2020). Perhaps a more general mechanism responds to stress, altering development to effect earlier maturation if growth rates (or even egg development) are suppressed.

Trans-generational impacts of ionizing radiation on maturation mass and growth rate may involve several mechanisms that could generalize across broad phylogenies. One of the main effects of ionizing radiation on organisms is a generalized stress response that negatively affects maturation size and growth rate. On the one hand, stress could simply disrupt normal functions resulting in reduced growth and altered maturation. Alternatively, mechanisms associated with stress resistance can alter heritable methylation and epigenetic markers that can impact both the adult and subsequent offspring. This calls for

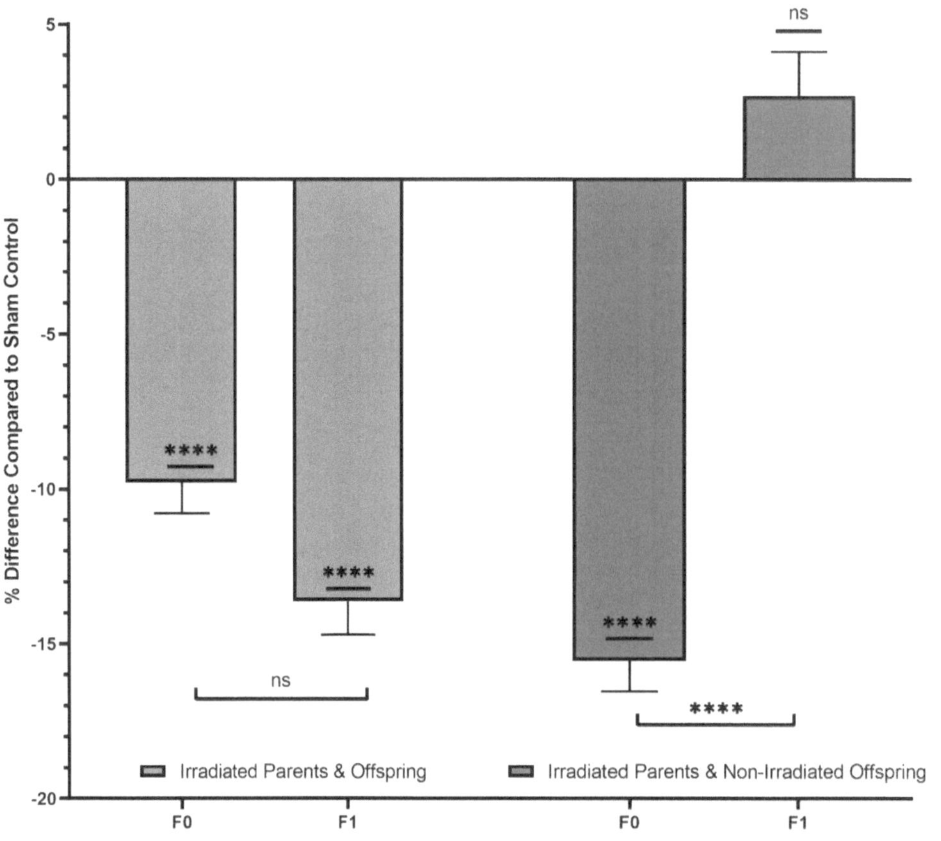

Figure 5. % Effects of early life ionizing radiation on average growth rate (mg/day) ± SEM of F0 and F1 female *Acheta domesticus* compared to Sham Controls. Percent difference in average growth rate of Irradiated Parents & Offspring (IPO) and Irradiated Parents & Non-Irradiated Offspring (IPNIO) compared to sham control. Sham crickets had no exposure to radiation but were exposed to same environmental stressors of being handled and traveling to the source. F0 and F1 Crickets in the IPO group were exposed to 13.92 Gy of ionizing radiation (dose rate = 0.58 Gy/min) at 14 d old. F0 crickets in IPNIO group were exposed to 13.92 Gy of radiation at 14 day old, F1 crickets were exposed to the same environmental stressors as the other groups but were not exposed to radiation. Radiation significantly decreased mean maturation size in crickets exposed to 13.92 Gy of radiation. F1 IPNIO crickets exhibits significant difference from the parents but no significant difference from the sham control. Dunnett's multiple comparison test indicating significant differences between IPO and IPNIO compared to sham, and significant differences within and between groups. $^{*}p < .05$, $^{**}p < .01$, $^{***}p < .001$, $^{****}p < .0001$.

understanding of the underlying physiological mechanisms inducing heritable epigenetic changes.

This work adds to a growing body of literature on trans-generational and paternal inheritance in both invertebrate and vertebrate systems. Due to their relatively small body size and short generational times, insects are an excellent model for examining trans-generational impacts of ionizing radiation. Understanding stress responses to radiation has important practical applications for nuclear power, exploration of space, medical applications, and possible accidents.

Understanding transgenerational consequences of stress is particularly relevant today, given the increased range of novel stressors from anthropogenic sources. Radiation is an excellent stressor for research given the ability to apply highly accurate doses, the wealth of scientific research and the range of topics and applications (energy generation, medicine, space travel, environmental pollution). Our results emphasize that parental and early-life exposure to ionizing radiation impacts development and can be heritable across generations.

Acknowledgements

This manuscript was improved from advice provided by Tamara Fuciarelli. Special thanks go to Johnathon Stone and Carmel Mothersill, who provided helpful advice over the course of the project. The authors acknowledge funding from NSERC, and Bruce Power awarded to CDR and XL.

Disclosure statement

No potential conflict of interest was reported by the author(s).

ORCID

Xiaobing Li http://orcid.org/0000-0002-2719-8593
C. D. Rollo http://orcid.org/0000-0001-6849-2003

References

Anway MD, Leathers C, Skinner MK. 2006. Endocrine disruptor vinclozolin induced epigenetic transgenerational adult-onset disease. Endocrinology. 147(12):5515–5523.

Bhatnagar PL, Rockstein M, Dauer M. 1965. X-irradiation of pupae of the house fly *Musca domestica* L. and adult survival. Exp Gerontol. 1(2):149–159.

Christensen DM, Iddins CJ, Sugarman SL. 2014. Ionizing radiation injuriesand illnesses. Emer Med Clin. 32(1):245–265.

Cordier S. 2008. Evidence for a role of paternal exposures in developmental toxicity. Basic Clin Pharmacol Toxicol. 102(2):176–181.

Crofton KM, Padilla S, Tilson HA, Anthony DC, Raymer JH, MacPhail RC. 1996. The impact of dose rate on the neurotoxicity of acrylamide: the interaction of administered dose, target tissue concentrations, tissue damage, and functional effects. Toxicol Appl Pharmacol. 139(1):163–176.

Curley JP, Mashoodh R, Champagne FA. 2011. Epigenetics and the origins of paternal effects. Horm Behav. 59(3):306–314.

Desouky O, Ding N, Zhou G. 2015. Targeted and non-targeted effects of ionizing radiation. J Radiat Res Appl Sci. 8(2):247–254.

Filkowski JN, Ilnytskyy Y, Tamminga J, Koturbash I, Golubov A, Bagnyukova T, Pogribny IP, Kovalchuk O. 2010. Hypomethylation and genome instability in the germline of exposed parents and their progeny is associated with altered miRNA expression. Carcinogenesis. 31(6): 1110–1115.

Franklin TB, Russig H, Weiss IC, Gräff J, Linder N, Michalon A, Vizi S, Mansuy IM. 2010. Epigenetic transmission of the impact of early stress across generations. Biol Psychiatry. 68(5):408–415.

Guerrero-Bosagna C, Covert TR, Haque MM, Settles M, Nilsson EE, Anway MD, Skinner MK. 2012. Epigenetic transgenerational inheritance of vinclozolin induced mouse adult onset disease and associated sperm epigenome biomarkers. Reprod Toxicol. 34(4): 694–707.

Imai M, Naraki Y, Tochinai S, Miura T. 2009. Elaborate regulations of the predator-induced polyphenism in the water flea *Daphnia pulex*: kairomone-sensitive periods and life-history tradeoffs. J Exp Zool A: Ecol Genet Physiol. 311(10):788–795.

Luckey TD. 1991. Radiation hormesis. Boca Raton (FL): CRC Press.

O'Brien CE, Jozet-Alves C, Mezrai N, Bellanger C, Darmaillacq AS, Dickel L. 2017. Maternal and embryonic stress influence offspring behavior in the cuttlefish *Sepia officinalis*. Front Physiol. 8:981.

Paithankar JG, Deeksha K, Patil RK. 2017. Gamma radiation tolerance in different life stages of the fruit fly *Drosophila melanogaster*. Int J Radiat Biol. 93(4):440–448.

Pembrey ME, Bygren LO, Kaati G, Edvinsson S, Northstone K, Sjöström M, Golding J. 2006. Sex-specific, male-line transgenerational responses in humans. Eur J Hum Genet. 14(2):159–166.

Romerius P, Ståhl O, Moëll C, Relander T, Cavallin-Ståhl E, Gustafsson H, Löfvander Thapper K, Jepson K, Spanò M, Wiebe T, et al. 2010. Sperm DNA integrity in men treated for childhood cancer. Clin Cancer Res. 16(15):3843–3850.

Rugh R, Duhamel L, Osborne AW, Varma A. 1964. Persistent stunting following x-irradiation of the fetus. Am J Anat. 115(1):185–197.

Shephard AM, Aksenov V, Tran J, Nelson CJ, Boreham DR, Rollo CD. 2018. Hormetic effects of early juvenile radiation exposure on adult reproduction and offspring performance in the cricket (*Acheta domesticus*). Dose Response. 16(3):1559325818797499.

Skinner MK. 2014. Environmental stress and epigenetic transgenerational inheritance. BMC Med. 12(1):153.

Soubry A, Hoyo C, Jirtle RL, Murphy SK. 2014. A paternal environmental legacy: evidence for epigenetic inheritance through the male germ line. Bioessays. 36(4):359–371.

Sperfeld E, Nilssen JP, Rinehart S, Schwenk K, Hessen DO. 2020. Ecology of predator-induced morphological defense traits in *Daphnia longispina* (Cladocera, Arthropoda). Oecologia. 192(3):687–698.

Timbadiya B, Sisodiya D, Sharma A. 2018. Gamma radiation: an important tool for pest management in agriculture. Trends Biosci. 47(11):4347–4349.

Evaluation of interactive effects of phosphorus-32 and copper on marine and freshwater bivalve mollusks

Emily L. Vernon, Michael N. Moore, Tim P. Bean, and Awadhesh N. Jha

ABSTRACT

Purpose: Contaminants seldom occur in isolation in the aquatic environment. While pollution of coastal and inland water bodies has received considerable attention to date, there is limited information on potential interactive effects between radionuclides and metals. Whether by accidental or controlled release, such contaminants co-exist in aquatic ecosystems and can pose an enhanced threat to biota. Using a range of biological responses, the study aimed to evaluate relative interactive effects on representative freshwater and marine bivalve species.

Methods: An integrated, multi-biomarker approach was adopted to investigate response to copper (Cu, 18 μg L^{-1}), a known environmentally relevant genotoxic metal and differing concentrations of phosphorus-32 (^{32}P; 0.1 and 1 mGy d^{-1}), alone and in combination in marine (*Mytilus galloprovincialis*) and freshwater (*Dreissena polymorpha*) mussels. Genetic and molecular biomarkers were determined post-exposure and included DNA damage (as measured by the comet assay), micronuclei (MN) formation, γ-H2AX foci induction and the expression of key stress-related genes (i.e. *hsp70/90, sod, cat, gst*).

Results: Overall, using a tissue-specific (i.e. gill and digestive gland) approach, genotoxic response was reflective of exposures where Cu had a slight additive effect on ^{32}P-induced damage across the species (but not all), cell types and dose rates. Multivariate analysis found significant correlations between comet and γ-H2AX assays, across both the tissues. Transcriptional expression of selected genes were generally unaltered in response to contaminant exposures, independent of species or tissues.

Conclusions: Our study is the first to explore the interactive effects of ionizing radiation (IR) and Cu on two bivalve species representing two ecological habitats. The complexity of IR-metal interactions demonstrate that extrapolation of findings obtained from single stressor studies into field conditions could be misrepresentative of real-world environments. In turn, environmental protective strategies deemed suitable in protecting biota from a single, isolated stressor may not be wholly adequate.

HIGHLIGHTS

- Adoption of an integrated, multi-biomarker approach in two bivalve species.
- Toxicity of combined mixtures of ^{32}P and Cu compared.
- Cu induced additive effects with ^{32}P in the tissues.
- DNA damage and DDR showed strong correlations.
- Multiple stressors should be considered in assessing the impact of ionizing radiations.

Abbreviations: μg: microgram; γ-H2AX: Gamma Histone 2AX; ^{60}Co: Cobalt 60; ^{137}Cs: Cesium 137; ^{3}H: Tritium; ^{32}P: Phosphorus 32; ^{32}S: Sulfur 32; ^{90}Sr: Strontium 90; ACT: Actin; ATP: Adenosine triphosphate; Bq: Becquerel; BHA: Butylated hydroxyanisole; CAT: Catalase; cDna: Complimentary DNA; CPM: Counts per minute; Cq: Threshold cycle; Cu: Copper; DAPI: 4′:6-diamidino-2-phenylindole dihydrochloride; DDR: DNA damage response; DG: Digestive gland; DI: Deionized water; DO: Dissolved oxygen; DOC: Dissolved organic carbon; DP: *Dreissena polymorpha*; DSB: Double strand break; EF1: Elongation factor 1; ERICA: Environmental Risk from Ionizing Contaminants: Assessment and Management; GST: Glutathione-S-Transferase; HCl: Hydrochloric acid; HSP: Heat Shock Protein; ICP-MS: Inductively coupled plasma mass spectrometry; IR: Ionizing radiation; LET: Linear energy transfer; LMA: Low melting point agarose; LSC: Liquid Scintillation Counting; MBq: Mega Becquerel; MG: *Mytilus galloprovincialis*; mGy: Milligray; MN: Micronuclei; MOA: Mechanism of action; MT-10: Metallothionein 10; NPP: Nuclear power plant; PCA: Principal component analysis;

PCR: Polymerase chain reaction; qPCR: Quantitative PCR; RER: Relative expression ratio; ROS: Reactive oxygen species; SOD: Superoxide Dismutase; SSB: Single strand break; TBHQ: Tertiary-butyl hydroquinone

1. Introduction

Radionuclides and metals co-exist and are often co-discharged in the aquatic environment. Their increased prominence results primarily from anthropogenic activities (e.g. industrial discharge, nuclear power generation and decommissioning, accidents or weapons tests, mining, wastewater treatment; Salbu et al. 2005; Hu et al. 2010; Skipperud and Salbu 2018). Nuclear Power Plants (NPPs) established near the aquatic environment, in general, co-dispose or discharge both radionuclides and non-radioactive wastes in the environment. The requirement for large volumes of water for cooling processes typically results in NPPs being located close to large water bodies (i.e. coasts, lakes), where waste products are disposed under set regulations. Whilst discharges of contaminants are largely controlled, little is known about potential interactive effects (Pearson et al. 2018). Exposure to a mixture of stressors which in isolation may not induce significant damage may cause deleterious effects on organism's health through additive or synergistic mechanisms when acting in combination (Eggen et al. 2004; Salbu et al. 2005; Mothersill et al. 2007; Olsvik et al. 2010; Heier et al. 2013; Dallas et al. 2012, 2016).

It is well accepted that radionuclides and metals can readily bioaccumulate and concentrate in the tissues of aquatic organisms, posing a threat to both humans via food consumption and biota by trophic level transfer (Carvalho 2018). When present within a cell, such contaminants have varying potential to cause significant damage to molecules such as DNA, both directly (direct interaction with DNA molecules or other cellular structures) or indirectly (generation of reactive oxidative species [ROS]), the extent dependent on concentration, linear energy transfer (LET), distribution and biological half-life. Short-lived radionuclide phosphorus-32 (^{32}P, half-life = 14.29 d) was utilized in this study as a cost-effective, accessible surrogate for more environmentally ubiquitous beta and gamma emitting radionuclides (e.g. cesium-137; ^{137}Cs, strontium 90; ^{90}Sr). In comparison to other radionuclides there is limited information regarding its environmental presence. ^{32}P reference levels, referring to concentrations that, when consumed at 2 L day^{-1} equate to a total ingested dose of 0.10 mSv y^{-1} (in humans) are fixed at 57 Bq L^{-1} (DWQR, 2014). Environmental background levels average 0.27 ± 0.21 Bq L^{-1} (2005–2013, River Clyde, Erskine Harbor, King George V Dock, SEPA 2013). Despite its short half-life it could readily be accumulated by organisms and concentrated in tissues, particularly following chronic exposures. In terms of biological impact, ^{32}P is chemically and radiologically distinct (Cheng et al. 2015). Subsequent to the incorporation of aqueous ^{32}P into the ribose-phosphate backbone of replicating DNA, isotopic decay (^{32}P to sulfur-32, ^{32}S) results in chemical breakage of DNA (single strand breaks, SSBs). More detrimental DNA damage (double strand breaks, DSBs) are induced via the release of high-energy beta particles. As demonstrated by us previously, ^{32}P readily accumulated in the tissues of mussels and detrimental ^{32}P-induced responses (DNA damage, micronuclei [MN] and γ-H2AX foci induction) have been noted at dose rates as low as 0.1 mGy d^{-1} (Vernon et al. 2018; 2020). Similarly, radionuclides such as ^{137}Cs and tritium (^{3}H) have been found to induce biological damage across multiple aquatic species including mollusks (*C. gigas*, *Mytilus spp.*), arthropods (*D. magna*) and fish (*D. rerio*, *O. mykiss*, *O. latipes*) (Hagger et al. 2005; Jha et al. 2005; Jha et al. 2006; Jaeschke et al. 2011; Pereira et al. 2011; Devos et al. 2015; Parisot et al. 2015; Dallas et al. 2016; Sayed et al. 2017; Pearson et al. 2018; Festarini et al. 2019). For example at low ^{3}H doses ($\sim$15.58 μGy h^{-1}), Dallas et al. (2016) noted upregulation of genes involved in protein folding, DNA DSB repair and cell cycle checkpoint control in *M. galloprovincialis*, transcriptome level effects were well correlated with genetic (DNA) damage.

Metals such as copper (Cu) can be highly toxic to organisms at concentrations present within marine and freshwater environments. Although biologically essential, playing an important role in numerous physiological and biochemical processes (i.e. growth, metabolism and enzymatic activities), Cu ions are redox-active and facilitate the production of ROS, leading to oxidative damage to biomolecules including DNA, proteins and lipids (Cid et al. 1995; Gaetke and Chow 2003; Bopp et al. 2008; Kim et al. 2008). Typically, permitted Cu levels in England and Wales are 5 μg L^{-1} in seawater and between 1–28 μg L^{-1} in freshwater (dependant on water hardness, DEFRA 2014), however total dissolved Cu in contaminated environments, such as around NPPs can reach concentrations of 689 μg L^{-1} (Bryan and Gibbs 1983). Cu-induced damaging effects in aquatic invertebrates are well studied (Al-Subiai et al. 2011; Vosloo et al. 2012; Brooks et al. 2015; Xu et al. 2016; Vernon and Jha 2019), and much research has focused on interactions between Cu and additional stressors such as metals, ocean acidification, temperature, pesticides, microplastics (Clayton et al. 2000; Bouskill et al. 2006; Trevisan et al. 2011; Maria et al. 2013; Lewis et al. 2016). However, to our knowledge no studies have determined the potential interactive effects of IR and Cu in mussels.

In this study, the effects of ^{32}P and Cu, alone and in combination were studied in two ecologically relevant adult bivalve species. *Mytilus galloprovincialis* (MG) and *Dreissena polymorpha* (DP) were selected as representatives of marine and freshwater environments (McDonald et al. 1991; Binelli et al. 2015; Vernon et al. 2018; Vernon and Jha, 2019). Mussels are well established model species in ecotoxicological studies and as such the physiology, anatomy and ecology of MG and DP are well understood. This multi-species exposure enables determination of relative sensitivity and is arguably a more robust, environmentally representative approach than single-species

Table 1. Table to show the expected dose rates in mGy d^{-1} and the average dose rate achieved in *M. galloprovincialis* and *D. polmorpha* whole-body, digestive gland and gill tissue (mGy d^{-1}).

		Av. Dose rate (mGy d^{-1})		
	Expected dose rate	Whole body	Digestive gland	Gill
M. galloprovincialis	0.1	0.11	**4.34**	0.09
	1	**0.96**	**38.76**	**0.66**
D. polymorpha	0.1	0.08	**1.53**	**0.8**
	1	**0.87**	**3.72**	**1.68**

Values in bold are those above the ERICA tool screening value of 0.24 mGy d^{-1}.

Table 2. Water quality parameters (pH, temperature, salinity and dissolved oxygen [DO]), copper concentrations in water (μg L^{-1}) and ^{32}P concentrations in water (Bq L^{-1}), for both species.

Water parameters	*M. galloprovincialis*	*D. polymorpha*
pH	8.1 ± 0.06	8.1 ± 0.08
Temp (°C)	14.6 ± 0.24	14.7 ± 0.23
Salinity	36.7 ± 3.90	0.3 ± 0.03
DO (%)	99.4 ± 2.15	93.2 ± 1.33
Copper water conc. (μg L^{-1})		
Control	1.9 ± 0.08	0.3 ± 0.02
0.1 mGy d^{-1} + Cu	14.0 ± 0.72	17.1 ± 0.21
0.1 mGy d^{-1}	2.4 ± 0.24	0.3 ± 0.07
1 mGy d^{-1} + Cu	15.8 ± 0.49	17 ± 0.14
1 mGy d^{-1}	4.1 ± 0.56	0.4 ± 0.16
56 μg L^{-1}	40.9 ± 2.09	56.0 ± 0.60
^{32}P water conc. (Bq L^{-1})		
Control	0.8 ± 1.04	0.2 ± 0.58
0.1 mGy d^{-1} – 993 / 579	1246.3 ± 360.94	573.2 ± 444.04
1 mGy d^{-1} – 9930 / 5786	9712.8 ± 1235.16	4641.0 ± 2128.19

Data is presented as mean ± standard deviation. $n = 3$.

exposures (Chapman 2002; Solomon and Sibley 2002; Schnug et al. 2014).

A suite of sub-lethal biological responses or biomarkers were measured in the digestive gland (DG) and gill cells of the selected mussel species, following exposure to a range of ^{32}P and Cu, alone and in combination. ^{32}P dose rates of 0.10 and 1 mGy d^{-1} (Table 1) were in line with a generic screening value (all species) of 10 μGy h^{-1} (0.24 mGy d^{-1}), where no significant negative effects are expected at the population level (Andersson et al. 2008; Andersson et al. 2009). Cu concentrations (18 μg L^{-1}) were reflective of environmental realistic values, and in line with previous studies (Al-Subiai et al. 2011; Vernon and Jha 2019). The overall aims and objectives of this study were (a) to adopt an integrated, multi-biomarker approach in investigating combined effects of Cu and ^{32}P in two bivalve species, (b) to determine potential correlations between different parameters (molecular, genetic) studied and (c) to determine relative sensitivity between different cell types (i.e. gill and digestive gland cells). Principal Component Analysis (PCA) combined with Hierarchical Cluster Analysis was utilized to integrate data from molecular and genetic biomarkers. With regards to species variation, we hypothesized firstly that little disparity in response will be evident. Secondly, Cu would have a modifying (i.e. additive or synergistic) effect on the ^{32}P-induced biological responses in mussels.

2. Materials and methods

2.1. Chemicals and suppliers

Radiolabelled-ATP (Adenosine triphosphate, γ-^{32}P) was purchased from Perkin Elmer (Perkin Elmer Ltd, Beaconsfield, UK, 9.25 MBq: specific activity: 370 MBq mL^{-1}) and diluted with DI water to form appropriate working stocks (Vernon et al. 2018). All working solutions were decay adjusted throughout the exposure. Where appropriate, product details for chemicals and reagents are noted within the text.

2.2. Mussel exposure conditions

Mussels were collected from a reference site and maintained as described in detail in other studies from our research group (Dallas et al. 2013; Dallas et al. 2016; Pearson et al. 2018; Vernon et al. 2018).

Ten-day exposures were performed between September and October 2017. MG and DP experiments were staggered by two weeks for logistical reasons. Following collection and after a 2-week acclimation, MG (mean wet weight: 12.17 ± 0.67 g) and DP (mean wet weight: 3.22 ± 0.27 g) individuals (total wet weight, 35 g L^{-1}) per labeled beaker (9 mussels treatment^{-1}; 3 mussels beaker^{-1}) were exposed to the following exposure scenarios in triplicate: (a) 0.1 mGy d^{-1}, (b) 0.1 mGy d^{-1} + Cu, (c) 1 mGy d^{-1} and (d) 1 mGy d^{-1} + Cu. The Cu (CuSO4.5H2O, 99% purity) concentration used in combination with ^{32}P was 18 μg L^{-1}. Control and positive control ($CuSO_4.5H_2O$, 56 μg L^{-1}) treatments were run alongside. ^{32}P activity levels in water were calculated in a previous study (Vernon et al. 2018). To meet expected nominal dose rates ^{32}P activity concentrations for 0.1 mGy d^{-1} were 993 Bq L^{-1} (MG) and 579 Bq L^{-1} (DP), and for 1 mGy d^{-1} 9930 Bq L^{-1} (MG) and 5786 Bq L^{-1} (DP).

Mussels were fed (2 h before 50% water change) on days 3, 5, 7 and 9, as described in earlier studies (Vernon et al. 2018). Water quality parameters (pH, temperature, salinity and dissolved oxygen [DO]) were measured before and after water changes. ^{32}P and Cu activity levels were determined using water samples (1 mL, in duplicate), taken ~30 min after each water change and processed for liquid scintillation counting (LSC) or ICP-MS, as described in Section 2.3. LSC counting and ICP-MS data confirmed that achieved values across all treatment groups were in line with nominal concentrations (Table 2).

2.3. Sampling procedures

After exposures, gill and digestive gland tissues were dissected from each individual, other soft tissue and shell was discarded. $^2/_3$ tissue was stored in tube (on ice, 4 °C) until isolation of cells and $^1/_3$ in RNAlater (1.5 mL, Fisher UK) at −20 °C.

2.4. Biological assays

2.4.1. Isolation of digestive gland and gill cells

Gill and digestive gland cells were extracted in accordance with Vincent-Hubert et al. (2011), with minor adjustments. Samples were incubated with 1 mL dispase II solution

(37 °C, 30 min) and shaken every 10 min. Following incubation, cell suspension was centrifuged (775 g, 5 min) and supernatant utilized in consequent assays. Viability of cells was greater than (<) 90% across all treatments, as determined using the Trypan Blue exclusion dye assay (Strober 2001, data not included).

2.4.2. Comet assay to determine DNA strand breaks, analysis of micronuclei (MN) and γ-H2AX foci induction

Biological assays are described in detail in previous studies from our laboratory (Jha et al. 2005; Cheung et al. 2006; Al-Subiai et al. 2011; Dallas et al. 2013; Pearson et al. 2018; Vernon et al. 2020). DNA strand breakage was determined using the single-cell gel-electrophoresis or comet assay. Comet IV imaging software (Perceptive Imaging, Bury St Edmunds, UK) was used for scoring the cells (100 cells per slide). Genotoxicity data are presented as % Tail DNA in accordance to previous studies (Kumaravel and Jha 2006; Vernon et al. 2020). MN induction and γ-H2AX expression analysis are described in full in Vernon et al. (2020). MN were scored following widely accepted criteria (Venier et al. 1997; Bolognesi and Fenech 2012). The results were reported as mean MN per 1000 cells in accordance with previously published data from our laboratory (Dallas et al. 2013). In terms of γ-H2AX expression analysis, number of γ-H2AX foci in each of 50 cells per individual/slide were counted using a fluorescence microscope (NIKON Epifluorescence 80i, 60x magnification). An appropriate DAPI filter was used to locate cell nuclei and a FITC filter set for the FITC signal of the primary antibody (anti-GamaH2H, Mouse Monoclonal Antibody, Sigma) for γ-H2AX foci (Festarini et al. 2015).

2.4.3. Determination of transcriptional expression of key genes

RNA extraction, cDNA synthesis and Real-time Polymerase chain reaction (qPCR) were performed in accordance to Vernon et al. (2020).

2.5. Water quality measurements and ^{32}P and Cu analyses

2.5.1. ^{32}p activity level analysis in water samples

Prior to liquid scintillation counting (Hidex 300SL), water samples were combined with 4 mL ScintLogic U scintillant (Labologic Systems Ltd, Sheffield, UK) in 20 mL borosilicate glass (Fisherbrand™) vials. Samples, including blanks were stored in dark (~2) and read for 10 s (in triplicate). Samples were decay adjusted (due to short half-life of ^{32}P) and background corrected by subtracting blanks. An activity of 0.000 was assigned to CPM (counts per min) sample values below the blank (Vernon et al. 2018; Jaeschke and Bradshaw 2013).

2.6. Cu concentration analysis in water samples

Immediately post extraction, water samples were spiked with 50 μl HCl, and stored until analysis (room temperature). In line with Dallas et al. (2013), indium (115-In) and iridium (193-Ir) were added as internal standards. In accordance with Al-Subiai et al. (2011), appropriate Cu standards were used to calibrate the instrument before and during analysis of samples. Following dilution of seawater samples (1:5, DI water), standards, samples and blanks (run every 10 samples) were analyzed using X Series II ICP-MS (Plasma Quad PQ2 Turbo, Thermo Elemental, Winsford, UK) with PQ Vision 4.1.2 software.

2.7. Dosimetry and the ERICA TOOL

The ERICA (Environmental Risk from Ionizing Contaminants: Assessment and Management) tool was used for dose estimation, as described in detail in Vernon et al. (2018, 2020). Dose rates for whole-body and tissue-specific ^{32}P concentrations are presented in Table 1.

2.8. Statistical analysis

LinRegPCR (version 11, Ramakers et al. 2003; Ruijter et al. 2009) and threshold cycle (Cq) was utilized to determine PCR efficiencies. From such values, the relative mRNA expression ratio (RER) of the five selected genes was quantified (REST software [v 2009, Qiagen Ltd], Dallas et al. 2013, 2016), relative to reference genes actin (*act*) and elongation factor 1 (*ef1*), using control values as calibrators.

Statistics were performed in R (RStudio, R 3.4.3 GUI 1.70 El Capitan build (7463), https://www.r-project.org/). Where applicable, data were checked for normality and homogeneity of variance, with visual inspection of residuals. Appropriate non-parametric (Kruskal–Wallis with Dunn's pairwise comparison with Bonferroni correction) or parametric tests (one-way ANOVA was run with Tukey's post hoc, with Wilcoxon rank sum test with Holm–Bonferroni correction). Significance level was fixed at $p < .05$ (*) and data (unless noted otherwise) shown as mean ± standard deviation.

2.8.1. Multivariate analysis

Biomarker data for gill and digestive gland tissues in both species were analyzed using multivariate analysis software, PRIMER v 6.1.5 (PRIMER-€ Ltd., U. Auckland, New Zealand; Clarke 1999; Clarke and Warwick 2001), in accordance with Sforzini et al. (2020). Possible 'significant biomarkers', those which capture the full PCA biological response pattern were identified using PRIMER v6 – BIO-ENV routine.

3. Results

Cu and ^{32}P concentrations, along with water quality measurements are displayed in Table 2. Whole body and specific tissue (i.e. gill and digestive gland) dose rates are illustrated in Table 1. Whole body dose rates were 0.11 and 0.96 for MG and 0.08, 0.87 mGy d^{-1} for DP, in line with predicted values.

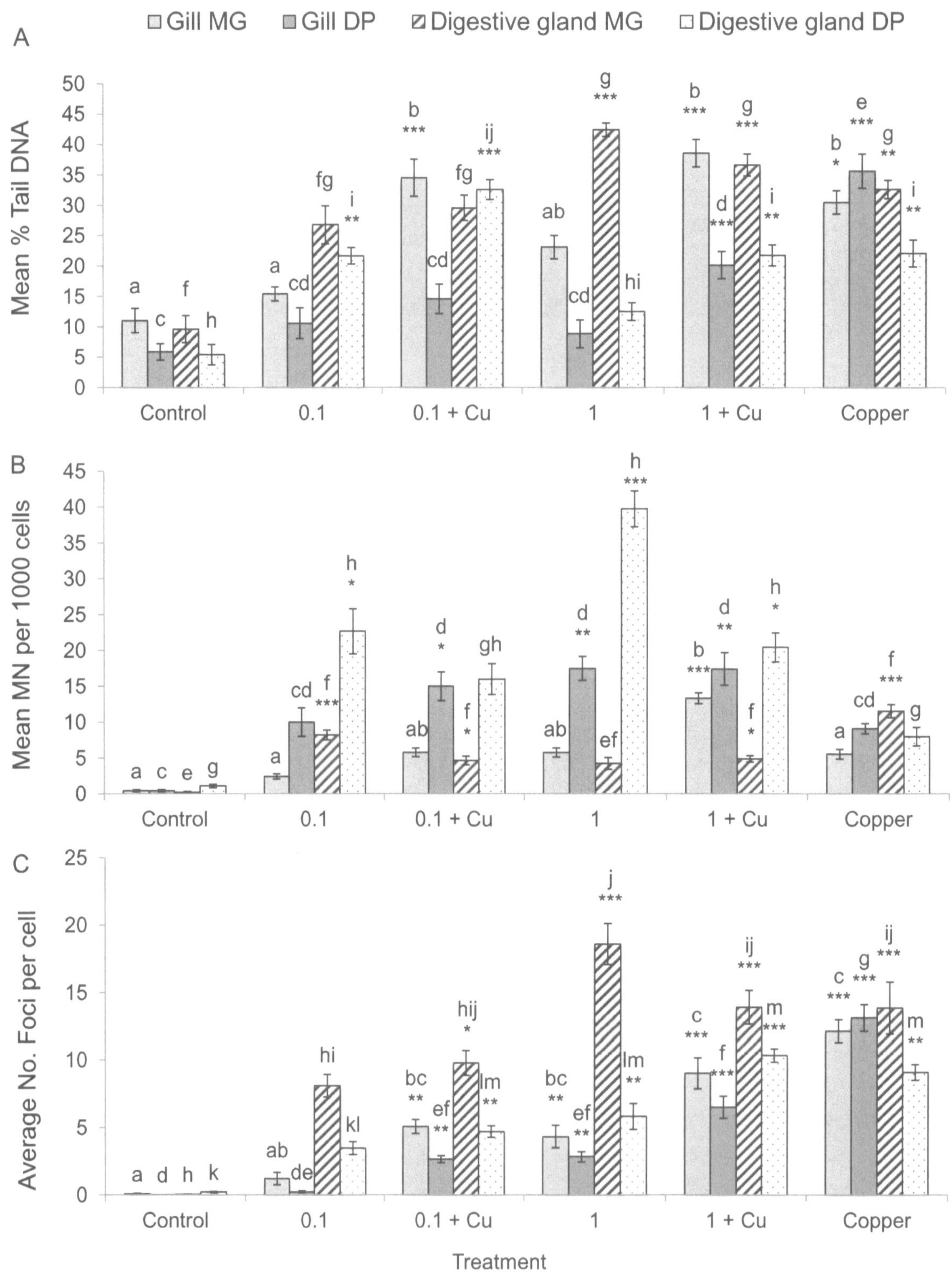

Figure 1. Genotoxic effects and subsequent repair in *M. galloprovincialis* and *D. polymorpha* gill and digestive gland cells following a 10 day exposure to ^{32}P and Cu, alone and in combination. Asterisks (*, ** or ***) are indicative of significant differences ($p < .05$, .01, .001) from the corresponding control. Letters are indicative of significant differences ($p < .05$) between species tissue (i.e. MG gill tissue). SD is standard deviation of mean data. $n = 9$. Copper = positive control.

3.1. Genotoxic response following in vivo exposures to ^{32}P and Cu

Figure 1 shows mean (A) % tail DNA damage, (B) MN per 1000 cells and (C) γ-H2AX induction in MG and DP gill and digestive gland cells, following exposure to varying concentrations of ^{32}P and Cu, alone and in combination for 10 days. Potential correlations within the biomarker data are displayed on Figure 2(a,b,A–D). Control (unexposed) treatments show a low degree of damage across all biomarkers, indicative of good health in both mussel species.

3.1.1. Comet assay to determine DNA strand breaks

While the only significant ($p < .01$) interaction between stressors was evident in MG gill, between the 0.1 and 0.1

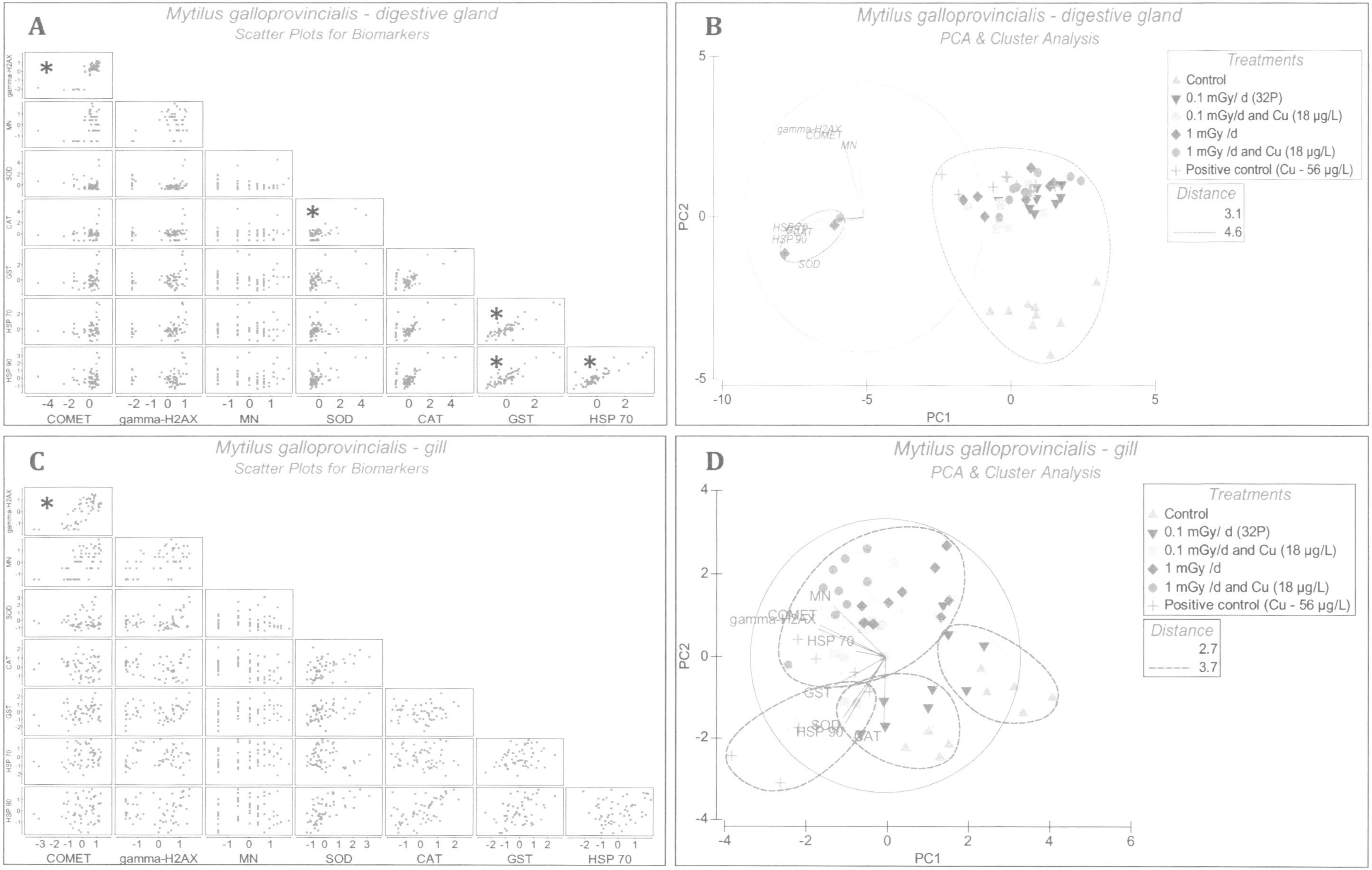

Figure 2. (a) Multivariate analysis (PRIMER 6.1.5) between treatments for the integrated biomarker data showing scatter plots (A, C), principal component (PCA) and cluster analysis of the biomarker data. Vectors indicate the directionality of specific biomarkers. (A, B) *M. galloprovincialis* digestive gland and (C, D) gill tissue. Asterisks (*) are indicative of significant differences ($p < .05$). PC1 captured (B) 47.5% and (D) 32.9% of the variation. (b) Multivariate analysis (PRIMER 6.1.5) between treatments for the integrated biomarker data showing scatter plots (A, C), principal component (PCA) and cluster analysis of the biomarker data. Vectors indicate the directionality of specific biomarkers. (A, B) *D. polymorpha* digestive gland and (C, D) gill tissue. Asterisks (*) are indicative of significant differences ($p < .05$). PC1 captured (B) 32.6% and (D) 33.2% of the variation.

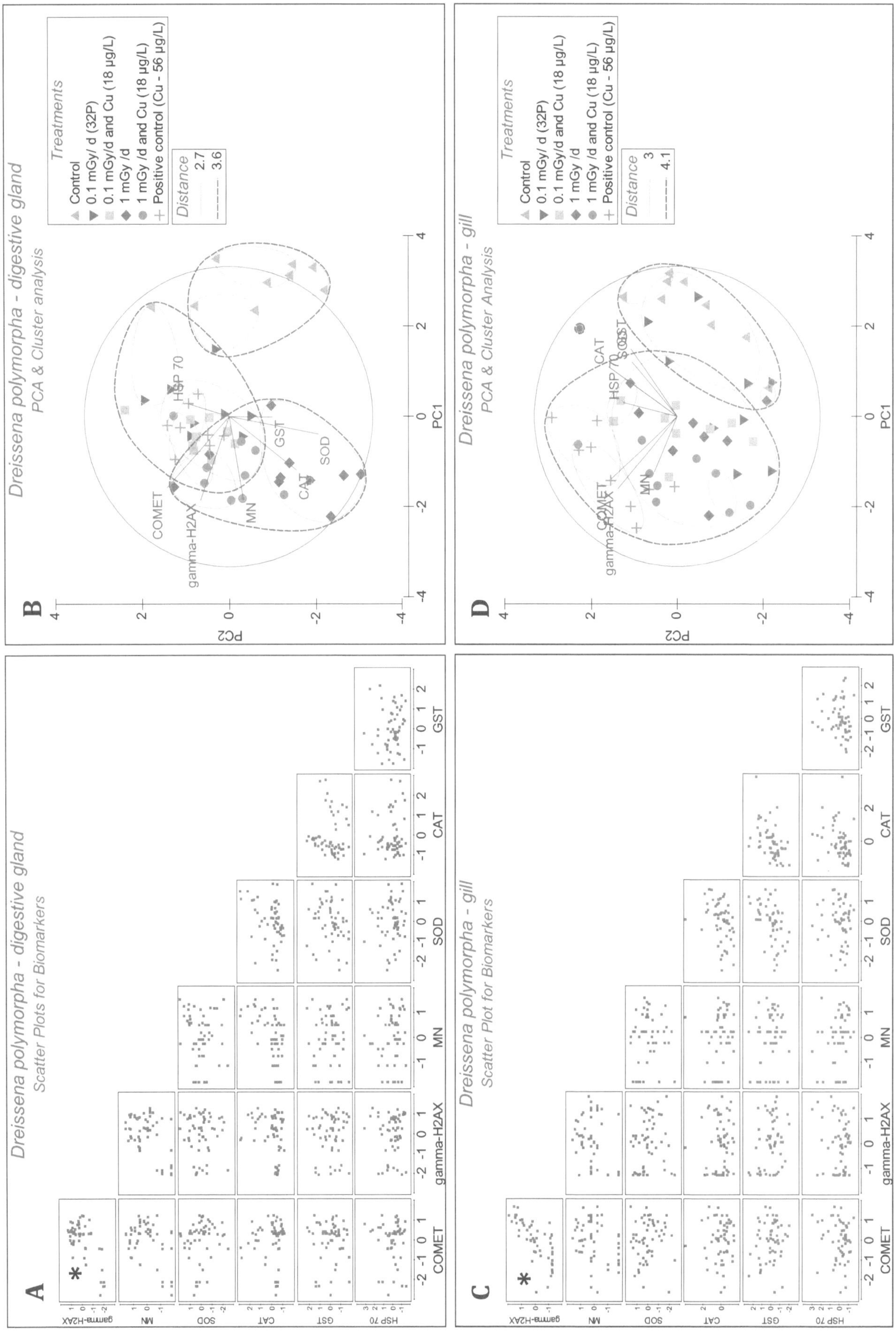

Figure 2. (Continued).

Table 3. Gene and primers used for *M. galloprovincialis* and *D. polymorpha* in RT-PCR.

Gene	Short name	Forward primer	Reverse primer	Accession No.
Mytilus galloprovincialis				
Actin	*act*	GGGAGTCATGGTTGGTATGG	TCAGAAGGACTGGGTGCTCT	AF157491
Elongation factor 1	*ef1*	CACCACGAGTCTCTCCCAGA	GCTGTCACCACAGACCATTCC	AF063420
Glutathione S-transferase	*GST*	ATGGCTCTTTTCCTGCTTCA	TTTGGCCAGTGTCCATGTTA	AF527010
Superoxide dismutase	*CuZn-sod*	CATTTCCCAGATCACCAACA	CATTTCCCAGATCACCAACA	FM177867
Catalase	*CAT*	CACCAGGTGTCCTTCCTGTT	CTTCCGAGATGGCGTTGTAT	AY743716
Heat-shock Protein 70	*HSP70*	GGGTGGTGGAACTTTTGATG	GCCGTTGAAAAAGTCCTGAA	AF172607
Heat-shock Protein 90	*HSP90*	TCAGTGATGATCCTAGATTAGGCA	CGTTCCTCTCTTTCCATCTGTAAC	AJ625655
Dreissena polymorpha				
Actin	*act*	CCTCACCCTCAAGTACCCCAT	TTGGCCTTTGGGTTGAGTG	AF082863
Elongation factor 1	*ef1*	CCACCAAAGGCAGCCAAGAG	TGGGACGAGGTCAGCCATAC	AJ250733
Glutathione S-transferase	*GST*	TCCGCTATATCTGCCTGGAC	GCTCCTTCAGACCTGCTTTC	EF194203
Superoxide dismutase	*SOD*	GACAGCATGGCTTCCATGTG	AGATTCTGGGCCAGTCAGAG	AY377970
Catalase	*CAT*	ACGGCTATGGAAGCCACACG	AGGTCGCGCATCGCATAGTC	EF681763
Heat-shock Protein 70	*HSP70*	TGTCCTGCTTGTGGATGTAG	CGTGGTGAATGTCTGTGTAG	EF526096
Heat-shock Protein 90	*HSP90*	TTGATCATCAATACTTTCTATTC	ACACCAAACTGTCCAATCAT	GU433881a

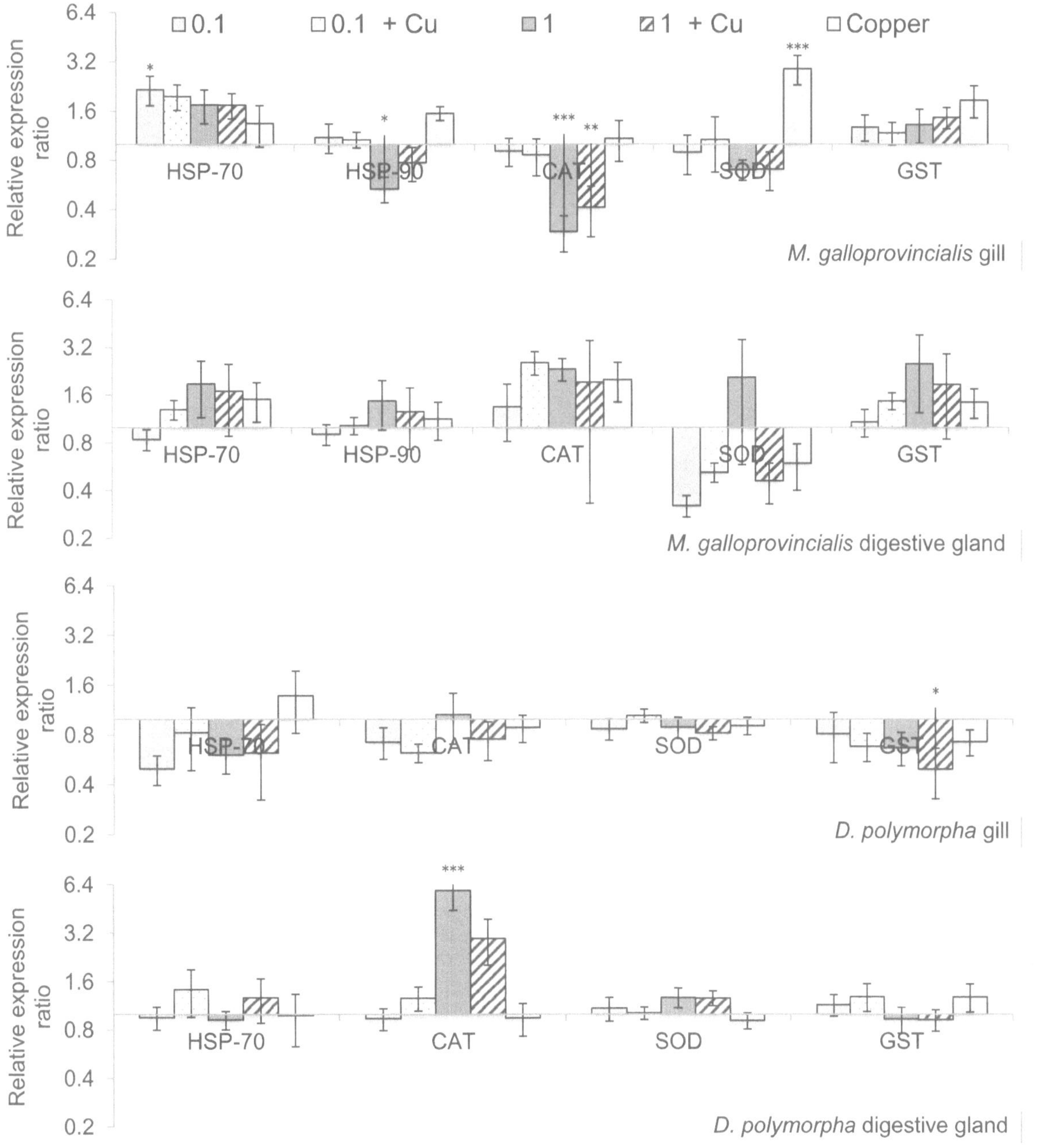

Figure 3. Relative expression ratios (RER) of key genes in *M. galloprovincialis* and *D. polymorpha* gill and digestive gland cells following a 10-day exposure to ^{32}P and Cu, alone and in combination. Data are normalized for reference genes (ef1, actin) and controls. Error bars indicate the 95% confidence intervals. Asterisks (*, ** or ***) are indicative of significant differences ($p < .05, .01, .001$) from the corresponding control. $n = 9$. Copper = positive control.

mGy d^{-1}+ Cu treatments, Cu had a weak interaction with ^{32}P across all treatments excluding MG digestive gland (1 and 1 mGy d^{-1}+ Cu), where the addition of Cu decreased the damaging effect of ^{32}P (Figure 1(A)). The lowest degree of damage was evident in DP gill tissue across all treatments, % Tail DNA was only significantly greater than controls with the addition of Cu (1 mGy d^{-1} + Cu, $p < .001$). This trend was continued in MG gill (both treatments). Cu appeared to have the greatest interaction with ^{32}P within this tissue, where % Tail DNA was increased by 2.3- and 1.7-fold (0.1 and 1 mGy d^{-1}, respectively).

3.1.2. Analysis of micronuclei (MN) formation

Trends observed in DNA damage response were not reflected for MN formation (Figure 1(B)), as Cu did not appear to interact with ^{32}P in a detrimental manner (excluding MG gill). No significant difference was noted between treatments in DP gill, MG and DP digestive gland. The greatest degree of damage was evident in DP digestive gland, particularly in the 1 mGy d^{-1} treatment at 40 MN/1000 cells. Cu appeared to have an antagonistic interaction with ^{32}P, where MN induction decreased when exposed to both stressors. In terms of relative sensitivity DP shows a higher MN frequency in gill and digestive gland cells over its marine counterpart, across all treatments (excluding negative/positive controls). The higher incidence of MN in DP, compared to MG, was however only significant at 1 mGY d^{-1} dose rate ($p < .05$).

3.1.3. Induction of γ-H2AX foci

Figure 1(C) shows the average number of γ-H2AX foci per cell, indicative of DSBs. In keeping with DNA damage, an enhanced number of γ-H2AX foci were evident in MG digestive gland, particularly in 1 mGy d^{-1} treatment at 19 foci/cell. On average, γ-H2AX foci in MG digestive gland was 2.3- to 3.1-fold higher than DP. In relation to the corresponding control, no significant increase in γ-H2AX foci/cell were noted at the lowest ^{32}P treatment of 0.1 mGy d^{-1} across both species and tissues, however, foci number significantly increased with the addition of Cu (MG digestive gland: $p < .05$, MG and DP gill, DP digestive gland: $p < .01$).

3.2. Transcriptional expression of key genes

Due to failure of the assay to amplify, *hsp90* data are not included for DP. For key genes studied in DP, PCR efficiencies were (primer details, Table 3): *actin* (*act*): 1.80, *elongation factor 1* (*ef1*): 1.79, *catalase* (*cat*): 1.82, *glutathione-s-transferase* (*gst*): 1.79, *superoxide dismutase* (*sod*): 1.78 and *heat shock protein 70* (*hsp70*): 1.81. In MG: *act:* 1.79, *ef1:* 1.79, *cat:* 1.81, *gst:* 1.81, *sod:* 1.80, *hsp70:* 1.75 and *heat shock protein 90* (*hsp90*): 1.83. Relative gene expression of key genes is presented in Figure 3. Overall there was limited variation across all biological tissue and species, with no change evident in MG digestive gland. At 1 mGy d^{-1}, *cat* was downregulated in MG gill, but upregulated in DP digestive gland ($p < .001$). Downregulation of *cat* was further noted in MG gill 1 + Cu treatment, but to a lesser extent ($p < .01$). In response to Cu, *sod* was significantly upregulated in MG gill ($p < .01$).

3.3. Multivariate analysis of biomarker reactions

Analysis of the potential correlations within the biomarker data indicated that most of the biological parameters were not strongly correlated (Figure 2(a,b,A,C)), with several exceptions (primarily comet [measure of DNA damage] and γ-H2AX assays, across all tissues, $p < .05$). PCA and MDS (MDS plots not shown) combined with cluster analysis showed that the experimental treatments were clearly distinct from the control groups (Figure 2(a,b,B,D)). The percentage of variance explained by PC1 and PC2 was as follows: DP digestive gland, 32.6% and 21.4%; DP gill, 33.2% and 24.3%; MG digestive gland, 47.5% and 27.1% and MG gill, 32.9% and 28.4%. ANOSIM analysis gave global significances of $p < .001$ for both tissues and species. In DP gill tissue, pairwise analysis showed that all treatments were significantly different from one another ($p < .05$), digestive gland tissue of this species showed similarly distinct differences between pairs of treatments except for 0.1 mGy d^{-1} and 0.1 mGy d^{-1} + Cu. MG gill tissue showed clear differences between pairs of treatments except for 0.1 mGy d^{-1} and 1 mGy d^{-1}, and 1 mGy d^{-1} and 1 mGy d^{-1} + Cu. However, the digestive gland tissue of MG showed greater overlap in treatments, with no significant differences between 0.1 mGy d^{-1} and 1 mGy d^{-1} + Cu, 0.1 mGy d^{-1} and positive Cu Control, 1 mGy d^{-1} and 1 mGy d^{-1} + Cu, and 1 mGy d^{-1} + Cu and positive Cu control. The PCA results for this tissue showed that the experimental treatments were more strongly grouped together than in any other tissues (Figure 2(B)).

As already stated, most biomarkers were not correlated with each other, and the BIO-ENV routine for various combinations of biomarkers indicated that there were no influential biomarkers among the various combinations capable of capturing the full PCA biomarker response pattern. Principal component (PCA) coupled with hierarchical cluster analysis and ANOSIM results for all treatments showed that both ^{32}P and Cu had a detrimental effect on the genetic integrity and oxidative stress status in the four tissues tested (Figure 2(a,b,A–D)).

4. Discussions

This novel study is the first to explore the interactive effects of ^{32}P and Cu on two aquatic bivalve species using a multi-biomarker approach. Qualities of radionuclides and speciation of metals (i.e. physical and chemical properties) in aquatic systems have a large influence on bioavailability and subsequent toxicity (Gunten and Beneš 1995; Richards et al. 2011). It is well documented that Cu, focusing on the toxic ionic form (Cu^{2+}) forms complexes with natural organic matter, decreasing bioavailability and at lower salinities (i.e. freshwater) becoming more abundant (Grosell et al. 2007). Despite this, little is known on the possible interactions between Cu and lesser studied radionuclides such as ^{32}P and the combined influence of water parameters (i.e. pH, salinity, dissolved organic carbon [DOC], alkalinity). Due to time and logistical constraints certain parameters (e.g. DOC) were not determined during this experiment. It is possible that differing water chemistry (i.e. salt and freshwater), along with chemical interactions between stressors could have affected the noted biological effects in each species.

4.1. Biomarker interactions

PCA can be used to integrate information from multiple biological parameters. Our study demonstrates that PCA can aid interpretation of multi-biomarker responses to combined environmental stressors (i.e. IR and metals), by reducing multi-dimensionality into a less complex two-dimensional representation (Chatfield and Collins 1980; Allen and Moore 2004; Moore et al. 2006). However, PCA and cluster analysis are limited in the inability to integrate biomarkers in a more functional manner, wherein, the endpoints (i.e. DNA damage, MN formation) adopted in this study are not necessarily reflective of 'overall health status'. It is likely that such biomarkers indicate whether a contaminant-induced response has taken place, or that health status within a narrow range has been altered, but they do not generally indicate an individual's health status for the whole range from healthy to permanent damage (Depledge et al. 1993; Köhler et al. 2002; Moore et al. 2004; Moore et al. 2006). Despite this, subsequent investigations could build on the information provided in order to develop such models in the future. Furthermore, this study is certainly useful in establishing relationships between stressor and response, PCA coupled with hierarchical cluster analysis and ANOSIM results for all treatments showed that both ^{32}P and Cu had deleterious effects on genetic integrity and oxidative stress status in species and tissues tested (Figure 2(a,b,A–D)).

4.2. ^{32}p induced genotoxic response in gill and digestive gland cells

The majority of IR-induced toxicity studies have focused on single radionuclide exposures. DNA damage as a biomarker has been noted at dose rates ranging from 0.8 to 41,666 μGy h^{-1} in a wide range of biota, including marine bivalves (*Mytilus* spp., *C. gigas*, *P. perna*, *C. fluminea*, *P. malabarica* & *M. casta;* Jha et al. 2005; Jha et al. 2006; Godoy et al. 2008; Farcy et al. 2011; AlAmri et al. 2012; Praveen Kumar et al. 2014). PCA and MDS, combined with cluster analysis demonstrate a clear distinction between controls and experimental groups, across both species, tissue and biomarker (Figure 2(a,b,B,D)). In keeping with previous work (Vernon et al. 2020), no significant change in DNA damage was noted at lower ^{32}P doses of 0.10 mGy d^{-1} (Excluding DP digestive gland, present study). However, where previous findings demonstrated significant increased % Tail DNA in both species and cell types at 1 mGy d^{-1}, only MG digestive gland was significantly increased. The present study had greater baseline levels of DNA damage of ~5–11%, compared to <5% in some previous literature, this along with individual differences may account for such variation.

Exposure of the mussels to ^{32}P alone caused no significant increase in % Tail DNA (Excluding DP [0.1] and MG digestive gland [1 mGy d^{-1}]), but when combined with Cu, % Tail DNA was statistically greater than controls in MG gill and DP digestive gland at 0.10 mGy d^{-1}, and all species and tissues at 1 mGy d^{-1}. Cu in isolation, at environmentally relevant concentrations (18 μg L^{-1}) has been found to have no significant effect on % Tail DNA in MG or DP gill cells (Vernon and Jha 2019), when in combination with ^{32}P there appears to be an additive effect on mussels. Whilst not significantly so the addition of Cu to IR exposures also increased γ-H2AX foci induction, across both species, tissue and dose rate (excluding MG digestive gland). This apparent additive effect of Cu on the genotoxicity of ^{32}P on marine and freshwater mussels is the first reported. Cu-contaminant induced effects have been noted in previous literature. In *M. edulis*, Cu-induced (0.1 μM) damage to DNA and lipids was noted as significantly greater under low pH conditions (reflective of ocean acidification), in comparison to controls (Lewis et al. 2016). Similarly, combined Cu (10 μg L^{-1}) and IR (^{60}Co, 70 mGy) exposure was found to induce significantly depleted glutathione compared to exposure to Cu alone, in presmolt salmon, *S. salar* (Heier et al. 2013).

Aquatic biota are continuously exposed to both endogenously and environmentally generated contaminants, giving cause for the development of highly effective biochemical mechanisms that afford the ability to protect and defend on multiple biological levels. H2AX is quickly phosphorylated to form γ-H2AX, a crucial factor in DSB repair response and therefore a relevant, useful technique in radiation biology, where the MoA (mechanism of action) of ^{32}P in particular is mediated by induction of DNA DSBs (Kuo and Yang 2008; Cheng et al. 2015). Pereira et al. (2011), Urushihara et al. (2012), Pereira et al. (2014 and Sayed et al. (2017) have successfully utilized this technique to assess DDR in fish (*O. latipes* and *D. rerio*) following acute exposures to ^{137}Cs, but to our knowledge, is yet to be fully utilized in mussels. The γ-H2AX assay is further validated when utilized alongside classical DNA damage techniques, such as comet and MN assays, relationships have been demonstrated between such biomarkers (Pereira et al. 2011; Sayed et al. 2017; Dallas et al. 2013, 2018). In our study, γ-H2AX was strongly correlated with DNA damage across the tissues ($p < .05$), correlation was not noted between other molecular or genetic biomarkers (Figure 2(a,b,A,C)). The greatest induction of γ-H2AX foci was noted in MG digestive gland, specifically at 1 mGy d^{-1} (~ 19 foci/cell), followed by the 1 mGy d^{-1} + Cu, positive Cu control, 0.1 mGy d^{-1}+ Cu and 0.1 mGy d^{-1} treatments. At significantly higher dose rates of 10 mGy d^{-1} (^{137}Cs), less than 5 foci were present per zebrafish ZF4 cell (Pereira et al. 2011). γ-H2AX foci occurs rapidly post-irradiation (30 min – 2 h) and decrease in number over time, as both studies measured damage immediately (~10–40 min) post exposure, varying radiosensitivity may be a result of species, radionuclide source or associated MoA (Sedelnikova et al. 2003;

Ivashkevich et al. 2011; Ivashkevich et al. 2012). Trends in γ-H2AX foci induction in MG digestive gland followed DNA damage, but not MN formation, where MN/1000 cells remained uniform respective of experimental treatment. The presence of MN is generally regarded as more long-lasting damage, suggesting that the induced DSBs were easily repairable. This was further noted in MG gill cells.

In line with previous studies (Vernon et al. 2020), a greater degree of MN formation is noted in DP gill and digestive gland cells, relative to associated MG tissue, across all treatments, suggesting a significant impact on DNA integrity. Unexpectedly, Cu showed weak antagonistic interaction with ^{32}P for MN induction, where it showed a decreased level. It was however not found to be statistically significant. It is possible that Cu-specific repair mechanisms, as a by-product buffer against the more permanent effects induced by IR as an isolated threat (Yu et al. 2012).

As described in a previous study (Vernon et al. 2020), ^{32}P concentrates differentially in gill and digestive gland tissues, with the latter accumulating 87% (MG) and 44% (DP) of ^{32}P (proportion of whole body ^{32}P concentration) at 10 mGy d^{-1}. Similarly, Cu uptake varies between tissue and species (Cai and Wang 2019; Vernon and Jha 2019). In turn, there are clear differences in tissue-specific dose rate (Table 1) and subsequent genotoxic response to IR and metals, highlighting the importance of a multiple-tissue approach in ecotoxicological studies. Overall, the addition of Cu arguably has the lowest impact on ^{32}P-induced damage in MG digestive gland, where MG gill tissue shows far less variation between treatments (excluding % Tail DNA, 0.1 and 0.1 + Cu). In keeping with previous findings, in terms of longer-term damage, high MN frequency in DP cells suggests a more long-lasting response in freshwater mussels, in relation to MG.

4.3. Transcriptional expression of key genes

In our study, transcriptional expression of key genes showed little variation between differing treatments. The findings are in agreement with previous work (Vernon et al. 2020). Excluding significant upregulation of *sod* in MG gill, the Cu positive control (56 μg L^{-1}) caused no variation in gene expression. This is in contrast to work by Xu et al. (2018) where exposure to much lower Cu concentrations (2 and 8 μg L^{-1}) significantly induced the expressions of stress responsive genes (*hsp70, hsp90, mt-10*) in *Mytilus coruscus* (haemocytes), particularly 12 d post exposure. Furthermore, significantly altered transcriptional profiles in fish (*D. rerio* and *S. salar*) have been noted following IR exposure (^{137}Cs or ^{60}Co). Common MoAs associated with low-dose gamma radiation included the induction of oxidative stress and DNA damage (Jaafar et al. 2013; Freeman et al. 2014; Song et al. 2014; Song et al. 2016). At a molecular level, lack of significant variation between response may be an indication of disparity in toxicity MoA, or different sensitivities between the end points (Devos et al. 2015). Noted genomic damage may result from the direction interaction of ^{32}P with DNA, through the release of high-energy beta particles or isotopic decay, and to a lesser extent via the generation of ROS.

Due to both logistical constraints and lack of readily available gene sequences for MG and DP, we were only able to assess five key genes in this study. This provides limited information pertaining to molecular mechanistic responses to IR and Cu. Furthermore, while identification of transcriptome variation is undoubtedly important and increasingly popular within radiobiological research, limited information can be acquired by studying genes in isolation. If this work were taken further, it would be logical to use microarrays, RNAseq and proteomics analyses (Banni et al. 2017; Dallas et al. 2018; Barranger, Langan, et al. 2019; Barranger, Rance, et al. 2019) and at multiple, earlier time points and at different environmentally relevant temperatures as performed in previous studies in our laboratory conditions in order to dissect out the mechanisms of toxicity (Dallas et al. 2013, 2016, 2018). Identifying known and potentially novel molecular targets (genes and proteins) involved, as well as activated signaling pathways in a range of biological systems under multi-stress exposure scenarios may be the first step in predicting potential larger scale impacts. These could eventually include detrimental effects at high levels of biological organization, such as community or population level (Maria et al. 2013; Gomes et al. 2014).

Along with other hydrological characteristics (e.g. pH, salinity, dissolved oxygen) which show temporal and seasonal variabilities in the natural environment, temperature is considered to be an important factor which is known to influence many biological responses (e.g. expression of genes, induction of DNA damage) in aquatic organisms (Jha 2008; Dallas and Jha 2015). It can also influence cell division and therefore cell-cycle dependent phenomenon such as induction of MN in tissue and species-specific manner. Furthermore, given that thermal discharge from nuclear power plants is considered to be an important environmental issue along with potential sea surface temperature rise by up to 3.5° C in the next 100 years (IPCC 2007), determining the interaction of elevated temperature with radionuclide and other environmental stressors (e.g. metals) could be considered highly relevant. Previous studies from our laboratory have elaborated potential interactive impacts of temperature and an environmentally relevant radionuclide (i.e. tritium) on these parameters in marine mussels (Dallas et al. 2016). In line with most of the studies carried out by us and other researchers, the experiments in this study were carried out at 15 °C, which is considered to be environmentally relevant for these species. In future, it would be helpful to elucidate the impact of temperature and other water quality parameters while comparing the relative sensitivities of different species following exposure to multiple stressors.

5. Conclusions

In conclusion, the results of this integrated, multi-biomarker study represent the most extensive data to date obtained on the combined effects of IR and metals (Cu) in two environmentally relevant bivalve species. The results indicate that (a) genotoxic response was reflective of exposure, where Cu had an overall additive effect on ^{32}P-induced damage across several (but not all) species, cell types and dose rates, (b) selected genes were generally unaltered in terms of transcriptional response to contaminants, independent of species and (c) in exclusion of comet and γ-H2AX assays biological parameters were not strongly correlated. Whilst it is difficult to extrapolate such findings to exposures in realistic environmental conditions, these data contribute to the limited amount of information available in the literature on the possible mechanisms involved in multi-stressor (i.e. IR and metals) induced biological responses. The study highlights the importance of investigating the interactive effects of pollutants on ecologically relevant species.

Acknowledgements

The authors thank Ms Charlotte Crowther, Mr Nick Crocker, Dr Alex Taylor and Professor William Blake, University of Plymouth, for their technical assistance and radiological protection advices.

Disclosure statement

The authors declare no conflict of interest.

Funding

This work was funded by the Natural Environment Research Council (NERC), the Environment Agency (EA) and Radioactive Waste Management Limited (RWM) under the Radioactivity and the Environment (RATE) programme [Grant no.: NE/L000393/1].

ORCID

Awadhesh N. Jha http://orcid.org/0000-0001-9660-4308

References

AlAmri OD, Cundy AB, Di Y, Jha AN, Rotchell JM. 2012. Ionizing radiation-induced DNA damage response identified in marine mussels, *Mytilus* sp. Environ Pollut. 168:107–112.

Allen JI, Moore MN. 2004. Environmental prognostics: is the current use of biomarkers appropriate for environmental risk evaluation? Mar Environ Res. 58:227–232.

Al-Subiai SN, Moody AJ, Mustafa SA, Jha AN. 2011. A multiple biomarker approach to investigate the effects of copper on the marine bivalve mollusc, *Mytilus edulis*. Ecotoxicol Environ Saf. 74:1913–1920.

Andersson P, Beaugelin-Seiller K, Beresford N, Copplestone D, Della Vedova C, Garnier-Laplace J, Howard B, Howe P, Oughton D, Wells Whitehouse CP. 2008. Numerical benchmarks for protecting biota from radiation in the environment: proposed levels, underlying reasoning and recommendation. NERC/Centre for Ecology & Hydrology.

Andersson P, Garnier-Laplace J, Beresford NA, Copplestone D, Howard BJ, Howe P, Oughton D, Whitehouse P. 2009. Protection of the environment from ionising radiation in a regulatory context (protect): proposed numerical benchmark values. J Environ Radioact. 100:1100–1108.

Banni M, Sforzini S, Arlt VM, Barranger A, Dallas LJ, Oliveri C, Aminot Y, Pacchioni B, Millino C, Lanfranchi G, et al. 2017. Assessing the impact of Benzo[*a*]pyrene on Marine Mussels: application of a novel targeted low density microarray complementing classical biomarker responses. PLoS One. 12:e0178460.

Barranger A, Langan LM, Sharma V, Rance GA, Aminot Y, Weston NJ, Akcha F, Moore MN, Arlt VM, Khlobystov AN, et al. 2019. Antagonistic interactions between benzo[a]pyrene and fullerene (C60) in toxicological response of marine mussels. Nanomaterials. 9:987.

Barranger A, Rance GA, Aminot Y, Dallas LJ, Sforzini S, Weston NJ, Lodge RW, Banni M, Arlt VM, Moore MN, et al. 2019. An integrated approach to determine interactive genotoxic and global gene expression effects of multiwalled carbon nanotubes (MWCNTs) and Benzo[*a*]pyrene (BaP) on marine mussels: evidence of reverse 'Trojan Horse' effects. Nanotoxicology. 13:1324–1343.

Binelli A, Della Torre C, Magni S, Parolini M. 2015. Does zebra mussel (*Dreissena polymorpha*) represent the freshwater counterpart of *Mytilus* in ecotoxicological studies? A critical review. Environ Pollut. 196:386–403.

Bolognesi C, Fenech M. 2012. Mussel micronucleus cytome assay. Nat Protoc. 7:1125–1137.

Bopp SK, Abicht HK, Knauer K. 2008. Copper-induced oxidative stress in rainbow trout gill cells. Aquat Toxicol. 86:197–204.

Bouskill NJ, Handy RD, Ford TE, Galloway TS. 2006. Differentiating copper and arsenic toxicity using biochemical biomarkers in *Asellus aquaticus* and *Dreissena polymorpha*. Ecotoxicol Environ Saf. 65:342–349.

Brooks SJ, Farmen E, Heier LS, Blanco-Rayón E, Izagirre U. 2015. Differences in copper bioaccumulation and biological responses in three *Mytilus* species. Aquat Toxicol. 160:1–12.

Bryan GW, Gibbs PE. 1983. Heavy Metals in the Fal Estuary, Cornwall: a study of long-term contamination by mining waste and its effects on estuarine organisms. Marine Biological Association of the United Kingdom

Cai C, Wang WX. 2019. Inter-species difference of copper accumulation in three species of marine mussels: implication for biomonitoring. Sci Total Environ. 692:1029–1036.

Carvalho FP. 2018. Radionuclide concentration processes in marine organisms: a comprehensive review. J Environ Radioact. 186:124–130.

Chapman PM. 2002. Integrating toxicology and ecology: putting the "eco" into ecotoxicology. Mar Pollut Bull. 44:7–15.

Chatfield C, Collins A. 1980. Introduction to multivariate analysis. New York (NY): Springer.

Cheng Y, Kiess AP, Herman JM, Pomper MG, Meltzer SJ, Abraham JM. 2015. Phosphorus-32, a clinically available drug, inhibits cancer growth by inducing DNA double-strand breakage. Plos One. 10: e0128152.

Cheung VV, Depledge MH, Jha AN. 2006. An evaluation of the relative sensitivity of two marine bivalve mollusc species using the Comet assay. Marine Environmental Research. 62:S301–S305.

Cid A, Herrero C, Torres E, Abalde J. 1995. Copper toxicity on the marine microalga *Phaeodactylum tricornutum*: effects on photosynthesis and related parameters. Aquat Toxicol. 31:165–174.

Clarke KR, Warwick RM. 2001. Change in marine communities: an approach to statistical analysis and interpretation. Plymouth (UK): PRIMER-є.

Clarke KR. 1999. Non-metric multivariate analysis in community-level ecotoxicology. Environ Toxicol Chem. 18:117–127.

Clayton ME, Steinmann R, Fent K. 2000. Different expression patterns of heat shock proteins hsp 60 and hsp 70 in zebra mussels (*Dreissena polymorpha*) exposed to copper and tributyltin. Aquat Toxicol. 47:213–226.

Dallas LJ, Bean TP, Turner A, Lyons BP, Jha AN. 2013. Oxidative DNA damage may not mediate Ni-induced genotoxicity in marine mussels: assessment of genotoxic biomarkers and transcriptional responses of key stress genes. Mutat Res. 754:22–31.

Dallas LJ, Bean TP, Turner A, Lyons BP, Jha AN. 2016. Exposure to tritiated water at an elevated temperature: genotoxic and transcriptomic effects in marine mussels (*M. galloprovincialis*). J Environ Radioact. 164:325–336.

Dallas LJ, Jha AN. 2015. Applications of biological tools or biomarkers in aquatic biota: a case study of the Tamar estuary, South West England. Mar Pollut Bull. 95:618–633.

Dallas LJ, Keith-Roach M, Lyons BP, Jha AN. 2012. Assessing the impact of ionizing radiation on aquatic invertebrates: a critical overview. Radiat Res . 177:693–716.

Dallas LJ, Turner A, Bean TP, Lyons BP, Jha AN. 2018. An integrated approach to assess the impacts of zinc pyrithione at different levels of biological organization in marine mussels. Chemosphere. 196: 531–539.

DEFRA. 2014. Water Framework Directive implementation in England and Wales: new and updated standards to protect the water environment. Environment FaRA.

Depledge MH, Amaral-Mendes JJ, Daniel B, Halbrook RS, Kloepper-Sams P, Moore MN, Peakall DB. 1993. The conceptual basis of the biomarker approach. In: Peakall DB, Shugart LR, editors. Biomarkers. Berlin: Springer; p. 15–29.

Devos A, Dallas LJ, Voiseux C, Lecomte-Pradines C, Jha AN, Fiévet B. 2015. Assessment of growth, genotoxic responses and expression of stress related genes in the Pacific oyster *Crassostrea gigas* following chronic exposure to ionizing radiation. Mar Pollut Bull. 95:688–698.

DWQR. 2014. Drinking Water Quality Regulator for Scotland: monitoring of radioactivity in Scottish drinking water. Millican D, editor. [accessed 2019 Oct 7]. http://dwqr.scot/media/12749/dwqr-information-letter-2014-1-monitoring-of-radioactivity-in-scottish-drinking-water.pdf

Eggen RIL, Behra R, Burkhardt-Holm P, Escher BI, Schweigert N. 2004. Challenges in ecotoxicology. Environ Sci Technol. 38: 58A–64A.

Farcy E, Voiseux C, Robbes I, Lebel JM, Fievet B. 2011. Effect of ionizing radiation on the transcription levels of cell stress marker genes in the Pacific oyster *Crassostrea gigas*. Radiat Res. 176:38–48.

Festarini A, Shultz C, Stuart M, Kim SB, Ferreri C. 2015. Cellular responses to tritium exposure in Rainbow trout: Hto- and obt-spiked feed exposure experiments. CNL Nuclear Review. 5:155–172.

Festarini A, Shultz C, Stuart M, Kim SB, Ferreri C. 2019. Cellular responses in rainbow trout (*Oncorhynchus mykiss*) reared in tritiated water and/or fed organically bound tritium. Appl Radiat Isot. 151: 217–225.

Freeman JL, Weber GJ, Peterson SM, Nie LH. 2014. Embryonic ionizing radiation exposure results in expression alterations of genes associated with cardiovascular and neurological development, function, and disease and modified cardiovascular function in zebrafish. Front Genet. 5:268.

Gaetke LM, Chow CK. 2003. Copper toxicity, oxidative stress, and antioxidant nutrients. Toxicology. 189:147–163.

Godoy JM, de Oliveira MS, de Almeida CE, de Carvalho ZL, da Silva ER, Fernandes Fda C, Pitanga FL, Danelon OM. 2008. ^{210}Po concentration in *Perna perna* mussels: looking for radiation effects. J Environ Radioact. 99:631–640.

Gomes T, Chora S, Pereira CG, Cardoso C, Bebianno MJ. 2014. Proteomic response of mussels *Mytilus galloprovincialis* exposed to CuO NPs and Cu2+: an exploratory biomarker discovery. Aquat Toxicol. 155:327–336.

Grosell M, Blanchard J, Brix KV, Gerdes R. 2007. Physiology is pivotal for interactions between salinity and acute copper toxicity to fish and invertebrates. Aquat Toxicol. 84:162–172.

Gunten HRV, Beneš P. 1995. Speciation of radionuclides in the environment. Radiochim Acta. 69:1–29.

Hagger JA, Atienzar FA, Jha AN. 2005. Genotoxic, cytotoxic, developmental and survival effects of tritiated water in the early life stages of the marine mollusc, *Mytilus edulis*. Aquat Toxicol. 74:205–217.

Heier LS, Teien HC, Oughton D, Tollefsen KE, Olsvik PA, Rosseland BO, Lind OC, Farmen E, Skipperud L, Salbu B. 2013. Sublethal effects in Atlantic salmon (*Salmo salar*) exposed to mixtures of copper, aluminium and gamma radiation. J Environ Radioact. 121: 33–42.

Hu QH, Weng JQ, Wang JS. 2010. Sources of anthropogenic radionuclides in the environment: a review. J Environ Radioact. 101: 426–437.

IPCC. 2007. Summary for policymakers. In: Solomon S, editor. Climate Change 2007: the Physical Science Basis. Contribution of Working Group I to the Fourth Assessment Report of the Intergovernmental Panel on Climate Change. Cambridge (UK): Cambridge University Press; p. 1–18.

Ivashkevich A, Redon CE, Nakamura AJ, Martin RF, Martin OA. 2012. Use of the γ-H2AX assay to monitor DNA damage and repair in translational cancer research. Cancer Lett. 327:123–133.

Ivashkevich AN, Martin OA, Smith AJ, Redon CE, Bonner WM, Martin RF, Lobachevsky PN. 2011. γH2AX foci as a measure of DNA damage: a computational approach to automatic analysis. Mutat Res. 711:49–60.

Jaafar L, Podolsky RH, Dynan WS. 2013. Long-term effects of ionizing radiation on gene expression in a zebrafish model. PLoS ONE. 8: e69445.

Jaeschke BC, Bradshaw C. 2013. Bioaccumulation of tritiated water in phytoplankton and trophic transfer of organically bound tritium to the blue mussel, *Mytilus edulis*. J Environ Radioact. 115:28–33.

Jaeschke BC, Millward GE, Moody AJ, Jha AN. 2011. Tissue-specific incorporation and genotoxicity of different forms of tritium in the marine mussel, *Mytilus edulis*. Environ Pollut. 159:274–280.

Jha AN, Dogra Y, Turner A, Millward GE. 2005. Impact of low doses of tritium on the marine mussel, *Mytilus edulis*: genotoxic effects and tissue-specific bioconcentration. Mutat Res. 586:47–57.

Jha AN, Dogra Y, Turner A, Millward GE. 2006. Are low doses of tritium genotoxic to *Mytilus edulis*? Mar Environ Res. 62:S297–S300.

Jha AN. 2008. Ecotoxicological applications and significance of the comet assay. Mutagenesis. 23:207–221.

Kim BE, Nevitt T, Thiele DJ. 2008. Mechanisms for copper acquisition, distribution and regulation. Nat Chem Biol. 4:176–185.

Köhler A, Wahl E, Soffker K. 2002. Functional and morphological changes of lysosomes as prognostic biomarkers of toxic liver injury in a marine flatfish (*Platichthys flesus* (L.)). Environ Toxicol Chem. 21:2434–2444.

Praveen Kumar MK, Shyama SK, Sonaye BS, Naik UR, Kadam SB, Bipin PD, D'costa A, Chaubey RC. 2014. Evaluation of γ-radiation-induced DNA damage in two species of bivalves and their relative sensitivity using comet assay. Aquat Toxicol. 150:1–8.

Kumaravel TS, Jha AN. 2006. Reliable Comet assay measurements for detecting DNA damage induced by ionising radiation and chemicals. Mutat Res. 605:7–16.

Kuo LJ, Yang LX. 2008. Gamma-H2AX – a novel biomarker for DNA double-strand breaks. In Vivo. 22:305–309.

Lewis C, Ellis RP, Vernon E, Elliot K, Newbatt S, Wilson RW. 2016. Ocean acidification increases copper toxicity differentially in two key marine invertebrates with distinct acid-base responses. Sci Rep. 6:21554.

Maria VL, Gomes T, Barreira L, Bebianno MJ. 2013. Impact of benzo(a)pyrene, Cu and their mixture on the proteomic response of *Mytilus galloprovincialis*. Aquat Toxicol. 144-145:284–295.

McDonald JH, Seed R, Koehn RK. 1991. Allozymes and morphometric characters of three species of *Mytilus* in the Northern and Southern Hemispheres. Mar Biol. 111:323–333.

Moore MN, Allen JI, McVeigh A. 2006. Environmental prognostics: an integrated model supporting lysosomal stress responses as predictive biomarkers of animal health status. Mar Environ Res. 61:278–304.

Moore MN, Depledge MH, Readman JW, Paul Leonard DR. 2004. An integrated biomarker-based strategy for ecotoxicological evaluation of risk in environmental management. Mutat Res. 552:247–268.

Mothersill C, Salbu B, Heier LS, Teien HC, Denbeigh J, Oughton D, Rosseland BO, Seymour CB. 2007. Multiple stressor effects of radiation and metals in salmon (*Salmo salar*). J Environ Radioact. 96: 20–31.

Olsvik PA, Heier LS, Rosseland BO, Teien HC, Salbu B. 2010. Effects of combined gamma-irradiation and metal (Al + Cd) exposures in Atlantic salmon (*Salmo salar* L.). J Environ Radioact. 101:230–236.

Parisot F, Bourdineaud JP, Plaire D, Adam-Guillermin C, Alonzo F. 2015. DNA alterations and effects on growth and reproduction in *Daphnia magna* during chronic exposure to gamma radiation over three successive generations. Aquat Toxicol. 163:27–36.

Pearson HBC, Dallas LJ, Comber SDW, Braungardt CB, Worsfold PJ, Jha AN. 2018. Mixtures of tritiated water, zinc and dissolved organic carbon: assessing interactive bioaccumulation and genotoxic effects in marine mussels, *Mytilus galloprovincialis*. J Environ Radioact. 187:133–143.

Pereira S, Bourrachot S, Cavalie I, Plaire D, Dutilleul M, Gilbin R, Adam-Guillermin C. 2011. Genotoxicity of acute and chronic gamma-irradiation on zebrafish cells and consequences for embryo development. Environ Toxicol Chem. 30:2831–2837.

Pereira S, Malard V, Ravanat JL, Davin AH, Armengaud J, Foray N, Adam-Guillermin C. 2014. Low Doses of Gamma-Irradiation Induce an Early Bystander Effect in Zebrafish Cells Which Is Sufficient to Radioprotect Cells. PLoS ONE. 9:e92974.

Ramakers C, Ruijter JM, Deprez RH, Moorman AF. 2003. Assumption-free analysis of quantitative real-time polymerase chain reaction (PCR) data. Neurosci Lett. 339:62–66.

Richards R, Chaloupka M, Sanò M, Tomlinson R. 2011. Modelling the effects of 'coastal' acidification on copper speciation. Ecol Modell. 222:3559–3567.

Ruijter JM, Ramakers C, Hoogaars WM, Karlen Y, Bakker O, van den Hoff MJ, Moorman AF. 2009. Amplification efficiency: linking baseline and bias in the analysis of quantitative PCR data. Nucleic Acids Res. 37:e45.

Salbu B, Rosseland B, Oughton D. 2005. Multiple stressors – a challenge for the future. J Environ Monit. 7:539.

Sayed AEDH, Igarashi K, Watanabe-Asaka T, Mitani H. 2017. Double strand break repair and γ-H2AX formation in erythrocytes of medaka (*Oryzias latipes*) after γ-irradiation. Environ Pollut. 224: 35–43.

Schnug L, Jensen J, Scott-Fordsmand JJ, Leinaas HP. 2014. Toxicity of three biocides to springtails and earthworms in a soil multi-species (SMS) test system. Soil Biol Biochem. 74:115–126.

Sedelnikova OA, Pilch DR, Redon C, Bonner WM. 2003. Histone H2AX in DNA damage and repair. Cancer Biol Ther. 2:233–235.

SEPA. 2013. Phosphorus-32 concentrations in the River Clyde, Scotland between 2005–2013. Scottish Environmental Protection Agency. Personal communication.

Sforzini S, Oliveri C, Barranger A, Jha AN, Banni M, Moore MN, Viarengo A. 2020. Effects of fullerene C_{60} in blue mussels: role of mTOR in autophagy related cellular/tissue alterations. Chemosphere. 246:125707.

Skipperud L, Salbu B. 2018. Radionuclides: sources, speciation, transfer and impacts in the aquatic and terrestrial environment. In: Dellasala DA, Goldstein MI, editors. Encyclopedia of the anthropocene. Oxford (UK): Elsevier; p. 195–206.

Solomon KR, Sibley P. 2002. New concepts in ecological risk assessment: where do we go from here? Mar Pollut Bull. 44:279–285.

Song Y, Salbu B, Teien HC, Evensen Ø, Lind OC, Rosseland BO, Tollefsen KE. 2016. Hepatic transcriptional responses in Atlantic salmon (*Salmo salar*) exposed to gamma radiation and depleted uranium singly and in combination. Sci Total Environ. 562:270–279.

Song Y, Salbu B, Teien HC, Heier LS, Rosseland BO, Tollefsen KE. 2014. Dose-dependent hepatic transcriptional responses in Atlantic salmon (*Salmo salar*) exposed to sublethal doses of gamma radiation. Aquat Toxicol. 156:52–64.

Strober W. 2001. Trypan blue exclusion test of cell viability. Curr Protoc Immunol. Appendix 3: Appendix 3B.

Trevisan R, Ferraz Mello D, Fisher AS, Schuwerack PM, Dafre AL, Moody AJ. 2011. Selenium in water enhances antioxidant defenses and protects against copper-induced DNA damage in the blue mussel *Mytilus edulis*. Aquat Toxicol. 101:64–71.

Urushihara Y, Kobayashi J, Matsumoto Y, Komatsu K, Oda S, Mitani H. 2012. DNA-PK inhibition causes a low level of H2AX phosphorylation and homologous recombination repair in Medaka (*Oryzias latipes*) cells. Biochem Biophys Res Commun. 429:131–136.

Venier P, Maron S, Canova S. 1997. Detection of micronuclei in gill cells and haemocytes of mussels exposed to benzo[a]pyrene. Mutat Res. 390:33–44.

Vernon EL, Bean TP, Jha AN. 2020. Assessing relative biomarker responses in marine and freshwater bivalve molluscs following exposure to phosphorus 32 (^{32}P): application of genotoxicological and molecular biomarkers. J Environ Radioact. 213:106120.

Vernon EL, Jha AN. 2019. Assessing relative sensitivity of marine and freshwater bivalves following exposure to copper: application of classical and novel genotoxicological biomarkers. Mutat Res. 842:60–71.

Vernon EL, Smith JT, Jha AN. 2018. Relative comparison of tissue specific bioaccumulation and radiation dose estimation in marine and freshwater bivalve molluscs following exposure to phosphorus-32. J Environ Radioact. 192:312–320.

Vincent-Hubert F, Arini A, Gourlay-Francé C. 2011. Early genotoxic effects in gill cells and haemocytes of *Dreissena polymorpha* exposed to cadmium, B[a]P and a combination of B[a]P and Cd. Mutat Res. 723:26–35.

Vosloo D, Sara J, Vosloo A. 2012. Acute responses of brown mussel (*Perna perna*) exposed to sub-lethal copper levels: integration of physiological and cellular responses. Aquat Toxicol. 106-107:1–8.

Xu K, Tang Z, Liu S, Liao Z, Xia H, Liu L, Wang Z, Qi P. 2018. Effects of low concentrations copper on antioxidant responses, DNA damage and genotoxicity in thick shell mussel *Mytilus coruscus*. Fish Shellfish Immunol. 82:77–83.

Xu M, Jiang L, Shen KN, Wu C, He G, Hsiao CD. 2016. Transcriptome response to copper heavy metal stress in hard-shelled mussel (*Mytilus coruscus*). Genom Data. 7:152–154.

Yu KN, Tung MMT, Choi VWY, Cheng SH. 2012. Alpha radiation exposure decreases apoptotic cells in zebrafish embryos subsequently exposed to the chemical stressor, Cd. Environ Sci Pollut Res Int. 19: 3831–3839.

Combined action of gamma radiation and exposure to copper ions on *Lemna minor L*

Irina S. Bodnar and Evgenia V. Cheban

ABSTRACT

Purpose: Under natural conditions, the reaction of living organisms to the action of acute gamma radiation depends on other stressors, including heavy metals. The aim of this work was to study changes in morphometric parameters, the content of photoassimilation pigments and the level of oxidative stress in irradiated duckweed at various copper concentrations in the culture medium.

Materials and methods: As a model organism, we used *Lemna minor* L. Duckweed was exposed to acute γ-radiation at doses of 18, 42, 63 Gy. After irradiation, the plants were transferred into a medium containing 3, 5, 6.3 µmol/L Cu. On the 4th day of exposure, the levels of chlorophyll, carotenoids, malondialdehyde (MDA) were measured; after 7 days, the specific growth rate, the level of damage, the change in the frond area, copper concentration in plant tissues were determined.

Results: The action of γ-radiation (18, 42, 63 Gy) and copper ions (3, 5, 6.3 µmol/L) reduced the growth rate, increased the membrane lipid peroxidation, reduced the area of the fronds more significantly than under the separate action of the factors. The factors acted antagonistically on the specific growth rate. The content of copper in the tissues of irradiated plants (42, 63 Gy) increased.

Conclusion: Irradiation of duckweed with acute doses of gamma radiation reduced the resistance of plants to excess copper in the environment.

Introduction

The impact of radiation on ecosystems depends on abiotic and biotic stressors (Gagnaire et al. 2017; Beresford et al. 2020). The study of the combined effect of radiation and non-radiation factors on living organisms is a priority in radiological research and is necessary to expand understanding of the effects of radiation on wildlife and ecosystem processes, as well as to improve the radiological assessment of environmental impact (Fevrier et al. 2014; Garnier-Laplace et al. 2018; Beresford et al. 2020). Radiation/chemical interactions are regulated completely separately but clearly occur together in the ecosystem and in the individual, formed combined toxicity (Mothersill et al. 2014).

There are several types of response of living organisms to the combined action of two or more factors. Additivity – the combined effect of the action of several factors is equal to the sum of the effects of their separate action. Synergism – the combined effect is not equal to a simple addition of the effects of separate action of the factors, but it is intensified. Antagonism – the combined effect is less than expected from a simple addition of their independent action. Approaches to assessing the effects of the combined action of chemical and non-chemical stressors are under development (Altenburger and Greco 2009). The combined effect of two or more chemicals can be calculated using several methods, including fractional effects analysis (Backhaus et al. 2004; Long et al. 2009; Altenburger et al. 2018), response surface approach (Kanzawa and Saijo 1997), median effect analysis (Bijnsdorp et al. 2011; De Liguoro et al. 2018). In this study, CompuSyn software (Chou and Martin 2005) was used to assess the combined effects of gamma radiation and copper ions, which uses median effect analysis. In ecotoxicology, the synergistic effect is expressed as an increase in the toxic effect on the organism (or the considered parameter, for example, the growth rate), and the antagonistic effect, on the contrary, as a decrease.

Heavy metals are among the most common pollutants. The main sources of heavy metal pollution of water bodies are waste from metallurgical enterprises, household wastewater, electronic waste (Akram et al. 2019; Breivik et al. 2016). Combined radiation and chemical contamination is typical for the Chernobyl region which has areas with high concentrations of lead that was dumped into the reactor during the decontamination of the territory after the Chernobyl accident, and also with an increased content of copper and zinc from the meltdown of the core of the nuclear power plant (Skesters et al. 2010; Ministry of Emergencies of Ukraine 2011). Copper is an essential trace element for plants, a part of a number of enzymes, including Cu, Zn-superoxide dismutase, oxidoreductases; it participates in biochemical and physiological processes in plant cells. However, a slight increase in the concentration of copper in the environment has a detrimental effect on the life of aquatic plants. Increased concentrations of copper in water bodies can be the result of the extraction and processing of the metal, as well as the operations of metalworking and agricultural enterprises (Yruela 2009). The redox cycle between its two ionic forms promotes the formation of reactive oxygen species (ROS), while the possibility of erroneous binding to nonspecific sites in the molecules of proteins and other compounds can lead to the loss of the function of key enzymes (Vidaković-Cifrek et al. 2015).

Since 2000, one of the requirements of the International Commission on Radiological Protection (ICRP) became the protection of the environment from radioactive releases (ICRP 2003). Human health-related environmental approach is to be replaced by an ecocentric approach, in which ecosystems are viewed as 'targets' of radiation exposure (Tsvetnova et al. 2014). In nature, plants exposed to other stressors may be more radiosensitive than those grown under experimental conditions (Garnier-Laplace et al. 2013). In this case, it is necessary to focus on the most vulnerable link in ecosystems, that is, on species with high radiosensitivity, when regulating the negative impact. Aquatic plants can be exposed to significant radiation levels when water bodies are contaminated with radionuclides (Kryshev and Sazykina 1998).

Lemna minor is a free-floating vascular macrophyte that develops roots and is capable of absorbing nutrients through both fronds and roots (Van Hoeck et al. 2015a). Systematically, the species belongs to monocotyledonous plants of the *Lemnaceae* family (Landolt 1980, Sree et al. 2016). Like most plants, duckweed is a producer of primary organic matter by photosynthesis, it serves as food for fish and birds and is an essential component of aquatic ecosystems (Landolt and Kandeler 1987). Duckweed is widely used as a model organism in ecotoxicological research of surface water and wastewater (Van Hoeck et al. 2017). Its small size and ease of cultivation allow it to be used as a test object when working with various environmental factors, including ionizing radiation (Bodnar et al. 2016; Xie et al. 2019).

Lemna minor consists of an individual 'leaf-like' structure (frond) and a single root. The frond comprises a meristematic tissue, by means of which vegetative reproduction is carried out. On each side of the proximal part of the frond, there are two pockets, in which vegetative buds are formed, giving rise to daughter plants. Most often, this plant reproduces vegetatively, which makes it possible to use a genetically homogeneous population throughout the experiment (Landolt 1980). Plants are connected via a hyaline thread, a thin outgrowth of the membrane, in colonies of 2–5 fronds.

The aim of this work was to study the response of irradiated *Lemna minor* L. to Cu stress.

Materials and methods

Plant material, growth conditions, gamma radiation and Cu exposure

A laboratory culture of duckweed from the collection of the Institute of Biology of the Komi Scientific Center of the Ural Branch of the Russian Academy of Sciences (IB KSC UB RAS, Russian Federation) was grown on the Steinberg's medium (Steinberg 1946) in a KBWF 240 climatic chamber (Binder, Germany) at a temperature of 24 ± 0.1°C, photoperiodicity of 16 h light/8 h dark, 70% humidity, light intensity of 8000 lux (cool white light of fluorescent lamps). The composition of Steinberg's medium does not contain copper ions.

Gamma irradiation experiments were conducted at the 'Researcher' experimental facility of IB KSC UB RAS. *Lemna minor* was exposed to external ^{137}Cs gamma irradiation at three acute doses (18, 42, 63 Gy). As we established earlier, 18 Gy is the minimum dose at which the growth of a laboratory culture of duckweed is inhibited and chlorosis appears (Bodnar et al. 2016). After exposure to ionizing radiation, the plants were transferred onto a medium with 3, 5, 6.3 μmol/L Cu. Cu was added to the Steinberg's medium as sterilized solution of $CuCl_2{\cdot}2H_20$ (dissolved in deionized water). The concentrations were chosen based on our previous experiments (Bodnar et al. 2018). As a control, we used plants grown on the Steinberg's medium (without copper ions). All experiments were performed in triplicates.

A scheme of the different conditions:

0 Gy + 0 μM Cu – (0 Gy + Cu0)
0 Gy + 3 μM Cu – (0 Gy + Cu3)
0 Gy + 5 μM Cu – (0 Gy + Cu5)
0 Gy + 6.3 μM Cu – (0 Gy + Cu6)
18 Gy + 0 μM Cu – (18 Gy + Cu0)
42 Gy + 0 μM Cu – (42 Gy + Cu0)
63 Gy + 0 μM Cu – (63 Gy + Cu0)
18 Gy + 3 μM Cu – (18 Gy + Cu3)
18 Gy + 5 μM Cu – (18 Gy + Cu5)
18 Gy + 6.3 μM Cu – (18 Gy + Cu6)
42 Gy + 3 μM Cu – (42 Gy + Cu3)
42 Gy + 5 μM Cu – (42 Gy + Cu5)
42 Gy + 6.3 μM Cu – (42 Gy + Cu6)
63 Gy + 3 μM Cu – (63 Gy + Cu3)
63 Gy + 5 μM Cu – (63 Gy + Cu5)
63 Gy + 6.3 μM Cu – (63 Gy + Cu6)

To calculate the specific growth rate, we selected colonies of 2–4 fronds and transferred to sterile glass containers, 9–12 plants per container. Irradiated and non-irradiated plants were transplanted simultaneously. After irradiation, the duckweed for seven days was maintained in an experimental medium. At the end of the experiment, the plants were counted. The growth of L. minor was adequate, i.e. doubling time of frond number in control conditions <2.5 days (60 h), as described as validity criteria in guideline (OECD 2006). The average specific growth rate for a specific period was calculated as the angle of the slope of the logarithmic growth curve from the equation:

$$\mu_{i-j} = \frac{\ln(N_j) - \ln(N_i)}{t_j - t_i}, \quad (1)$$

where μ_{i-j} – the average value of the specific growth rate obtained over time from i to j; N_i – using the number of fronds observed in an experimental or control container at the moment I; N_j – the number of fronds observed in an experimental or control container at the moment j; t_i – the time of the beginning the experiment; t_j – the time of the end of the experiment (OCDE 2006).

Copper accumulation

Copper concentration in plant tissues was measured using inductively coupled plasma atomic emission spectroscopy. Microwave mineralization with nitric acid and peroxide was used in the digestion of the samples. The analysis was carried out by the Ecoanalytical Laboratory of the Institute of Biology of the Federal Research Center of the Komi Scientific Center of the Ural Branch of the Russian Academy of Sciences.

Frond surface area and damage

On the first and seventh days, the plants were photographed. To calculate their surface areas, the obtained images were analyzed using the Image-J software program version 1.48 (National Institutes of Health, Bethesda, MD, USA). The ratio of plant areas at the beginning and at the end of the experiment (S2/S1) was determined. On the seventh day of the experiment, the number of damaged fronds (with chlorosis, necrosis) was counted.

Assessment of lipid peroxidation (malondialdehyde content)

To assess the level of lipid peroxidation, four days after the start of the experiment we measured the content of malondialdehyde (MDA) formed during the destruction of membrane lipids by reactive oxygen species (ROS). The plants were weighed, 50 mg of the wet mass was triturated with quartz sand, then 1.5 ml of 20% trichloroacetic acid was added and the mixture was centrifuged at 10,000 *g* for 15 min. 0.3 ml of the supernatant was mixed with 1.2 ml of 0.5% thiobarbituric acid dissolved in 20% trichloroacetic acid. The mixture was incubated at 95°C for 30 min, then quickly cooled and centrifuged at 10,000 g for 15 min. The absorbance was determined at 532 nm and 600 nm using a spectrophotometer Spectrumlab SS2107 (LEKI Instruments, Finland); as a control, a solution of thiobarbituric acid in trichloroacetic acid was used. The MDA content was calculated from the equation:

$$C_x = \frac{(E_{532} - E_{600}) \times V_e}{K \times m_s \times V_a}, \quad (2)$$

where Cx – the MDA content, nmol/g of fresh weight; E_{532} is the maximum absorbance of the TBA-MDA complex; E_{600} is the nonspecific turbidity, Ve – the volume of the extract, ml; Va – the volume of the extract taken for analysis, ml; k – the molar extinction coefficient for MDA: 156 $mM^{-1} \times cm^{-1}$; m_s – the mass of the sample for extraction (Heath and Packer 1968).

Assessment of pigment concentration

Four days after the experiment started, we assessed the level of photoassimilating pigments. Plants (25 mg) were ground with the addition of a small amount of quartz sand (for better grinding) and calcium carbonate (to neutralize endogenous acids), then 0.7 ml of 96% ethanol was added and the resulting homogenate was incubated for 30 min at 4 °C. Then it was centrifuged 13,000 g for 10 min and the extract was transferred into a clean test tube. Ethanol was added to the precipitate and the tube was centrifuged again. The procedure was repeated until the precipitate became gray. The absorbance was determined using a spectrophotometer 470, 649 and 664 nm. The concentration of pigments was calculated in accordance with experimental equations as described in (Lichtenthaler 1987). Each experiment was carried out in triplicate.

Statistical analysis

CompuSyn software was used to quantify the interaction of factors (Chou and Martin 2005). CompuSyn uses a median effect analysis method whereby the combination index (CI) is calculated from growth inhibition curves (Bijnsdorp et al. 2011). This software was originally developed for the analysis of drug cytotoxicity (Chou 2006). De Liguoro et al. (2018) successfully used CompuSyn to evaluate a mixture of sulfonylamides in an aqueous medium for *Daphnia magna*. In this study, the inhibition of the specific growth rate of duckweed for individual copper concentrations, radiation doses, and their combination served as the initial data for the calculation.

The Shapiro–Wilk tests were used to determine whether the data had a normal distribution. One-way ANOVA, Student's and Mann–Whitney's tests were used. Statistical analysis was performed by computer software Statistica 6.0 (StatSoft, USA).

Results

Growth inhibition

Gamma radiation caused a dose dependent linear reduction in duckweed growth rate, including in all variants of the experiment with copper ($p \leq .05$, ANOVA). 7 days after irradiation at a dose of 18 Gy, the duckweed growth rates decreased by 20% compared to control plants, 42 Gy – by 36%, 63 Gy – by 55%. After irradiation and 7 days of exposure to copper, the growth rates decreased compared to control in variant 18 Gy + Cu3 by 40%, 18 Gy + Cu5 – by 50%, 18 Gy + Cu6 – by 56.5% (Figure 1). After irradiation at a dose of 18 Gy and exposure to 5 and 6.3 µmol/L Cu, the growth rate coincided with 0 Gy + Cu5 and 0 Gy + Cu6, at 42 and 63 Gy it was lower ($p \leq .05$). Data analysis using CompuSyn software showed that a combination of factors from gamma radiation and subsequent exposure to different copper concentrations most often resulted to antagonistic effects. Additivism was present with a combination of doses of 18 and 42 Gy and 5 µmol/L copper (Table 1). The lowest

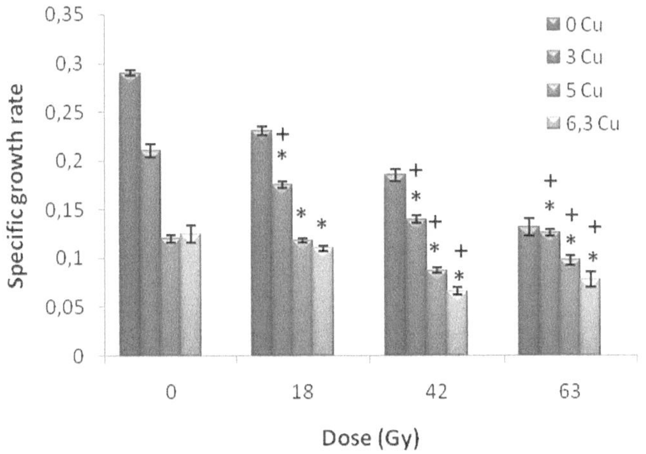

Figure 1. Dependence of the specific growth rate of the irradiated culture of duckweed on the concentration of copper in the medium. *Differences relative to irradiated plants are significant, +differences relative to non-irradiated plants grown on a medium with excess copper are significant ($p \leq .05$, Student's test).

Table 1. Combination index (CI), obtained from CompuSyn Report.

option	CI	Synergism/antagonism
18Gy + Cu3	1.19	Slight antagonism
18Gy + Cu5	1.07	Nearly additive
18Gy + Cu6	1.21	Moderate antagonism
42Gy + Cu3	1.26	Moderate antagonism
42Gy + Cu5	1.0	Nearly additive
42Gy + Cu6	1.47	Antagonism
63Gy + Cu3	1.37	Moderate antagonism
63Gy + Cu5	1.33	Moderate antagonism
63Gy + Cu6	1.24	Moderate antagonism

growth rate was observed upon simultaneous exposure to 6.3 µmol/L Cu and irradiation at doses of 42 and 63 Gy. At a dose of 18 Gy, the chemical action of copper had the most significant effect on the decrease in the growth rate; the growth rate under the action of irradiation decreased only in variant 18 Gy + Cu3 ($p \leq .05$). Upon irradiation at a dose of 42 Gy, the specific growth rate was lower compared to 18 Gy for all Cu concentrations.

Cu accumulation

The tissue copper concentrations in control plants were 0.89 ± 0.18 mg kg^{-1} fresh weight (FW) (Table 2). As the copper concentration increased, plant tissue copper concentration increased and ranged from 28 to 81 $mgkg^{-1}$ FW. Plants irradiated at doses of 42 and 63 Gy accumulated more copper than non-irradiated plants grown with the addition of copper to the medium ($p \leq .05$).The copper content in the irradiated duckweed was from 36 to 150 mg/kg fresh weight. The largest accumulation of copper was at 63 Gy + Cu6.

Frond surface area

Upon exposure to ionizing radiation and copper, the frond surface area decreased ($p \leq .05$, ANOVA). However, if the pressure from stressors was increased even more, the average ratio of the areas did not change due to the low growth rate of duckweed and the appearance of a small number of plants with a smaller frond area. The most significant decrease in the average area occurred when the plants were irradiated at a dose of 18 Gy and at a copper concentration of 3 µmol/L; it was 40% lower than the initial frond area (Figure 2).

Table 2. Copper content in plants after 7 days of treatment.

option	Cu, $mgkg^{-1}$FW
Control (0Gy + 0Cu)	0.89 ± 0.18
0Gy + Cu3	28 ± 6
0Gy + Cu5	45 ± 9
0Gy + Cu6	81 ± 16
18Gy + Cu3	36 ± 7
18Gy + Cu5	87 ± 17
18Gy + Cu6	74 ± 15
42Gy + Cu3	47 ± 9
42Gy + Cu5	72 ± 14
42Gy + Cu6	100 ± 20
63Gy + Cu3	51 ± 10
63Gy + Cu5	108 ± 22
63Gy + Cu6	132 ± 26

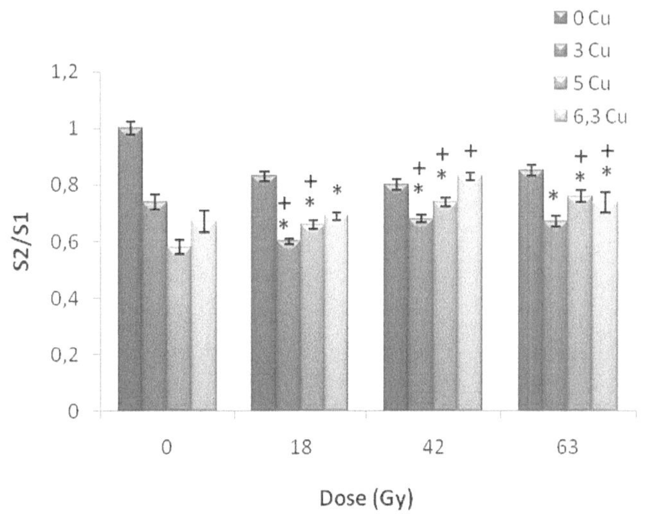

Figure 2. Change in the average frond area (S2/S1, where S1 is the initial area, S2 is the area after 7 days). *Differences relative to to irradiated plants are significant, +differences relative to non-irradiated plants grown on a medium with excess copper are significant ($p \leq .05$, Student's test).

Damage to fronds

Chemical exposure caused more significant damage to the surface of the fronds than ionizing radiation. The highest level of damage was observed at a copper concentration of 6.3 µmol/L, reaching 100% with all variants of plant irradiation (Figure 3). At 42 Gy + Cu5 µmol/L over 90% of plants were damaged. After irradiation, the plants of positive irradiated control had a dark green color. Upon the combined action of copper and radiation, duckweed turned yellow, chlorosis appeared, in variants 18 Gy + Cu5, 42 Gy + Cu5, 18 Gy + Cu5, 18Cy + Cu6, as well as early death of plants, while the fronds became white. At higher levels of exposure, necrosis appeared, the color of the plants became white-green, and in variants 42 Gy + Cu6, 63 Gy + Cu6 – white with growth points.

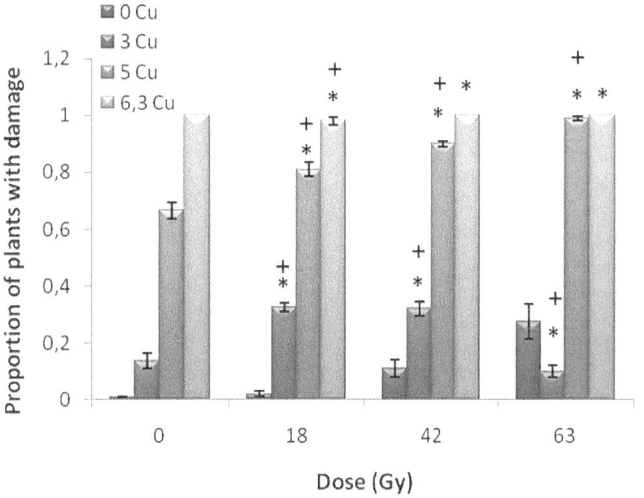

Figure 3. Change in the proportion of damaged fronds. *Differences relative to irradiated plants are significant, +differences relative to non-irradiated plants grown on a medium with excess copper are significant ($p \leq .05$, Mann–Whitney's test).

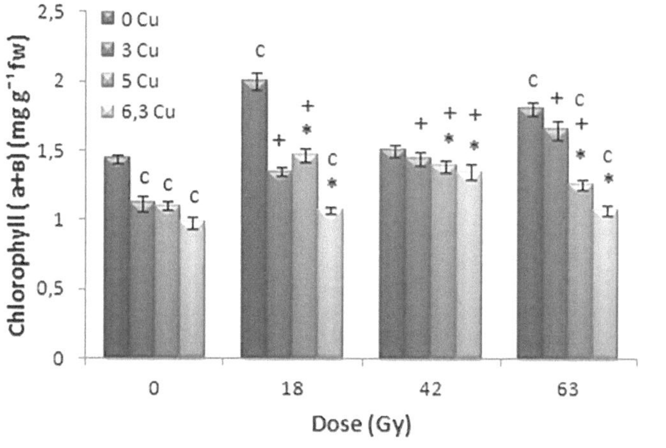

Figure 5. Change in chlorophyll content a + b. *Differences relative to irradiated plants are significant, +differences relative to non-irradiated plants grown on a medium with excess copper are significant, [c]difference relative to control are significant (0 Gy + Cu0) ($p \leq .05$, Student's test).

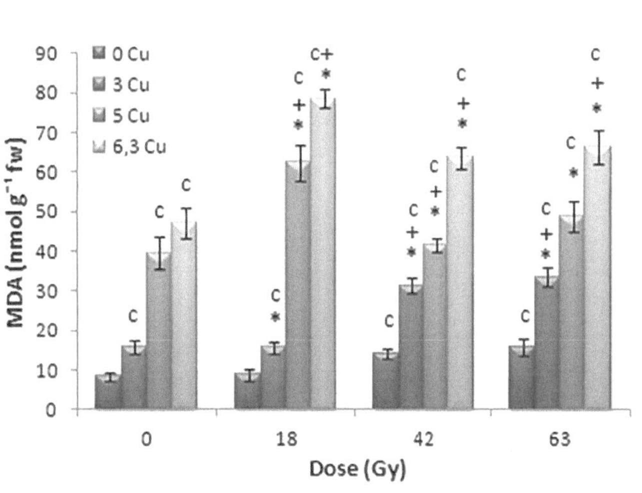

Figure 4. Change in the content of malondialdehyde. *Differences relative to irradiated plants are significant, +differences relative to non-irradiated plants grown on a medium with excess copper are significant, [c]difference relative to control are significant (0 Gy + Cu0) ($p \leq .05$, Mann–Whitney's test).

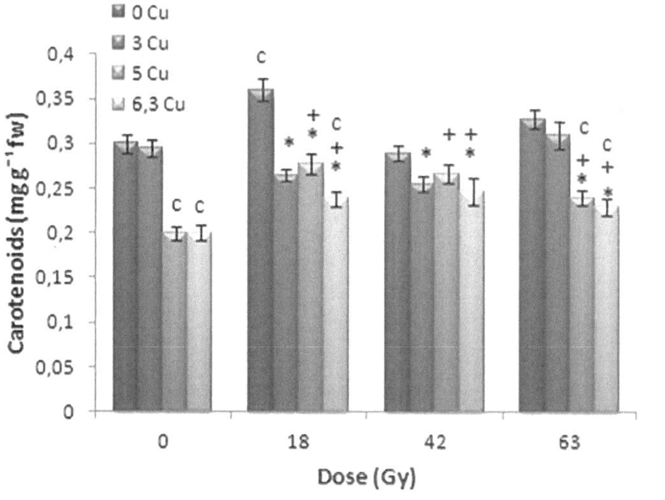

Figure 6. Change in total carotenoid content. *Differences relative to irradiated plants are significant, +differences relative to non-irradiated plants grown on a medium with excess copper are significant, [c]difference relative to control are significant (0 Gy + Cu0) ($p \leq .05$, Student's test).

Lipid peroxidation

Ionizing radiation provoked peroxidation of membrane lipids; a statistically significant increase in the MDA concentration occurred at doses of 42 and 63 Gy (Figure 4). Copper is a redox-active metal; starting at 3 μmol/L Cu, the MDA content increased linearly. Compared to radiation, copper provokes the development of stronger oxidative stress in plant cells. But the highest MDA level was observed upon the combined action of irradiation at a dose of 18 Gy and exposure to copper at concentrations of 5 and 6.3 μmol/L: 7 and 9 times higher than in the control, respectively.

Chlorophyll and carotenoid content

Under the action of ionizing radiation (at 18 and 63 Gy), the content of chlorophylls a and b increased, while copper, on the contrary, caused a decrease in the content of pigments (Figure 5). The factors acted as antagonists: copper destroyed chlorophyll, while radiation stimulated its production. In variants 18 Gy + Cu6, 63 Gy + Cu5, 63 Gy + Cu6, the chlorophyll content was lower than in the control ($p \leq .05$). Compared with 0 Gy + Cu3, 0 Gy + Cu5, 0 Gy + Cu6, the content of chlorophylls a and b increased when exposed to gamma radiation, except 18 Gy + Cu6. The content of chlorophylls a and b in plants exposed to both irradiation and copper was lower than in only irradiated plants. In contrast, the chlorophyll content was higher in plants exposed to both factors than in non-irradiated plants cultured with copper.

As seen in the Figure 6, the level of carotenoids in plants exposed to 5 or 6.3 μmol/L Cu is reduced by one third. Irradiation at a dose of 18 Gy caused an increase in the carotenoid content. The response to the combined action of both factors in most variants remained at the level of control values, except for variants 18 Gy + Cu6, 63 Gy + Cu5, 63 Gy + Cu6, where the levels of carotenoids were lower than in the control.

A characteristic sign of the action of copper ions on duckweed was the death of roots. At a copper concentration of 3 μmol/L, the roots were small, often white. Since a plant of *Lemna minor* has only one green root capable of photosynthesis, a decrease in the level of photo-assimilating pigments in duckweed at this copper concentration in comparison with control plants is associated with the destruction of chlorophyll in the roots. At copper concentrations of 5 and 6 μmol/L, the roots died off.

Discussion

The effects of ionizing radiation on plants, animals, and bacteria are being actively studied, however, data on the radiosensitivity of species obtained at laboratory conditions most often differs from what is observed in nature. Changes in stability of biological systems of different levels in response to an increased level of environmental radioactivity in the wild may be caused by a complex of other stressors acting in addition to ionizing radiation. (Beresford et al. 2020). This study made it possible to expand the existing understanding of the responses of irradiated plants to Cu stress. In this work, we analyzed the changes in morphological and biochemical parameters under separate and combined action of stressors and quantitatively determined the nature of their interaction.

The growth rate of a laboratory culture of duckweed is an integral characteristic of biochemical and molecular changes in response to stressors. Under the action of acute gamma radiation, a dose-dependent decrease in the specific growth rate was observed, which is consistent with previous studies (Van Hoeck et al. 2015b). Copper at the considered concentrations (3–6 μmol/L) also led to a decrease in the growth rate of the laboratory culture. Similar results were obtained by Dumont et al. 2019, when exposed to copper, the growth rate at 3.8 μmol/L decreased at any light intensity; the study was carried out according to OECD (2006) and on Steinberg's medium. The sensitivity of duckweed to copper has been shown to vary with environmental conditions, highlighting the importance of strict guidelines in standardized protocols (Dumont et al. 2019).

The toxicity of copper to agricultural plants and was manifested a decrease in biomass, a change in water content, the appearance of chlorosis and necrosis, the inhibition of shoot growth, a decrease in the rate of seed germination, a decrease in chlorophyll content and a change in the structure of chloroplasts and thylakoid membranes in spinach, rice, wheat (Baszynski et al. 1988; Lidon and Henriques 1993; Ciscato et al. 1997; Pätsikkä et al. 1998; Quartacci et al. 2000; Panou-Filotheou et al. 2001; Ahsan et al. 2007).

In irradiated duckweed grown in a medium with excess copper, the growth rate is lower than in the case of independent action of stressors, although the factors acted as antagonists. Environmental variations might greatly affect the vigor of a species, its sensitivity to chemicals (Vitasse et al. 2010; Dumont et al. 2019). After irradiation, the accumulation of copper in plant tissues changed, which is reflected in morphological and biochemical parameters. When cultivated with excess copper, the pre-irradiated plants at doses 42 and 63 Gy had a higher copper accumulation than the non-irradiated. This increased toxicity and suppressed the growth rate. Saha et al. (2010) have shown that irradiation changed the elemental profile of plants. The combined action of radiation and copper induced metallothionein proteins, which are homeostatic regulators and radical scavengers against oxidative stress (Saha et al. 2008).

Irradiation at a dose of 42 Gy and exposure to copper at 3 μmol/L led to a 50% reduction in the growth rate compared to a control. Thus, the EC50 became now not 63 Gy, as in the irradiated plants without copper exposure, but 42 Gy at a relatively low copper concentration of 3 μmol/L (0.2 mg/L) in the culture medium. For comparison: according to European standards, the maximum admissible concentration of copper in water is 3 mg/L (The Council of the European Union 1998).

An increase in toxicity in response to exposure to copper and ionizing radiation was also observed when analyzing data on changes in the surface area of fronds. A decrease in the frond area was more pronounced after exposure to copper than after irradiation compared to an independent action of the factors; the minimum values were at 5 μmol/L Cu and at 42 Gy, which is lower than the control by 42% and 20%, respectively. Under the combined action of the factors, the minimum of the indicator was at 18 Gy and 3 μmol/L. A dose-dependent change in the area of fronds upon exposure to copper was shown in Yue et al. (2018). Microscopy imaging showed that the epidermis was severely damaged in fronds, edges were severely sloughed off (Yue et al. 2018). The rupture of the epidermal layers occurs due to the binding of Cu^{2+} ions to the cell walls (Kopittke et al. 2007). Under an independent and combined action of the factors, the relationship between them and the average change in the frond area is U-shaped: at first, the area under the action of the factors decreases, and with an increase in the dose and concentration, it increases, since there are few new plants and their area is less than the initial one, in the calculation the mean values included area of the parental individuals. The exposure acts with a delay in time. As we showed earlier, after fourteen days, due to its small size, the frond becomes like a ball (Bodnar et al. 2016). The smaller the leaf area, the lower its ability for photosynthesis and therefore for further growth and development. The cause of the delayed action of the irradiation has been established by Grodzinskii et al. in experiments on *Spirodela polyrhiza* (Grodzinskii et al. 1984). They have proven that the radiosensitivity increases linearly with the generation number of the daughter fronds: in plants of 1–2 generations, LD50 comprised 60–70 Gy, and in the seventh generation – 10 Gy. Because of the delayed action of radiation, the first generations of fronds had minimal damage, while the subsequent ones that were more affected by the action of radiation had multiple morphological disorders. Van Hoeck et al. (2017) determined that although no visible signs of cell death for irradiated plants were observed at the end of the exposure period, cell death was triggered as a delayed response to DNA damage during the recovery period. 13 of the 56 genes double-strand DNA breaks repair genes were significantly downregulated, including LIG4, RFC1, RPA1B, KU70, and KU80, which are all key players in DNA repair mechanisms (Van Hoeck et al. 2017).

Under irradiation at a dose of 63 Gy, the proportion of damaged fronds reaches 30%, and under exposure to copper at 6.3 μmol/L – 100%. In the population of control plants, after 7 days, the proportion of damaged fronds was less than 1%. Copper, due to its high reactivity, acts directly and in a 7-day test produces a greater proportion of plants with chlorosis, necrosis and leaf depigmentation. At the molecular level, copper binds to sulfhydryl groups of proteins, which negatively affects the absorption of essential elements and the transfer processes in the cell (Vidacović-Cifrek et al. 2015). A specific feature of the combined action of the factors is accelerated aging of duckweed.

Copper catalyzes the formation of hydroxyl radicals (OH*) in the Haber–Weiss chemical reaction between superoxide (O_2*) and hydrogen peroxide (H_2O_2) (Halliwell and Gutteridge 1984). An excess of heavy metal leads to oxidative stress in plants, an increase in the content of highly toxic free radicals, and changes in the activity and content of some antioxidants (Luna et al. 1994; Stohs and Bagchi 1995; Gupta et al. 1999; Tang et al. 2016; Saleem et al. 2020). In addition to the direct degenerative effect, ionizing radiation causes the development of free radical reactions in biological objects. Chemically aggressive radicals interact with lipid and protein molecules of cell membranes with subsequent destruction of cells (Sisakht et al. 2020). Oxidative stress is the main component of radioadaptive reactions in cells exposed to gamma radiation. Ionizing radiation causes an increased production of ROS due to water radiolysis. In plants, membranes of chloroplasts and mitochondria are the main targets for gamma-induced ROS due to membrane lipid peroxidation (LPO) (Shadyro et al. 2002). A recent study (Xie et al. 2019) showed that a significant increase in LPO in duckweed upon exposure to gamma radiation positively correlated with the ROS formation.

The breakdown of fatty acids accompanying LPO results in the formation of diene conjugates at first, and then such metabolites as malondialdehyde (Zagoskina and Nazarenko 2016). Radiation and copper ions shift the oxidant/antioxidant balance in favor of oxidation. The content of such oxidation marker of lipids as MDA is significantly increased (Deshmukh et al. 2013). Copper ions at the considered concentrations caused a stronger oxidative stress than γ-radiation. Copper caused an increase in the MDA content even at a minimum concentration affecting the growth rate of plants, and irradiation caused oxidative stress at doses of 42 and 63 Gy. Irradiation and Cu resulted to higher concentrations of MDA in plants than under the independent action of factors. Plants have an established system to reduce the consequences of such influences. It is known that small doses of radiation can lead to the stimulation of the growth and development of plants, including those under stress from heavy metals. At high and medium doses of radiation, not only synergistic and additive, but also antagonistic effects can be observed (Wang et al. 2017). An analysis of our data on the change in the growth rate of the laboratory culture of duckweed showed that the interaction of the factors most often had the antagonistic nature. Therefore, the toxic effect under the combined action of the factors was lower than expected. At these dose levels, ionizing radiation can have a protective effect. The explanation may be the changes in the antioxidant system and in the expression profile of genes, including those associated with the transport of heavy metals and the metabolism of abscisic acid under conditions of stress due to exposure to Cd or Pb (Wang et al. 2017).

The content of chlorophylls a and b under the action of the stressors on plants depended on the type and intensity of the action. After acute γ-radiation at doses of 18 and 63 Gy, the chlorophyll concentration was higher than in unirradiated plants, and at 42 Gy it was similar to that in the control samples. In the literature, a similar increase in the concentration of pigment after irradiation of plants is described. When exposed to an ion beam at doses of 10 and 20 Gy, the total chlorophyll content in *O. sativa* seedlings was similar to that in the control, while irradiation at a dose of 40 Gy had a stimulating effect and the level of chlorophyll increased (Ling et al. 2013). The work (Creanga et al. 2005) has shown that upon irradiation of seeds there is a stimulating effect of low doses of radiation on the chlorophyll content in bean leaves. In 1971, Singh has also found an increase in the chlorophyll content in maize leaves after exposure to γ-radiation at 86.96–347.80 Gy (Singh 1971). The work (Vanhoudt et al. 2014) has confirmed that the photosynthetic capacity of photosystem II (PSII) does not change after 7-day γ-radiation at a total dose from 3.9 to 58.8 Gy.

Copper has a destructive effect on chlorophyll, interfering with the synthesis of the photosynthetic apparatus, protein components of photosynthetic membranes and modifying pigments (Lidon and Henriques 1991; Maksymiec et al. 1994; Pätsikkä et al. 2002); it also negatively affects photosystem II (PSII) Therefore, the excess of copper results in the inhibition of growth and disruption of such cellular processes as photosynthesis and respiration (Marschner 1995; Prasad and Strazalka 1999; Yruela 2005). In addition, the central magnesium ion of the chlorophyll is replaced by copper ion, which leads to inhibition of photosynthesis (Küpper et al. 2003; Küpper and Kroneck 2005). The level of chlorophyll *a* in duckweed in the positive copper control decreased by 27–30%. Pre-irradiation at an acute dose led to an increase in the chlorophyll content in plants compared to the positive copper control in all variants, except 63 Gy + Cu6. This means that the irradiation had an adaptive effect for the rudiments of fronds of subsequent generations. A similar effect has been observed in agricultural species subjected to pre-irradiation of seeds. The positive effect of irradiation before cultivation in a medium with copper may be responsible for the fact that only in variants 18 Gy + Cu6, 63 Gy + Cu5, 63 Gy + Cu6 the content of chlorophyll *a* and *b* was lower than in the control. Upon pre-irradiation of highland barley seeds at low doses (50 Gy), a similar adaptive effect has been observed; there has been a decrease in toxicity under the exposure to Pb, Cd compared to non-irradiated seedlings due to an increase in the proline content and an improvement of the ultrastructure of chloroplasts. γ-radiation can significantly affect the expression of genes involved in the transport of heavy metals and the metabolism of abscisic acid under stress conditions of exposure to Cd, Pb (Wang et al. 2017).

In addition to the photoassimilatory function, carotenoids are involved in the antioxidant defense of the organism against oxidative stress caused by chemical and physical factors (Hashimoto et al. 2016). Carotenoids are considered the first line of plant defense against the formation of ROS in chloroplasts; they can extinguish singlet oxygen ($^1O_2$), as well as triplet excited chlorophyll (3Chl) through any thermal deactivation or oxidation (Ramel et al. 2012). A decrease in the level of carotenoids can result in the inhibition of the ability to collect light and reduce the efficiency of photosynthesis. Irradiation of duckweed at a dose of 18 Gy stimulated the production of carotenoids, contributing to the minimization of the effects of oxidative stress on the cell, which is evident from the fact that the concentration of MDA remained at a level similar to the control values. When plants are cultivated on a medium with the addition of copper, the level of carotenoids is higher in pre-irradiated plants as compared to non-irradiated plants. Stimulation of carotenoid production may be caused by antioxidant reactions in chloroplasts (Lind et al. 2019).

Conclusions

Acute doses of gamma radiation reduced the resistance of duckweed to excess copper in the environment. The accumulation of copper in the tissues of irradiated plants increased. This should be taken into account when analyzing the risk for aquatic ecosystems with increased radioactivity and heavy metal contamination. At the minimum effective concentration of the metal and dose of γ-radiation, the growth rate of the laboratory culture decreased more than under the separate action of the factors. However, the nature of the interaction of the factors was antagonistic due to changes at the biochemical level. Irradiation of duckweed contributed to the formation of an adaptive response in plants under the action of increased copper concentrations in the environment, which stimulated the production of photoassimilating pigments. The combined action of the factors intensified the processes of membrane lipid peroxidation in cells, contributed to the reduction in the area and the formation of damage to the fronds.

Disclosure statement

The authors declare that they have no conflict of interest.

Funding

The work was carried out as a part of the state project 'Mechanisms of Biogenic Migration of Radionuclides and Long-Term Effects Induced by Chronic Radiation and Chemical Exposure in Plants and Animals' [number 0414-2018-0002, AAAA-A18-118011190102-7].

References

Ahsan N, Lee DG, Lee SH, Kang KY, Lee JJ, Kim PJ, Yoon HS, Kim JS, Lee BH. 2007. Excess copper induced physiological and proteomic changes in germinating rice seeds. Chemosphere. 67:1182–1193.

Akram R, Natasha Fahad S, Hashmi MZ, Wahid A, Adnan M, Mubeen M, Khan N, Rehmani M, Awais M, Abbas M, et al. 2019. Trends of electronic waste pollution and its impact on the global environment and ecosystem. Environ Sci Pollut Res Int. 26:16923–16938.

Altenburger R, Greco WR. 2009. Extrapolation concepts for dealing with multiple contamination in environmental risk assessment. Integr Environ Assess Manag. 5:62–68.

Altenburger R, Scholze M, Busch W, Escher BI, Jakobs G, Krauss M, Krüger J, Neale PA, Ait-Aissa S, Almeida AC, et al. 2018. Mixture effects in samples of multiple contaminants – an inter-laboratory study with manifold bioassays. Environ Int. 114:95–106.

Backhaus T, Faust M, Scholze M, Gramatica P, Vighi M, Grimme LH. 2004. Joint algal toxicity of phenylurea herbicides is equally predictable by concentration addition and independent action. Environ Toxicol Chem. 23:258–264.

Baszynski T, Tukendorf A, Ruszkowska M, Skórzynska E, Maksymiec W. 1988. Characteristics of the photosynthetic apparatus of copper nontolerant spinach exposed to excess copper. J Plant Physiol. 132:708–713.

Beresford NA, Horemans N, Copplestone D, Raines KE, Orizaola G, Wood MD, Laanen P, Whitehead HC, Burrows JE, Tinsley MC, et al. 2020. Towards solving a scientific controversy – the effects of ionising radiation on the environment. J Environ Radioact. 211:106033.

Bijnsdorp I, Giovannetti E, Peters GJ. 2011. Analysis of drug interactions. Methods Mol Biol. 731:431–434.

Bodnar IS, Cheban EV, Zainullin VG. 2018. Features of the effect of copper and strontium ions on the laboratory culture of duckweed (*Lemna minor* L.). Princ Ecol. 2:3–21. Russian.

Bodnar IS, Yushkova EA, Zainullin VG. 2016. Influence γ-radiation on morphometric characteristics of duckweed (*Lemna minor* L.). Radiat Biol Radioecol. 56:617–622. Russian.

Breivik K, Armitage JM, Wania F, Sweetman AJ, Jones KC. 2016. Tracking the global distribution of persistent organic pollutants accounting for E-waste exports to developing regions. Environ Sci Technol. 50:798–805.

Chou TC. 2006. Theoretical basis, experimental design, and computerized simulation of synergism and antagonism in drug combination studies. Pharmacol Rev. 58:621–681.

Chou TC, Martin N. 2005. CompuSyn for drug combinations: PC software and user's guide: a computer program for quantitation of synergism and antagonism in drug combinations, and the determination of IC50 and ED50 and LD50 values. Paramus (NJ): ComboSyn.

Ciscato M, Valcke R, Loven K, Clijsters H, Navari-Izzo F. 1997. Effects of in vivo copper treatment on the photosynthetic apparatus of two *Triticum durum* cultivars with different stress sensitivity. Physiol Plant. 100:901–908.

Creanga D, Mantale AM, Ecaterina F. 2005. The radiosensitivity of photosynthetic processes in young maize plants. Analele Ştiinţifice ale Universităţii "Al.I.Cuza". Biofizică, Fizică medicală şi Fizica mediului. 1:81–86.

De Liguoro M, Riga A, Fariselli P. 2018. Synergistic toxicity of some sulfonamide mixtures on *Daphnia magna*. Ecotoxicol Environ Saf. 164:84–91.

Deshmukh PS, Banerjee BD, Abegaonkar MP, Megha K, Ahmed RS, Tripathi AK, Mediratta PK. 2013. Effect of low level microwave radiation exposure on cognitive function and oxidative stress in rats. Indian J Biochem Biophys. 50:114–119.

Dumont ER, Larue C, Pujol B, Lamaze T, Elger A. 2019. Environmental variations mediate duckweed (*Lemna minor* L.) sensitivity to copper exposure through phenotypic plasticity. Environ Sci Pollut Res Int. 26:14106–14115.

Fevrier L, Oughton D, Hinton TG. 2014. Strategic research agenda for radioecology - an updated version with stakeholder input. Star.

Gagnaire B, Adam-Guillermin Festarini A, Cavalie I, Della-Vedova C, Shultz C, Kim SB, Ilkert H, Dubois C, Walsh S, Farrow F, et al. 2017. Effects of in situ exposure to tritiated natural environments: a multi-biomarker approach using the fathead minnow, Pimephales promelas. Sci Total Environ. 599-600:597–611.

Garnier-Laplace J, Vandenhove H, Beresford N, Muikku M, Real A. 2018. COMET strongly supported the development and implementation of medium-term topical research roadmaps consistent with the ALLIANCE Strategic Research Agenda. J Radiol Prot. 38: 164–174.

Garnier-Laplace J, Geras'kin S, Della-Vedova C, Beaugelin-Seiller K, Hinton TG, Real A, Oudalova A. 2013. Are radiosensitivity data derived from natural field conditions consistent with data from controlled exposures? A case study of Chernobyl wildlife chronically exposed to low dose rates. J Environ Radioact. 121:12–21.

Grodzinskii DM, Kolomiez KD, Gudkov IN, Kutlakhmedov YA, Bulakh AA. 1984. Formation of the radiobiological reaction of plants. Kiev: Nauka. Russian.

Gupta M, Cuypers A, Vangronsveld J, Clijsters H. 1999. Copper affects the enzymes of the ascorbate-glutathione cycle and its related metabolites in the roots of *Phaseolus vulgaris*. Physiol Plant. 106: 262–267.

Halliwell B, Gutteridge JMC. 1984. Oxygen toxicity, oxygen radicals, transition metals and disease. Biochem J. 219:1–14.

Hashimoto H, Uragami C, Cogdell RJ. 2016. Carotenoids and Photosynthesis. Subcell Biochem. 79:111–141.

Heath RL, Packer L. 1968. Photoperoxidation in isolated chloroplasts. Kinetics and stoichiometry of fatty acid peroxidation. Archives of Biochem and Biophys. 125:189–198.

ICRP [International Commission on Radiological Protection]. 2003. A framework for assessing the impact of ioni-zing radiation on non-human species. ICRP Publication 91. Ann IRCP. 33:201–226.

Kanzawa F, Saijo N. 1997. In vitro interaction between gemcitabine and other anticancer drugs using a novel three-dimensional model. Semin. Oncol. 24:8–16.

Kopittke PM, Dart PJ, Menzies NW. 2007. Toxic effects of low concentrations of Cu on nodulation of cowpea (*Vigna unguiculata*). Environ Pollut. 145:309–315.

Kryshev I, Sazykina T. 1998. Radioecological effects on aquatic organisms in the areas with high levels of radioactive contamination: environmental protection criteria. Radiar Prot Dosim. 75:187–191.

Küpper H, Kroneck PMH. 2005. Heavy metal uptake by plants and cyanobacteria. Met Ions Biol Syst. 44:97–144.

Küpper H, Šetlík I, Šetliková E, Ferimazova N, Spiller M, Küpper FC. 2003. Copper-induced inhibition of photosynthesis: limiting steps of in vivo copper chlorophyll formation in *Scenedesmus quadricauda*. Funct Plant Biol. 30:1187–1196.

Landolt E. 1980. Key to the determination of taxa witnin the family of Lemnaceae. Zürich: Veroeffentlichungen des Geobotanischen Instituts der ETH, Stiftung Rübel.

Landolt E, Kandeler R. 1987. Biosystematic investigations in the family of duckweeds (Lemnaceae). The Family of Lemnaceae. Vol. 4. Zürich: Veroeffentlichungen Des Geobotanischen Instituts Der ETH, Stiftung Rübel.

Lichtenthaler HK. 1987. Chlorophylls and carotenoids; pigments of photosynthetic biomembranes. Method Enzymol. 148:350–382.

Lidon FC, Henriques FS. 1991. Limiting step in photosynthesis of rise plants treated with varying copper levels. J Plant Physiol. 138: 115–118.

Lidon FC, Henriques FS. 1993. Changes in the thylakoid membrane polypeptide patterns triggered by excess Cu in rice. Photosynthetica. 28:109–117.

Lind OC, Oughton DH, Salbu B. 2019. The NMBU FIGARO low dose irradiation facility. Int J Radiat Biol. 95:76–81.

Ling APK, Ung YC, Hussein S, Harun AR, Tanaka A, Yoshihiro H. 2013. Morphological and biochemical responses of *Oryza sativa* L. (cultivar MR219) to ion beam irradiation. J Zhejiang Univ Sci B. 14: 1132–1143.

Long SM, Reichenberg F, Lister LJ, Hankard PK, Townsend J, Mayer P, Wright J, Holmstrup M, Svendsen C, Spurgeon DJ. 2009. Combined chemical (fluoranthene) and drought effects on *Lumbricus rubellus* demonstrate the applicability of the independent action model for multiple stressor assessment. Environ Toxicol Chem. 28:629–636.

Luna CM, González CA, Trippi VS. 1994. Oxidative damage caused by excess of copper in oat leaves. PCP. 35:11–15.

Maksymiec W, Russa R, Urbanik-Sypniewska T, Baszynski T. 1994. Effect of excess Cu on the photosynthetic apparantus of runner bean leaves treated at two different growth stages. Physiol Plant. 91: 715–721.

Marschner H. 1995. Mineral nutrition of higher plants. London (UK): Academic Press; p. 333–347.

Ministry of Emergencies of Ukraine. 2011. Twenty-five years Chornobyl after accident: safety for the future. National report of Ukraine. Kyiv: KIM.

Mothersill C, Smith RW, Heier LS, Teien HC, Lind OC, Land OC, Seymour CB, Oughton D, Salbu B. 2014. Radiation-induced bystander effects in the Atlantic salmon (*Salmo salar* L.) following mixed exposure to copper and aluminum combined with low-dose gamma radiation. Radiat Environ Biophys. 53:103–114.

[OECD] Organisation for Economic Co-operation and Development. 2006. Guidelines for the testing chemicals. *Lemna* sp. growth inhibition test. Paris: OECD.

Panou-Filotheou H, Bosabalidis AM, Karataglis S. 2001. Effects of copper toxicity on leaves of oregano (*Origanum vulgare* subsp. *hirtum*). Ann Bot. 88:207–214.

Pätsikkä E, Aro EM, Tyystjärvi E. 1998. Increase in the quantum yield of photoinhibition contributes to copper toxicity in vivo. Plant Physiol. 117:619–627.

Pätsikkä E, Kairavuo M, Sersen F, Aro EN, Tyystjärvi E. 2002. Excess copper predisposes photosystem II to photoinhibition in vivo by outcompeting iron and causing decrease in leaf chlorophyll. Plant Physiol. 129:1359–1367.

Prasad MNV, Strazalka K. 1999. Impact of heavy metals on photosynthesis. In: Prasad MNV, Hagemeyer J, editors. Heavy metal stress in plants. Berlin: Springer; p. 117–178.

Quartacci MF, Pinzino C, Sgherri CLM, Dalla Vecchia F, Navari-Izzo F. 2000. Growth in excess copper induces changes in the lipid composition and fluidity of PSII-enriched membranes in wheat. Physiol Plantarum. 108:87–93.

Ramel F, Birtic S, Cuine S, Triantaphylides C, Ravanat JL, Havaux M. 2012. Chemical quenching of singlet oxygen by carotenoids in plants. Plant Physiol. 158:1267–1278.

Saha P, Mishra D, Chakraborty A, Sudarshan M, Raychaudhuri SS. 2008. In vitro radiation induced alterations in heavy metals and metallothionein content in *Plantago ovata* Forsk. Biol Trace Elem Res. 124:251–261.

Saha P, Raychaudhuri SS, Chakraborty A, Sudarshan M. 2010. PIXE analysis of trace elements in relation to chlorophyll concentration in *Plantago ovata* Forsk. Appl Radiat Isot. 68:444–449.

Saleem MH, Fahad S, Khan SU, Din M, Ullah A, Sabagh AE, Hossain A, Llanes A, Liu L. 2020. Copper-induced oxidative stress, initiation of antioxidants and phytoremediation potential of flax (Linum usitatissimum L.) seedlings grown under the mixing of two different soils of China. Environ Sci Pollut Res Int. 27:5211–5221.

Shadyro OI, Yurkova IL, Kisel MA. 2002. Radiation-induced peroxidation and fragmentation of lipids in a model membrane. Int J Radiat Biol. 78:211–217.

Singh BB. 1971. Effect of gamma-irradiation on chlorophyll content of maize leaves. Radiat Bot. 11:243–244.

Sisakht M, Darabian M, Mahmoodzadeh A, Bazi A, Shafiee SM, Mokarram P, Khoshdel Z. 2020. The role of radiation induced oxidative stress as a regulator of radio-adaptive responses. Int J Radiat Biol. 96:561–576.

Skesters A, Zvagule T, Silova A, Rusakova N, Larmane L, Reste J, Eglite M, Rainsford KD, Callingham BA, Bake MA, et al. 2010. Biochemical observations relating to oxidant stress injury in Chernobyl clean-up workers ("liquidators") from Latvia. Inflammopharmacology. 18:17–23.

Sree KS, Bog M, Appenroth KJ. 2016. Taxonomy of duckweeds (Lemnaceae), potential new crop plants. Emir J Food Agric. 28: 291–302.

Steinberg R. 1946. Mineral requirements of *Lemna minor*. Plant Physiol. 21:42–48.

Stohs SJ, Bagchi D. 1995. Oxidative mechanisms in the toxicity of metal ions. Free Radic Biol Med. 18:321–336.

Tang Y, He R, Zhao J, Nie G, Xu L, Xing B. 2016. Oxidative stress-induced toxicity of CuO nanoparticles and related toxicogenomic responses in *Arabidopsis thaliana*. Environ Pollut. 212:605–614.

The Council of the European Union. 1998. Council Directive 98/83/EC of 3 November 1998 on the quality of water intended for human consumption. OJ L. 330:32–54.

Tsvetnova OB, Shcheglov AI, Stolbova VV. 2014. On the question of biodiagnostic methods. Radiat Biol Radioecol. 54:423–431. Russian.

Van Hoeck A, Horemans N, Hees M, Nauts R, Knapen D, Vandenhove H, Blust R. 2015a. β-Radiation stress responses on growth and antioxidative defense system in plants: a study with strontium-90 in *Lemna minor*. Int J Mol Sci. 16:15309–15327.

Van Hoeck A, Horemans N, Hees M, Nauts R, Knapen D, Vandenhove H, Blust R. 2015b. Characterizing dose response relationships: chronic gamma radiation in *Lemna minor* induces oxidative stress and altered polyploidy level. J Environ Radioact. 150: 195–202.

Van Hoeck A, Horemans N, Nauts R, Hees MV, Vandenhove H, Blust R. 2017. *Lemna minor* plants chronically exposed to ionising radiation: RNA-seq analysis indicates a dose rate dependent shift from acclimation to survival strategies. Plant Sci. 257:84–95.

Vanhoudt N, Horemans N, Wannijn J, Nauts R, Van Hees M, Vandenhove H. 2014. Primary stress responses in *Arabidopsis thaliana* exposed to gamma radiation. J Environ Radioact. 129:1–6.

Vidaković-Cifrek Ž, Tkalec M, Šikić S, Tolić S, Lepeduš H, Pevalek-Kozlina B. 2015. Growth and photosynthetic responses of *Lemna minor* L. exposed to cadmium in combination with zinc or copper. Arh Hig Rada Toksikol. 66:141–152.

Vitasse Y, Bresson C, Kremer A, Michalet R, Delzon S. 2010. Quantifying phenological plasticity to temperature in two temperate tree species. Funct Ecol. 24:1211–1218.

Wang X, Ma R, Cui D, Cao Q, Shan Z, Jiao Z. 2017. Physio-biochemical and molecular mechanism underlying the enhanced heavy metal tolerance in highland barley seedlings pre-treated with low-dose gamma irradiation. Sci Rep. 7:14233.

Xie L, Solhaug KA, Song Y, Brede DA, Lind OC, Salbu B, Tollefsen KE. 2019. Modes of action and adverse effects of gamma radiation in an aquatic macrophyte *Lemna minor*. Sci Total Environ. 680: 23–34.

Yruela I. 2005. Copper in plants. Braz J Plant Physiol. 17:145–146.

Yruela I. 2009. Copper in plants: acquisition, transport and interactions. Funct Plant Biol. 36:409–430.

Yue L, Zhao J, Yu X, Lv K, Wang Z, Xing B. 2018. Interaction of CuO nanoparticles with duckweed (*Lemna minor* L): uptake, distribution and ROS production sites. Environ Pollut. 243:543–552.

Zagoskina NV, Nazarenko LV. 2016. Reactive oxygen species and antioxidant system of plants. Vestnik of Moscow City University. Natural Sciences. 2:9–23.

Copper uptake in adult rainbow trout irradiated during early life stages and in non-irradiated bystander trout which swam with the irradiated fish

Richard Smith, Sunita Nadella, Richard Moccia, Colin Seymour, and Carmel Mothersill

ABSTRACT

Purpose: This investigation forms part of a wider study into the legacy effects of exposure of rainbow trout eggs 38 h after fertilization, eyed eggs, yolk sac larvae (YSL) or first feeders to a single 0.5 Gy X-ray dose, including the induction of a bystander effect, by the irradiated fish, to non-irradiated fish. Fish may be exposed to multiple environmental stressors, including waterborne metals, during their lifespan and, while there are data on how the legacy of early life stage irradiation and bystander effect induction is affected by waterborne aluminum and cadmium, there are no studies into the effects radiation or the radiation induced bystander effect on metal uptake. Therefore the aim of this investigation was to determine if the legacy of early life stage irradiation included an effect on copper uptake by adult fish and by non-irradiated bystander adult trout which swam with the irradiated fish.

Methods: The four early life stages mentioned above were exposed to a single 0.5 Gy X-ray dose and then maintained, for two years with no further irradiation. At two years old the irradiated fish were allowed to swim, for 2 h with non-irradiated bystander trout (also two years old). After this time copper uptake was determined using ^{64}Cu.

Results: Copper uptake was increased in adult trout irradiated as eggs at 48 h after fertilization and as first feeders but eyed egg or YSL irradiation had no effect. Copper uptake was also increased in the bystander trout which swam with trout irradiated as eggs at 48 h after fertilization and as eyed eggs but there was no effect on non-irradiated adult trout which swam with trout irradiated as YSL or first feeders.

Conclusions: When put in context with the proteomic changes observed in these fish we propose the increased copper uptake in adult trout irradiated as eggs at 48 h after fertilization could be part of an anti-tumorigenic response and the increase in copper uptake in adult trout irradiated as first feeders could be part of a potentially protective pro-apoptotic response. Similarly we propose the increase in copper uptake in non-irradiated adult trout, induced by trout irradiated as eggs at 48 h after fertilization or as eyed eggs, was part of the universally anti-tumorigenic nature of the X-ray induced bystander effect in fish. However this was exclusive to embryonic irradiation.

Abbreviations: AARS: Alma Aquaculture Research Station; HPV-G: human papilloma virus transfected keratinocytes; YSL: yolk sac larvae

Introduction

The radiation induced bystander effect occurs when cells which have not been exposed to radiation receive and respond to signals emitted from irradiated cells (reviewed by Mothersill and Seymour 2012; Mothersill et al. 2018). This effect has now been shown in rainbow trout (*Oncorhynchus mykiss*), (Mothersill et al. 2006), zebrafish (*Danio rerio*) (Mothersill et al. 2007; Saroya et al. 2010; Mothersill et al. 2012; Smith, Seymour, Moccia et al. 2013) and medaka (*Oryzias latipes*) (Mothersill et al. 2009; Smith, Seymour, Moccia et al. 2013), irradiated by X-rays, Atlantic salmon (*Salmo salar*), irradiated by gamma radiation (Salbu et al. 2008; Mothersill et al. 2014), and in fathead minnows (*Pimphales promelas*) irradiated by alpha radiation from ^{226}Ra (Smith, Seymour, Mothersill et al. 2013). In rainbow trout it has also been shown that early life stage irradiation, by a single X-ray dose applied to eggs 48 h after fertilization, eyed eggs, yolk sac larvae (YSL) or first feeders, can result in a legacy effect in two year old adult fish, including the ability of these irradiated trout to induce a bystander effect in non-irradiated trout (Mothersill et al. 2010).

Apart from radiation it is quite possible that fish may face exposure to multiple stressors, either simultaneously or sequentially, and some studies have been made on the interaction between exposure to metals and radiation. In Atlantic salmon, exposed to gamma radiation, both the direct effects (Salbu et al. 2008; Mothersill et al. 2014) and the direct and bystander effects (Mothersill et al. 2014) are modified by simultaneous exposure to copper and aluminum (Mothersill et al. 2014) or aluminum and cadmium (Salbu et al. 2008) and in rainbow trout both the direct effects and bystander effects of X-rays are modified by exposure to aluminum immediately prior to irradiation (Smith, Seymour et al. 2015). Similarly the long term legacy of early life stage irradiation (including bystander effect induction) has been shown to be modified, in the adult trout, by exposure to aluminum (Smith, Saroya et al. 2015). However, to date, no experiments have been made to address the question of whether radiation exposure can affect the uptake of water-borne metals by fish.

Copper is a natural component of freshwater (e.g. Farag et al. 1994) but it is also a potentially hazardous pollutant which is known to increase permeability to water (reviewed by Wendelaar Bonga and Lock 1991) and compromise Na^+/K^+ ATPase function (Laurén and McDonald 1987a) thus disrupting whole body ion balance (Laurén and McDonald 1987b). Therefore the aim of this investigation was to determine if early life stage irradiation had an effect on copper uptake in adult rainbow trout and in non-irradiated bystander fish which swam with the trout which had been irradiated.

Materials and methods

Fish and irradiation

This investigation involves two year old adult trout which had been irradiated with a single 0.5 Gy X-ray dose at one of four early life history stages; eggs 48 h after fertilization, eyed eggs, YSL, or first feeders (Mothersill et al. 2010). This group of animals has already been the subject of five published investigations so more details of the source of the fish, the early life stage radiation exposure protocol and the subsequent rearing and husbandry can be found elsewhere (Mothersill et al. 2010; Smith, Saroya et al. 2015; Smith et al. 2016; Smith, Moccia, Mothersill et al. 2018; Smith, Moccia, Seymour et al. 2018).

Briefly; eggs and sperm were taken from six females and six males from the established breeding stock at the Alma Aquaculture Research Station (AARS), University of Guelph. The AARS is a purpose built fish husbandry center and all husbandry practices are carried out according to a number of well-established standard operating procedures. Water is drawn from a local, highly protected aquifer, from a depth of 25–30 m below ground level, filtered and aerated and then supplied to the husbandry tanks. Prior to delivery to the husbandry tanks dissolved oxygen was within 95% of saturation and each tank was also independently aerated. The water has a year round temperature of 8.5 ± 0.5 C and a pH range of 7.8–8.1, and is routinely sampled and regularly subjected to a full chemical analysis by an ISO 17025 accredited laboratory (ALS Environmental, Waterloo, Ontario), using Environmental Protection Agency (EPA) or American Public Health Association (APHA) certified tests. The AARS retains a 30 year archive of the results of this analysis and a recent set of data are provided in Table 1. All of the fish used in this study were held in a flow through system such that each tank received a minimum of two complete turnovers of water per hour, irrespective of tank size.

At the four early life stages stated above the fertilized eggs, YSL or first feeders were transported to McMaster University and exposed to a single 0.5 Gy X-ray dose using a portable X-ray device (Faxitron X-ray Corporation, Wheeling, Il, USA). This dose was given by irradiation of 100 kVp (kilovolts peak) for 5 mins; i.e. 0.1 Gy min-1 (Mothersill et al. 2006; 2007). To ensure uniformity of radiation dose the eggs were irradiated in two groups of approximately 200–250, spread in a single layer in 500 ml water, the YSL were irradiated in two groups of 80–100, also in a single layer in 1000 ml water, and the first feeders were irradiated in ten groups of 25, each in 750 ml water. In all cases the containers used for irradiation were placed in the center of the X-ray beam target. Following irradiation the eggs, YSL or first feeders were returned to the AARS and reared, with no additional treatment, for two years.

At the first feeder stage the fish from each X-ray groups were divided into two husbandry tanks. And were all fed with an appropriate (according to size/age) commercial trout diet (Martin Mills, Elmira, Ontario, Canada). Note this diet contains approximately 2.5 mg kg^{-1} Cu (i.e. 2.5 ppm). Growth and general health were monitored and, as required, moved to larger tanks and thinned in accordance with

Table 1. Influent water chemistry at the Alma Aquaculture Research Station.

Physical parameters[a]							
Colour, apparent (CU)	<2.0	Conductivity (umhos cm^{-1})	506	Hardness[b] (mg l^{-1})	241	pH	8.21
Total dissolved solids (mg l^{-1})	311	Turbidity (NTU)	0.13				
Anions and total organic carbon (mg l^{-1})[a]							
Alkalinity[b]	242	Ammonia[b]	<0.02	Chloride	7.72	Fluoride	0.199
Nitrate[c]	0.482	Nitrite[c]	<0.01	Orthophosphate[d]	<0.003	Sulphate	31.6
Total organic carbon	1.4						
Total metals and minerals (mg l^{-1})[a]							
Aluminum	<0.01	Chromium	<0.0005	Phosphorous	<0.05	Thallium	0.000012
Antimony	<0.0001	Cobalt	<0.0001	Potassium	0.952	Thorium	<0.0001
Arsenic	0.00016	Copper	<0.001	Rubidium	<0.0002	Tin	<0.0001
Barium	0.836	Iron	<0.05	Selenium	0.000088	Titanium	<0.0003
Beryllium	<0.0001	Lead	<0.0001	Silicon	5.72	Tungsten	<0.0001
Bismuth	<0.00005	Magnesium	24.4	Silver	<0.00005	Uranium	0.000289
Boron	0.033	Manganese	0.0942	Sodium	13.2	Vanadium	<0.0005
Cadmium	<0.00001	Mercury	<0.0001	Strontium	0.946	Zinc	<0.003
Calcium	54.5	Molybdenum	0.00185	Sulphur	10.5	Zirconium	<0.0003
Caesium	<0.00001	Nickel	<0.0005	Tellurium	<0.0002		

[a]All values preceded by < (i.e. less than) indicate concentrations below assay detection limit.
[b]Hardness and alkalinity expressed as $CaCO_3$.
[c]Nitrate and nitrite expressed as N.
[d]Orthophosphate expressed as P.

AARS procedures, to prevent overcrowding. At the time of experimentation the fish ranged from approximately 1.8–2.0 kg and were held at a stocking densities of 41–49 kg m^{-3} (50 kg m^{-3} being the maximum density employed at AARS). Each tank contained between 66 (min) -80 (max) trout from which the individual fish used in the present study were randomly taken.

Bystander effect induction

Full details on the induction of a bystander effect in these fish can also be found elsewhere (Mothersill et al. 2010) but, in summary, each adult fish, which had been irradiated at one of the four early life stages, were placed, in one side of a 75 l, mesh-partitioned container and a non-irradiated two year old trout (also taken from the AARS stock) was then placed on the other side of the partition. Four irradiated fish, from each irradiated group, and four bystander fish were used in this study. Since there were two tanks of fish, from each X-ray group, two trout were taken from each husbandry tank. Full aeration was provided at all times. After 2 h each irradiated and bystander fish were transferred to a series of 10 l containers for copper influx measurement.

Determination of copper influx in irradiated and bystander fish

Copper uptake was measured in both the irradiated and the bystander fish using ^{64}Cu generated at the McMaster University Nuclear Reactor. The trout were placed in 10 l containers, ^{64}Cu (as $CuSO_4$ dissolved in 0.5% HNO_3) was added to the water and uptake was calculated from the specific activities of the water at 15 mins after addition (initial specific activity) and 2 h (final specific activity) after ^{64}Cu addition, and the specific activity of the fish (also at 2 h after ^{64}Cu addition). For more details, including ^{64}Cu preparation, displacement of surface bound ^{64}Cu and measurement of ^{64}Cu activity refer to; Grosell and Wood (2002) and Sloman et al. (2003).

Statistical analysis

All data are expressed as mean ± standard deviation ($n = 4$ individual fish). The effects of early life stage irradiation and the radiation induced bystander effect, induced in non-irradiated fish by the irradiated trout, were analyzed by Analysis of Variance followed by analysis of Least Square Difference; a $p < 0.05$ was considered statistically significant.

Results and discussion

Experimental controls

Waterborne copper in the influent water at the AARS was less than 0.001 mg l^{-1} (i.e. <1.57 nM) (Table 1). To put this in context the US EPA has determined that 0.2 μM is the maximum acceptable concentration for the chronic protection of aquatic life. Thus we conclude the trout used in this study were not exposed to detrimental waterborne copper concentrations prior to the measurement of copper uptake. The mean (± Standard Deviation) copper influx rates measured in non-irradiated trout was 0.043 (± 0.004) μmol kg^{-1} h^{-1} (Figure 1). This compares to 0.3–1.6 ng g^{-1} h^{-1} (i.e. 0.005–0.025 μmol kg^{-1} h^{-1}), depending on dietary sodium (Pyle et al. 2003), and 20–65 pmol g^{-1} h^{-1} (i.e. 0.02–0.065 μmol kg^{-1} h^{-1}), depending on ambient sodium, potassium or calcium (Grosell and Wood 2002) previously measured in rainbow trout (Note; these are estimates from graphical presentation in the articles cited).

Early life stage irradiation and increased copper influx in the irradiated trout and in non-irradiated bystander fish which swam with the irradiated trout

There was an increase in copper influx in two year old trout irradiated as eggs 48 h after fertilization (Figure 1). Although there was no change in two year old trout irradiated as eyed eggs or as YSL there was also an increase in copper influx in two year old trout irradiated as first feeders, albeit less than the increase seen in adult trout irradiated as eggs 48 h after fertilization (Figure 1). These results suggest that, for increased copper uptake to be increased in adult trout, irradiation had to occur at the very earliest stages of embryonic development or immediately after the onset of exogenous feeding. Irradiation at a later embryonic developmental stage or during the endogenous feeding stage had no effect on copper uptake by the adult fish.

Copper influx was also increased in the non-irradiated bystander fish which swam with two year old trout irradiated as eggs 48 h after fertilization and in non-irradiated bystander fish which swam with two year old trout irradiated as eyed eggs (Figure 2). There was no change in copper influx in non-irradiated bystander fish which swam with two year old trout irradiated as YSL or as first feeders (Figure 2). This meant that with irradiation as eggs 48 h after fertilization the directly irradiated and bystander fish showed the same response. However, in the case of eyed egg irradiation, the response in non-irradiated bystander fish occurred despite not being seen in the irradiated trout (refer to; Increased copper influx and early life stage irradiation).

This is the first investigation to demonstrate that early life stage irradiation can result in increased metal uptake in adult fish. It is also the first investigation to demonstrate that metal uptake can increase due to the radiation induced bystander effect induced. We recognize such findings require confirmation and we therefore propose that additional experiments are now required; (a) to independently confirm the preliminary results, on copper uptake, presented here and (b) to expand this line of research to other metals and to other radiation types and exposure scenarios.

Increased copper influx; a pro- or anti-tumorigenic response?

Aside from compromising ion and water balance (as previously outlined in the Introduction) the pathological action

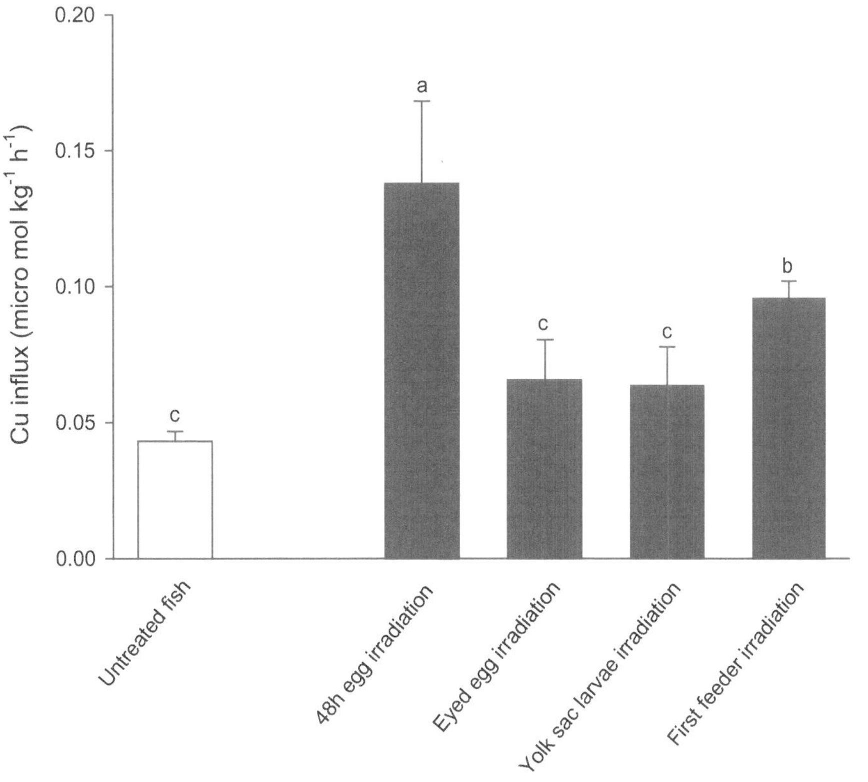

Figure 1. Copper influx rates in non-irradiated two year old rainbow trout (*Oncorhynchus mykiss*) which had been irradiated with a single 0.5 Gy X-ray dose at one of four early life stages. Data are shown as the mean value ± standard deviation. Analysis was by Analysis of Variance (ANOVA) followed by Least Square Difference (LSD). Bars annotated with the same letter indicate a statistical similarity in the data and bars annotated with different letters indicate a statistical difference in the data; $p = 1.3 \times 10^{-5}$.

of copper extends to tumorigenesis and cancer progression. Cancer cells contain higher levels of copper than nonmalignant cells (Ebadi and Swanson 1988; Nasulewicz et al. 2004; Lowndes and Harris 2005) and the copper chaperone/transcription factor Atox1 is instrumental in cancer cell migration (Blockhuys and Wittung-Stafshede 2017). Consequently lowering cellular copper (i.e. the opposite of the increases reported here) may therefore have potential as an anti-cancer treatment (reviewed by Brewer 2014). However, in cancer cells, copper has also been shown to exert a specific anti-proliferative action (Koňarikova et al. 2016) and promote the pro-oxidant cytotoxicity of flavonoids (Arif et al 2018). Consequently copper induced oxidative stress (e.g. Valko et al. 2005) offers a possibility for cancer treatment (Gupte and Mumper 2009). Additionally copper enhances the action of anti-cancer chemotherapeutics, such as curcumin (Lee et al. 2016), oleuropein (Capo et al. 2017), cisplatin (Liu et al. 2016) and disulfiram (Wang et al. 2018).

Increased copper influx in adult trout irradiated as eggs at 48 h after fertilization and as first feeders

There is clearly a complex relationship between copper influx and tumorigenesis (refer to; Increased copper influx; a pro- or anti- tumorigenic response?). Therefore, taken by itself, it is not clear whether an increase in copper influx in the trout irradiated as eggs at 48 h after fertilization or as first feeders could be interpreted as detrimental or beneficial; i.e. potentially tumorigenic or anti-tumorigenic. However the data presented here were generated by using fish that were taken (at random) from larger homogeneous groups. Therefore placing the copper influx results presented here in context with the proteomic changes to the gills (Smith, Moccia, Seymour et al. 2018), which were also measured in individuals taken from the exact same groups of fish, may best explain the consequences of increased copper uptake in adult trout irradiated at an early life stage.

The proteome changes in the gills from adult trout irradiated as eggs 48 h after fertilization were characterized by a number of anti-tumorigenic responses (Smith, Moccia, Seymour et al. 2018), specifically the elevation of hemoglobin, which has been found to improve the prognosis of esophageal cancer treatment (Rades et al. 2006), and the downregulation of pyruvate dehydrogenase, which has been found to increase resistance to cancer treatment (Roh et al. 2016). Significantly this is in direct contrast to the potentially tumorigenic proteomic responses from the gills of adult trout and also medaka (Smith et al. 2011), immediately following exposure to the same 0.5 Gy X-ray dose; trout showed an increase in annexin II, which is associated with a variety of cancers (Singh 2007), while medaka showed an increase in annexin max 3, which is has been observed in pancreatic, breast, lung and kidney cancer (Chuthapisith et al. 2009) and a decrease in annexin A4, which has been observed in prostate cancer (Hayes and Moss 2004). This does suggest it is possible the same radiation type and dose may elicit a different response depending on when, during the animal's life, exposure occurred. With this in mind, we

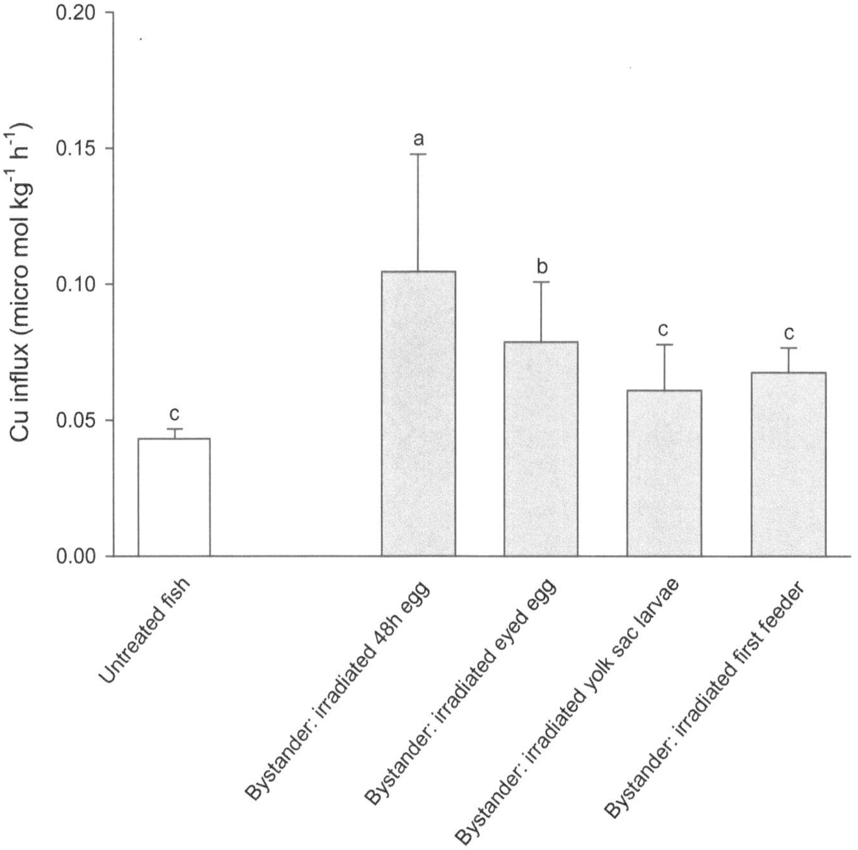

Figure 2. Copper influx rates in non-irradi ated two year old rainbow trout (*Oncorhynchus mykiss*) which had swam with adult trout irradiated with a single 0.5 Gy X-ray dose at one of four early life stages. Data are shown as the mean value ± standard deviation. Analysis was by Analysis of Variance (ANOVA) followed by Least Square Difference (LSD). Bars annotated with the same letter indicate a statistical similarity in the data and bars annotated with different letters indicate a statistical difference in the data; $p = 4.8 \times 10^{-3}$.

propose the increased influx of copper in adult trout irradiated as eggs at 48 h after fertilization was more likely to be part of a wider anti-tumorigenic legacy of early life stage irradiation.

The fact that an increase in copper uptake observed was not observed in in adult trout irradiated as eyed eggs suggests a degree of exclusivity in long term the response to embryonic irradiation depending on when radiation exposure occurred. This result continues an emerging trend of differences between the effects of 48 h egg and eyed egg irradiation, which has been observed in previous studies conducted on the same stocks of fish as those used here. No change in clonogenic survival of the HPV-G (human papilloma virus transfected keratinocytes) reporter cells was observed following treatment with media from gill explants of adult trout irradiated as eggs at 48 h after fertilization yet there was decrease in survival following treatment with gill explant media from adult trout irradiated as eyed eggs (Mothersill et al. 2010) and this response was differently modified by acute exposure of the adult trout to waterborne aluminum (Smith, Saroya et al. 2015). Transgerational effects, in F1 trout bred from these irradiated fish, also measured by changes in HPV-G cell survival, differed depending on whether F0 irradiation had occurred at the egg 48 h after fertilization or at the eyed egg stage (Smith et al. 2016). Finally differences in proteomic responses in the gills have been recorded between adult trout irradiated as eggs at 48 h after fertilization or as eyed eggs (Smith, Moccia, Mothersill et al. 2018; Smith Moccia, Seymour et al. 2018).

Although there were some anti-tumorigenic proteomic responses in the gill proteome of adult trout irradiated as first feeders there were also changes to two proteins which could be described as "dual function" (He et al. 2015) since their relationship with tumorigenesis is more complicated. This term was originally applied to nucleoside diphosphate kinase (He et al. 2015) which was downregulated in adult trout irradiated as first feeders (Smith, Moccia, Seymour et al. 2018). However the tumor suppressive properties of nucleoside diphosphate kinase are dependent on redox state (reviewed by Gellert et al. 2015). Consequently, despite being a metastasis inhibitor, nucleoside diphosphate kinase over expression has also been measured during hepatocarcinogenesis (He et al. 2015). In addition it has been proposed that hyperosmotic glycine rich protein (upregulated in adult trout irradiated as first feeders; Smith, Moccia, Seymour et al 2018), which is overexpressed in cancer cells lines (Calderón-González et al. 2015), yet downregulated during cancer progression and metastasis in breast cancer (Đermadi et al. 2014), could also be considered a dual function protein

(Smith, Moccia, Seymour et al. 2018). As summarized above (refer to; Increased copper influx; a pro- or anti- tumorigenic response?) copper could also be considered as having a dual function with regard to tumorigenesis or cancer progression.

Aside from proteomic responses the results of the present investigation can also be considered in context with apoptosis. Gill explant media from adult trout irradiated as eggs 48 h after fertilization had no effect on HPV-G cell survival (Mothersill et al. 2010). However copper increases the pro-apoptotic inhibition of deubiquitinase (Chen et al. 2017) and the increased copper influx in adult trout irradiated as first feeders corresponds to the lowest levels of clonogenic survival in the apoptosis sensitive HPV-G reporter cell line following treatment with media from gill explants from these fish (Mothersill et al. 2010).

We therefore propose that there may be some differences in the consequences of increased copper uptake depending on whether irradiation occurred at the earliest embryonic stage (i.e. eggs at 48 h after fertilization) or at the onset of exogenous feeding. Specifically we hypothesize that increased copper uptake arising from embryonic irradiation is more likely to be part of an anti-tumorigenic response whereas an increase in uptake arising from irradiation at the start of exogenous feeding is part of a potentially protective pro-apoptotic response.

Although the effects of early life stage irradiation on copper influx are long lived, from this study, we cannot say conclusively as to whether this constitutes an epigenetic change. Also it is worth noting that none of the early life stage irradiations resulted in a reduction in copper influx in adult trout. However, depending on dose and the tissue analyzed, both increased and decreased copper influx has been recorded in rats exposed to gamma radiation, (Hawas 2013). Similarly alpha radiation, from naturally occurring thorium (^{232}Th), also increases or decreases metal influx in earthworms (Mrdakovic Popic et al. 2012).

Increased copper influx and the bystander effect in adult trout induced by early life stage irradiation

This is the first investigation to show that the radiation induced bystander effect can result in an increase in copper (or any metal) influx in an organism which was not been directly irradiated and therefore extend this effect of early life stage irradiation outside of the organisms which had been irradiated. As was the case with increased copper uptake in the irradiated trout (refer to; Increased copper influx in adult trout irradiated as eggs at 48 h after fertilization and as first feeders) it is not clear from the copper uptake data alone whether the increase in uptake in non-irradiated bystander fish, which swam with the irradiated trout, was beneficial or detrimental. Therefore, again, we propose placing this result in context with the proteomic changes observed in comparable bystander fish (Smith, Moccia, Seymour et al. 2018. Note; an identical bystander effect induction method was used in the present investigation).

Proteomic analysis of the gill of non-irradiated adult bystander trout, which swam with the early life stage irradiated trout, has revealed an anti-tumorigenic and protective response in the gills of trout which swam with trout irradiated as eggs at 48 h after fertilization and as eyed eggs (Smith, Moccia, Seymour et al. 2018). In the case of trout irradiated as eggs at 48 h after fertilization, the bystander effect increase in copper influx reported here was associated with the same increase in hemoglobin as seen in the irradiated fish (Smith, Moccia, Seymour et al. 2018) whereas, in the case of bystander fish which swam with trout irradiated as eyed eggs, the increase in copper influx reported here was associated with the downregulation of triosephosphate isomerase (TPI) and ubiquitin-40S ribosomal protein S2 (Smith, Moccia, Seymour et al. 2018). Ubiquitin-40S ribosomal protein S2 downregulation corresponds to ubiquitin knockdown resulting in reduced colon cancer cell migration and invasion (Yun et al. 2015) and TPI downregulation is the opposite of the increase seen lung cancer cases (Dakubo et al. 2006).

Similar anti-tumourigenic responses have been seen in the gill proteome of bystander trout (Smith et al. 2007) and also medaka (Smith et al. 2011), which swam with adult fish immediately after they were irradiated with the same 0.5 Gy X-ray dose and also in the brain proteome of non-irradiated rats housed with irradiated rats, again immediately after irradiation (Smith, Wang et al. 2018). Therefore there seems little doubt that a bystander effect resulting from a single radiation dose may be beneficial, irrespective of when the irradiation occurred and how long after irradiation a bystander effect was induced, by the irradiated fish, in non-irradiated fish.

Consequently we propose the increased copper influx in non-irradiated fish which swam with adult trout irradiated as eggs at 48 h after fertilization or as eyed eggs should be considered as another aspect of the beneficial nature of the radiation induced bystander effect (refer to: Increased copper influx; a pro- or anti- tumorigenic response?). However, since there was only a bystander effect induced increase in copper influx in non-irradiated fish swimming with trout irradiated as eggs at 48 h after fertilization or as eyed eggs, this investigation strongly suggests this was exclusively the result of a bystander effect induced following embryonic irradiation. Obviously we recognize this hypothesis should now be confirmed or refuted with additional experiments.

Given that copper influx was increased in the bystander fish only we also propose that focused experiments on embryonic (particularly eyed egg) irradiation are now required to fully investigate responses exclusive to the bystander effect and those seen in both directly irradiated and non-irradiated bystander fish. For example, in another study of the same stocks of fish as those used here, the bystander effect only induction of increased copper influx, shown in the present investigation, contrasts to the production of pro-apoptotic signals (as detected by HPV-G clonogenic survival) by both the adult trout irradiated as eyed eggs and by the non-irradiated bystander fish which swam with these trout (Mothersill et al. 2010). However the results presented here mirror the findings of a transgenic study (again on the same stocks of fish as were used in this investigation) which showed that F1 and F2 fish, bred from both male and female trout irradiated as eyed eggs, induced an increase in pro-apoptotic signal production (again detected in HPV-G cells) from the skin and spleen of non-irradiated bystander fish whereas no signals (from the same tissues) were detected in the F1 and F2 fish themselves (Smith et al. 2016).

Given that copper is a known generator of Reactive Oxygen Species (ROS) (e.g. reviewed by Linder 2012) there is one more remark to be made, regarding early life stage irradiation, the bystander effect and copper influx. It is of interest to note that the unchanged copper influx in bystander fish which swam with trout irradiated as YSL or as first feeders coincided with the downregulation of cold-inducible RNA binding protein B-like isoform X3 (CIRBP) in bystander fish which swam with trout irradiated as YSL and superoxide dismutase (SOD) in bystander fish which swam with trout irradiated as first feeders (Smith, Moccia, Seymour et al. 2018). ROS is well documented in terms of reducing cancer cell proliferation (e.g. Feng et al. 2016) apoptosis and cancer treatment (e.g. Chen et al. 2016; Gong et al. 2016; Sun et al. 2016; Zhang et al. 2016) yet CIRBP counteracts the effects of ROS and SOD downregulation (and any subsequent alteration to ROS) can result in cellular senescence and the deterioration of biological function (reviewed by Lleonart 2010). One could therefore speculate as to whether this observation does in fact provides more detail on the differences in the beneficial nature between a bystander effect resulting from embryonic or post-embryonic irradiation.

Conclusions

The results presented here constitute a preliminary study into the effects of early life stage irradiation on the ability of trout regulate copper influx. Such studies are becoming more important as the interactions of multiple stressors become better understood. However, we concede that further investigations, involving more radiation types and doses, additional metals and other species, and larger datasets, including those derived from field sampling as well as laboratory based studies, are now required to support or refute the findings and conclusions presented here.

Nevertheless, this study has demonstrated that exposure to a single radiation dose at an early life stage can result in an increase in copper influx in rainbow trout. However, this is specific to the irradiation occurring either within 48 h of fertilization of the egg or immediately following the onset of exogenous feeding. Irradiation of early life stages in between had no effect.

When put into context with proteomic or apoptotic responses in similarly irradiated fish it is proposed increased that copper influx may be beneficial, specifically as part of a potentially anti-tumorigenic mechanism, if irradiation occurred during the earliest embryonic stages or, as part of a more complex response involving apoptosis, if irradiation occurred after the onset of exogenous feeding.

A parallel can be drawn between fetal irradiation of Heligenberger mice and early life stage irradiation of trout. Genetic instability has been proposed as a possible mechanism (Streffer and Müller 1996) for the deleterious effects observed in the mice (Hillebrandt and Streffer 1994; Müller et al. 1999). It is therefore important to note that, depending on the exact cross involving the early life stage irradiated trout, there is a transgenerational effect in the F1 generation, and a bystander effect in non-irradiated fish paired with the F1 and F2 generation (Smith et al. 2016). Thus, despite early life stage irradiation in trout appearing to be largely beneficial (Smith, Moccia, Seymour et al. 2018) and therefore in contrast to the effects fetal irradiation on mice, it is possible there might be similarities in terms of an epigenetic effect of radiation exposure at the earliest developmental stages.

This study also demonstrated that similar increases in copper influx was caused in non-irradiated trout which swam with two year old trout irradiated as eggs at 48 h after fertilization or as eyed eggs. This is therefore the first investigation to demonstrate a radiation-induced bystander effect increase in metal uptake in fish (or indeed any organism) and raises important questions regarding the extension of radiation exposure effects beyond the organism which had been irradiated as well as with regard to the beneficial nature of the bystander effect induced by embryonic irradiation.

Acknowledgements

The authors express our grateful thanks to Michael Burke, David Bevan, Michael Kirk and Neil MacBeth, Alma Aquaculture Research Station, University of Guelph, for the initial breeding, rearing and husbandry of all the fish used in this investigation.

Disclosure statement

The authors report no conflict of interest. The authors alone are responsible for the content and writing of this paper.

Funding

Funding for this investigation was provided by; Natural Sciences and Engineering Research Council (NSERC) (RGPIN06223-17 and RPGIN05542-14), Industrial Chairs Programme, Canada Deuterium Uranium (CANDU) Owners Group, Bruce Power and The Canada Chairs Programme. All procedures used in this investigation were carried out in accordance with the animal care protocols and regulatory guidelines imposed by McMaster University Central Animal Care Facility and by the University of Guelph, and were specifically covered by McMaster Animal Utilisation Protocol (AUP) 06-12-65.

References

Arif H, Sohail A, Farhan M, Abdur Rehman A, Ahmad A, Hadi SM. 2018. Flavonoids-induced redox cycling of copper ions leads to generation of reactive oxygen species: a potential role in cancer chemoprevention. Int J Biol Macromol. 106:569–578.

Blockhuys S, Wittung-Stafshede P. 2017. Copper chaperone Atox1 plays role in breast cancer cell migration. Biochem Biophys Res Commun. 483(1):301–304.

Brewer GJ. 2014. The promise of copper lowering therapy with tetrathiomolybdate in the cure of cancer and in the treatment of inflammatory disease. J Trace Elem Med Biol. 28(4):372–378.

Calderón-González KG, Rustarazo MLV, Labra-Barrios ML, Bazán-Méndez CI, Tavera-Tapia A, Herrera-Aguirre ME, Sánchez del Pino MM, Gellegos-Pérez JL, González-Márquez H, Hernández-Hernández JM, León-Ávila G, et al. 2015. Determination of the protein expression profiles of breast cancer cell lines by quantitative proteomics using ITRAQ labelling and tandem mass spectrometry. J Proteomics. 124:50–78.

Capo CR, Pedersen JZ, Falconi M, Rossi L. 2017. Oleuropein shows copper complexing properties and noxious effect on cultured SH-SY5Y neuroblastoma cells depending on cell copper content. J Trace Elem Med Biol. 44:225–232.

Chen W, Zou P, Zhao Z, Chen X, Fan X, Vinothkumar R, Cui R, Wu F, Zhang Q, Liang G, et al. 2016. Synergistic antitumor activity of rapamycin and EF24 via increasing ROS for the treatment of gastric cancer. Redox Biol. 10:78–89.

Chen X, Zhang X, Chen J, Yang Q, Yang L, Xu D, Zhang P, Wang X, Liu J. 2017. Hinokitiol copper complex inhibits proteasomal deubiquitination and induces paraptosis-like cell death in human cancer cells . Eur J Pharmacol. 815:147–155.

Chuthapisith S, Bean BE, Cowley G, Eremin JM, Samphao S, Layfield R, Kerr LD, Wiseman J, El-Sheemy M, Sreenivasan T, et al. 2009. Annexins in human breast cancer: possible predictors of pathological response to neoadjuvant chemotherapy. Eur J Cancer. 45(7): 1274–1281.

Dakubo GD, Parr RL, Costello LC, Franklin RB, Thayer RE. 2006. Altered metabolism and mitochondrial genome in prostate cancer. J Clin Pathol. 59(1):10–16.

Đermadi D, Valo S, Pussila M, Reyhani N, Sarantaus L, Lalowski M, Baumann M, Nyström M. 2014. Inherited cancer predisposition sensitizes colonic mucosa to address western diet effects and putative cancer-predisposing changes on mouse proteome. J Nutr Biochem. 25(11):1196–1047.

Ebadi M, Swanson S. 1988. The status of zinc, copper and metallothionein in cancer patients. Prog Clin Biol Res. 259:161–175.

Farag AM, Boese CJ, Woodward DF, Bergman HL. 1994. Physiological changes and tissue metal accumulation in rainbow trout exposed to foodborne and waterborne metals. Environ Toxicol Chem. 13(12): 2021–2029.

Feng X, Yu W, Zhou F, Chen J, Shen P. 2016. A novel small molecule compound diaporine inhibits breast cancer cell proliferation via promoting ROS generation. Biomed Pharmacother. 83:1038–1047.

Gellert M, Hanschmann E-M, Lepka K, Berndt C, Lillig CH. 2015. Redox regulation of cytoskeletal dynamics during differentiation and de-differentiation. Biochim Biophys Acta. 1850(8):1575–1587.

Gong E-Y, Shin YJ, Hwang L-Y, Kim JH, Kim S-M, Moon J-H, Shin J-S, Lee D-H, Hur DY, Jin D-H, et al. 2016. Combined treatment with vitamin C and sulindac synergistically induces p53- and ROS-dependent apoptosis in human colon cancer cells . Toxicol Lett. 258: 126–133.

Grosell M, Wood CM. 2002. Copper uptake across rainbow trout gills: mechanisms of apical entry. J Exp Biol. 205(8):1179–1188.

Gupte A, Mumper RJ. 2009. Elevated copper and oxidative stress in cancer cells as a target for cancer treatment. Cancer Treat Rev. 35(1):32–46.

Hawas AH. 2013. The biosensitivity of certain organs in rats exposed to low doses of γ-radiation. J Rad Res Appl Sci. 6(2):56–62.

Hayes MJ, Moss SE. 2004. Annexins and disease. Biochem Biophys Acta. 1762:164–180.

He F, York JP, Burroughs SG, Qin L, Xia J, Chen D, Quigley EM, Webb P, LeSage GD, Xia X. 2015. Recruited metastasis suppressor NM23-H2 attenuates expression and activity of peroxisome proliferator-activated receptor δ (PPARδ) in human cholangiocarcinoma. Dig Liver Dis. 47(1):62–67.

Hillebrandt S, Streffer C. 1994. Protein patterns in tissues of foetuses with radiation-induced gastroschisis. Mutat Res. 308(1):11–22.

Koňarikova K, Perdikaris GA, Gbelcová H, Andrezálová L, Švéda M, Ruml T, Laubertová L, Režnáková S, Žitňanová I. 2016. Autophagy in MCF-7 cancer cells induced by copper complexes. Pharmacol Rep. 68(6):1221–1224.

Laurén DL, McDonald DG. 1987a. Acclimation to copper by rainbow trout, *Salmo gairdneri*: biochemistry. Can J Fish Aquat Sci. 44(1): 105–111.

Laurén DL, McDonald DG. 1987b. Acclimation to copper by rainbow trout, *Salmo gairdneri*: physiology. Can J Fish Aquat Sci. 44(1): 99–104.

Lee H-M, Patel V, Shyur L-F, Lee W-L. 2016. Copper supplementation amplifies the anti-tumor effect of curcumin in oral cancer cells. Phytomedicine. 23(12):1535–1544.

Linder MC. 2012. Review: The relationship of copper to DNA damage and damage prevention in humans. Mutat Res. 733(1-2):83–91.

Liu H, Zhang Y, Zheng S, Weng Z, Ma J, Li Y, Xie X, Zheng W. 2016. Detention of copper by sulfur nanoparticles inhibits the proliferation of A375 malignant melanoma and MCF-7 breast cancer cells. Biochem Biophys Res Commun. 477(4):1031–1037.

Lleonart ME. 2010. A new generation of proto-oncogenes: cold-inducible RNA binding proteins . Biochim Biophys Acta. 1805(1):43–52.

Lowndes SA, Harris AL. 2005. The role of copper in tumour angiogenesis. J Mammary Gland Biol Neoplasia. 10(4):299–310.

Mothersill C, Bucking C, Smith RW, Agnihotri N, Oneill A, Kilemade M, Seymour CB. 2006. Communication of radiation-induced stress or bystander signals between fish *in vivo*. Environ Sci Technol. 40(21):6859–6864.

Mothersill C, Rusin A, Fernandez-Palomo C, Seymour C. 2018. History of bystander effects research 1905-present; what is in a name? Int J Radiat Biol. 94(8):696–707.

Mothersill C, Seymour C. 2012. Changing paradigms in radiobiology. Mutat Res. 750(2):85–95.

Mothersill C, Smith RW, Agnihotri N, Seymour CB. 2007. Characterization of a radiation-induced stress response communicated *in vivo* between zebrafish. Environ Sci Technol. 41(9): 3382–3387.

Mothersill C, Smith RW, Fazzari J, McNeill F, Prestwich W, Seymour CB. 2012. Evidence for a physical component to the radiation-induced bystander effect? Int J Radiat Biol. 88(8):583–591.

Mothersill C, Smith RW, Heier LS, Teien H-C, Lind OC, Land OC, Seymour CB, Oughton D, Salbu B. 2014. Radiation-induced bystander effects in the Atlantic salmon (salmo salar L.) following mixed exposure to copper and aluminum combined with low-dose gamma radiation. Radiat Environ Biophys. 53(1):103–114.

Mothersill C, Smith RW, Hinton TG, Aizawa K, Seymour CB. 2009. Communication of radiation-induced signals *in vivo* between DNA repair deficient and proficient medaka (*Oryzias latipes*). Environ Sci Technol. 43(9):3335–3342.

Mothersill C, Smith RW, Saroya R, Denbeigh J, Rowe B, Banevicius L, Timmins R, Moccia R, Seymour CB. 2010. Irradiation of rainbow trout at early life stages results in legacy effects in adults. Int J Radiat Biol. 86(10):817–828.

Mrdakovic Popic J, Salbu B, Skipperud L. 2012. Ecological transfer of radionuclides and metals to free-living earthworm species in natural habitats rich in NORM. Sci Total Environ. 414:167–176.

Müller W-U, Streffer C, Wojcik A, Niedereichholz F. 1999. Radiation-induced malformations after exposure of murin germ cells in various stages of spermatogenesis. Mutat Res. 425(1):99–106.

Nasulewicz A, Mazur A, Opolski A. 2004. Role of copper in tumour angiogenesis-clinical implications. J Trace Elem Med Biol. 18(1):1–8.

Pyle GG, Kamunde CN, McDonald DG, Wood CM. 2003. Dietary sodium inhibits aqueous copper uptake in rainbow trout (*Oncorhynchus mykiss*). J Exp Biol. 206(Pt 3):609–618.

Rades D, Lang S, Schild SE, Alberti W. 2006. Prognostic value of haemoglobin levels during concurrent radio-chemotherapy in the treatment of oesophageal cancer. Clin Oncol. 18(2):139–144.

Roh J-L, Park JY, Kim EH, Jang HJ, Kwon M. 2016. Activation of mitochondrial oxidation by PDK2 inhibition reverses cisplatin resistance in head and neck cancer. Cancer Lett. 371(1):20–29.

Salbu B, Denbeigh J, Smith RW, Heier LS, Teien HC, Rosseland BO, Oughton D, Seymour CB, Mothersill C. 2008. Environmentally relevant mixed exposures to radiation and heavy metals induce measurable stress responses in Atlantic salmon. Environ Sci Technol. 42(9): 3441–3446.

Saroya R, Smith R, Seymour C, Mothersill C. 2010. Injection of reserpine into zebrafish prevents fish to fish communication of radiation-induced bystander signals: confirmation *in vivo* of a role for serotonin in the mechanism. Dose Response. 8:317–330.

Singh P. 2007. Role of Annexin-II in GI cancers: interaction with gastrins/progastrins. Cancer Lett. 252(1):19–35.

Sloman KA, Morgan TP, McDonald DG, Wood CM. 2003. Socially-induced changes in sodium regulation affect the uptake of water-borne copper and silver in rainbow trout, *Oncorhynchus mykiss*. Comparat Biochem Physiol C. 135(4):393–403.

Smith R, Wang J, Seymour C, Fernandez-Palomo C, Fazzari J, Schültke E, Bräuer-Krisch E, Laissue J, Schroll C, Mothersill C. 2018. Homogenous and microbeam X-ray radiation induces proteomic changes in the brains of irradiated rats and in the brains of non-irradiated cage mate rats. Dose Response. 16(1):1559325817750068.

Smith RW, Moccia RD, Mothersill CE, Seymour CB. 2018. Irradiation of rainbow trout at early life stages results in a proteomic legacy in adult gills. Part B; the effect of a second radiation dose, after one year, on the proteomic responses in the irradiated and non-irradiated bystander fish. Environ Res. 163:307–313.

Smith RW, Moccia RD, Seymour CB, Mothersill CE. 2018. Irradiation of rainbow trout at early life stages results in a proteomic legacy in adult gills. Part A; proteomic responses in the irradiated fish and in non-irradiated bystander fish. Environ Res. 163:297–306.

Smith RW, Saroya R, Seymour CB, Moccia RD, Mothersill CE. 2015. Exposure to acute levels of waterborne aluminium modifies the legacy of early life stage irradiation, including the communication of radiation-induced bystander signals, in adult rainbow trout. Int J Radiat Biol. 91(11):878–890.

Smith RW, Seymour CB, Moccia RD, Hinton TG, Mothersill CE. 2013. The induction of a radiation-induced bystander effect in fish transcends taxonomic group and trophic level. Int J Radiat Biol. 89(4): 225–233.

Smith RW, Seymour CB, Moccia RD, Mothersill CE. 2015. Tissue-specific effects of acute aluminium exposure on the radiation-induced bystander effect in rainbow trout (*Onchorynchus mykiss*, Walbaum). Int J Radiat Biol. 91(9):715–723.

Smith RW, Seymour CB, Moccia RD, Mothersill CE. 2016. Irradiation of rainbow trout at early life stages results in trans-generational effects including the induction of a bystander effect in non-irradiated fish. Environ Res. 145:26–38.

Smith RW, Seymour CB, Mothersill CE. 2013. Short and long term bystander effect induction by fathead minnows (*Pimphales promelas*, Rafinesque, 1820) injected with environmentally relevant whole body doses of ^{226}Ra. J Environ Radioact. 126:133–136.

Smith RW, Wang J, Bucking CP, Mothersill CE, Seymour CB. 2007. Evidence for a protective response by the gill proteome of rainbow trout exposed to X-ray induced bystander signals. Proteomics. 7(22): 4171–4180.

Smith RW, Wang J, Mothersill CE, Hinton TG, Aizawa K, Seymour CB. 2011. Proteomic changes in the gills of wild-type and transgenic radiosensitive medaka following exposure to direct irradiation and to X-ray induced bystander signals. Biochim Biophys Acta. 1814(2): 290–298.

Streffer C, Müller W-U. 1996. Malformations after radiation exposure of preimplantation stages. International Journal of Biology. 40: 355–360.

Sun B, Gao L, Ahsan A, Chu P, Song Y, Li H, Zhang Z, Lin Y, Peng J, Song Z, et al. 2016. Anticancer effect of SZC015 on lung cancer cells through ROS-dependent apoptosis and autophagy induction mechanisms in vitro . Int Immunopharmacol. 40:400–409.

Valko M, Morris H, Cronin MT. 2005. Metals, toxicity and oxidative stress. Curr Med Chem. 12(10):1161–1208.

Wang N-N, Wang L-H, Li Y, Fu S-Y, Xue X, Jia L-N, Yuan X-Z, Wang Y-T, Tang X, Yang J-Y, et al. 2018. Targeting ALDH2 with disulfiram/copper reverses the resistance of cancer cells to microtubule inhibitors. Exp Cell Res. 362(1):72–82.

Wendelaar Bonga SE, Lock RAC. 1991. Toxicants and osmoregulation in fish. Neth J Zool. 42(2-3):478–493.

Yun S-L, Kim HH, Yoon JH, Park WS, Hahn M-J, Kim HC, Chung CH, Kim KK. 2015. Ubiquitin specific protease 4 positively regulates the WNT/β-catenin signaling in colorectal cancer. Mol Oncol. 9(9): 1834–1851.

Zhang S, Li T, Zhang Y, Xu H, Li Y, Zi X, Yu H, Li J, Jin C-Y, Liu H-M. 2016. A new brominated chalcone derivative suppresses the growth of gastric cancer cells in vitro and in vivo involving ROS mediated up-regulation of DR5 and 4 expression and apoptosis . Toxicol Appl Pharmacol. 309:77–86.

A pilot study of radiation-induced bystander effect in radio-adapting frogs at a radiologically contaminated site located on the chalk river laboratories property

Nguyen T. K. Vo, Harleen Singh, Marilyne Stuart, Colin B. Seymour, and Carmel E. Mothersill

ABSTRACT

Purpose: To measure medium borne bystander effects, to study the influence of radioadaptive response (RAR) on bystander response, and to discover reliable radioresponsive biomarkers in radio-adapting frogs from Duke Swamp contaminated with an above-background radiation level and in naïve frogs from Twin Lake as the background control site.

Materials and methods: Frogs were captured at Duke Swamp and Twin Lake and brought to the lab at the Canadian Nuclear Laboratories facility. Half of the frogs from each site were irradiated with 4 Gy while the other half of the frogs were left with no further radiation treatment. Frog bladders were removed and placed in sterile culture media. Upon arrival at McMaster University, the bladders were processed for tissue cultures. After 48 h, the culture media conditioned by the bladder explants were harvested for clonogenic reporter survival assay and calcium flux measurements for assessing bystander effects. HPV-G cells were used as bystander reporter cells in all radiation-induced bystander effect (RIBE) assays. The frog bladder cultures were incubated for another 10–12 days followed by immunochemical staining for bcl-2 and c-myc expressions to analyze cellular anti-apoptotic (pro-survival) and pro-apoptotic (pro-death) responses, respectively.

Results: Only culture media conditioned by bladders from 4-Gy-irradiated naïve frogs from Twin Lake induced bystander effects (reduction of HPV-G reporter cells' clonogenic survival and presence of strong calcium flux activities). The 4 Gy irradiation dose increased pro-apoptotic c-myc expression in naïve frogs' bladder explants. Culture media conditioned by bladders from radio-adapting frogs from Duke Swamp enhanced HPV-G's clonogenic survival and a 4 Gy irradiation challenge did not change the enhanced clonogenic survival nature nor induce calcium flux. In bladder explants from both control and 4-Gy-irradiated radio-adapting frogs, anti-apoptotic bcl-2 expression for pro-survival responses was ubiquitous while c-myc expression for pro-death responses was limited to a small fraction of cells.

Conclusion: The clonogenic RIBE reporter assay using HPV-G and calcium flux measurements are useful diagnostic tools for RIBE assessment of field biological samples, specifically those from frogs. RAR induced by environmentally relevant low-dose radiation induces protective bystander response. Bcl-2 and c-myc are reliable biomarkers for evaluating low dose radiation responses in wild populations of amphibians. Overall, this pilot study emphasizes the importance of looking at non-targeted effects (NTEs) in natural populations of non-human biota that could be vulnerable to chronic low-dose radiation exposures.

Introduction

As secondary consumers in both aquatic and terrestrial environments, amphibians play a pivotal role in many ecosystems and are relevant to radioecology. Field and laboratory data on environmental radiation effects on amphibians are currently seriously insufficient (Vo 2021). Biological mechanisms operating at low-dose regions relevant to environmental levels of radiation contamination are largely attributed to cellular responses mediated by non-targeted effects (NTEs) such as radiation-induced bystander effect (RIBE) and radioadaptive response (RAR) (Seymour and Mothersill 2000; Mothersill et al. 2006, 2007, 2020; Ryan et al. 2008; Smith et al. 2013; Cohen et al. 2019; Vo et al. 2019). In RIBE, unirradiated cells receive signals from irradiated counterparts and develop characteristics or defects typically caused by direct irradiation (Mothersill et al. 2006, 2007, 2017). In RAR, cells can tolerate radiation insults better than normally anticipated due to genomic/extragenomic changes that confer radioresistance (Ryan et al. 2008; Audette-Stuart et al. 2011, Audette-Stuart and Yankovich, 2011; Cohen et al. 2019; Hancock et al. 2019). While RIBE and RAR have been studied in a number of animals of medical and ecological importance, studies in amphibians are sporadic and limited to only a few (Vo 2021). This presents a major setback when trying to understand the interaction between environmental radiation and overall health and dynamics of affected amphibian communities.

Of particular interest is studying NTEs on frogs exposed to radioactive wastewater generated by nuclear reactors that are currently operational or in various stages of construction or decommissioning on five continents. Tritium is a major common by-product found in radioactive wastewater from CANDU nuclear reactors (IAEA 2004; Audette-Stuart et al. 2011). An example of an increased tritium level in aquatic bodies near a nuclear reactor site in Canada is the Duke Swamp (DS) pond, which receives the contaminating radionuclides via the groundwater from the nearby waste management area (Ontario, Canada) (Audette-Stuart et al. 2011). In 2000s, Stuart and colleagues at the Canadian Nuclear Laboratories designed field studies to examine how native frogs were affected by the above-background radiation level in DS. The tritium level in DS was 5,000-35,000 Bq/L in 2003 and 2,800 Bq/mL in 2008 (Audette-Stuart et al. 2011) and was above-background with respect to the radiation level. The tritium level at Twin Lake (TL) was 707 Bq/mL in 2003 and 214 Bq/mL in 2008 and Twin Lake was the background control site (Audette-Stuart et al. 2011). The natural background radiation level in Chalk River was estimated to be 2.4 mGy/year (Audette-Stuart et al. 2011). The elevated radiation level in DS contributed to less than 1 mGy/year to the natural background radiation (Audette-Stuart et al. 2011). It was found that DS frogs radio-adapted and were more radioresistant than TL frogs (Audette-Stuart et al. 2011); thus, elevated environmental radiation exposure invoked RAR in frogs. In a separate study, Audette-Stuart and Yankovich (2011) further showed that like tritiated frogs, control frogs that did not receive an experimental low dose of tritium radiation developed RAR as well when kept in contact with tritiated frogs. This suggests that tritiated frogs transferred bystander signals to and induced control frogs to radio-adapt.

Questions about how to measure medium borne bystander effects, the influence of RAR on RIBE biology, and possible molecular mechanisms involved in frogs, especially from natural habitats, remain open. The experimental set-up (Table 1) in the field study by Audette-Stuart et al. (2011) conducted in 2008 in combination with the robust RIBE reporter assays developed in the Mothersill and Seymour's lab at McMaster University presents a unique opportunity to address these questions. Mink frogs (*Rana Septentrionalis*) were caught at the above-background DS site and the background control TL site and brought back to the Canadian Nuclear Laboratories. As previously noted, DS frogs, but not TL frogs, naturally developed RAR; therefore, in this paper, DS frogs were referred to as radio-adapting frogs and TL frogs as naïve frogs. Half of the frogs caught at each site received no more radiation upon capture while the other half were given a whole-body irradiation (WBI) dose of 4 Gy (dose rate was 8 Gy/min) (Table 1). Bladders were chosen to evaluate bystander signal production because the vertebrate urothelium is a radiosensitive tissue and harbors a robust RIBE production following a wide range of irradiation conditions (Mothersill et al. 2001; Vines et al. 2009; Singh et al. 2011; Fernandez-Palomo et al. 2013, 2016). RIBEs were assessed for the induction of clonogenic cell

Table 1. Results of the study show bystander reporter's percent clonogenic survival, calcium flux activities in RIBE reporter cells, percent of bcl-2 and c-myc protein expressions in bladder explant cultures from the background control Twin Lake (TL) site and above-background Duke Swamp (DS) site.

	Frog ID	Sex	Site	Challenge dose (Gy)	Bystander reporter's percent clonogenic survival (%)	Calcium flux (difference between baseline and peak calcium)	bcl-2 (% positive cells in bladder explant)	c-myc (% positive cells in bladder explant)
	MF-1	M	TL	0	N/A	N/A	N/A	N/A
	MF-2	F	TL	0	N/A	N/A	N/A	N/A
Group 1	MF-5	M	TL	0	N/A	N/A	N/A	N/A
	MF-17	F	TL	0	75 ± 4.7	Weak signal [0.015]	26	15
	MF-19	F	TL	0	85 ± 10.1	No signal [0]	31	21
	MF-21	M	TL	0	98	No signal [0]	21	6
	MF-23	F	TL	0	84 ± 9.1	No signal [0]	18	10
				Average	85.5 ± 4.7		24 ± 2.9	13 ± 3.2
	MF-16	F	DS	0	N/A	N/A	N/A	N/A
	MF-18	F	DS	0	N/A	N/A	N/A	N/A
Group 2	MF-20	M	DS	0	157 ± 3.9	No signal [0]	100	5
	MF-22	M	DS	0	133 ± 7.4	No signal [0]	100	3
	MF-24	F	DS	0	N/A	N/A	N/A	N/A
				Average	145 ± 12		100 ± 0.0	4.0 ± 1.0
	MF-3	M	TL	4	N/A	N/A	N/A	N/A
	MF-4	F	TL	4	52.1 ± 4.8	Strong signal [0.1]	2	87
	MF-6	M	TL	4	54.3 ± 5.1	Strong signal [0.1]	11	76
Group 3	MF-7	F	TL	4	N/A	N/A	N/A	N/A
	MF-11	F	TL	4	N/A	N/A	N/A	N/A
	MF-13	F	TL	4	N/A	N/A	N/A	N/A
	MF-15	F	TL	4	N/A	N/A	N/A	N/A
				Average	53.2 ± 1.1		6.5 ± 4.5	81.5 ± 5.5
	MF-8	M	DS	4	120 ± 3.7	No signal [0]	100	2
	MF-9	F	DS	4	N/A	N/A	N/A	N/A
Group 4	MF-10	M	DS	4	127 ± 2.6	No signal [0]	100	4
	MF-12	M	DS	4	N/A	N/A	N/A	N/A
	MF-14	M	DS	4	145 ± 8.8	No signal [0]	100	1
				Average	130.7 ± 7.4		100 ± 0.0	2.3 ± 0.9
Reporter control					100 ± 10	No signal [0]	10	4

Average values were calculated to include standard error of the mean. N/A (gray shaded) indicates that frog bladders were collected and tissue-cultured but the cell cultures were microbially contaminated and discarded and thus data were not available.

death (Maguire et al. 2005) and calcium flux activities (Lyng et al. 2002). Protein signals of radio-responsive oncogenes bcl-2 (B-cell lymphoma 2) associated with survival responses (Lyng et al. 2006) and c-myc (cellular myelocytomatosis oncogene) associated with death responses (Vines et al. 2009) were also studied.

Materials and methods

RIBE reporter cell line

The RIBE reporter assay employed the HPV-G cell line, which was originally derived from human foreskin (Pirisi et al. 1987). HPV-G cells were generously supplied by Dr. J. DiPaolo (National Institutes of Health, Bethesda, MD). HPV-G cells produced a robust and sensitive response to RIBEs following different low-dose irradiation treatments in a number of murine and fish studies (Mothersill et al. 2006, 2007; Singh et al. 2011; Fernandez-Palomo et al. 2013; Smith et al. 2013). Two hallmarks of RIBEs in HPV-G reporter cells are induction of clonogenic cell death and induction of calcium flux into cell cytoplasm (Singh et al. 2011). HPV-G cells were maintained in RPMI 1640 supplemented with 100 U/mL penicillin, 100 μg/mL streptomycin, 25 mM HEPES, 2.05 mM L-glutamine (Gibco, Burlington, ON), 10% fetal bovine serum (FBS) (Invitrogen, Burlington, ON), and 0.5 μg/mL hydrocortisone (Sigma-Aldrich, Oakville, ON). Routine cell maintenance at 37 °C and 5% CO_2 was performed as previously described (Singh et al. 2011). A 0.25% trypsin solution containing 1 mM EDTA (Gibco, Burlington, ON) was used to subculture cells.

Frog bladder samples and bladder explant cultures

Bladders were removed from euthanized frogs and placed in sterilized transport media and shipped overnight to McMaster University. We used the GammaBeam 150 C Cobalt-60 irradiator and irradiation was performed at 10 mGy/min. For frogs that received 4 Gy irradiation, bladders were excised 1 h post-irradiation. Upon arrival on the campus of McMaster University, bladder tissues were placed in sterile Petri dishes (VWR, Burlington, ON), and chopped into three pieces of approximately 2 mm^2, and washed with Dulbecco's Phosphate Buffered Saline (Gibco, Burlington, ON) three times. The explants were then transferred into 25-cm^2 tissue culture flasks (Falcon, Franklin Lanes, NJ) with 2 mL of the RPMI 1640 (Gibco, Burlington, ON) complete medium described above. The explant culture flasks were kept in the 22 °C incubator equilibrated at 5% CO_2. After 48 h, the media from the explant culture flasks were harvested, filtered through sterile 0.22-μm Acrodisc® syringe filters (Pall Corporation, Port Washington, NY) and frozen at −20 °C. Collected media from the bladder tissue explant cultures derived from control and 4-Gy-irradiated groups of naïve frogs from TL (groups 1 and 2 in Table 1) were referred to as control tissue conditioned medium (TL-CTCM) and irradiated tissue conditioned medium (TL-ITCM), respectively. Similarly, DS-CTCM and DS-ITCM were from the corresponding control and 4-Gy-irradiated groups or radio-adapting frogs from DS (groups 3 and 4 in Table 1). In the explant culture flasks, 2 mL of serum-free KGM medium (Clonetics Corporation, San Diego, CA) was added and the explant cultures were incubated for 10–12 days. Subsequently, the explants and surrounding cell growth were fixed in 2 mL of 10% formalin (Fischer Scientific, Kalamazoo, MI) for immunochemistry.

In summary, for the clonogenics experiments, three replicate flasks were plated for each of the frogs caught. After the media were collected for clonogenic assays, one flask was used for bcl-2 immunostaining, one flask for c-myc immunostaining, and the last flask as a negative control for immunocytochemistry. In the calcium flux assays, one dish was used for imagining and data were collected from 10 randomly selected cells in the field of imaging as previously described in detail by our group (Singh et al. 2011; Fernandez-Palomo et al. 2013, 2016).

Immunostaining for bcl-2 and c-myc expressions in explant cultures

Fixed cultures were probed with primary anti-bcl-2 and anti-c-myc antibodies purchased from DAKOCytomation and Novocastra, respectively. The primary antibodies have been previously validated to work on invertebrates, fish, and mouse (Mothersill et al. 2006; Smith et al 2011). Immunochemistry was performed with the Vectastain Elite Avidin/Biotinylated enzyme Complex (ABC) HRP kit (Vector Laboratories, Burlingame, CA) as previously described by our group (Mothersill et al. 2006, 2007, 2009) and diaminobendizine as a detection substrate. Over 200 cells were scored over 5 randomly selected fields using an Olympus image analysis system (ProDiscovery).

RIBE clonogenic reporter assay

Clonogenic reporter flasks designated to receive CTCM or ITCM from bladder explants had 500 or 1000 HPV-G cells initially seeded as previously described (Singh et al. 2011). CTCM or ITCM of the same group were thawed and pooled to over 5 mL of volume. Six hours after cell seeding, the old media in the reporter flasks were replaced with 5 mL of CTCM or ITCM. A set of plating efficiency control flasks does not require a medium change. All flasks had 5 mL of media and were incubated at 37 °C with 5% CO_2. After 10 days, cultures are stained with carbol fuchsin (VWR, Bridgeport, NJ) to visualize surviving colonies and score surviving cell fraction.

RIBE-induced calcium flux measurement assay

One of the well-known RIBEs is to induce calcium influx in bystander cells (Lyng et al. 2002). This biomarker was investigated in this study to see if ITCM and CTCM-induced calcium influx in HPV-G reporter cells. Calcium measurement was measured using the ratiometric fluorescent dye Fura 2-AM as previously described by our group (Singh et al.

2011). About half a million HPV-G cells were seeded per plate (MatTek, Ashland, MA) and cultured for 72 h. Cells are rinsed three times with a buffer solution containing 130 mM NaCl, 5 mM KCl, 1 mM Na_2HPO_4, 1 mM $CaCl_2$, 1 mM $MgCl_2$, and 25 mM HEPES (pH 7.4). Cells were then loaded with 1 mL of 4.2 µM Fura 2-AM (Sigma-Aldrich, Milwaukee, WI) for 30 min at 37 °C. Excess dye was removed and cells were rinsed with the same buffer three times. Then 300 µL of buffer was added to each plate for measurement purposes. The detection of the fluorescence from the HPV-G cells was temporally measured with an automated Olympus 1 × 81 microscope using the 40× oil objective and a Fura filter cube with 510 nm emission. Image acquisition was achieved by Photon Technology International ImageMaster 5.0 software system (Photon Technology International Inc., Birmingham, NJ). Once Fura-2 was excited, measurements at 380 nm and 345 nm were recorded every 4 seconds for 8–9 min. After 45 s to 1 min of recording, 100 µL of ITCM or CTCM was added to the cells. The pre-event time frame of 45–60 s was chosen so as to firmly establish a baseline reading. All measurements were made in the dark at room temperature. Acquired ratios of fluorescence at the two wavelengths versus time were then graphed using the Sigma Plus software system to visualize free cytosolic calcium induction in real time. The calcium flux data are presented as a mean ± SEM from measurements taken from five trials.

Statistical analysis

Significance of differences was tested using Student's *t*-test with Welch's correction in the GraphPad Prism 9 software (GraphPad Software, La Jolla, CA). $p < .05$ was deemed to be statistically significant.

Results

Clonogenic survival of RIBE reporter cells

RIBE reporters receiving TL-CTCM from control naïve frogs (group 1) and TL-ITCM from irradiated naïve frogs (group 3) had cell survival at 85.5 ± 4.7% and 53.2 ± 1.1%, respectively (Table 1) and the reduction in cell survival was statistically significant ($p = .005$, Figure 1). Four Gy WBI of naïve frogs (group 3) living in the background radiation level (Twin Lake) resulted in the bladders inducing bystander effects.

RIBE reporters receiving DS-CTCM from control radio-adapting frogs (group 2) had cell survival at 145 ± 12%, which was higher than that (85.5 ± 4.7%) by TL-CTCM from control naïve frogs (group 1). The difference in cell survival between DS-CTCM and TL-CTCM was, however, not statistically significant ($p = .09$, Figure 2), likely due to the limited number of bladder samples usable for testing.

RIBE reporters receiving DS-ITCM from irradiated radio-adapting frogs (group 4) had cell survival at 130.7 ± 7.4%, which was statistically similar to that by DS-CTCM from control counterparts (group 2) ($p = .43$, Figure 3). Therefore, in contrast to bladder samples from naïve frogs from the

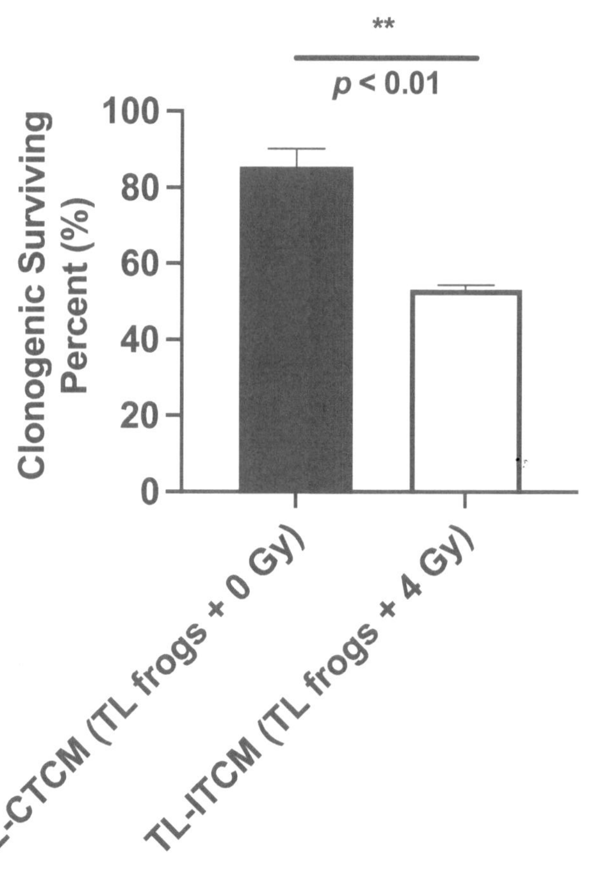

Figure 1. Clonogenic survival percent of RIBE reporter cells in receipt of control tissue conditioned media (CTCM) or irradiated tissue conditioned media (ITCM) from bladder tissue cultures derived from control or 4-Gy-irradiated naïve frogs from the background control site Twin Lake (TL), respectively. $^{**}p < .01$ from Student's *t*-test with Welch's correction.

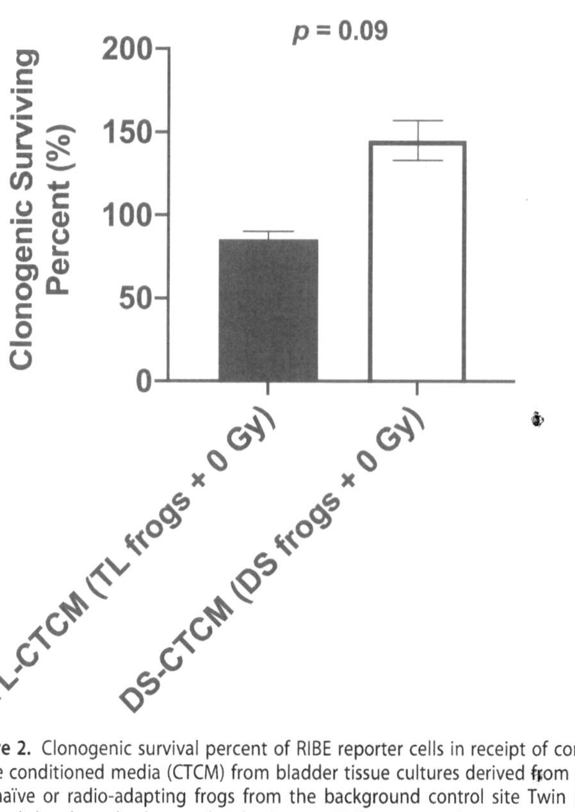

Figure 2. Clonogenic survival percent of RIBE reporter cells in receipt of control tissue conditioned media (CTCM) from bladder tissue cultures derived from control naïve or radio-adapting frogs from the background control site Twin Lake (TL) and the above-background Duke Swamp (DS), respectively.

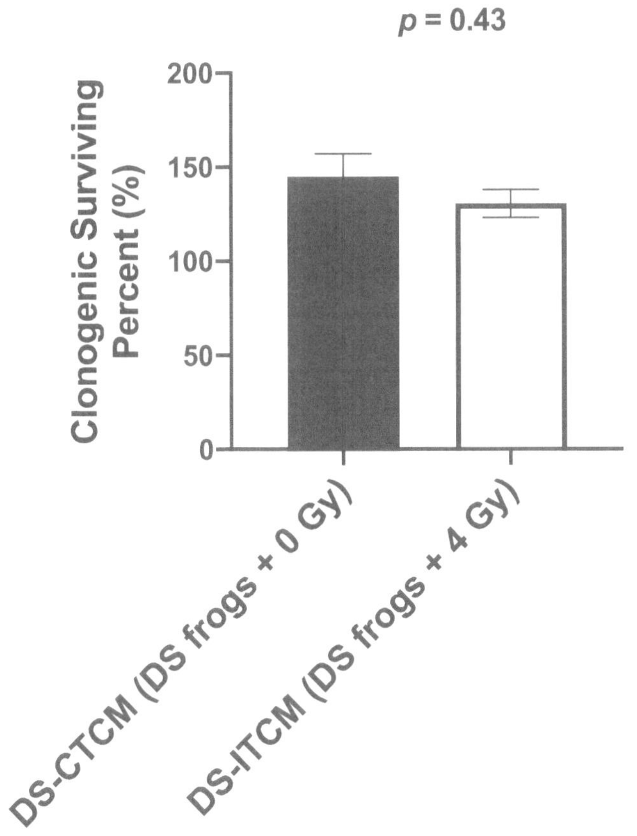

Figure 3. Clonogenic survival percent of RIBE reporter cells in receipt of control tissue conditioned media (CTCM) or irradiated tissue conditioned media (ITCM) from bladder tissue cultures derived from control or 4-Gy-irradiated radio-adapting frogs from the above-background Duke Swamp (DS), respectively.

background control Twin Lake site, 4 Gy WBI of radio-adapting frogs in the above-background DS site did not induce bladders to produce bystander effects.

Bladder samples from DS frogs (groups 2 and 4, Table 1), despite being chronically exposed to elevated levels of contaminating radionuclides such as tritium, did not produce bystander effects.

Calcium flux activities in RIBE reporter cells

The calcium flux results corroborated with the clonogenic survival results. Only bladder samples from group 3 (irradiated naïve frogs from TL) showing bystander effects induced strong calcium flux activities in RIBE reporter cells (Table 1 and Figure 4). The strong calcium flux activity was typically judged by a prominent peak in calcium signal measurements as shown in the right panel of Figure 4. The calcium flux lasted for approximately 150 s, before returning to baseline levels (Figure 4).

No or weak calcium flux activities were observed in reporter cells in receipt of conditioned media from bladder samples from groups 1 and 2 (control naïve frogs from TL and control radio-adapting frogs from DS) and group 4 (irradiated radio-adapting frogs from DS).

Bcl-2 and c-myc expressions in frog bladder explants

In bladder explant cultures from control naïve frogs from TL (group 1), 24% of the cells were positive for bcl-2 and

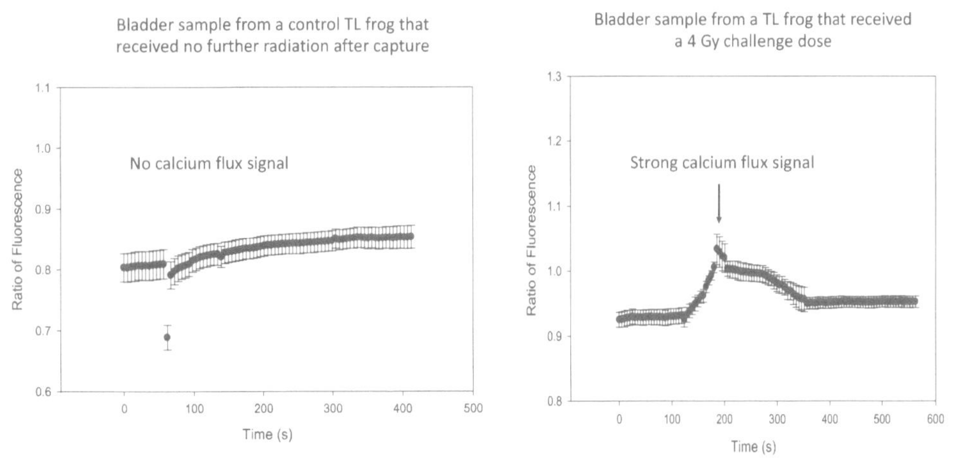

Figure 4. Measurement of calcium flux activities in RIBE reporter cells. An example is shown below showing real-time calcium flux measurement using the fluorescent Fura-2 AM dye in response to exposure to CTCM (left panel) and ITCM (right panel). Ony ITCM induced a calcium flux, which resulted in a peak (arrow) in signal recordings shown on the right panel.

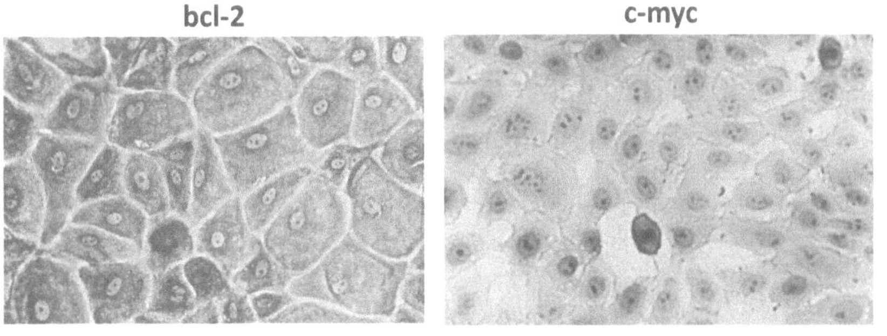

Figure 5. Representative images of frog bladder cells stained immunopositive for bcl-2 and c-myc.

13% of the cells positive for c-myc (Table 1). In bladder explant cultures from control radio-adapting frogs from DS (group 2), all cells (100%) were positive for bcl-2 and about 4% of the cells were positive for c-myc (Table 1). The immunopositivity patterns for bcl-2, but not for c-myc, between groups 1 and 2 was significantly different ($p < .001$). These results show that bladder cells in radio-adapting frogs chronically exposed to elevated radiation levels in DS fully activated the bcl-2-mediated defense response well-known to promote cell survival (Lyng et al. 2006).

The bladder explant cultures from radio-adaptive frogs irradiated with 4 Gy (group 4) showed statistically similar immunopositivity patterns ($p > .05$) with those from control radio-adaptive cells with no 4 Gy challenge. In cultures from group 4, all cells (100%) were positive for bcl-2 and about 2.3% of the cells were positive for c-myc (Table 1). The challenge dose of 4 Gy did not increase c-myc activities, which are associated with death responses, in cells of radio-adapting frogs.

By contrast, 4 Gy irradiation significantly increased pro-apoptotic c-myc activities in cells from naïve frogs from TL from 13% (from control animals, group 1) to 81.5% (in irradiated animals, group 3) ($p = .014$) while the anti-apoptotic bcl-2 expression decreased from 24% to 6.5% ($p = .088$). These results show that bladder cells in naïve frogs, in contrast to radio-adapting frogs, activated death responses when exposed to a high and potential lethal dose of 4 Gy. Representative images of frog bladder cells that were stained immuno-positive for bcl-2 and c-myc are shown in Figure 5.

Discussion

One of the objectives of this study was to test our robust RIBE reporter assay using HPV-G cells to detect and quantify bystander effects from field biological samples (e.g. frog bladders in this case). The applicability of our RIBE reporter assay was confirmed as reduction of clonogenic survival of reporter cells and induction of a strong calcium flux activity, two known characteristics of RIBEs (Maguire et al. 2007), were observed upon exposing reporter cells to TL-ITCM from bladders of naïve frogs (group 3) irradiated with a 4 Gy challenge dose (Figure 1).

Media conditioned by bladders from radio-adapting frogs from DS did not reduce clonogenic survival of reporter cells and did not induce calcium flux. The conditioned media increased clonogenic potential of reporter cells, suggesting that the conditioned media contained soluble factors that enhanced the proliferation or clonogenic potential of reporter cells. Enhancement of clonogenic potential of HPV-G cells in receipt of medium borne bystander signals has been previously noted for acute doses of less than 2 mGy (Liu et al. 2006). The DS frogs received an additional dose of <1 mGy per year (Audette-Stuart et al. 2011). Therefore, it would be likely that the DS frogs received an additional cumulative dose of less than 2 mGy, which could stimulate the release of cell-derived factors that increased the clonogenic survival of HPV-G cells. However, a point of caution should be stressed due to known differences in biological responses to chronic vs acute exposures and low vs. high dose rates. At this moment, it is, therefore, not possible to make further speculations on this point. But the enhancement of clonogenic potential of HPV-G cells had to be due to factors in bladder tissue conditioned media specifically from DS frogs and not from TL frogs as TL-CTCM did not enhance clonogenic survival of HPV-G (Figure 2). What we also know is that the DS frogs radio-adapted in response to the elevated radiation levels at the DS site (Audette-Stuart et al. 2011). RAR involves a number of defense mechanisms that are turned on and enable animals or cells to withstand deleterious radiation insults more effectively and prevent cell death. The lack of mechanisms to induce calcium flux, which initiates signaling cascades to cell death (Lyng et al. 2002) is indicative of a cellular defense to inhibit cell death. Our results further show that radio-adapting frogs exposed to elevated radiation levels fully switched on key anti-apoptotic programs such as those mediated by the cellular survival protein bcl-2, whose expression persisted in bladder cells despite cells being removed from frogs and cultured in a radiation-free environment for about two weeks. Sustained release of cell-derived mitogenic factors has been documented in fish cells following ionizing radiation (Vo et al. 2017). Collective evidence suggests that radio-adaptive programs activated in DS frogs exposed to low dose radiation were likely responsible for the production of soluble cell-derived factors with beneficial effects. This observation would agree with Audette-Stuart and Yankovich (2011)'s study showing that factors produced by radiation-induced radio-adapting frogs protected unirradiated bystander frogs from radiation damage.

Much to our surprise and amazement, 4 Gy irradiation led to bystander signal production by naïve frogs from TL but not by radio-adapting frogs from DS (as judged by both clonogenic survival and calcium flux results). While 4 Gy invoked increased c-myc expression in naïve frogs, the same high dose did little to induce c-myc expression in radio-adapting frogs with sustained widespread strength of pro-survival bcl-2 responses. Previous works with cultured mouse bladder cells (Vines et al. 2009) and in fish models (Mothersill et al. 2007, 2009; Smith et al. 2011) showed that c-myc expression induced by low dose radiation induced bystander effects was generally associated with pro-apoptotic programs. However, c-myc expression is part of double-strand break repair response (Cui et al. 2015) and, under normal circumstances, this is an immediate early gene induced in response to cytokine stimulation. It is difficult to assess if double-strand break repair would still be going on at the time the immunocytochemistry assay was carried out, as we had no additional data regarding the timescale. It is only chronically induced c-myc that is associated with apoptosis; however, it is not known at the present time that c-myc was chronically elevated. The inability to express c-myc in response to 4-Gy irradiation suggests that DNA repair might be compromised in radio-adapting frogs or that in these animals c-myc has become uncoupled for the double-strand damage response. In the absence of a time course and measurement of p53, c-myc, and bax expression together with measures of cell death, such as annexin 5 expression it is challenging to interpret differential Bcl2 expression. Therefore, more experiments need to be conducted in the future to further understand this.

The results of this study have important implications in scenarios where low-dose radiation contamination potentially raises environmental concerns. Mobilization of bystander signals in the environment, especially via the water medium, has been previously demonstrated (Mothersill et al. 2006, 2007). The present work shows that bystander signals could be released by amphibian bladder epithelial cells, which line the anatomical channel that will lead to the release of these signals into the environment via urination. These signals could enter bystander animals and elicit biological effects. When signal-producing amphibians have radio-adapted, the releasing bystander signals could have radioprotective effects on bystander animals. However, the degree and legacy effect of radioprotection mediated by these bystander signals require further experimentation. It is also noted that radiobiological effects are dependent on cell types, which have different intrinsic genetic and biochemical programs activated throughout development. Bystander signals that induce radio-protective effects on one cell type may not show the same beneficial effects on another. Therefore, a larger study with more tissues involved is essential to construct a fuller picture of bystander effects that can display.

We acknowledge that a small sample size is a major limitation of this study. Ecological wildlife research in the environmental radiobiology field that is not as well-funded always faces this challenge. We are encouraged by our results with novel assays and a validated molecular toolbox that can help push forward this important work.

This pilot study emphasizes the importance of looking at NTEs in natural populations of non-human biota that could be vulnerable to chronic low-dose radiation exposures. The clonogenic reporter assay and calcium flux measurement proved to be useful reporter assays to assess RIBEs in the environment. Bcl-2 and c-myc proved to be reliable biomarkers for chronic low-dose radiation exposures in wild populations of amphibians. Efforts to include samples and data from amphibians living in radiologically contaminated habitats in Chernobyl and Fukushima would help better understand RIBE and RAR in amphibians in particular and in ecosystem health in general.

Disclosure statement

The authors report no conflicts of interest.

Funding

The research was supported by the Canada Research Chairs Program, the NSERC Industrial Chairs Program, the NSERC Discovery grant program, the NOTE EU Integrated Project, The CANDU Owners Group, and Atomic Energy Canada Ltd. Canada Research Chair in Sexual and Gender Minority Health; Natural Sciences and Engineering Research Council of Canada; Sixth Framework Programme.

ORCID

Nguyen T. K. Vo http://orcid.org/0000-0002-4709-3078

References

Audette-Stuart M, Kim SB, McMullin D, Festarini A, Yankovich TL, Carr J, Mulpuru S. 2011. Adaptive response in frogs chronically exposed to low doses of ionizing radiation in the environment. J Environ Radioact. 102(6):566–573.

Audette-Stuart M, Yankovich T. 2011. Bystander effects in bullfrog tadpoles. Radioprotection. 46(6):S497–S502.

Cohen J, Vo NTK, Seymour CB, Mothersill CE. 2019. Parallel comparison of pre-conditioning and post-conditioning effects in human cancers and keratinocytes upon acute gamma irradiation. Int J Radiat Biol. 95(2):170–178.

Cui F, Fan R, Chen Q, He Y, Song M, Shang Z, Zhang S, Zhu W, Cao J, Guan H, et al. 2015. The involvement of c-Myc in the DNA double-strand break repair via regulating radiation-induced phosphorylation of ATM and DNA-PKcs activity. Mol Cell Biochem. 406(1-2):43–51.

Fernandez-Palomo C, Schültke E, Bräuer-Krisch E, Laissue JA, Blattmann H, Seymour C, Mothersill C. 2016. Investigation of abscopal and bystander effects in immunocompromised mice after exposure to pencilbeam and microbeam synchrotron radiation. Health Phys. 111(2):149–159.

Fernandez-Palomo C, Schültke E, Smith R, Bräuer-Krisch E, Laissue J, Schroll C, Fazzari J, Seymour C, Mothersill C. 2013. Bystander effects in tumor-free and tumor-bearing rat brains following irradiation by synchrotron X-rays. Int J Radiat Biol. 89(6):445–453.

Hancock S, Vo NTK, Byun SH, Zainullin VG, Seymour CB, Mothersill CE. 2019. Effects of historic radiation dose on the frequency of sex-linked recessive lethals in Drosophila populations following the Chernobyl nuclear accident. Environ Res. 172:333–337.

[IAEA] International Atomic Energy Agency. 2004. Management of Waste Containing Tritium and Carbon-14, Technical Reports Series No. 421, IAEA, Vienna.

Liu Z, Mothersill CE, McNeill FE, Lyng FM, Byun SH, Seymour CB, Prestwich WV. 2006. A dose threshold for a medium transfer bystander effect for a human skin cell line. Radiat Res. 166 (1 Pt 1):19–23.

Lyng FM, Maguire P, Kilmurray N, Mothersill C, Shao C, Folkard M, Prise KM. 2006. Apoptosis is initiated in human keratinocytes exposed to signalling factors from microbeam irradiated cells. Int J Radiat Biol. 82(6):393–399.

Lyng FM, Seymour CB, Mothersill C. 2002. Initiation of apoptosis in cells exposed to medium from the progeny of irradiated cells: a possible mechanism for bystander-induced genomic instability. Radiat Res. 157(4):365–370.

Maguire P, Mothersill C, McClean B, Seymour C, Lyng FM. 2007. Modulation of radiation responses by pre-exposure to irradiated cell conditioned medium. Radiat Res. 167(4):485–492.

Maguire P, Mothersill C, Seymour C, Lyng FM. 2005. Medium from irradiated cells induces dose-dependent mitochondrial changes and BCL2 responses in unirradiated human keratinocytes. Radiat Res. 163(4):384–390.

Mothersill C, Bucking C, Smith RW, Agnihotri N, Oneill A, Kilemade M, Seymour CB. 2006. Communication of radiation-induced stress or bystander signals between fish in vivo. Environ Sci Technol. 40(21):6859–6864.

Mothersill C, Rea D, Wright EG, Lorimore SA, Murphy D, Seymour CB, O'Malley K. 2001. Individual variation in the production of a 'bystander signal' following irradiation of primary cultures of normal human urothelium. Carcinogenesis. 22(9):1465–1471.

Mothersill C, Rusin A, Seymour C. 2017. Low doses and non-targeted effects in environmental radiation protection; where are we now and where should we go? Environ Res. 159:484–490.

Mothersill C, Smith RW, Agnihotri N, Seymour CB. 2007. Characterization of a radiation-induced stress response communicated in vivo between zebrafish. Environ Sci Technol. 41(9): 3382–3387.

Mothersill C, Smith RW, Hinton TG, Aizawa K, Seymour CB. 2009. Communication of radiation-induced signals in vivo between DNA repair deficient and proficient medaka (Oryzias latipes). Environ Sci Technol. 43(9):3335–3342.

Mothersill CE, Oughton DH, Schofield PN, Abend M, Adam-Guillermin C, Ariyoshi K, Beresford NA, Bonisoli-Alquati A, Cohen J, Dubrova Y, et al. 2020. From tangled banks to toxic bunnies; a reflection on the issues involved in developing an ecosystem approach for environmental radiation protection. Int J Radiat Biol. 1–16.

Pirisi L, Yasumoto S, Feller M, Doniger J, DiPaolo JA. 1987. Transformation of human fibroblasts and keratinocytes with human papillomavirus type 16 DNA. J Virol. 61(4):1061–1066.

Ryan LA, Seymour CB, O'Neill-Mehlenbacher A, Mothersill CE. 2008. Radiation-induced adaptive response in fish cell lines. J Environ Radioact. 99(4):739–747.

Seymour CB, Mothersill C. 2000. Relative contribution of bystander and targeted cell killing to the low-dose region of the radiation dose-response curve. Radiat Res. 153(5 Pt 1):508–511.

Singh H, Saroya R, Smith R, Mantha R, Guindon L, Mitchel RE, Seymour C, Mothersill C. 2011. Radiation induced bystander effects in mice given low doses of radiation in vivo. Dose Response. 9(2): 225–242.

Smith RW, Mothersill C, Hinton T, Seymour CB. 2011. Exposure to low level chronic radiation leads to adaptation to a subsequent acute X-ray dose and communication of modified acute X-ray induced bystander signals in medaka (Japanese rice fish, Oryzias latipes). Int J Radiat Biol. 87(10):1011–1022.

Smith RW, Seymour CB, Mothersill CE. 2013. Short and long term bystander effect induction by fathead minnows (Pimephales promelas, Rafinesque, 1820) injected with environmentally relevant whole body doses of 226Ra. J Environ Radioact. 126:133–136.

Vines AM, Lyng FM, McClean B, Seymour C, Mothersill CE. 2009. Bystander effect induced changes in apoptosis related proteins and terminal differentiation in in vitro murine bladder cultures. Int J Radiat Biol. 85(1):48–56.

Vo NTK, Seymour CB, Mothersill CE. 2019. Radiobiological characteristics of descendant progeny of fish and amphibian cells that survive the initial ionizing radiation dose. Environ Res. 169:494–500.

Vo NTK, Sokeechand BSH, Seymour CB, Mothersill CE. 2017. Characterizing responses to gamma radiation by a highly clonogenic fish brain endothelial cell line. Environ Res. 156:297–305.

Vo NTK. 2021. Environmental radiobiology of amphibians – knowledge gaps to be filled using cell lines. Int J Radiat Biol. 1–38.

Environmental radiation on large Japanese field mice in Fukushima reduced colony forming potential in hematopoietic progenitor cells without inducing genomic instability

Kentaro Ariyoshi, Tomisato Miura, Kosuke Kasai, Valerie Swee Ting Goh, Yohei Fujishima, Akifumi Nakata, Atsushi Takahashi, Yoshinaka Shimizu, Hisashi Shinoda, Hideaki Yamashiro, Colin Seymour, Carmel Mothersill, and Mitsuaki A. Yoshida

ABSTRACT

Purpose: To study the environmental radiation effects of wild animals after the Fukushima Daiichi nuclear power plant accident, we assessed effects on hematopoietic progenitor cells (HPCs) in large Japanese field mice (*Apodemus speciosus*).

Materials and methods: A. speciosus were collected from three contaminated sites and control area. The air dose-rates at the control and contaminated areas were 0.96 ± 0.05 μGy/d (Hirosaki), 14.4 ± 2.4 μGy/d (Tanashio), 208.8 ± 31.2 μGy/d (Ide), 470.4 ± 93.6 μGy/d (Omaru), respectively. We investigated possible DNA damage and pro-inflammatory markers in the bone marrow (BM) cells. The colony-forming potential of BM cells was estimated by the number of HPC colony-forming cells. Radiation-induced genomic instability (RIGI) in HPCs was also analyzed by quantifying delayed DNA damage in CFU-GM clones.

Results: Although no significant differences in DNA damage and inflammation markers in BM cells from control and contaminated areas, the number of HPC colonies exhibited an inverse correlation with air dose-rate. With regard to RIGI, no significant differences in DNA damage of CFU-GM clones between the mice from the control and the three contaminated areas.

Conclusions: Our study suggests that low dose-rate radiation of more than 200 Gy/d reduced HPCs, possibly eliminating genomically unstable HPCs.

Introduction

After the Fukushima Daiichi Nuclear Power Plant (FDNPP) accident, there has been a concern about the effects of dispersed radionuclides on the environment. Although nine years have passed since the FDNPP disaster and the air dose-rates and radionuclide concentrations have been gradually decreasing, the biological effects caused by long-term exposure to radioactive substances remain a major concern. Several studies have assessed the biological effects of exposure to artificial radionuclides after the FDNPP accident (Hiyama et al. 2012; Yamashiro et al. 2013; Ochiai et al. 2014; Bonisoli-Alquati et al. 2015; Hiyama et al. 2015; Kubota et al. 2015; Suzuki 2015; Okano et al. 2016; Urushihara et al. 2016; Kawagoshi et al. 2017; Morimoto et al. 2017; Takino et al. 2017; Giraudeau et al. 2018; Itoh et al. 2018; Urushihara et al. 2018; Taira et al. 2019). There was no induction of inversion-type chromosomal abnormalities in *Drosophila melanogaster* (Itoh et al. 2018). Furthermore, there was no significant effect on the tissue (blood, liver, and skin) carotenoid levels in Japanese tree frogs (*Hyla japonica*) (Giraudeau et al. 2018). There were no obvious radiation-induced effects in bull testes (Yamashiro et al. 2013) nor in large Japanese field mouse testes (Okano et al. 2016). On the contrary, Hiyama et al. reported morphological abnormalities in pale grass blue butterflies (*Pseudozizeeria maha*) (Hiyama et al. 2012; Hiyama et al. 2015). Moreover, white butterflies that were fed cabbage grown in Fukushima showed morphological abnormalities and decreased percentage of granulocytes (Taira et al. 2019). Abnormal hyperplasia of macrophages was observed in multiple organs (spleen, kidney, liver, and pancreas) of carp (Suzuki 2015). Although the radioactive contamination levels did not correlate with DNA fragmentation in nestlings, higher contamination level decreased the number of barn swallows and the fraction of juveniles (Bonisoli-Alquati et al. 2015). DNA damage, measured by γ-H2AX foci, was significantly increased in lymphocytes of cattle in the evacuation zone (Nakamura et al. 2017). Decreased number of blood cells (Ochiai et al. 2014), small head size, and delayed body weight growth (Hayama et al. 2017) were observed in wild Japanese monkeys. Urushihara et al. also reported that the number of myeloid, megakaryocyte and hematopoietic cells were inversely correlated and the occupancy of adipose tissue was positively correlated with the internal dose-rate in the bone marrow of wild Japanese monkeys (Urushihara et al. 2018). Despite these extensive studies on the biological effects on wild animals after the FDNPP accident, to the best of our knowledge, the potential effects of accidental release of radionuclides on stem cells or progenitor cells of wild animals have never been studied. The object of our study is to assess the radiation effect on hematopoietic progenitor cells of wild animal.

The large Japanese field mouse (*Apodemus speciosus*) is endemic to Japan and can be found across the country. At approximately 1 and 3 mGy/d, similar to that of locations near FDNPP, an increase in the frequency of dicentric chromosomes in splenic lymphocytes was observed in *A. speciosus* and other wild rodents (house mice; *Mus musculus*; the small Japanese field mouse, *Apodemus argenteus*) (Kubota et al. 2015; Kawagoshi et al. 2017). In agreement with these results, Tanaka et al. reported that in laboratory mice (*M. musculus*), chronic exposure to very low dose-rate radiation (1 mGy/d) caused chromosomal aberrations in splenic lymphocytes, characterized by a high sensitivity to radiation, at doses of 400 and 8000 mGy (Tanaka et al. 2009; Tanaka et al. 2014). It is noteworthy that no increase in chromosome aberrations was observed after 1 mGy/d chronic exposure to a total dose of 20 mGy (Braga-Tanaka et al. 2018). Our previous in vitro experiments suggest that embryonic fibroblast cells of *A. speciosus* are more sensitive to radiation than mouse embryonic fibroblast (MEF) cells (Ariyoshi, Miura, Kasai, Akifumi, Fujishima, and Yoshida 2018). Based on these results, greater harmful effects are expected in radiosensitive tissues or cells of *A. speciosus* than those of laboratory mice.

Hematopoietic stem cells (HSCs) and hematopoietic progenitor cells (HPCs) have high proliferative potential, which makes them highly sensitive to extracellular stresses including radiation stress (Heylmann et al. 2014). Low dose radiation effect on HSCs or HPCs has been investigated in vitro or ex vivo, and also in vivo). Yagunov et al. reviewed low dose and low dose-rate radiation effects on HSCs or HPCs and suggested that low dose-rate radiation results in a deficiency of HSCs (Yagunov et al. 1998). The effect of low dose-rate radiation (0.5 mGy/min) on the recovery of murine hemopoietic and stromal progenitor cells also suggests a decrease in HSCs in mouse (Gallini et al. 1988). Similar cellular damage was observed in HSCs and HPCs (CFU-granulocyte macrophages CFU-GMs) after exposure to 750 cGy by fractioned radiation (25 cGy/min) or low dose-rate radiation (4 cGy/min) (Girinski et al. 1990). Nothdurft et al. reported that low dose-rate radiation exposure (18.8 mGy/d) resulted in a strong decrease in HPCs (CFU-GMs) within the first 150 d of exposure in dogs (Nothdurft et al. 1995). These results suggest that chronic low dose-rate radiation could cause a decrease in HSCs/HPCs in wild animals in contaminated areas in Fukushima.

Another concern is the non-targeted effects of radiation, radiation-induced genomic instability (RIGI), and radiation-induced bystander effect (RIBE), of chronic exposure to low dose-rate radiation in wild animals, which have been pointed out and reviewed by Mothersill et al. (Mothersill et al. 2017). In our previous study, we observed RIBE in embryonic fibroblast cells of *A. speciosus* after experimental radiation exposure (Ariyoshi, Miura, Kasai, Akifumi, Fujishima, and Yoshida 2018). While our results indicated that RIBE could occur in cells of *A. speciosus*, little is known about the RIGI in wild animals. Non-clonal cytogenetic aberrations in the clonal descendants of irradiated human and mouse HSCs in vitro have been reported (Kadhim et al. 1992; Kadhim et al. 1994; Kadhim et al. 1995). In a mouse model, accumulating evidence suggests that some genetic component determines the susceptibility to RIGI in HSCs (Watson et al. 1997; Gowans et al. 2005). Furthermore, our previous study suggests that age-at-exposure determines the susceptibility of mouse HSCs to RIGI (Ariyoshi, Miura, Kasai, Akifumi, Fujishima, Shinagawa, et al. 2018). However, the effect of low-dose-rate radiation on RIGI in *A. speciosus* HSCs and/or HPCs remains unclear.

In this study, to fill the gap, we assessed the biological effects after long-term low-dose-rate radiation on HPCs from the bone marrow of large Japanese field mice, *A. speciosus*, in three contaminated areas around FDNPP with low, intermediate and high air dose-rates. To reduce the biological effect of synthetic compounds, we captured *A. speciosus* in forests 50 to 100 m away from farmland. We investigated (i) the frequency of DNA damage and pro-inflammatory markers in bone marrow cells, (ii) performed a quantitative analysis of HPC colonies from bone marrow, and (iii) evaluated RIGI by analyzing delayed DNA damage in HPCs.

Materials and methods

Capture of large Japanese field mice

Large Japanese field mice (*A. speciosus*) were captured using Sherman traps (H. B. Sherman Traps, Inc.) at Koguriyama area in Hirosaki City (non-contaminated control site, $n = 5$), Aomori in May 2015. *Apodemus speciosus* were also captured at three contaminated areas, Omaru area (high dose-rate, air: 470.4 ± 93.6 μGy/d, soil: 471.2 ± 192.9 μGy/d, $n = 5$), Ide area (intermediate dose-rate, air: 208.8 ± 31.2 μGy/d, soil: 211.7 ± 41.3 μGy/d, $n = 14$) and Tanashio area (low dose-rate, 14.4 ± 2.4 μGy/d, 15.5 ± 4.4 μGy/d, $n = 6$) at Namie Town, Fukushima in April 2015 (Figure 1(A)).

The capture of wild rodents in sampling areas was permitted by the Fukushima Prefectural Office and Aomori Prefectural Office. Age in *A. speciosus* was estimated based on the dry weight of the optic lens (Takada et al. 2006). All animal experiments were conducted in accordance with the

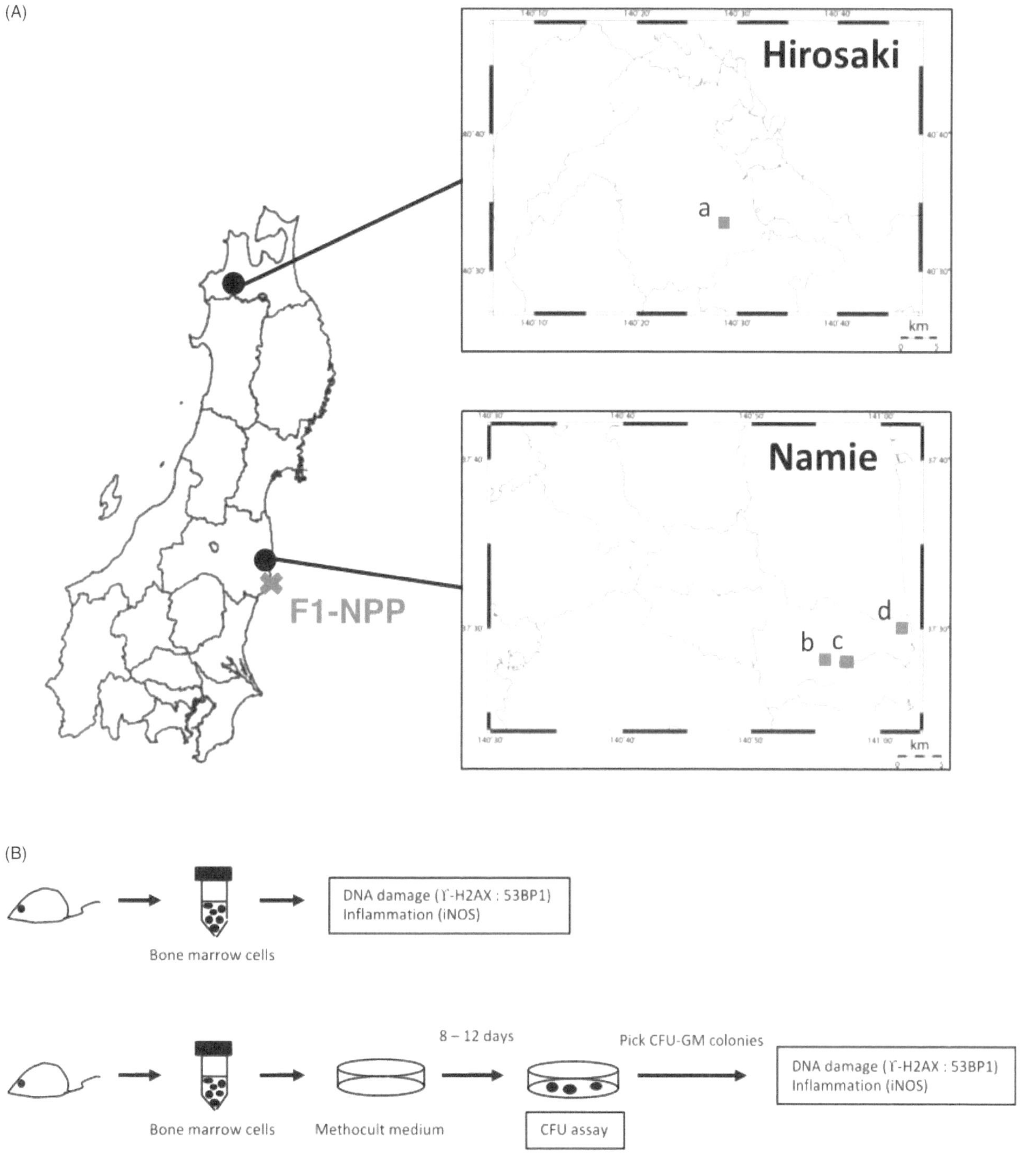

Figure 1. Sampling areas of *A. speciosus* in Fukushima and Aomori Prefectures, and a schematic of the experimental protocol. (A) Detailed map represents the control area in Hirosaki, Aomori (a: Koguriyama), and contaminated areas in Namie Town, Fukushima (b: Omaru, c: Ide, d: Tanashio). FDNPP: The TEPCO Fukushima Daiichi Nuclear power plant. (B) Schematic of CFU assay performed in *A. speciosus* bone marrow cells to obtain HPC colonies. IF staining was performed in both bone marrow cells and HPC colonies.

laboratory animal care guidelines of the Ethics Committee for Care and Use of Laboratory Animals for Research of Hirosaki University, Japan (Approval No. R12001).

Measurements of air dose-rate

The air dose-rates 1 m above the ground and dose-rates from the soil of sampling points were measured with a NaI (Tl) scintillation survey meter TCS-171B (Hitachi Aloka Medical, Ltd., Wallingford, CT). Measurements were made at μGy per hour, time-constant set to 10 s, and recorded when the readings were stabilized after a minimum of 30 s.

Colony-forming cell (CFU) assay

The experimental procedure for the CFU assay is shown in Figure 1(B). The flushed bone marrow cells were dissociated into single-cell suspensions by pipetting. Cells were then suspended in CELLBANKER 1plus (Nippon Zenyaku Kogyo) and stored at −80 °C until use (typically from 1 to 2 months). Thawed cells were diluted with Iscove's MDM containing 2% fetal bovine serum (FBS) (Thermo Scientific) to determine total cell count and cell viability using the Muse Count and Viability kit (Muse Cell Analyzer; Millipore) according to the manufacturer's instructions. Over 85% of viable cell stocks were used for the CFC assay. CFC assay was performed with MethoCult GF M3434 methylcellulose medium (Stem Cell Technologies). Cells ($0.5–1 \times 10^5$ cells) were seeded per dish and maintained at 37 °C in a humidified atmosphere with 5% CO_2. Colonies of burst-forming unit-erythroid cells (BFU-E) and colony-forming-unit-megakaryocytes (CFU-Meg) were scored on day 7, while CFU–granulocyte macrophages (CFU-GM) were scored on day 12 of the incubation according to the manufacturer's protocol. Colonies of CFU-GM were pooled on day 12 of the incubation and used in the following experiments.

Immunofluorescence staining for DNA damage and pro-inflammatory markers in bone marrow cells and HPCs

The experimental procedure for the immunofluorescence staining is shown in Figure 1(B). The tibias of wild rodents were removed and placed in ice-cold PBS, and bone marrow cells were flushed out. To investigate the induction of DSBs in bone marrow cells, cells were washed twice in PBS(−) and fixed with 4% formaldehyde in PBS(−) for 15 min. To analyze the induction of DSBs in HPCs, pooled CFU-GM colonies were washed twice in PBS(−) and fixed with 4% formaldehyde in PBS(−) for 15 min. Fixed bone marrow cells and CFU-GM were washed twice in PBS(−) and then centrifuged onto glass slides at 600 rpm for 8 min in a Shandon Cytospin 4 cytocentrifuge (Thermo Scientific). Air-dried cells were then permeabilized in PBS(−) containing 0.1% Triton X-100 (Nacalai Tesque) for 5 min on ice. For DSB analysis using γH2AX/53BP1 foci, cells were labeled with anti-γH2AX (1:200; 05-636, Millipore) and anti-53BP1 (1:100; ab36823, Abcam) antibodies in TBS-T (a mixture of tris-buffered saline and Tween 20) supplemented with 5% nonfat dry milk (Wako Pure Chemical Industries, Ltd.) for 4 h at 37 °C. Labeled cells were washed in PBS(−) twice and stained with Alexa Fluor 488-conjugated secondary antibodies (1:500; ab150113, Abcam) or Alexa Fluor 647-conjugated secondary antibodies (1:500; ab150079, Abcam) for 30 min at 37 °C. Expression of pro-inflammatory proteins in bone marrow cells and CFU-GM was measured by staining inducible nitric oxide synthase (iNOS). Fixed samples were cytospun onto slides and permeabilized in PBS(−) containing 0.5% Triton X-100 for 3 min on ice, then labeled with anti-iNOS antibodies (1:100; ab36823, Abcam) in TBS-T supplemented with 5% nonfat dry milk for 6 h at 37 °C. Labeled cells were washed in PBS(−) twice and stained with Alexa Fluor 647-conjugated secondary antibodies for 30 min at 37 °C. The nuclei were counterstained with 4′,6-diamidino-2-phenylindole (DAPI) (Nacalai Tesque). Fluorescent images were captured and recorded using a fluorescent microscope and a CCD camera (Olympus, Shinjuku, Japan).

Ex vivo X-ray exposure on bone marrow cells of the control area

Although the radionuclides spread by the FDNPP accident emit gamma rays and beta rays, we do not have a gamma-ray irradiator. So, we used an X-ray generator for analyzing the radiation sensitivity of HPCs. Bone marrow cells were collected from wild rodents captured at Koguriyama area at Hirosaki City, and cell suspension samples were irradiated 0, 50, 100, 200, 500, 1000 and 2000 mGy at a dose rate of 0.5 Gy/min with an X-ray generator (MBR-1520R-3; Hitachi Medical), then immediately cultured for CFU assay. During X-ray exposure, the total dose and dose rate were monitored with a thimble ionization chamber placed next to the sample. Average D_0 values (the dose required to reduce cellular survival to $e^{-1} = 0.37$ on the linear portion of the dose survival curve) were evaluated.

Statistical analysis

The statistical significance of the difference between groups was assessed using the statistical tests indicated in the figure legends. All the analyses were performed using StatMate III software, ver. 3.08 (ATMS Co., Ltd., Tokyo, Japan). Statistical significance was set at $p < .01$ unless otherwise stated.

Results

DNA damage and pro-inflammatory protein analyses in bone marrow cells of A. speciosus

We examined the frequency of DNA damage in bone marrow cells of *A. speciosus* by scoring co-localized γH2AX/53BP1 foci per cell in positive cells (Figure 2(A)). All the γH2AX foci were co-localized with 53BP1. DNA damage foci per cell of *A. speciosus* captured in Hirosaki (control area, air: 0.96 ± 0.05 μGy/d, soil: 1.01 ± 0.05 μGy/d), Tanashio (low dose-rate, air: 14.4 ± 2.4 μGy/d, soil: 15.5 ± 4.4 μGy/d), Ide (intermediate dose-rate, air: 208.8 ± 31.2 μGy/d, soil: 211.7 ± 41.3 μGy/d), and Omaru (high dose-rate, air: 470.4 ± 93.6 μGy/d, soil: 471.2 ± 192.9 μGy/d) were 0.047 ± 0.02, 0.046 ± 0.04, 0.058 ± 0.03, and 0.042 ± 0.02, respectively (Figure 2(B)). No significant differences were observed between the Hirosaki control area and the three contaminated areas. Furthermore, a similar tendency was observed in the number of DNA damage focus in the Hirosaki control area and the three contaminated areas (Figure 2(C)). Regardless of the air dose-rate, we observed DNA damage-harboring cells (four foci per cell) (Figure 2(C)). The relationship of DNA damage induction and other factors was further assessed in age, body weight, sex, and activated pro-inflammatory responses using iNOS expression (Figure S1; Figure 3). No correlation was observed between DNA damage and age or sex, and a weak negative

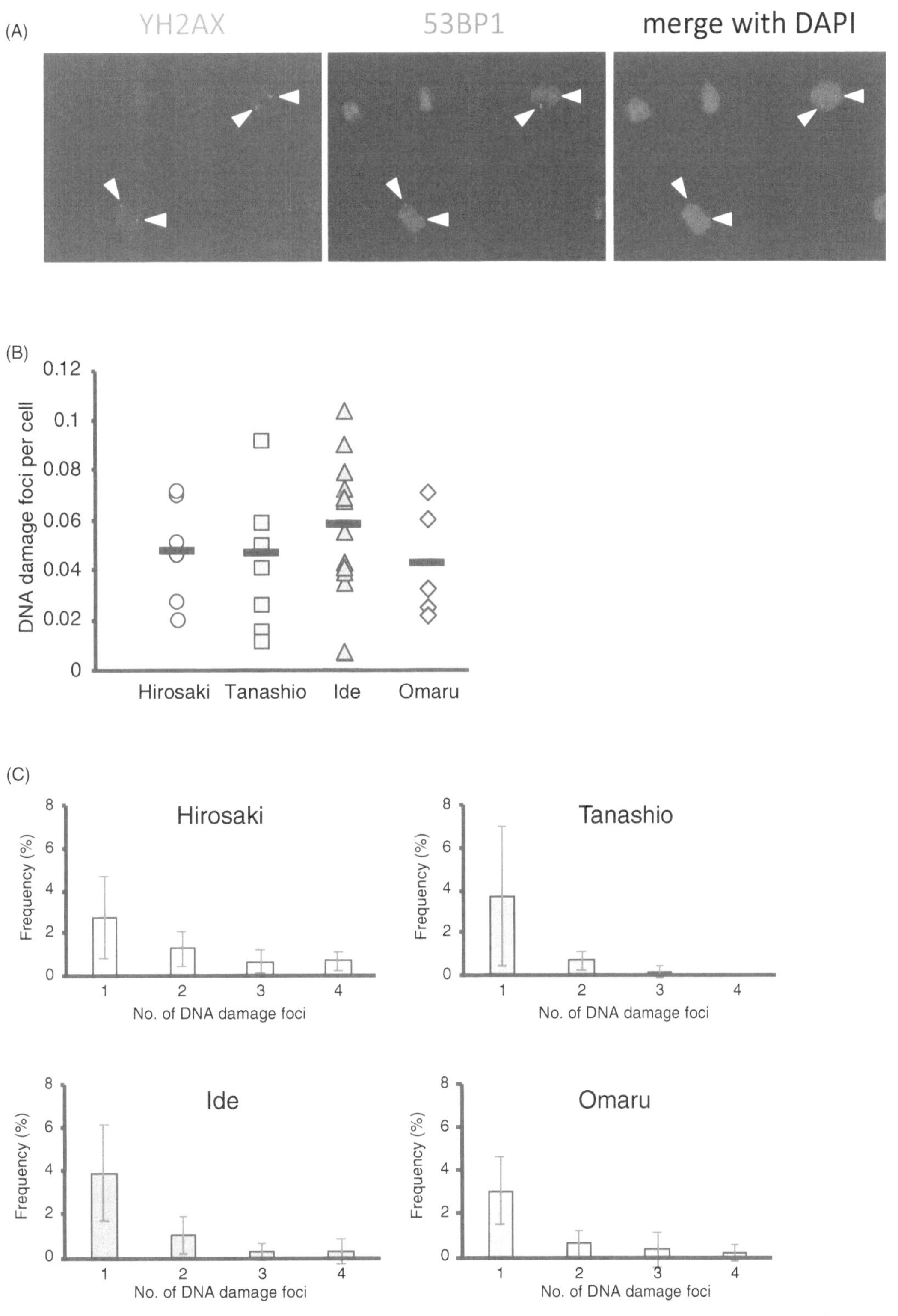

Figure 2. DNA damage analysis in bone marrow cells of *A. speciosus* captured at control and contaminated areas. (A) Representative photomicrographs of γ-H2AX and 53BP1 immunofluorescence staining and nucleic counterstaining with DAPI in bone marrow cells. (B) The frequency of DNA damage foci in *A. speciosus* bone marrow cells captured in Hirosaki (control area: 0.96 ± 0.05 μGy/d, $n = 6$), and three areas in Namie Town, Tanashio (low dose-rate, 14.4 ± 2.4 μGy/d, $n = 7$), Ide (intermediate dose-rate, 208.8 ± 31.2 μGy/d, $n = 14$) and Omaru (high dose-rate, 470.4 ± 93.6 μGy/d, $n = 5$). All data are expressed as Mean ± SD. No significant differences were found after Welch's *t*-test ($p > .01$). (C) The proportion of *A. speciosus* bone marrow cells with different number of DNA damage foci captured at Hirosaki (control area, $n = 6$), Tanashio (low dose-rate, $n = 7$), Ide (intermediate dose-rate, $n = 14$) and Omaru (high dose-rate, $n = 5$).

correlation was seen between DNA damage and body weight (Figure S1). However, we could observe a clear correlation between DNA damage and iNOS positive cells (Figure 3(C)). These results suggested that activated inflammation responses, not activated by low dose rate radiation, induced DNA damage in *A. speciosus* bone marrow cells.

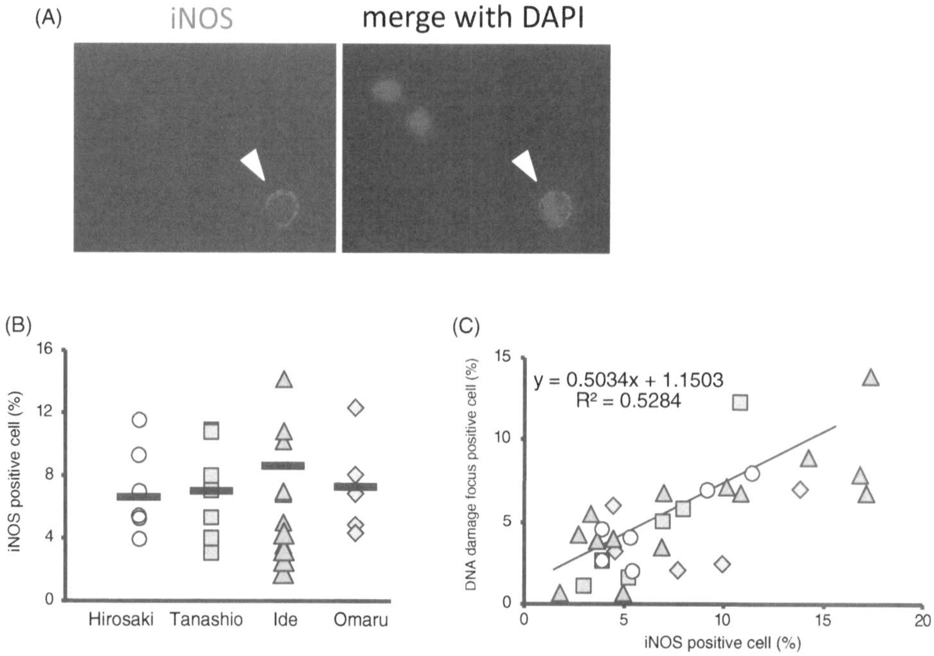

Figure 3. Pro-inflammatory response in bone marrow cells of *A. speciosus* captured at control and contaminated areas. (A) Representative photomicrographs of iNOS immunofluorescence staining and nucleic counterstaining with DAPI of bone marrow cells. (B) The proportion of iNOS positive *A. speciosus* bone marrow cells in Hirosaki (control area, $n = 6$), Tanashio (low dose-rate, $n = 7$), Ide (intermediate dose-rate, $n = 14$) and Omaru (high dose-rate, $n = 5$). No significant differences were found after Welch's *t*-test ($p > .01$). (C) Linear regression analysis between iNOS positive cells (*X*-axis) and DNA damage focus positive cells (*Y*-axis).

Number of HPC colonies in A. speciosus

While we examined flow cytometric analysis, we could not detect CD34$^-$ c-Kit$^+$ Sca-1$^+$ Lin$^-$ (CD34$^-$ KSL) cell populations (data not shown). It is probable that anti-mouse antibodies could not detect the epitope of *A. speciosus*. So, we evaluated the number of HPCs based on the results of the CFU assay in Methocult GF medium (Figure 4). We identified CFU-GM, BFU-E, and CFU-Meg based on size, distribution, and cell morphology after 12 d of culture (Figure 4(A)). Because we could not find antibodies for c-Kit or Sca-1, we could not examine positive selection and could not detect CFU-granulocytes, -erythrocytes, -monocytes, and -megakaryocytes (CFU-GEMM). The total number of bone marrow cells in the tibia was mostly similar among the control and contaminated areas, and the average number of cells was as follows: Hirosaki (5.08×10^6 cells), Tanashio (5.82×10^6 cells), Ide (7.02×10^6 cells) and Omaru (4.90×10^6 cells) (Figure 4(B)). A significantly higher number of bone marrow cells was observed in *A. speciosus* captured at Ide, possibly due to their larger body size compared with that of mice in the other areas (Table 1). To determine the radiosensitivity of different types of HPC colony-forming cells, bone marrow cells of *A. speciosus* captured in Hirosaki were exposed to various X-ray doses from 0 to 2000 mGy at 0.5 Gy/min and cultured up to 12 d for CFU assay. (Figure 4(C)). All colony-forming cells decreased in a dose-dependent manner (Figure 4(C)) and the D_0 values of CFU-GM, BFU-E, and CFU-Meg colonies were 2.1, 0.5, and 0.9 Gy respectively. The CFU assay was also performed in *A. speciosus* captured in both control and contaminated areas without additional X-ray irradiation. The number of CFU-GM colonies in mice from Ide (38.8 ± 19.3) and Omaru (29.5 ± 9.33) was significantly lesser than that of mice from Tanashio (57.3 ± 22.5) and Hirosaki (63.9 ± 14.6) (Figure 4(D)). A significant reduction in BFU-E and CFU-Meg colonies in mice from Ide and Tanashio was also observed (Figure 4(D)). The number of BFU-E colonies in mice from Ide, Omaru, Tanashio, and Hirosaki was 10.1 ± 3.9, 8.6 ± 3.0, 16.1 ± 7.0 and 17.4 ± 3.8 respectively, and the number of CFU-Meg colonies in mice from Ide, Omaru, Tanashio, and Hirosaki was 2.7 ± 1.6, 3.5 ± 2.3, 4.6 ± 2.7 and 5.9 ± 3.1, respectively (Figure 4(D)). No significant differences were observed in all HPC clones between Hirosaki and Tanashio, and between Ide and Omaru (Figure 4(C)). There was no correlation between the number of CFUs (CFU-GM, BFU-E, and CFU-Meg) and age, body weight, DNA damage, and pro-inflammatory response (Figure S2). On the contrary, the number of HPC colonies (CFU-GM and BFU-E) weakly correlated with dose-rate in the air (Figure S3).

Delayed induction of DNA damage in CFU-GM colonies

DNA damage in CFU colonies was evaluated to estimate the induction of RIGI by radiation. We first examined the induction of RIGI under experimental conditions by pooling and analyzing CFU-GM colonies, which we could obtain more than 50 clones per experiment, after bone marrow cells of *A. speciosus* captured at Hirosaki. The bone marrow cells were irradiated with 0, 50, 100, 200, 500, and 1000 mGy of X-rays at 0.5 Gy/min and CFU-GM colonies of similar sizes were selected (eight colonies per dose), then delayed DNA damage was measured with γ-H2AX/53BP1 foci per cell (Figure S4(A)). We observed no differences between 0 Gy control and 50, 100, 200, and 500 mGy-irradiated CFU-GM (Figure S4(B)). However, RIGI was observed after 1000 mGy irradiation (Figure S4(B)). These results

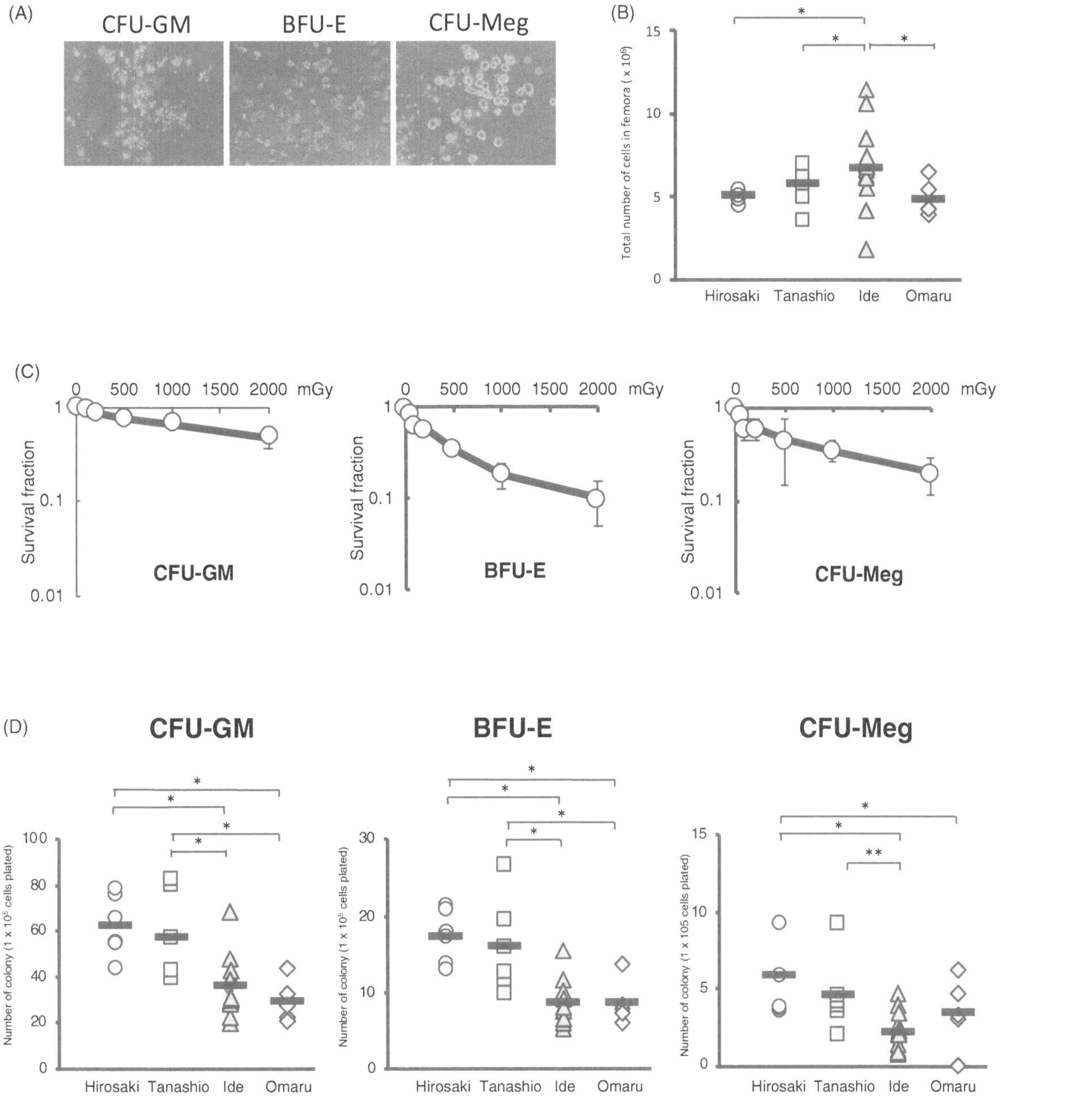

Figure 4. Number of bone marrow cells and HPC colonies of *A. speciosus* captured at control and contaminated areas. (A) Representative photomicrographs of CFU-GM, BFU-E and CFU-Meg of *A. speciosus*. Colonies for CFU-GM and BFU-E were scored 7 d after incubation, and CFU-Meg were scored 12 d after incubation; 0.5–1 $\times 10^5$ cells were plated for each experiment. (B) The number of bone marrow cells per femur in Hirosaki (control area, $n = 6$), Tanashio (low dose-rate, $n = 7$), Ide (intermediate dose-rate, $n = 14$) and Omaru (high dose-rate, $n = 5$). All data are expressed as Mean ± SD. *The differences are significant based on Welch's *t*-test ($p < .01$). (C) Survival fraction of CFU-GM, BFU-E and CFU-Meg in *A. speciosus* captured at Hirosaki (control area, $n = 4$) after X-ray exposure to 0, 50, 100, 200, 500, 1000 and 2000 mGy at a dose rate of 0.5 Gy/min. (D) The number of CFU-GM, BFU-E and CFU-Meg colonies of *A. specious* captured at Hirosaki (control area, $n = 6$), Tanashio (low dose-rate, $n = 7$), Ide (intermediate dose-rate, $n = 14$) and Omaru (high dose-rate, $n = 5$). All data are expressed as Mean ± SD. *The differences are significant based on Student's *t*-test ($p < .01$).

suggest that RIGI in CFU-GM of *A. speciosus* is radiation dose-dependent. We then measured delayed DNA damage in CFU-GM colonies of *A. speciosus* captured in the contaminated areas (10 colonies per *A. speciosus*). Among the CFU-GM colonies of Hirosaki and the other contaminated areas, the frequency of DNA damage in CFU-GM in Hirosaki, Tanashio, Ide and Omaru was 0.39 ± 0.17, 0.44 ± 0.19, 0.43 ± 0.15 and 0.41 ± 0.16 respectively, with no significant differences seen (Figure 5(A)). A similar tendency was observed in the number of DNA damage foci in Hirosaki control area and the three contaminated areas (Figure 5(B)).

Discussion

The effects of artificial radionuclide exposure on wildlife are a major concern after the FDNPP accident. The report of the United Nations Scientific Committee on the Effects of Atomic Radiation (UNSCEAR) about the effects of ionizing radiation on wildlife suggested that at below the chronic dose rate of 100 μGy/h, significant effects are unlikely to be observed at the population level (United Nations Scientific Committee on the Effects of Atomic Radiation 2008). In addition, the International Committee of Radiation Protection (ICRP) suggested the concept of reference

Table 1. Individual information of *A. speciosus* caught in control (Hirosaki) and contaminated areas (Tanashio, Ide, Omaru).

Individual ID	Location	Sex	Body weight (g)	Expected age by lens (d)*
447	Tanashio	M	27.0	229.94
448	Tanashio	F	21.4	217.17
449	Tanashio	F	22.6	229.94
450	Tanashio	F	21.5	265.24
451	Tanashio	M	28.6	236.60
452	Tanashio	M	31.3	154.15
453	Tanashio	M	35.0	N.D.**
445	Ide	F	25.1	177.86
446	Ide	F	33.7	384.52
454	Ide	M	22.3	182.97
456	Ide	M	29.2	217.17
457	Ide	F	22.6	211.06
458	Ide	F	17.8	32.96
459	Ide	M	22.0	126.21
460	Ide	F	29.7	288.98
461	Ide	M	30.5	211.06
466	Ide	M	35.2	236.61
467	Ide	F	24.0	122.66
468	Ide	M	43.2	808.08
477	Ide	F	31.5	257.77
478	Ide	F	35.5	352.94
462	Omaru	M	31.8	265.24
469	Omaru	F	27.4	172.81
470	Omaru	M	37.6	643.00
471	Omaru	M	31.6	229.94
472	Omaru	F	33.8	831.50
486	Hirosaki	F	21.23	39.13
487	Hirosaki	F	20.30	250.52
490	Hirosaki	M	15.17	28.58
491	Hirosaki	F	47.68	456.40
493	Hirosaki	M	23.40	373.69
495	Hirosaki	F	15.30	27.77

*Age in *A. speciosus* was estimated based on the dry weight of optic lens (Takada et al. 2006); **Lens of 453 was damaged before estimating age.

animals for radiation environmental protection studies and reported that the derived consideration reference level (DCRL) in small terrestrial animals is 0.1–1.0 mGy/d (ICRP 2008). To assess in-depth biological consequences of long-term low dose-rate radiation on *A. speciosus* after the FDNPP accident, we investigated DNA damage and pro-inflammatory markers in bone marrow cells of *A. specious*, evaluated the number of HPC colonies and assessed RIGI induction in HPCs.

We found no significant DNA damage increase in bone marrow cells of *A. speciosus* captured from contaminated areas (Figure 2) but a significant positive correlation between DNA damage and the pro-inflammatory response was seen (Figure 3). We found no significant DNA damage increase in bone marrow cells of *A. speciosus* captured from contaminated areas (max 0.5 mGy/d) as compared to those caught in the control area (Figure 2). This was consistently seen in a previous study by Kawagoshi et al. (2017) where no *A. speciosus* splenic chromosomal aberrations were observed in contaminated areas below 0.7 mGy/d (30.7 ± 7.4 μGv/h). In more severely contaminated areas ($\sim$1.9 mGy/d; 80.0 ± 13.6 μGv/h), splenic chromosomal aberrations were significantly higher (Kubota et al. 2015; Kawagoshi et al. 2017), which could suggest a higher bone marrow DNA damage seen in *A. speciosus* from those areas. Interestingly, a significant positive correlation between DNA damage and the pro-inflammatory response was seen (Figure 3). As no air dose-rate dependent relationship was seen in both DNA damage and pro-inflammatory markers, inflammation seems to play a bigger role in increased DNA damage in bone marrow cells. However, pro-inflammatory responses may be affected by fur mite infestation in several *A. speciosus* captured from all four sites (Johnston et al. 2009).

We then evaluated the colony-forming potential of *A. speciosus* bone marrow cells by quantifying the number of HPC colonies using the CFU assay with Methocult medium (Figure 4). Unexpectedly, we observed a dose-dependent decrease in colony-forming cells by ex vivo irradiation and estimated the HPC colonies' (CFU-GM, BFU-E, CFU-Meg) sensitivity to acute X-ray irradiation (Figure 4(C)). In our previous study, we reported that the D_0 values of CFU-GM colonies from 1- and 8-week-old C3H/He mouse were 4.5 and 3.9 Gy respectively, and that of BFU-E from 1- and 8-week-old were 5.0 and 4.6 Gy, respectively (Ariyoshi et al. 2014). Compared with these D_0 values, CFU-GM and BFU-E of *A. speciosus* showed greater radiation sensitivity (D_0 values of CFU-GM and BFU-E colonies were 2.1 and 0.5 Gy respectively). The D_0 value of femoral CFU-Meg colonies has been reported to be 1.48 Gy in B6D2F1 mouse strain (Scheding et al. 1996), suggesting CFU-Meg of *A. speciosus* exhibits a greater radiation sensitivity (D_0 value of CFU-Meg colonies was 0.9 Gy) than B6D2F1 mice. In addition, embryonic fibroblasts of *A. speciosus* are also more sensitive to radiation than laboratory mouse (ICR) embryonic fibroblasts (Ariyoshi, Miura, Kasai, Akifumi, Fujishima, and Yoshida 2018). Furthermore, we observed a significant decrease in the number of colony-forming cells (CFU-GM, BFU-E, and CFU-Meg), which exhibited an inverse correlation with air dose-rate (Figure 4(D); Figure S3). Indeed, several experimental studies have suggested that HSCs and HPCs exhibit higher radiation sensitivity after low dose-rate irradiation (Gallini et al. 1988; Girinski et al. 1990; Nothdurft et al. 1995; Yagunov et al. 1998). However, our results showed that *A. speciosus* HPC colonies only decreased when exposed to at least 208.8 ± 31.2 μGy/d radiation (Figure 4(D)). Interestingly, HPC colony frequency exhibited no correlation with age, indicating that the accumulating radiation dose might not affect the number of HPCs (Figure S2). These results seem to imply that the number of HPCs of *A. speciosus* decreased *in utero* or soon after birth in contaminated areas. Garnier-Laplace et al. compared the chronic radiotoxicity data of non-human species in laboratory conditions and in the Chernobyl Exclusion Zone and concluded that organisms in their natural environment were more, about eight times, sensitive to radiation (Garnier-Laplace et al. 2013). Our data support their results.

Finally, we analyzed delayed DNA damage in CFU-GM colonies to investigate RIGI in *A. speciosus* captured in contaminated areas. RIGI can lead accumulation of gene mutations and chromosomal rearrangements, and has been considered to play a role in radiation-induced carcinogenesis (Huang et al. 2003; Suzuki et al. 2003). Although the mechanisms underlying RIGI remain unclear, accumulating

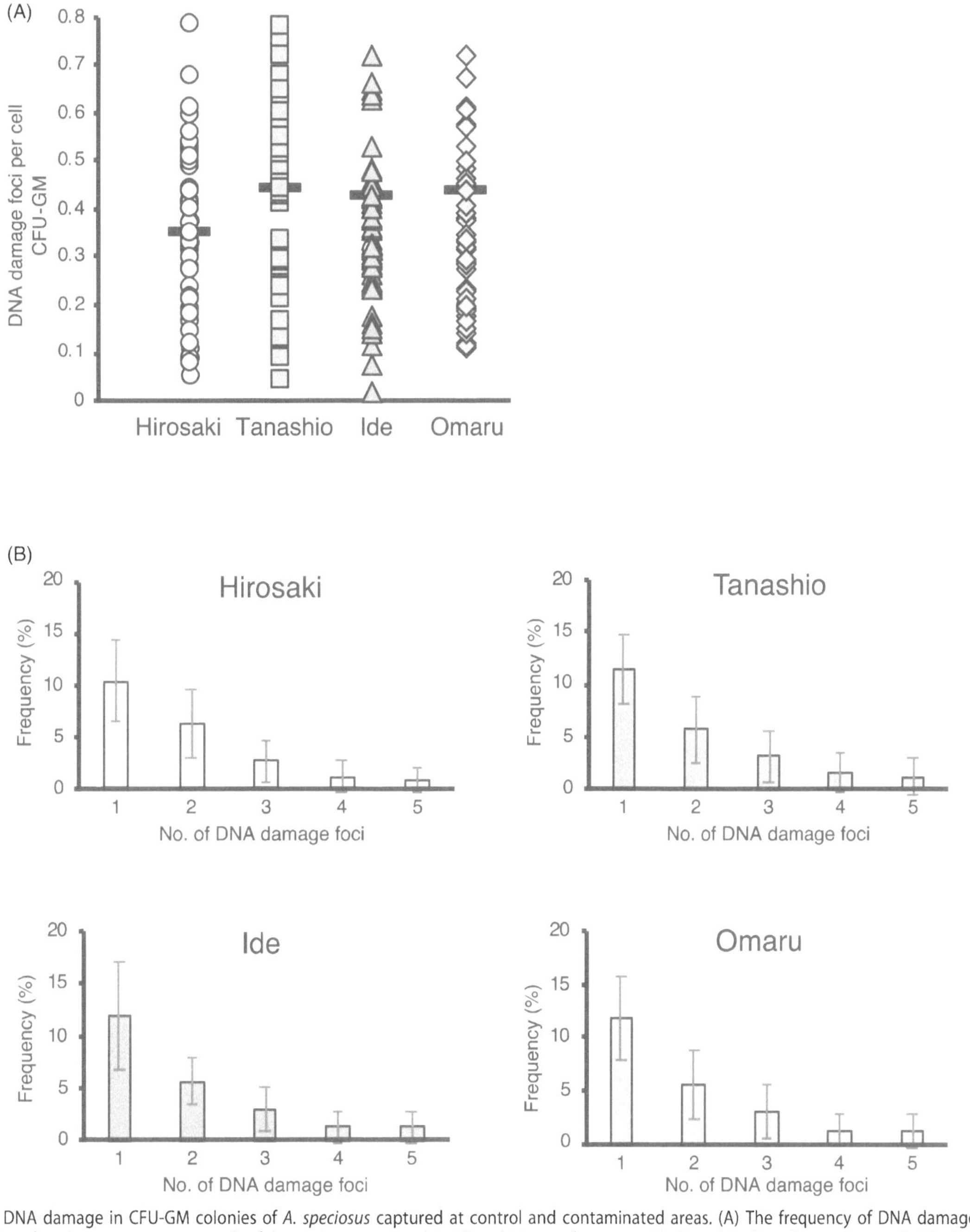

Figure 5. Delayed DNA damage in CFU-GM colonies of *A. speciosus* captured at control and contaminated areas. (A) The frequency of DNA damage foci in CFU-GM colonies (50 colonies of each area were measured) of *A. speciosus* captured in Hirosaki (control area, $n = 5$), Tanashio (low dose-rate, $n = 5$), Ide (intermediate dose-rate, $n = 5$) and Omaru (high dose-rate, $n = 5$). All data are expressed as Mean ± SD. No significant differences were seen between groups after Welch's *t*-test ($p > .01$). (B) Proportion of CFU-GM colonies with different number of DNA damage foci of *A. speciosus* captured at Hirosaki, Tanashio, Ide, and Omaru.

evidence suggests the induction of RIGI in HSCs and HPCs in both human and mouse models (Kadhim et al. 1992; Kadhim et al. 1994; Kadhim et al. 1995). Wang et al. reported that irradiated mice exhibited persistently increasing ROS in HSCs possibly caused by the up-regulation of NOX4 (Wang et al. 2010). Furthermore, several studies have shown that CBA mice are susceptible and C57BL/6 mice are resistant to RIGI, suggesting that the genomic background affects RIGI susceptibility (Watson et al. 1997); (Gowans et al. 2005). To examine the induction of RIGI in HPCs of *A. speciosus*, we evaluated delayed DNA damage in CFU-GM colonies after acute X-ray irradiation in bone marrow cells from *A. speciosus* caught in the control area (1000 mGy, 0.5 Gy/min). Although we could not observe the induction of RIGI at up to 500 mGy, we observed a significant increase in DNA damage in CFU-GM when irradiated at 1000 mGy (Figure S4). To the best of our knowledge, this is the first report on RIGI in HPCs from wild animals. We then analyzed RIGI in CFU-GM of *A. speciosus* caught in contaminated areas without additional X-ray irradiation and observed no delayed DNA damage (Figure 5). In a C3H/He mouse model, we previously reported that age-at-exposure is one of the factors affecting the induction of RIGI in HSCs, as seen from the more radioresistant adult bone marrow HSCs (Ariyoshi, Miura, Kasai, Akifumi, Fujishima, Shinagawa, et al. 2018). However, our data of HPC colonies in the present study suggests that the number of HPCs is affected by air dose-rates, not by age (Figure 4; Figures S2 and S3).

In addition, mouse HSCs and HPCs are sensitive in newborns (Ariyoshi et al. 2014), suggesting that genomically unstable HSCs, in the pro-inflammatory state, can be eliminated in newborns after exposure (Ariyoshi, Miura, Kasai, Akifumi, Fujishima, Shinagawa, et al. 2018). Consistent with our previous study on RIGI (Ariyoshi, Miura, Kasai, Akifumi, Fujishima, Shinagawa, et al. 2018), we did not observe significantly higher iNOS positive bone marrow cells in contaminated areas, suggesting pro-inflammatory response by environmental radiation was negligible.

In summary, these results suggest that low dose-rate radiation, of over 200 μGy/d, in the contaminated area reduced the number of HPCs, at least in CFU-GM, BFU-E, and CFU-Meg. Thus, genomically unstable HPCs might be eliminated or radiation damages were too low to induce RIGI in *A. speciosus* inhabiting contaminated areas with radiation levels below 500 μGy/d.

Author contributions

K.A. designed and directed the project; T.M., K.K., Y.F., V.S.T.G., A.N., A.T., Y.S., H.S., H.Y., and M.A.Y. performed capturing and preparation of samples; K.A. and Y.F. processed the experimental data; C.S. and C.M. provided critical discussions; K.A., V.S.T.G. wrote the paper with input from all authors.

Acknowledgments

The authors express their deep appreciation of Namie Town's local government in Fukushima Prefecture, Japan. The authors thank the laboratory staff for technical and secretarial assistance. The authors would also like to thank Editage (www.editage.jp) for English language editing.

Disclosure statement

No potential conflict of interest was reported by the author(s).

Funding

This work was supported in part by the Japan Society for the Promotion of Science Grant-in-Aid for Scientific Research (C) [19K12354].

References

Ariyoshi K, Miura T, Kasai K, Akifumi N, Fujishima Y, Shinagawa M, Kadono K, Nishimura M, Kakinuma S, Yoshida MA. 2018. Age dependence of radiation-induced genomic instability in mouse hematopoietic stem cells. Radiat Res. 190(6):623–633.

Ariyoshi K, Miura T, Kasai K, Akifumi N, Fujishima Y, Yoshida MA. 2018. Radiation-induced bystander effect in large japanese field mouse (*Apodemus speciosus*) embryonic cells. Radiat Environ Biophys. 57(3):223–231.

Ariyoshi K, Takabatake T, Shinagawa M, Kadono K, Daino K, Imaoka T, Kakinuma S, Nishimura M, Shimada Y. 2014. Age dependence of hematopoietic progenitor survival and chemokine family gene induction after gamma irradiation in bone marrow tissue in C3H/He mice. Radiat Res. 181(3):301–313.

Bonisoli-Alquati A, Koyama K, Tedeschi DJ, Kitamura W, Sukuzi H, Ostermiller S, Arai E, Møller AP, Mousseau TA. 2015. Abundance and genetic damage of barn swallows from Fukushima. Sci Rep. 5:9432.

Braga-Tanaka I, Tanaka S, Kohda A, Takai D, Nakamura S, Ono T, Tanaka K, Komura J, Ichiro 2018. Experimental studies on the biological effects of chronic low dose-rate radiation exposure in mice: overview of the studies at the Institute for Environmental Sciences. Int J Radiat Biol. 94(5):423–433.

Gallini R, Hendry JH, Molineux G, Testa NG. 1988. The effect of low dose rate on recovery of hemopoietic and stromal progenitor cells in gamma-irradiated mouse bone marrow. Radiat Res. 115(3):481–487.

Garnier-Laplace J, Geras'kin S, Della-Vedova C, Beaugelin-Seiller K, Hinton TG, Real A, Oudalova A. 2013. Are radiosensitivity data derived from natural field conditions consistent with data from controlled exposures? A case study of Chernobyl wildlife chronically exposed to low dose rates. J Environ Radioact. 121:12–21.

Giraudeau M, Bonzom JM, Ducatez S, Beaugelin-Seiller K, Deviche P, Lengagne T, Cavalie I, Camilleri V, Adam-Guillermin C, McGraw KJ. 2018. Carotenoid distribution in wild Japanese tree frogs (*Hyla japonica*) exposed to ionizing radiation in Fukushima. Sci Rep. 8(1): 7438.

Girinski T, Socie G, Cosset JM, Dutreix J, Chassagne D. 1990. Similar effects on murine haemopoietic compartment of low dose rate single dose and high dose rate fractionated total body irradiation. Preliminary results after a unique dose of 750 cGy. Br J Cancer. 61(6):797–800.

Gowans ID, Lorimore SA, McIlrath JM, Wright EG. 2005. Genotype-dependent induction of transmissible chromosomal instability by γ-radiation and the benzene metabolite hydroquinone. Cancer Res. 65(9):3527–3530.

Hayama SI, Tsuchiya M, Ochiai K, Nakiri S, Nakanishi S, Ishii N, Kato T, Tanaka A, Konno F, Kawamoto Y, Omi T. 2017. Small head size and delayed body weight growth in wild Japanese monkey fetuses after the Fukushima Daiichi nuclear disaster. Sci Rep. 7(1):3528.

Heylmann D, Rödel F, Kindler T, Kaina B. 2014. Radiation sensitivity of human and murine peripheral blood lymphocytes, stem and progenitor cells. Biochim Biophys Acta - Rev Cancer. 1846(1):121–129.

Hiyama A, Nohara C, Kinjo S, Taira W, Gima S, Tanahara A, Otaki JM. 2012. The biological impacts of the Fukushima nuclear accident on the pale grass blue butterfly. Sci Rep. 2:570.

Hiyama A, Taira W, Nohara C, Iwasaki M, Kinjo S, Iwata M, Otaki JM. 2015. Spatiotemporal abnormality dynamics of the pale grass blue butterfly: three years of monitoring (2011–2013) after the Fukushima nuclear accident. BMC Evol Biol. 15:15.

Huang L, Snyder AR, Morgan WF. 2003. Radiation-induced genomic instability and its implications for radiation carcinogenesis. Oncogene. 22(37):5848–5854.

ICRP. 2008. ICRP Publication 108. Environmental Protection: the concept and use of reference animals and plants. Annals of the ICRP. 38(4–6), 1–242.

Itoh M, Kajihara R, Kato Y, Takano-Shimizu T, Inoue Y. 2018. Frequencies of chromosomal inversions in *Drosophila melanogaster* in Fukushima after the nuclear power plant accident. PLoS One. 13(2):e0192096.

Johnston NA, Trammell RA, Ball-Kell S, Verhulst S, Toth LA. 2009. Assessment of immune activation in mice before and after eradication of mite infestation. J Am Assoc Lab Anim Sci. 48(4):371–377.

Kadhim MA, Lorimore SA, Goodhead DT, Wright EG, Hepburn MD, Buckle VJ. 1994. α-particle-induced chromosomal instability in human bone marrow cells. Lancet. 344(8928):987–988.

Kadhim MA, Lorimore SA, Townsend KMS, Goodhead DT, Buckle VJ, Wright EG. 1995. Radiation-induced genomic instability: delayed cytogenetic aberrations and apoptosis in primary human bone marrow cells. Int J Radiat Biol. 67(3):287–293.

Kadhim MA, Macdonald DA, Goodhead DT, Lorimore SA, Marsden SJ, Wright EG. 1992. Transmission of chromosomal instability after plutonium α-particle irradiation. Nature. 355(6362):738–740.

Kawagoshi T, Shiomi N, Takahashi H, Watanabe Y, Fuma S, Doi K, Kawaguchi I, Aoki M, Kubota M, Furuhata Y, et al. 2017. Chromosomal aberrations in large Japanese field mice (*Apodemus speciosus*) Captured near Fukushima Dai-ichi nuclear power plant. Environ Sci Technol. 51(8):4632–4641.

Kubota Y, Tsuji H, Kawagoshi T, Shiomi N, Takahashi H, Watanabe Y, Fuma S, Doi K, Kawaguchi I, Aoki M, et al. 2015. Chromosomal aberrations in wild mice captured in areas differentially contaminated by the Fukushima Dai-Ichi nuclear power plant accident. Environ Sci Technol. 49(16):10074–10083.

Morimoto M, Kato A, Kobayashi J, Okuda K, Kuwahara Y, Kino Y, Abe Y, Sekine T, Fukuda T, Isogai E, et al. 2017. Gene expression analyses of the small intestine of pigs in the ex-evacuation zone of the Fukushima Daiichi nuclear power plant. BMC Vet Res. 13(1): 337.

Mothersill C, Rusin A, Seymour C. 2017. Low doses and non-targeted effects in environmental radiation protection; where are we now and where should we go? Environ Res. 159:484–490.

Nakamura AJ, Suzuki M, Redon CE Kuwahara Y, Yamashiro H, Abe Y, Takahashi S, Fukuda T, Isogai E, Bonner W, Fukumoto M. 2017. The causal relationship between DNA damage induction in bovine lymphocytes and the Fukushima nuclear power plant accident. Radiat Res. 187(5):630–636.

Nothdurft W, Fliedner TM, Fritz TE, Seed TM. 1995. Cell and cell system responses: response of hemopoiesis in dogs to continuous low dose rate total body irradiation. Stem Cells. 13(S1): 261–267.

Ochiai K, Hayama SI, Nakiri S, Nakanishi S, Ishii N, Uno T, Kato T, Konno F, Kawamoto Y, Tsuchida S, et al. 2014. Low blood cell counts in wild Japanese monkeys after the Fukushima Daiichi nuclear disaster. Sci Rep. 4:5793.

Okano T, Ishiniwa H, Onuma M, Shindo J, Yokohata Y, Tamaoki M. 2016. Effects of environmental radiation on testes and spermatogenesis in wild large Japanese field mice (*Apodemus speciosus*) from Fukushima. Sci Rep. 6:23601.

Scheding S, Media JE, KuKuruga MA, Nakeff A. 1996. In situ radiation sensitivity of recombinant human granulocyte colony-stimulating factor-recruited murine circulating blood and bone marrow progenitors (colony-forming unit [CFU]-granulocyte-macrophage and CFU-megakaryocyte): evidence for possible biologi. Blood. 88(2):472–478.

Suzuki K, Ojima M, Kodama S, Watanabe M. 2003. Radiation-induced DNA damage and delayed induced genomic instability. Oncogene. 22(45):6988–6993.

Suzuki Y. 2015. Influences of radiation on carp from farm ponds in Fukushima. J Radiat Res. 56(1):i19–i23.

Taira W, Toki M, Kakinohana K, Sakauchi K, Otaki JM. 2019. Developmental and hemocytological effects of ingesting Fukushima's radiocesium on the cabbage white butterfly *Pieris rapae*. Sci Rep. 9(1):2625.

Takada Y, Sakai E, Uematsu Y, Tateishi T. 2006. Morphological variation of large Japanese field mice, *Apodemus speciosus* on the Izu and Oki Islands. Mammal Study. 31(1):29–40.

Takino S, Yamashiro H, Sugano Y, Fujishima Y, Nakata A, Kasai K, Hayashi G, Urushihara Y, Suzuki M, Shinoda H, et al. 2017. Analysis of the effect of chronic and low-dose radiation exposure on spermatogenic cells of male large Japanese field mice (*Apodemus speciosus*) after the Fukushima Daiichi nuclear power plant accident. Radiat Res. 187(2):161–168.

Tanaka K, Kohda A, Satoh K, Toyokawa T, Ichinohe K, Ohtaki M, Oghiso Y. 2009. Dose-rate effectiveness for unstable-type chromosome aberrations detected in mice after continuous irradiation with low-dose-rate γ rays. Radiat Res. 171(3):290–301.

Tanaka K, Satoh K, Kohda A. 2014. Dose and dose-rate response of lymphocyte chromosome aberrations in mice chronically irradiated within a low-dose-rate range after age adjustment. Radiat Prot Dosimetry. 159(1–4):38–45.

United Nations Scientific Committee on the Effects of Atomic Radiation (UNSCEAR). 2008. Sources and effects of ionizing radiation. Vol. II: Effects of Ionizing Radiation. Annex E: Effects of Ionizing Radiation on Non-human Biota.

Urushihara Y, Kawasumi K, Endo S, Tanaka K, Hirakawa Y, Hayashi G, Sekine T, Kino Y, Kuwahara Y, Suzuki M, et al. 2016. Analysis of plasma protein concentrations and enzyme activities in cattle within the ex-evacuation zone of the Fukushima Daiichi nuclear plant accident. PLoS One. 11(5):e0155069.

Urushihara Y, Suzuki T, Shimizu Y, Ohtaki M, Kuwahara Y, Suzuki M, Uno T, Fujita S, Saito A, Yamashiro H, et al. 2018. Haematological analysis of Japanese macaques (*Macaca fuscata*) in the area affected by the Fukushima Daiichi nuclear power plant accident. Sci Rep. 8(1):16748.

Wang Y, Liu L, Pazhanisamy SK, Li H, Meng A, Zhou D. 2010. Total body irradiation causes residual bone marrow injury by induction of persistent oxidative stress in murine hematopoietic stem cells. Free Radic Biol Med. 48(2):348.

Watson GE, Lorimore SA, Clutton SM, Kadhim MA, Wright EG. 1997. Genetic factors influencing α-particle-induced chromosomal instability. Int J Radiat Biol. 71(5):497–503.

Yagunov AS, Tokalov SV, Chukhlovin AB, Afanassiev BV. 1998. Animal studies of residual hematopoietic and immune system injury from low dose/low dose rate radiation and heavy metals. Bethesda (MD): Armed Forces Radiology Research Institute.

Yamashiro H, Abe Y, Fukuda T, Kino Y, Kawaguchi I, Kuwahara Y, Fukumoto M, Takahashi S, Suzuki M, Kobayashi J, et al. 2013. Effects of radioactive caesium on bull testes after the Fukushima nuclear plant accident. Sci Rep. 3:2850.

Assessment of chromosome aberrations in large Japanese field mice (*Apodemus speciosus*) in Namie Town, Fukushima

Yohei Fujishima, Akifumi Nakata, Risa Ujiie, Kosuke Kasai, Kentaro Ariyoshi, Valerie Swee Ting Goh, Kojun Suzuki, Hirofumi Tazoe, Masatoshi Yamada, Mitsuaki A. Yoshida, and Tomisato Miura

ABSTRACT

Purpose: After the Fukushima Daiichi Nuclear Power Plant (FDNPP) accident in Japan on March 11 2011, the surroundings became contaminated with radionuclides. To understand the possible biological effects after chronic low dose-rate radiation in contaminated areas of Fukushima, we assessed the effects in large Japanese field mice (*Apodemus speciosus*) by means of chromosome aberration analysis.

Materials and methods: We collected *A. speciosus* in five sites around Namie Town, Fukushima (contaminated areas) and in two sites in Hirosaki City, Aomori (control areas, 350 km north of FDNPP) from autumn 2011 to 2013. The number of mice captured and ambient dose-rates were as follows: high ($n = 11$, 10.1–30.0 μGy h^{-1}), moderate ($n = 10$, 5.7–15.6 μGy h^{-1}), low ($n = 12$, 0.23–1.14 μGy h^{-1}) and control ($n = 20$, 0.04–0.07 μGy h^{-1}). After spleen extraction from rodents, spleen cell culture was performed to obtain metaphase spreads. Chromosome aberrations were assessed on Giemsa-stained metaphase spreads.

Results: Although the mice in the contaminated areas were chronically exposed, there was no radiation-specific chromosome aberrations observed, such as dicentric chromosomes and rings. Some structural aberrations such as gaps and breaks were observed, and these frequencies decreased annually in mice from Namie Town.

Conclusion: These findings suggest that chromosome aberration analysis is useful to evaluate and monitor radiation effects in wild animals.

Introduction

After the accident in the TEPCO Fukushima Daiichi Nuclear Power Plant (FDNPP), the surrounding areas were contaminated by large amounts of released radionuclides, which could affect wild animals and plants living around the FDNPP. Although the ambient dose-rates are gradually decreasing, prolonged effects of chronic low-dose exposure on animals are expected to be seen in the coming years. Therefore, by assessing biological effects from radioactive substances, we are able to understand any possible health effects caused by radiation. To date, several species of animals have been evaluated (livestock bulls (Yamashiro et al. 2013), wild boars (Steinhauser and Saey 2016; Tomoko and Nakanishi 2016), wild Japanese monkeys (Hayama et al. 2013; Ochiai et al. 2014), wild rodents (Kubota et al. 2015) and freshwater fish (Mizuno and Kubo 2013; Fuma et al. 2016; Wada et al. 2016)) around the evacuation zone of the FDNPP. Furthermore, some studies suggest some negative biological effects seen in wild animals were likely caused by the release of radionuclides from the FDNPP accident, where an increased rate of abnormal wing development in butterflies (Hiyama et al. 2012, 2013, 2015) and the decline of bird populations (Møller et al. 2012) had been observed.

It is well-known that chromosome aberrations can be seen after radiation exposure (Edwards et al. 1979), and it is highly likely FDNPP-related radionuclide substances will cause some chromosome damage in wild animals. In published studies after the Chernobyl accident, most studies reported on accumulated radioactive substances in wildlife while comparatively fewer reports of chromosomal aberration analysis on wild animals were made. Bank voles (*Myodes glareolus*, formerly *Clethrionomys glareolus*) living in areas with high contamination (90 and 1524 kBq m^{-2} for ^{137}Cs) were reported to have a high frequency of chromosome aberrations (Goncharova and Ryabokon 1995). However, natural factors such as location, population size and viral infection had been shown to affect chromosome instability (Bol'shakov et al. 2003). Careful analysis should be made to ensure the increase in chromosome aberrations seen is radiation-specific and not caused by the species or inhabiting environment. As described above, reports on chromosome aberrations and genetic effects on the next generation in wildlife are limited. The influence of chronic low dose-rate exposure from the FDNPP nuclear accident toward wild animals and humans still remains unclear.

The International Commission on Radiological Protection (ICRP) listed nine species of animals and three species of plants as references for radiation environmental protection studies, with rodents being a representative model for small terrestrial animals (ICRP 2008). Specifically, the large Japanese field mouse (*Apodemus speciosus*) is endemic to Japan and widely distributed throughout the country. It inhabits a wide range of environments including forests, plantations, riversides, paddy and cultivated fields. This animal had been used in environmental toxicological studies throughout Japan (Kyoya et al. 2008) and radiation exposure studies caused by the FDNPP accident (Kubota et al. 2015; Kawagoshi et al. 2017). Hence in this study, we assessed chromosome aberrations in *A. speciosus* spleen cells to evaluate possible biological effects after chronic low dose-rate exposure in Fukushima.

Figure 1. A photograph of large Japanese field mice (*A. speciosus*) caught in Namie Town, Fukushima.

Materials and methods

Collection of specimens

The target wild rodents (large Japanese field mice—*A. speciosus* (Figure 1)) were captured using Sherman traps (H. B. Sherman Traps, Inc., Tallahassee, FL, USA) at five sites (Akougi, Ide, Murohara, Tatsuno, and Tanashio areas) in Namie Town, Fukushima, from 2011 to 2013 twice annually (Figure 2). Control rodents were also captured at Owasawa and Sakamoto areas in Hirosaki City, Aomori (Figure 2). The number of mice captured and ambient dose-rates were as follows: high (Akougi and Ide areas, $n = 11$, 10.1–30.0

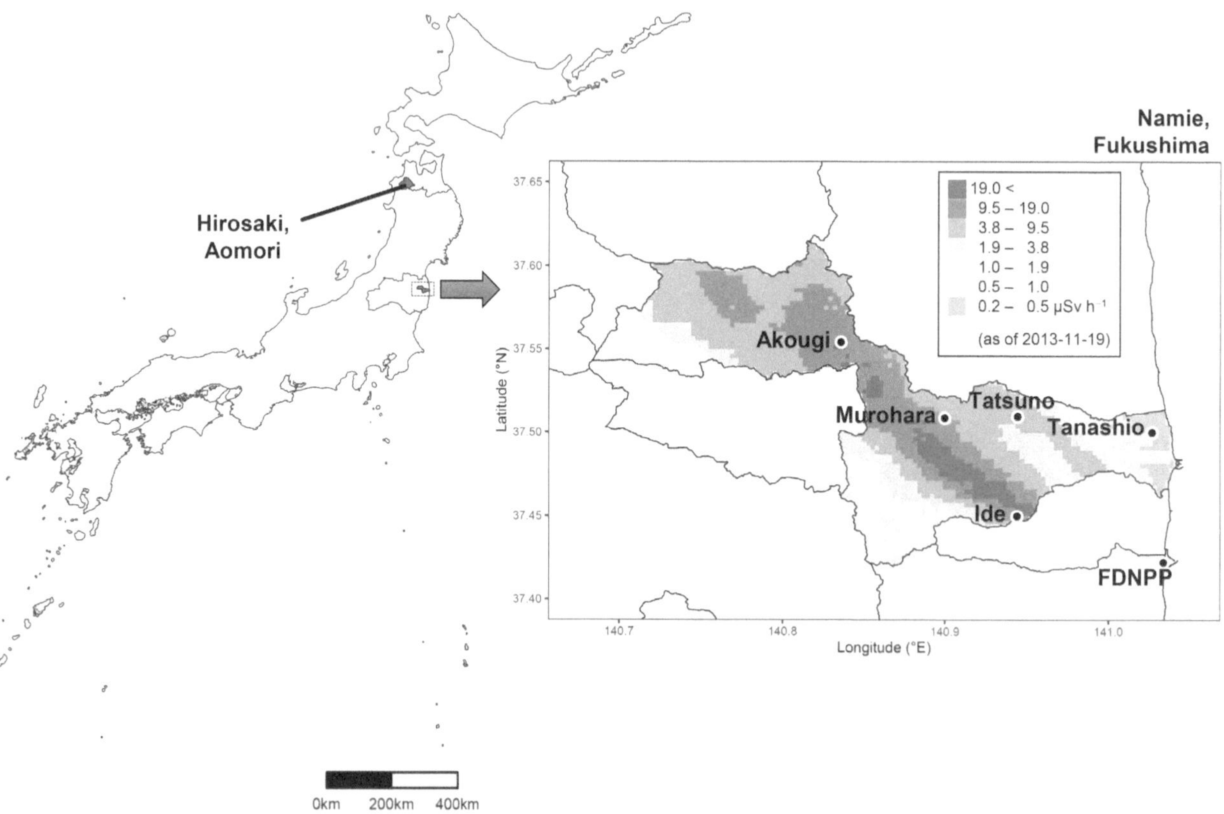

Figure 2. Sampling areas of wild rodent capture in Namie Town, Fukushima and Hirosaki City, Aomori. The magnified map shows Namie Town and Akougi, Ide, Murohara, Tatsuno and Tanashio areas, as well as the TEPCO Fukushima Daiichi Nuclear Power Plant (FDNPP). The maps were modified shapefiles from the National Land Numerical Information download service (data was retrieved from http://nlftp.mlit.go.jp/ksj/index.html on February 5th, 2016). The heat map shows ambient dose-rates on November 19th, 2013 provided by the Nuclear Regulation Authority 10th airborne monitoring survey (data were retrieved from http://radioactivity.nsr.go.jp/ja/list/362/list-1.html on February 5th, 2016).

μGy h^{-1}), moderate (Murohara and Tatsuno areas, $n = 10$, 5.7–15.6 μGy h^{-1}), low (Tanashio area, $n = 12$, 0.23–1.14 μGy h^{-1}) and control (Owasawa and Sakamoto areas, $n = 20$, 0.04–0.07 μGy h^{-1}).

The age of captured wild rodents was estimated with tooth wear status (Hikida and Murakami 1980) and eye lens weight (Takada 1982). The collection of specimens in sampling areas was authorized by the Fukushima Prefectural Office and Aomori Prefectural Office. All rodent experiments were conducted in accordance with the laboratory animal care guidelines of the Ethics Committee for Care and Use of Laboratory Animals for Research of Hirosaki University, Japan (Approval No. R12001).

Spleen cell culture and harvest

Spleen cell culture was performed to obtain metaphase spreads for chromosome aberration analysis. After spleen removal, the cells were isolated by gentle homogenization with RPMI 1640 medium, centrifugation (300 × g, 5 min, 4 °C) and resuspension to a concentration of 1.0×10^7 cells with RPMI 1640 medium (Thermo Fisher Scientific, Waltham, MA, USA) supplemented with 20% heat-inactivated fetal bovine serum (Sigma-Aldrich, MO, USA), 60 mg/ml kanamycin sulfate (Thermo Fisher Scientific, Waltham, MA, USA) in a 15 ml centrifuge tube. For lymphocyte mitogenesis induction, lipopolysaccharide (10 μg/ml, Sigma-Aldrich, MO, USA), concanavalin A (3 μg/ml, Sigma-Aldrich, MO, USA) and phytohemagglutinin (25 μg/ml, Remel Europe, Dartford, UK) were added. To obtain first-cycle metaphases, KaryoMAX colcemid solution (Thermo Fisher Scientific, Waltham, MA, USA) at a final concentration of 0.02 μg/ml was added 22 h before harvesting. Spleen cell cultures were maintained in a humidified 37 °C incubator with 5% CO_2 for 46 h. Cultured cells were harvested with hypotonic treatment using 75 mM potassium chloride and fixed in cold fixative (3:1 methanol/glacial acetic acid). Fixed cell suspensions were stored at −30 °C.

Chromosome aberration analysis

The cell pellets were suspended in fresh cold fixative to appropriate concentrations depending on the size of the pellets. Air-dried slides were prepared by in a humidified chamber HANABI-PIV (ADStec Co., Chiba, Japan) on pre-cleaned glass slides (Matsunami Glass Industry Ltd., Osaka, Japan). Each chromosome slide was immersed in 4% Giemsa solution (Merck KGaA, Darmstadt, Germany) with Gurr buffer (pH 6.8, Thermo Fisher Scientific, Waltham, MA, USA) for 10 min, then washed with tap water and air-dried. Giemsa-stained metaphase spreads were captured with the Metafer 4 software (version 3.9.6, MetaSystems GmbH, Altlussheim, Germany) combined with a microscope (Axio Imager.Z2, Carl Zeiss Microscopy GmbH, Jena, Germany) embedded with a CCD camera (CoolCube 1, MetaSystems GmbH, Altlussheim, Germany). Chromosome aberration analysis was performed according to the IAEA recommendations (IAEA 2011). All observable aberrations such as dicentric chromosomes (Dic), rings and acentric fragments were scored on captured metaphase images. Other chromosome- or chromatid-type aberrations, such as chromatid breaks (chtb), chromatid gaps (chtg), iso-chromatid breaks (ichtb), iso-chromatid gaps (ichtg) and chromosome exchanges, were also recorded (Supplementary Table S1). In general, *A. speciosus* inhabiting in Eastern Japan, including Fukushima and Aomori, has 48 chromosomes (Saitoh and Obara 1986; Saitoh et al. 1989). Therefore, metaphases with less than 47 centromeres were omitted from analysis.

Measurement of ambient dose-rate

Ambient dose-rates were measured at the sampling sites using a NaI(Tl) scintillation survey meter (TCS-171B, Hitachi Aloka Medical, Ltd., Tokyo, Japan). The measurements were expressed as micro-grays per hour at 1 m above the ground with the time constant of the survey meter set to 10 s. Measurements were recorded after a minimum wait of 30 s for the readings to be stabilized.

Measurement of cesium activity concentration in soil samples

As for the soil samples, five surface soil samples of 5 cm depth were collected at each sampling point. The soil samples were completely dried at 120 °C for 20 h before measurement. Each sample was transferred into separate plastic containers (U-8 container, SANPLATEC Co., Ltd., Osaka, Japan). The activity concentration for the collected samples was determined by gamma-ray spectrometry using a hyper-pure germanium (HPGe) detector (ORTEC GEM-40190, SEIKO-EG&G Co., Ltd., Tokyo, Japan). ^{134}Cs and ^{137}Cs were detected using 604.6 and 661.6 keV gamma-ray energies respectively for 1,000–80,000 s to satisfy measurement errors from counting to below 5% of the corresponding activity concentration. Activity was decay corrected to the sampling date, and activity concentration was calculated as per kilogram of dry weight of the soil samples.

Statistical analysis

Correlation analysis was based on the Pearson's product-moment correlation coefficient, and Wilcoxon rank sum test was applied to assess differences between two groups. The results were considered statistically significant if *p*-values below .05 were obtained. All statistical analyses were performed using the R version 3.6.1 (R Core Team 2019).

Results

Ambient dose-rates and accumulation of radioactive cesium in soil

The characteristics of the specimen collection sites (contaminated areas: Akougi, Ide, Murohara, Tatsuno and Tanashio in Namie Town, Fukushima; control area: Owasawa and

Table 1. Locations and ambient dose-rates of sampling areas from autumn 2011 to 2013.

	Coordinates			Ambient dose-rates (μGy h^{-1})*				
			Distance from FDNPP (km)	2011	2012		2013	
Location	Longitude (°)	Latitude (°)		Autumn	Spring	Autumn	Spring	Autumn
Namie, Fukushima								
High dose rate-areas								
Akougi	37°33′10.4″N	140°50′06.7″E	22.7	29.1^{a}	28.9 (27.8–30.0)	24.2 (20.5–30.0)	15.1 (11.2–21.9)	15.1 (10.3–24.1)
Ide	37°28′06.8″N	140°55′50.1″E	8.4	–	24.4 (19.0–30.0)	23.7 (17.8–25.9)	15.8 (10.1–28.4)	–
Moderate dose-rate areas								
Murohara	37°30′30.6″N	140°53′57.1″E	15.2	–	15.2 (13.3–15.6)	13.3 (7.6–14.3)	9.8 (5.7–13.1)	9.3 (7.8–14.7)
Tatsuno	37°30′33.1″N	140°56′39.8″E	12.4	8.79^{a}	–	–	–	–
Low dose-rate area								
Tanashio	37°29′58.6″N	141°01′32.2″E	8.5	0.59^{a}	0.56 (0.35–0.73)	0.52 (0.30–0.56)	0.61 (0.23–1.14)	0.51 (0.35–0.61)
Hirosaki, Aomori								
Owasawa	40°33′11.9″N	140°28′13.1″E	350.9	0.05^{a}	0.05 (0.04–0.06)	–	–	–
Sakamoto	40°31′25.3″N	140°26′24.0″E	348.0	0.06^{a}	0.07 (0.06–0.07)	–	–	–

*Ambient dose-rates are expressed as median, with their ranges indicated within parentheses.
[a]Data was recorded at only one point for each sampling area.

Sakamoto in Hirosaki City, Aomori), ambient dose-rates and activity concentrations in soil were listed in Table 1 and Table 2 respectively. According to the Ministry of Economy, Trade and Industry of Japan, evacuees are still unable to return back to the sampling areas in Namie Town except for Tatsuno and Tanashio because the ambient dose-rates and resulting annual radiation doses are higher than the allowed levels. Although the air dose-rates in Fukushima are declining over time (Takino et al. 2017), our results still show high levels of contamination at the sampling areas. In contrast, the evacuation order has been lifted from Tanashio area, where the ambient dose-rates and $^{134}Cs + ^{137}Cs$ activity concentrations in the soil were relatively low, as confirmed in our study.

In soil samples, the main radionuclides detected by HPGe detectors were ^{134}Cs and ^{137}Cs. Naturally occurring radionuclides such as ^{40}K were also detected, but their signals were either below the detection limit or slightly above it, as the counting time of 1,000 s was enough to measure the activity concentration of ^{134}Cs and ^{137}Cs. The distribution of radioactive cesium was extremely heterogeneous in Akougi, Ide and Murohara areas, as indicated by their wide range of the measurements (Table 2).

Owasawa and Sakamoto areas in Hirosaki, Aomori were chosen as control areas as their distance from the FDNPP is approximately 350 km, hence FDNPP-related radionuclides can be considered negligible. The ambient dose-rates (Owasawa area: 0.04 – 0.06 μGy h^{-1}; Sakamoto area: 0.06 – 0.07 μGy h^{-1}) and activity concentrations of ^{134}Cs and ^{137}Cs in soil (Owasawa area: 9.3 – 17.7 Bq kg^{-1}; Sakamoto area: 4.3 – 35.8 Bq kg^{-1}) of the control areas were significantly lower than those at the sampling areas in Fukushima.

Kinetics of observed chromosome aberrations

Spleen cells obtained from wild rodents were cultured for chromosome aberration analysis (Supplementary Table S1). There were no radiation-specific chromosome aberrations observed, such as Dics and rings, in wild rodents collected from autumn 2011 to 2013 at both Namie Town, Fukushima and Hirosaki City, Aomori. On the other hand, some chromosome structural aberrations such as gaps and breaks were observed (Figure 3, Supplementary Table S1). The combined frequencies of chromosome aberrations gradually decreased over time ($\rho = -0.57$, $p < .01$, Figure 4) in populations from Fukushima. In the control areas of Hirosaki City, however, the frequency of chromosome aberrations was significantly higher ($p < .05$, Wilcoxon rank sum test) only in autumn (Figure 4).

Regardless of the ambient dose-rates in both contaminated and control areas, we observed a similar frequency of chromosome aberrations. We further investigated the relationship of chromosome aberration frequency with other factors such as age (Figure 5(A), estimated from tooth wear status; Figure 5(B), estimated from eye lens weight), body weight (Figure 5(C)) and sex (Figure 5(D)). There were no significant correlations or differences observed between chromosome aberrations and age, body weight or sex in wild rodents collected both in Namie Town and Hirosaki City.

Discussion

The possibility of radiological effects on human/non-human biota is a major and lasting concern in the public since the occurrence of the FDNPP accident in 2011. To investigate biological consequences of long-term chronic low dose-rate radiation after the accident, we assessed chromosome aberrations in spleen cells of *A. speciosus*, which are endemic to Japan. We observed no radiation-related chromosome aberrations, such as Dics or ring chromosomes. In addition, some structural aberrations were both observed in mice from Namie Town, Fukushima and Hirosaki City, Aomori. There was a gradual annual decrease of aberration frequencies in wild rodents from Fukushima when the data from all contaminated areas was combined.

Table 2. $^{134}Cs+^{137}Cs$ activity in the soil of sampling areas from autumn 2011 to 2013.

Location	Coordinates		Distance from FDNPP (km)	$^{134}Cs + ^{137}Cs$ activities in soil (Bq kg^{-1} dry)*				
				2011	2012		2013	
	Longitude (°)	Latitude (°)		Autumn	Spring	Autumn	Spring	Autumn
Namie, Fukushima								
High dose-rate areas								
Akougi	37°33′10.4″N	140°50′06.7″E	22.7	138,404 (71,499–150,111)	91,932 (64,451–330,931)	104,313 (68,056–208,342)	130,664 (85,860–687,010)	45,923 (32,566–92,084)
Ide	37°28′06.8″N	140°55′50.1″E	8.4	–	64,594 (52,059–142,705)	39,541 (38,507–54,766)	170,577 (5,932–432,601)	–
Moderate dose-rate areas								
Murohara	37°30′30.6″N	140°53′57.1″E	15.2	–	28,595 (7,325–81,795)	12,827 (9,169–252,539)	406,127 (99,264–792,419)	58,757 (10,285–68,417)
Tatsuno	37°30′33.1″N	140°56′39.8″E	12.4	11,213 (7,665–19,659)	–	–	–	–
Low dose-rate area								
Tanashio	37°29′58.6″N	141°01′32.2″E	8.5	3,926 (3,524–3,932)	3,382 (2,508–4,613)	2,541 (1,656–3,080)	16,863 (4,552–23,875)	13,386 (6,292–26,463)
Hirosaki, Aomori								
Owasawa	40°33′11.9″N	140°28′13.1″E	350.9	17.6[a] (16.5–17.7)	11.0[a] (9.3–15.8)	–	–	–
Sakamoto	40°31′25.3″N	140°26′24.0″E	348.0	4.5 (4.3–4.7)	30.2[a] (25.9–35.8)	–	–	–

*Activity concentrations are expressed as median, with their ranges indicated within parentheses.
[a]As ^{134}Cs activity concentrations were below the detection limit, only ^{137}Cs activity concentrations were shown.

It is well known that Dics are possibly observed in acute radiation exposure exceeding 100–200 mGy, which is generally regarded as the detection limit (IAEA 2011). In high natural radiation background areas such as China, India, Brazil and Iran (Hendry et al. 2009), the frequency of Dics increased depending on the individual cumulative dose when residing in such areas (Jiang et al. 2000; Zakeri et al. 2011). The report of the United Nations Scientific Committee on the Effects of Atomic Radiation (UNSCEAR) on the effects of ionizing radiation on non-human biota (United Nations Scientific Committee on the Effects of Atomic Radiation 2008) suggests a chronic dose-rate below the threshold of 100 μGy h^{-1} for animals and 400 μGy h^{-1} for plants would most likely not induce significant detrimental health effects at the population level. Furthermore, the derived consideration reference level (DCRL) in small terrestrial animals (ICRP 2008), which indicates a range of dose-rates for possible occurrences of deleterious effects, is 0.1−1.0 mGy day^{-1}. With regards to FDNPP accident, Kubota et al. reported a significant increase in Dic frequency in the small Japanese field mouse (*Apodemus argenteus*) captured at points where ambient dose-rates were 30.9 ± 5.4 μSv h^{-1} and 80.5 ± 7.7 μSv h^{-1} (estimated 2.7 ± 0.4 mGy day^{-1} and 1.1 ± 0.2 mGy day^{-1} respectively) (Kubota et al. 2015). However, no significant difference was observed in Dic frequency between *A. argenteus* captured at an ambient dose-rate of 0.5 ± 0.2 μSv h^{-1} (estimated 0.03 ± 0.01 μGy day^{-1}) and the control area. In laboratory-controlled experiments, Tanaka et al. focused on the effects of low dose-rate chronic exposure (for 400 days) on chromosomal abnormalities in C3H mice and reported that irradiation at 1 mGy day^{-1} and 20 mGy day^{-1} induced a significant increase in Dic frequency. Mice irradiated at 0.05 mGy day^{-1} were similar to that of non-irradiated control mice (Tanaka et al. 2009; 2013). From these reports, it is possible that the radiation dose levels in the contaminated capture sites did not cause the formation of radiation-specific chromosomal abnormalities such as Dic, despite the exposure to chronic low dose-rate radiation with a maximum of 29.1 μGy h^{-1} (0.698 mGy d^{-1}) recorded in Akougi area, Fukushima in our study.

In addition, we found that the frequency of non-radiation-specific structural abnormalities such as gaps and breaks decreased annually in *A. speciosus* spleen cells living in radiation-contaminated areas. Several reports (Gileva et al. 2006; Perucatti et al. 2006; Kyoya et al. 2008) have used chromosome aberrations such as gaps and breaks for environmental monitoring in several species of mammals. Similarly, our results show that the analysis of chromosomal aberrations less sensitive to radiation might also be useful for environmental monitoring. However, *A. speciosus* captured in control areas in Hirosaki City showed a significant increase in gaps and breaks only in autumn. The increase could be attributed to some type of pesticide, rodenticide, or fertilizer as the control areas are near to farms. Previous studies have also showed an increase by such agents (Salvadori et al. 1988; Garry et al. 1989; Amer et al. 1996; Giri, Giri, et al. 2002; Giri, Prasad, et al. 2002; Lyubimov and Garry 2010).

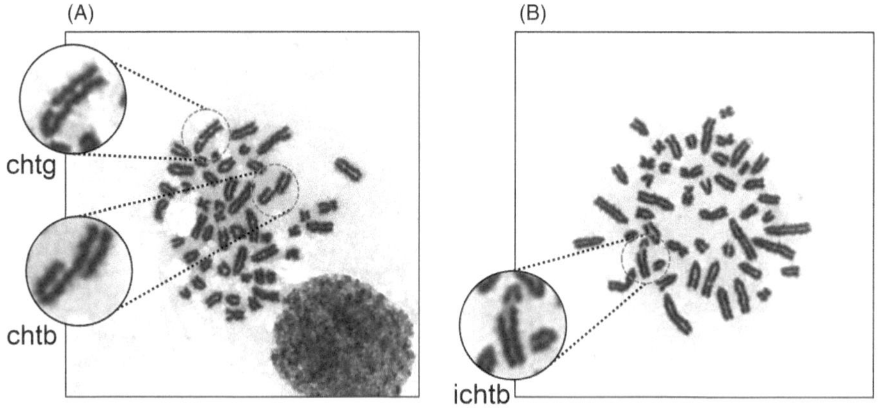

Figure 3. Representative images of *A. speciosus* chromosome spreads. (A) A metaphase spread with chromatid gap (chtg) and chromatid break (chtb). (B) A metaphase spread with iso-chromatid break (ichtb).

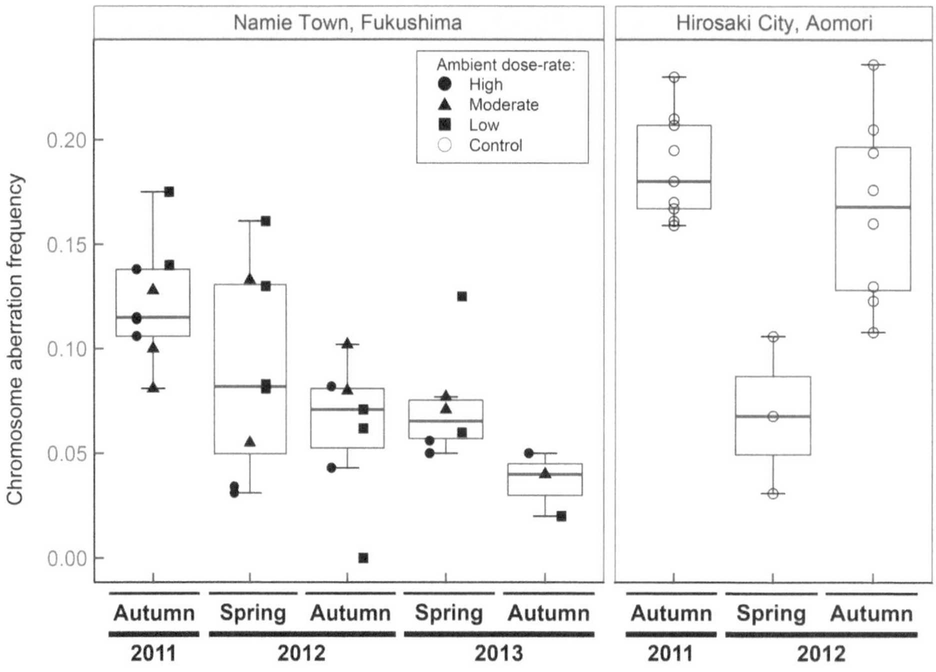

Figure 4. Transition of total chromosome structural aberration (ctg, ctb, ichtg, ichtb) frequency from autumn 2011 to 2013. Solid circles, triangles and squares represent wild rodents from contaminated areas with high, moderate, and low dose-rate radiation respectively. Open circles represent wild rodents from the control areas. In autumn 2011 and 2012, chromosome aberration frequency in Hirosaki City, Aomori was significantly higher than Namie Town, Fukushima ($p < .05$, Wilcoxon rank sum test).

Surprisingly, there was no correlation between frequency of chromosome structural aberrations and ambient dose-rates or activity concentrations in the soil. Radioactive substances in the soil were heterogeneously distributed as seen in Table 2. The home range of *A. speciosus* was estimated to be 30–40 m estimated from acorn hoarding (Soné and Kohno 1996, 1999), although acorns buried by *A. speciosus* were also found at approximately 10 cm below soil surface (Soné and Kohno 1999; Hirata et al. 2007). It is likely that the wild rodents were exposed to non-uniform radiation due to the type of food consumed and the depths in which they inhabit. Thus, ambient dose-rates and soil radioactivity concentrations may not provide the most accurate estimation of dose exposure to *A. speciosus* and hence no correlation was seen.

Overall, we collected spleen cells from *A. speciosus* inhabiting in both Namie Town, Fukushima (contaminated area) and in Hirosaki City, Aomori (control area). Although the wild rodents in contaminated areas were chronically exposed, there was no radiation-specific chromosomal aberrations observed, such as Dics and rings in mice collected. Some structural aberrations such as gaps and breaks were observed, and these frequencies were decreased annually in mice captured in Namie Town. These findings suggest that the chromosome analysis is useful to evaluate and to monitor the radiation effects on wild animals.

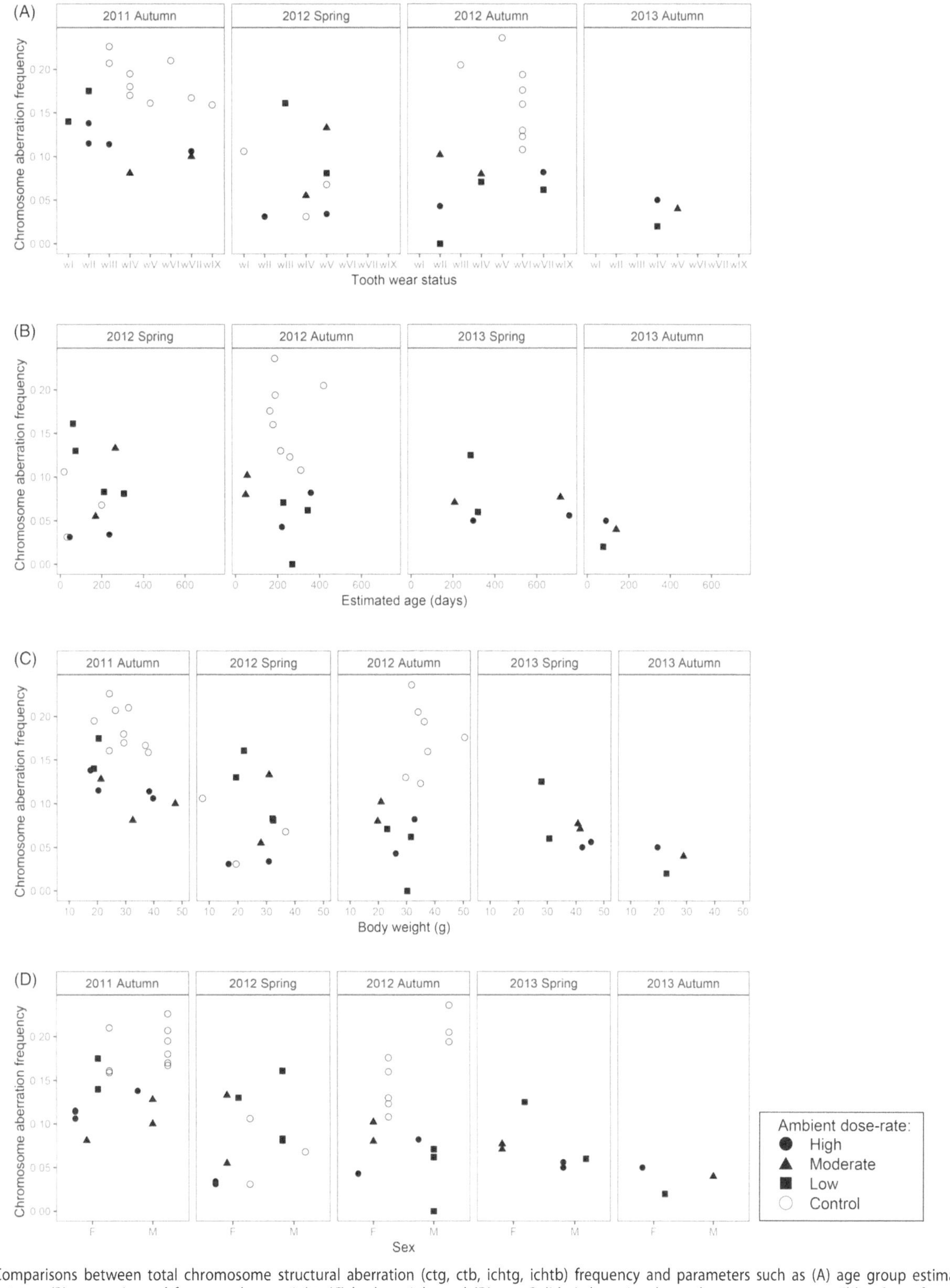

Figure 5. Comparisons between total chromosome structural aberration (ctg, ctb, ichtg, ichtb) frequency and parameters such as (A) age group estimated from tooth wear status, (B) age estimated from eye lens weight, (C) body weight and (D) sex. Solid circles, triangles and squares represent wild rodents from contaminated areas with high, moderate and low dose radiation respectively. Open circles represent wild rodents from the control areas.

Acknowledgements

We express deep appreciation to the local government of Namie Town in Fukushima Prefecture, Japan for facilitating our study. We also thank Ikue Asari, Naomi Sasaki, Yuki Sato and the other members of the laboratory for their technical and secretarial support. Lastly, we would like to extend their gratitude to the reviewers for their constructive comments.

Disclosure statement

No potential conflict of interest was reported by the author(s).

Funding

This work was supported in part by the Ministry of Education, Culture, Sports, Science and Technology (MEXT) [Nuclear Energy S&T and Human Resource Development Project through concentrating wisdom, Grant Number JPMX 17S17942991].

ORCID

Valerie Swee Ting Goh http://orcid.org/0000-0002-0016-609X
Tomisato Miura http://orcid.org/0000-0003-3104-2134

Reference

Amer SM, Fahmy MA, Donya SM. 1996. Cytogenetic effect of some insecticides in mouse spleen. J Appl Toxicol. 16(1):1–3.

Bol'shakov VN, Gileva EA, Yalkovskaya LE. 2003. Species-specific features of inter- and intrapopulation variation in the level of chromosome instability in rodents. Russ J Ecol. 34(5):314–319.

Edwards AA, Lloyd DC, Purrott RJ. 1979. Radiation induced chromosome aberrations and the poisson distribution. Radiat Environ Biophys. 16(2):89–100.

Fuma S, Kubota Y, Ihara S, Takahashi H, Watanabe Y, Aono T, Soeda H, Yoshida S. 2016. Radiocaesium contamination of wild boars in Fukushima and surrounding regions after the Fukushima nuclear accident. J Environ Radioact. 164:60–64.

Garry VF, Griffith J, Danzl TJ, Nelson RL, Whorton EB, Krueger LA, Cervenka J. 1989. Human genotoxicity: pesticide applicators and phosphine. Science. 246(4927):251–255.

Gileva EA, Rakitin SB, Cheprakov MI. 2006. Genomic instability in the bank vole: population-ecological aspects. Russ J Ecol. 37(4):271–277.

Giri S, Giri A, Sharma GD, Prasad SB. 2002. Mutagenic effects of carbosulfan, a carbamate pesticide. Mutat Res – Genet Toxicol Environ Mutagen. 519(1–2):75–82.

Giri S, Prasad SB, Giri A, Sharma GD. 2002. Genotoxic effects of malathion: an organophosphorus insecticide, using three mammalian bioassays in vivo. Mutat Res – Genet Toxicol Environ Mutagen. 514(1-2):223–231.

Goncharova RI, Ryabokon NI. 1995. Dynamics of cytogenetic injuries in natural populations of bank vole in the republic of belarus. Radiat Prot Dosimetry. 62(1–2):37–40.

Hayama S, Nakiri S, Nakanishi S, Ishii N, Uno T, Kato T, Konno F, Kawamoto Y, Tsuchida S, Ochiai K, et al. 2013. Concentration of radiocesium in the Wild Japanese Monkey (*Macaca fuscata*) over the First 15 months after the fukushima daiichi nuclear disaster. PLoS One. 8(7):e68530.

Hendry JH, Simon SL, Wojcik A, Sohrabi M, Burkart W, Cardis E, Laurier D, Tirmarche M, Hayata I. 2009. Human exposure to high natural background radiation: what can it teach us about radiation risks? J Radiol Prot. 29(2): A29–A42.

Hikida T, Murakami O. 1980. Age determination of the japanese wood mouse, *Apodemus speciosus*. Japanese J Ecol. 30(2):109–116.

Hirata R, Takamatsu N, Nakamura M, Fuchigami M, Hata K, Sone K. 2007. Secondary dispersal of pasania edulis acorns by *Apodemus speciosus* into a coniferous plantation. J Jpn for Soc. 89(2):113–120.

Hiyama A, Nohara C, Kinjo S, Taira W, Gima S, Tanahara A, Otaki JM. 2012. The biological impacts of the Fukushima nuclear accident on the pale grass blue butterfly. Sci Rep. 2:570.

Hiyama A, Nohara C, Taira W, Kinjo S, Iwata M, Otaki JM. 2013. The Fukushima nuclear accident and the pale grass blue butterfly: evaluating biological effects of long-term low-dose exposures. BMC Evol Biol. 13(1):168.

Hiyama A, Taira W, Nohara C, Iwasaki M, Kinjo S, Iwata M, Otaki JM. 2015. Spatiotemporal abnormality dynamics of the pale grass blue butterfly: three years of monitoring (2011–2013) after the Fukushima nuclear accident. BMC Evol Biol. 15(1):15–16.

IAEA. 2011. Cytogenetic dosimetry: applications in preparedness for and response to radiation emergencies. Vienna: International Atomic Energy Agency.

ICRP. 2008. ICRP Publication 108: environmental protection – the concept and use of reference animals and plants. Ann ICRP. 38:4–6.

Jiang T, Hayata I, Wang C, Nakai S, Yao S, Yuan Y, Dai L, Liu Q, Chen D, Wei L, et al. 2000. Dose-effect relationship of dicentric and ring chromosomes in lymphocytes of individuals living in the high background radiation areas in China. J Radiat Res. 41(Suppl):63–68.

Kawagoshi T, Shiomi N, Takahashi H, Watanabe Y, Fuma S, Doi K, Kawaguchi I, Aoki M, Kubota M, Furuhata Y, et al. 2017. Chromosomal aberrations in large Japanese field mice (*Apodemus speciosus*) captured near Fukushima Dai-Ichi nuclear power plant. Environ Sci Technol. 51(8):4632–4641.

Kubota Y, Takahashi H, Watanabe Y, Fuma S, Kawaguchi I, Aoki M, Kubota M, Furuhata Y, Shigemura Y, Yamada F, et al. 2015. Estimation of absorbed radiation dose rates in wild rodents inhabiting a site severely contaminated by the Fukushima Dai-Ichi nuclear power plant accident. J Environ Radioact. 142:124–131.

Kubota Y, Tsuji H, Kawagoshi T, Shiomi N, Takahashi H, Watanabe Y, Fuma S, Doi K, Kawaguchi I, Aoki M, et al. 2015. Chromosomal aberrations in wild mice captured in areas differentially contaminated by the Fukushima Dai-Ichi nuclear power plant accident. Environ Sci Technol. 49(16):10074–10083.

Kyoya T, Obara Y, Nakata A. 2008. Chromosomal aberrations in Japanese grass voles in and around an illegal dumpsite at the aomori-iwate prefectural boundary. Zoolog Sci. 25(3):307–312.

Lyubimov AV, Garry VF. 2010. Phosphine. In: Krieger R, editor. Hayes' handbook of pesticide toxicology. 3rd ed. San Diego (CA): Academic Press; p. 2259–2266.

Mizuno T, Kubo H. 2013. Overview of active cesium contamination of freshwater fish in Fukushima and Eastern. Sci Rep. 3(1):7–10.

Møller AP, Hagiwara A, Matsui S, Kasahara S, Kawatsu K, Nishiumi I, Suzuki H, Ueda K, Mousseau TA. 2012. Abundance of birds in Fukushima as judged from Chernobyl. Environ Pollut. 164:36–39.

Ochiai K, Hayama S-I, Nakiri S, Nakanishi S, Ishii N, Uno T, Kato T, Konno F, Kawamoto Y, Tsuchida S, et al. 2014. Low blood cell counts in wild Japanese monkeys after the Fukushima Daiichi nuclear disaster. Sci Rep. 4:5793.

Perucatti A, Di Meo GP, Albarella S, Ciotola F, Incarnato D, Caputi Jambrenghi A, Peretti V, Vonghia G, Iannuzzi L. 2006. Increased frequencies of both chromosome abnormalities and SCEs in two sheep flocks exposed to high dioxin levels during pasturage. Mutagenesis. 21(1):67–75.

R Core Team. 2019. R: A language and environment for statistical computing. R Foundation for Statistical Computing, Vienna, Austria; [accessed 2020 July 01]. https://www.R-project.org/.

Saitoh M, Matsuoka N, Obara Y. 1989. Biochemical systematics of three species of the Japanese long-tailed field mice; *Apodemus speciosus*, *A. giliacus* and *A. argenteus*: taxnomy and systematics. Zoolog Sci. 6(5):1005–1018.

Saitoh M, Obara Y. 1986. Chromosome banding patterns in five intraspecific taxa of the large Japanese field mouse *Apodemus speciosus*. Zoolog Sci. 3(10):785–792.

Salvadori DMF, Ribeiro LR, Pereira CAB, Beçak W. 1988. Cytogenetic effects of malathion insecticide on somatic and germ cells of mice. Mutat Res - Genet Toxicol. 204(2):283–287.

Soné K, Kohno A. 1996. Application of radiotelemetry to the survey of acorn dispersal by *Apodemus* mice. Ecol Res. 11(2):187–192.

Soné K, Kohno A. 1999. Acorn hoarding by the field mouse, *Apodemus speciosus* Temminck (Rodentia: Muridae). J for Res. 4(2): 167–175.

Steinhauser G, Saey P. 2016. ^{137}Cs in the meat of wild boars: a comparison of the impacts of Chernobyl and Fukushima. J Radioanal Nucl Chem. 307(3):1801–1806.

Takada Y. 1982. Life history of small mammals in fallow fields 4. The eye lens weight as an age indicator in the large Japanese field mouse, *Apodemus speciosus*. J Growth. 21(1–2):8–11.

Takino S, Yamashiro H, Sugano Y, Fujishima Y, Nakata A, Kasai K, Hayashi G, Urushihara Y, Suzuki M, Shinoda H, et al. 2017. Analysis of the effect of chronic and low-dose radiation exposure on spermatogenic cells of male large japanese field mice (*Apodemus speciosus*) after the Fukushima Daiichi nuclear power plant accident. Radiat Res. 187(2):161–168.

Tanaka K, Kohda A, Satoh K. 2013. Dose-rate effects and dose and dose-rate effectiveness factor on frequencies of chromosome aberrations in splenic lymphocytes from mice continuously exposed to low-dose-rate gamma-radiation. J Radiol Prot. 33(1):61–70.

Tanaka K, Kohda A, Satoh K, Toyokawa T, Ichinohe K, Ohtaki M, Oghiso Y. 2009. Dose-rate effectiveness for unstable-type chromosome aberrations detected in mice after continuous irradiation with low-dose-rate gamma rays. Radiat Res. 171(3):290–301.

Tomoko M, Nakanishi KT. 2016. Agricultural Implications of the Fukushima Nuclear Accident - The First Three Years. Nakanishi TM, Tanoi K, editors. Tokyo: Springer Japan; [accessed 2020 July 01].

United Nations Scientific Committee on the Effects of Atomic Radiation. 2008. UNSCEAR 2008 Report to the General Assembly, with scientific annexes, Volume II: Annex E: Effects of ionizing radiation on non-human biota; [accessed 2020 July 01]. http://www.unscear.org/docs/reports/2008/11-80076_Report_2008_Annex_E.pdf.

Wada T, Tomiya A, Enomoto M, Sato T, Morishita D, Izumi S, Niizeki K, Suzuki S, Morita T, Kawata G. 2016. Radiological impact of the nuclear power plant accident on freshwater fish in Fukushima: an overview of monitoring results. J Environ Radioact. 151:144–155.

Yamashiro H, Abe Y, Fukuda T, Kino Y, Kawaguchi I, Kuwahara Y, Fukumoto M, Takahashi S, Suzuki M, Kobayashi J, et al. 2013. Effects of radioactive caesium on bull testes after the Fukushima nuclear plant accident. Sci Rep. 3(1):2850.

Zakeri F, Rajabpour MR, Haeri SA, Kanda R, Hayata I, Nakamura S, Sugahara T, Ahmadpour MJ. 2011. Chromosome aberrations in peripheral blood lymphocytes of individuals living in high background radiation areas of Ramsar, Iran. Radiat Environ Biophys. 50(4): 571–578.

Isolation of the effects of alpha-related components from total effects of radium at low doses

Chandula Fernando, Soo Hyun Byun, Xiaopei Shi, Colin B. Seymour, and Carmel E. Mothersill

ABSTRACT

Purpose: Radium is the most common source of alpha radiation exposure to humans and non-human species in the environment but the dosimetry is complicated by the decay chain which involves gamma exposure due to radon daughters. This paper seeks to determine the separate contributions of alpha and gamma doses to the total dose and total direct and non-targeted effect in a fish and a human cell line.

Materials and methods: This study aimed to isolate the effect of alpha particles following exposure to low doses of radium in cells, and their progeny which received no further exposure. This was initially done by comparing the survival values of a human keratinocyte cell line (HaCaT) and an embryonic Chinook salmon cell line (CHSE-214) exposed to gamma radiation, from survival of the same cell lines exposed to mixed alpha and gamma radiation through exposure to Ra-226 and its decay products. A Monte Carlo simulation was later performed to determine the contributions of radium decay products including radon daughters.

Results: The human cell line showed increased radioresistance when exposed to low doses of alpha particles. In contrast the fish cell line, which demonstrated radioresistance to low dose gamma radiation, showed increased lethality when exposed to low doses of alpha particles. Significant and complex levels of non-targeted effects were induced in progeny of irradiated cells. The simulation showed that gamma and beta decay products did not contribute significant dose and the highest beta dose was below the threshold for inducing non-targeted effects.

Conclusions: The results confirm the need to consider the dose-response relationship when developing radiation weighting factors for low dose exposures, as well as the need to be aware of possible cell line and species differences.

Introduction

In dosimetry, linear energy transfer (LET) is the amount of energy (keV) that an ionizing particle transfers to the material transversed per unit distance (μm). It describes the quality of different types of radiation. It describes the action of radiation into matter. Photons such as gamma rays are able to traverse great distances unchanged before being entirely absorbed, however monoenergetic ions such as alpha particles cause frequent direct ionizations within a smaller range. Due to this, exposures to alpha particles are often considered as producing a significantly higher concentration of damage given the same amount of absorbed energy when compared to gamma radiation (Goodhead 1994). The derivation of concepts such as relative biological effectiveness (RBE) and the assignment of weighting factors have further supported this notion, aligning dose with biological effect in one seamless system. In vitro studies of alpha radiation effects have returned a wide range of RBE values, from <2 for the induction of double strand breaks, to 3.5–10 for cell lethality and transformation in different cell lines, to >25 for other endpoints assessed (Thomas et al. 2007). Most countries accept the single recommended radiation weighting factor of 20 for alpha particles when specifying exposure limits (ICRP 2007). A simplified relationship between dose and effect is useful in developing an all-encompassing regulatory policy for protecting against radioactive exposure, however, not all cases are alike. While existing RBE values and weighting factors may be true for high doses, there are many laboratory studies casting doubt that such relationships hold after exposure to low doses and low dose rates. This is particularly relevant in situations with risks of prolonged low dose exposure, including the mining and milling of uranium for nuclear power plants, and the handling of nuclear waste.

It has been shown that if a single alpha particle traverses a cell, it causes a significantly lower risk of oncogenic transformation than expected (Miller et al. 1999). Further, the work of Nagasawa and Little has shown significantly higher frequencies of mutation than would be expected through linear extrapolation from data for high doses, at doses where the *mean* number of alpha particle traversals per nucleus was significantly less than one, i.e. many cells in the flask got no traversal (Nagasawa and Little 1999). At low doses, both alpha and gamma rays can cause non-targeted effects (NTE) like genomic instability where damage does not cause direct mortality and cells appear completely normal but de novo con-clonal effects are seen in distant progeny and lethality occurs generations later, characterized by a reduction in plating efficiency or biochemical changes in the progeny of irradiated cells (Seymour et al. 1986). Cell survival at sub-lethal doses of gamma irradiation has also been observed to differ from what is expected by the traditional linear-quadratic model, instead displaying a region of low-dose hyper-radiosensitivity (HRS) followed by increased radioresistance (IRR) in both cells surviving exposure to ionizing radiation and the progeny of these cells (Marples et al. 1997).

Of particular interest to this study is whether NTE amplify low dose effects such that they are higher than what would be expected from the established linear no-threshold model (LNT) following exposure to low doses of an environmental alpha emitter: radium-226. Average natural radium-226 concentration in soil can vary from 3.7 to 126 Bq/kg around the world, however areas of high natural background can present concentrations up to three orders of magnitude higher than average. Natural activities in surface water generally range between 0.5 and 20 mBq/L, rising to 3000 mBq/L surrounding mining/milling sites. Further, monitoring data following human activities such as hydrofracking have observed values up to 620 Bq/L of radium-226 (Fesenko et al. 2014). Despite being an alpha emitter by itself, it is known that the uranium decay chain of which radium is part of involves many gamma and beta emissions, thereby making it difficult to measure pure alpha effects. To approach this problem two methods were considered. In the first approach survival curve data generated by our group for gamma irradiated cells exposed to acute Cs-137 or chronic Ra-226 were analyzed. This study measured survival and the lethal mutation phenotype assayed as reduced cloning efficiency in cultures of a human keratinocyte cell line (HaCaT). In addition, due to the increasing relevance of protecting non-human biota from radium in hydrogeologic contaminations from mining, etc., the study also investigated relative alpha exposure effects in the embryonic Chinook salmon cell line (CHSE-214) because to the importance of radium exposure in aquatic ecosystems left behind after uranium mining activities .

In the analysis data from the gamma irradiation experiments (through acute exposure to Cs-137) were subtracted from data generated from experiments with mixed alpha, beta and gamma irradiation (through chronic exposure to Ra-226 and its progeny). The data were analyzed to see the actual effect of caesium-137 exposure in directly exposed and distant progeny versus chronic exposure to radium assuming a dose and dose rate effectiveness factor (DDREF) value of 1 (Rühm et al 2016; Hoel 2018). In the second approach a Monte Carlo model was used to determine the contributions of alpha, beta and gamma decays to the total effect of the radium exposure in the directly exposed and progeny cells.

Materials and methods

Cell culture

The HaCaT cell line used in the study is an immortalized human keratinocyte cell line originally derived and characterized by Boukamp et al (Boukamp et al. 1988). The cell line used in this study was obtained as a gift from Dr. Orla Howe (Dublin, Ireland). The cell line was routinely maintained with RPMI-1640 medium supplemented with 10% fetal bovine serum (Invitrogen, Burlington, Canada), 5 ml of 200 mM L-Glutamine (Gibco, Burlington, Canada), 0.5 g/ml hydrocortisone (Sigma-Aldrich, Oakville, Canada), 25 mM Hepes buffer (Gibco), 10 U/ml penicillin (Gibco) and 10 mg/ml streptomycin sulfate, (Gibco). These cells were grown at 37 °C in an incubator with 5% CO_2.

The CHSE-214 is an embryonic cell line derived from Chinook salmon obtained as a gift from Dr. Neils Bols (Waterloo, Canada). CHSE-214 cells were cultured in Leibovitz's L-15 medium supplemented with 12% fetal bovine serum (Invitrogen), 5 ml of 200 mM L-Glutamine (Gibco), 25 mM Hepes buffer (Gibco), 10 U/ml penicillin (Gibco) and 10 mg/ml streptomycin sulfate (Gibco). These cells were grown at 19 °C in an incubator without CO2.

Reduction in cloning efficiency was observed using the clonogenic assay technique developed by Puck and Marcus (Puck and Marcus 1955). Cell stocks were maintained in T75 flasks with 30 ml medium. For clonogenic assay the cells were seeded into 4 replicate T25 flasks with 5 ml fresh culture media at the required cell density such that at least 100 viable colonies could be expected to form in control flasks.

Reporter T25 flasks were maintained in the incubator for 9 days (HaCaT cells) or 40 days (CHSE-214 cells). Following the incubation period, colonies in sham irradiated (control) flasks were visible to the naked eye. Flasks were stained using a 1:4 (v/v) dilution of Fuchsin-Carbol (Ricca Chemical Co., Arlington, TX) in water, and macroscopically visible colonies (confirmed to have more than 50 cells when observed under a microscope) were scored as survivors. The forth flask in each group was grown to confluence and subcultured to determine progeny 2 and 3 survival using the method in Seymour et al (1986).

Chronic irradiation using Ra-226 in medium

Stock solutions of medium containing the radioisotope ^{226}Ra were prepared using neutralized radium nitrate (Eckert & Ziegler 2010, Valencia, USA). 100 ml L-15 or RPMI medium was mixed with 1000 Bq of ^{226}Ra solution. The concentration of ^{226}Ra in this stock medium was 10,000 mBq/ml. After filtering into storage tubes, serial dilutions were made to give the required final concentrations.

500 cells were initially seeded into T25 flasks containing 5 ml of medium with Ra-226 or control medium. 4 flasks were prepared for each respective concentration: 0, 0.1, 1, 10, 100, 200 or 500 Bq/ml Ra-226. Flasks were maintained in the incubator for 9 days after which the radioactive medium was removed from both cell lines, and the cells were gently rinsed with calcium and magnesium-free DPBS. Ra-226

residues in the flasks were assumed to be insignificant. Flasks then received 5 ml of fresh culture medium without Ra-226 and returned to the incubator. 3 flasks from each concentration were deemed reporter flasks, incubated for 9 or 40 days, and stained as described above. Cloning efficiencies observed in these reporter flasks represented the initial plating efficiencies from direct chronic irradiation. The remaining fourth flask of each concentration was left to incubate until 80–90% confluency, after which it was subcultured as described above seeding 500 cells into a fresh flask. From here on however no further irradiation was done and all flasks received fresh culture medium containing 0 Bq/ml Ra-226. The process was repeated as before, and cloning efficiencies observed in these reporter flasks represented survival fractions of the progeny (P2). The process was repeated once more to observe further change in the cloning efficiency in subsequent generations (P3).

Acute irradiation using a Cs-137 source

As with the ^{226}Ra experiments, 4 T25 flasks were seeded with 500 cells for each respective dose: 0, 0.05, 0.1, 0.25, 0.5, 0.75 or 1 Gy. The flasks were incubated for six hours to allow for cells to adhere to the flask, after which they were exposed to their respective γ-ray dose using a ^{137}Cs source (Taylor source, McMaster University, Hamilton, Canada). Flasks were placed at 26 cm from the radiation source, irradiated at a dose rate of 0.273 Gy/min and the room temperature was around 26 °C.

All flasks were placed back in the incubator immediately after irradiation. Similar to the ^{226}Ra experiments, 3 flasks were deemed reporter flasks and incubated for approximately 9 days before being stained as described above (initial). The remaining fourth flask of each dose was incubated until cells became 80–90% confluent, after which they were subcultured as described above with fresh culture medium. This process was also repeated twice as above (P2 and P3).

Determining γ and electron absorbed doses from ^{226}Ra and its daughters

To compute the γ and electron absorbed doses from ^{226}Ra and its daughters, Monte Carlo simulations were carried out using the MCNP6 code. As a simplified geometry for the t-25 flask, a 1 mm thick plastic layer of $10 \times 5\,cm^2$ was used in MCNP. A culture medium was represented as a 3 mm thick layer on top of the plastic layer. The cell layer was assumed to be a 0.1 mm thick sub-layer within the culture medium, on contact with the plastic layer. The air volume of the flask was represented by a 2.2 cm thick layer on top of the culture medium.

As the source particles from the ^{226}Ra decay, two source cards, one for conversion electrons and the other for X-ray and gamma photon. The X-ray and gamma photon energies and emission probabilities were retrieved from Table of Radioactive Isotopes.

(WWW Table of Radioactive Isotopes, http://nucleardata.nuclear.lu.se/toi/) and the conversion electron data were retrieved from Browne et al (1986). The positions of the conversion electrons and the photons were uniformly sampled within the culture medium. Figure 1 shows an example of the source position sampling.

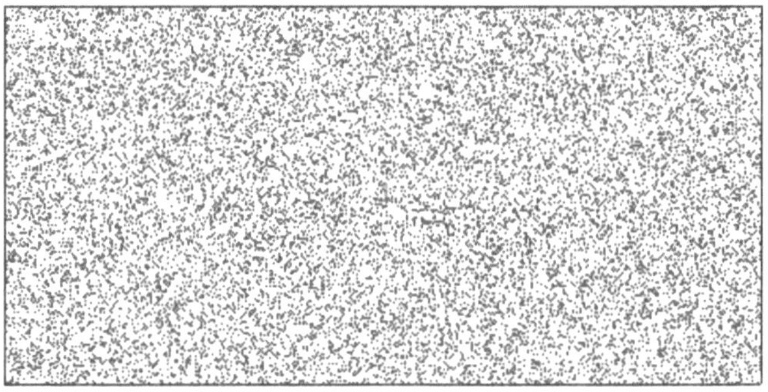

Figure 1. An example of the MCNP source position sampling for the X-ray and gamma photon on the horizontal plane. A culture medium was represented as a 3 mm thick layer on top of a plastic layer ($10 \times 5\,cm^2$, 1 mm thick). The position of each photon source particle was uniformly sampled within the culture medium.

From the list daughter nuclides, beta decay electrons from 214Pb and 214 Bi can make significant contribution to the absorbed dose and therefore, two source cards were created for the beta decay electrons and the conversion electron data were retrieved from Browne et al.'s data table (Browne et al. 1986). Unlike the conversion electron and photon sources, the positions of the beta decay electrons were uniformly sampled from the entire flask volume of $10 \times 5 \times 2.5\,cm^3$, given a large fraction of 222Rn is in the air region of the flask volume.

For each type of source card, the absorbed dose was computed using both F6 and *F8 tallies for either the primary or the secondary electrons, depending on the source card. The absorbed dose values from both tallies were in excellent agreement for each source card.

Table 1 compares the MCNP simulation result for each source card. From the 226Ra decay, the absorbed dose by conversion electrons are eighteen times higher than the dose from the X-ray and gamma photons, which is obvious since electrons always deposit energies to the cell layer as long as their tracks pass through the cell volume while the thickness of the cell layer is much thinner than the mean free path of the most photon energies. For the 214Pb and 214 Bi beta electrons, 214 Bi beta dose is about six times higher than the 214Pb dose, which may stem from the fact that the mean beta energy of 214 Bi is higher, hence a larger percentage of 214 Bi beta particles can penetrate the cell medium and arrive to the cell volume. Between the 226Ra dose and the daughters' beta dose, the daughters' dose rate is about 70 times higher than the 226Ra dose rate under the equilibrium condition. Therefore, the daughters' beta dose is the main low LET dose contributor in any case the exposure time is longer than 10 percent of the half-life of ^{222}Rn.

To compute the absorbed dose for an experimental set using the table MCNP simulation result, the total number of each decay must be computed. For ^{226}Ra decay, its computation is straightforward from the ^{226}Ra activity and exposure time while the integral of the time dependent activity must

Table 1. Absorbed dose computed by MCNP for each type of the source card.

226Ra α decay			
Conversion electron	X and gamma	214Pb β^- decay electrons	214Bi β^- decay electrons
2.36×10^{-14} Gy/decay	1.27×10^{-15} Gy/decay	2.51×10^{-13} Gy/decay	1.47×10^{-12} Gy/decay

Table 2. Survival of HaCaT cells exposed to Cs-137 (2a and Ra-226 (2 b) expressed as percentage surviving fraction ± SD after acute exposure to Cs-137 at a dose rate of 0.273 Gy/min or after chronic exposure to radium-226 for 9 days.

(a) Survival fraction (%) HaCaT exposed to ^{137}Cs

Dose (Gy)	0	0.05	0.1	0.25	0.5	0.75	1
P0	100 ± 2	93 ± 4	90 ± 3	92 ± 6	86 ± 3	76 ± 2	68 ± 8
P1	100 ± 3	79 ± 7	75 ± 4	85 ± 7	83 ± 7	78 ± 4	92 ± 4
P2	100 ± 3	77 ± 8	79 ± 3	90 ± 3	94 ± 6	96 ± 5	97 ± 9

(b) Survival fraction (%) HaCaT exposed to ^{226}Ra

Ra-226 (Bq/ml)	0	0.01	1	10	100	200	500
P0	100 ± 1.5	109 ± 3.0	107 ± 3.8	107 ± 3.0	104 ± 2.7	107 ± 3.3	92 ± 1.4
P1	100 ± 2.6	108 ± 5.4	109 ± 8.2	107 ± 3.4	99 ± 6.0	106 ± 4.4	84 ± 3.0
P2	100 ± 2.1	101 ± 2.5	106 ± 3.3	99 ± 2.0	102 ± 2.7	95 ± 2.5	101 ± 2.4
P3	100 ± 2.8	99 ± 2.8	101 ± 3.6	105 ± 4.3	106 ± 3.5	105 ± 1.7	102 ± 2.7

P0 refers to the directly exposed cells and P1, 2 and 3 refer to the passage numbers at which the lethal mutation burden was measured in progeny without further exposure to radiation.

be done in the initial activity buildup phase for ^{214}Pb and ^{214}Bi.

Determination of surviving fractions and progeny survivals

Survival fractions/plating efficiencies of cells are determined as cloning efficiency observed through staining, which visualizes the fraction of colonies formed from the 500 cells plated. Progeny survival fractions were calculated at each observed interval and for each dose through following the recorded cell numbers at the start and end of each interval, in accordance to previous delayed lethal effect assays by (Seymour et al. 1986; O'Reilly et al. 1994; Lyng et al. 1996; Mothersill et al. 2000). Here, the product of the cells observed at the end of the current passage, with the total cell number at the end of the preceding passage, was divided by the initial number of cells seeded per passage corrected for plating efficiency.

To isolate the effect of α-particles on the survival and genomic instability of cells, this study examined the effects observed after γ irradiation (through acute exposure to ^{137}Cs) and from mixed α,β and γ irradiation (through chronic exposure to 226Ra).

Results

Human keratinocyte cell line (HaCaT)

The data obtained for the initial and progeny survival fractions for HaCaT cells exposed to ^{226}Ra and ^{137}Cs are shown in Table 2. Data for acute ^{137}Cs exposure show a marked hypersensitivity in the low dose region followed by a region of increased radioresistance (IRR). The progeny of these cells continue to demonstrate such HRS/IRR behavior with further decrease in cloning efficiency thereby observing lethal mutations in those generations. In particular significant decreases in cloning efficiencies are observed in the first observation of progeny of cells (P2, 16 population doublings) irradiated at 0.05 Gy by 19% ($p = .007$), 0.1 Gy by 23% ($p = .00007$), 0.25 Gy by 14% ($p = .02$), 0.5 Gy by 15% ($p = .01$) and 0.75 Gy by 17% ($p = .0004$). Further significant decreases are observed in the second observation of progeny (P3, approximately 25 population doublings) at 0.05 Gy by 17% ($p = .01$) and at 0.1 Gy by 14% ($p = .001$).

In comparison, progenitor cells exposed to ^{226}Ra show significantly greater survival with many observations of higher cloning efficiency compared to sham irradiated control flasks (denoted as survival values greater than 100% i.e. more colonies survived in the treated flasks compared to controls). As such no HRS region is observed, with little to no change in survival compared to control in cells exposed to concentrations greater than 0.1 Bq/ml of Ra-226. At P2, survival of progeny in concentrations up to 10 Bq/ml of Ra-226 show significantly higher survival compared to control, while observations at P3 show similar survival values to what was observed in the progenitors

Embryonic chinook salmon cell line (CHSE-214)

In contrast to the studied human cell line, there was no significant cell death observed in the directly exposed cells of the CHSE-214 fish cell line and their progeny to acute Cs-137 exposure over the dose range tested (Table 3). This indicates low dose radioresistance when compared to human cell culture and confirms previous data from this group (Ryan et al. 2008). When CHSE-214 cells were exposed to ^{226}Ra however, progenitor cells show a marked response with decreasing cell survival following an almost linear trend with respect to dose. Residual survival observed at P2 (8 doubling periods after removal of radium) show increased lethal mutation however residual survival observed in the subsequent progeny at P3 (16 doubling periods after removal of radium) demonstrate a return of radioresistance with survival values similar to initial values.

Table 3. Survival of CHSE-214 cells exposed to Cs-137 (3a) and Ra-226 (3 b) expressed as percentage surviving fraction ± SD after acute exposure to Cs-137 at a dose rate of 0.273 Gy/min or after chronic exposure to radium-226 for 28 days.

(a) Survival fraction (%) CHSE-214 exposed to ^{137}Cs

Dose (Gy)	0	0.05	0.1	0.25	0.5	0.75	1
P0	100 ± 3	100 ± 2	101 ± 4	104 ± 3	100 ± 5	125 ± 3	99 ± 3
P1	100 ± 4	97 ± 3	105 ± 6	114 ± 4	118 ± 4	160 ± 6	119 ± 5
P2	100 ± 3	93 ± 2	111 ± 4	113 ± 4	140 ± 5	160 ± 5	139 ± 4

(b) Survival fraction (%) CHSE-214 exposed to ^{226}Ra

Ra-226 (Bq/ml)	0	0.1	1	10	100	200	500
P0	100 ± 5	69 ± 8	71 ± 13	70 ± 7	55 ± 6	31 ± 6	<0.2
P1	100 ± 16	72 ± 12	78 ± 9	81 ± 13	72 ± 4	46 ± 5	
P2	100 ± 10	91 ± 9	93 ± 10	99 ± 9	89 ± 6	97 ± 9	
P3	100 ± 7	82 ± 4	84 ± 8	72 ± 10	76 ± 8	89 ± 15	

P0 refers to the directly exposed cells and P1, 2 and 3 refer to the passage numbers at which the lethal mutation burden was measured in progeny without further exposure to radiation.

Table 4. Calculated contributions to the total dose of gamma and beta components of Ra-226 decay over the nine day exposure period for HaCaT cells (4a) or the 28 day period for CHSE-214 (4b).

(a) HaCaT

	^{226}Ra decay			Emission by daughters	
^{226}Ra Activity (Bq/ml)	α particles (mGy)	internal conversion (mGy)	X & γ photons (mGy)	^{214}Pb β^- decay (mGy)	^{214}Bi β^- decay (mGy)
0	0	0	0	0	0
0.1	0.06	9.18E − 06	4.94E − 07	0.0001	0.001
1	0.59	9.18E − 05	4.94E − 06	0.001	0.01
10	5.94	0.0009	4.94E − 05	0.01	0.06
100	59.37	0.009	0.0005	0.11	0.64
200	118.74	0.02	0.001	0.22	1.28
500	296.84	0.05	0.002	0.55	3.20

(b) CHSE-214

	^{226}Ra decay			Emission by daughters	
^{226}Ra Activity (Bq/ml)	α particles (mGy)	internal conversion (mGy)	X & γ photons (mGy)	^{214}Pb β^- decay (mGy)	^{214}Bi β^- decay (mGy)
0	0	0	0	0	0
0.1	0.18	2.85E − 05	1.54E − 06	0.0002	0.001
1	1.85	0.0003	1.54E − 05	0.002	0.01
10	18.47	0.003	0.0002	0.02	0.14
100	184.70	0.03	0.002	0.24	1.43
200	369.40	0.06	0.003	0.49	2.87
500	923.50	0.14	0.008	1.22	7.17

These results suggest that the approach of subtracting acute gamma exposure data from acute radium exposure data even in a range where biological effects are expected is too simplistic and does not yield meaningful information regarding the contribution of alpha radiation to the effect for either cell line.

The Monte Carlo model described in the methods section was then applied to the radium results. The results are shown in Table 4 and Figure 2(a,b). These data show the separate contributions in mGy of the alpha, beta and gamma components. It can be seen that the gamma component of dose is negligible and below that which could trigger lethal mutations or other NTE - 2mGy (Liu et al. 2006). The dose from beta decay of radon daughters is higher but only exceeds the 2mGy threshold for lethal mutation induction in the 500 Bq/ml exposure situation.

Discussion

At sub-lethal doses of gamma irradiation through acute exposure to ^{137}Cs, the HaCaT cell line displayed a region of low-dose hyper-radiosensitivity (HRS) followed by increased radioresistance (IRR). In addition, lethality was observed in subsequent generations (lethal mutation phenotype) with significant decreases in cloning efficiencies observed in unirradiated progeny cells. In contrast, only radioresistance was observed in the progenitor cells exposed to chronic ^{226}Ra with significantly higher survival and no observable region of HRS. Further, the observed progeny of these cells showed increased survival and lowered lethal mutation. In comparison, the results of experiments using the non-mammalian embryonic fish cell line showed the reverse of what was observed in human cell culture. Survival data following exposure to acute gamma irradiation confirmed existing radioresistance in the CHSE-214 cell line compared to human cell culture, with no significant lethality. However, survival data for cells exposed to ^{226}Ra suggested that alpha particles promoted lethality at doses otherwise known to have no significant effect. An aim of this paper was to determine to what extent the observed effects of ^{226}radium were due to radon decay chain elements contributing beta and gamma radiation to the biological material. Given the very different patterns of survival in directly irradiated cells and their progeny, it became apparent that comparison of

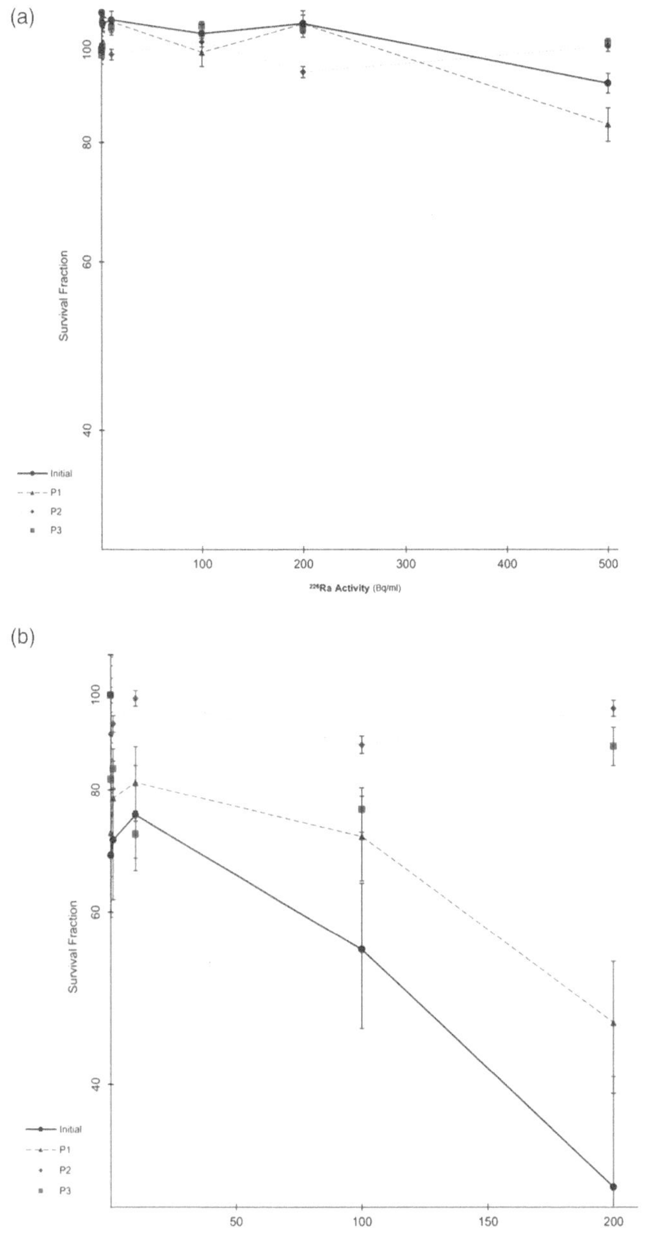

Figure 2. Surviving fraction (%) of initially irradiated and passage 1,2 and 3 progeny of HaCaT cells exposed to Ra-226 for 9 days (2a) and CHSE-214 cells exposed to Ra-226 for 28 days (2 b). The activity is shown in Bq/ml. The contributions of the decay products including radon daughters are shown on Table 4.

survival data for acute gamma exposure with mixed alpha/beta/gamma exposure would not yield useful data even if DDREF values were discounted. The Monte Carlo approach was used therefore to determine the individual components of dose in the culture flasks over the time of exposure to radium. Given the very low doses attributable to beta and gamma decay we conclude that the biological effects seen following ^{226}radium exposure in the flasks are most likely due to alpha radiation with a small contribution from beta radiation due to radon daughters. The results shows a marked difference in response of the human cell line (virtually no effect and a possible hormetic response), compared with the fish cell line (marked low dose hypersensitivity/induced radioresistance and persistent lethal mutation effects in progeny).

There are two important caveats when comparing these data:

1. There are significant differences in cell doubling times between CHSE cells and HaCaT cells. CHSE are a fish cell line and grow at 18 °C in a cooled incubator. They have a doubling time of approximately 127hrs (Plating efficiency 41.4 ± 2.2%) while HaCaT grow at 37 °C and have a doubling time of 23.6hrs (Plating efficiency 50.4 ± 0.8%). This difference will impact the average radiobiological effects (survival). It is probably best to regard the data as indicative of the likely complexities when extrapolating laboratory data to field situations or comparing likely effects of environmental exposures across cold and warm blooded species rather than as an absolute difference. Both the temperature difference and the doubling time difference would change DNA damage the repair kinetics (Mayer et al. 1987; MacLeod et al. 1990), The metabolic rate would also be different (Wheeler et al. 1992) although the Q10 of enzymes in poikilothermic versus homeothermic species show a remarkable ability of poikilothermic species to adapt to temperature swings while maintaining stable functions. (Robinson 2010). Either way the reality is that in the environment the ability of organisms to respond to radiation exposure will vary greatly depending on species and physical environment as well as dose.
2. In the experiments described above, an acute gamma exposure was used to try to determine the contribution of gamma alone to the total radium effect but this did not take into account dose rate effects, which are known to be important in low dose exposure scenarios. This means that the contribution of gamma should be taken as the maximum possible contribution rather than the actual contribution. An important point is that when the dose rate effect is examined for production of non-targeted effects (NTE) such as bystander effects or as examined here for lethal mutation induction, the literature suggests minimal impact of dose rate on NTE (Amundson et al. 2003; Belli and Tabocchini 2020). This probably relates to the fact that these are 'turned on' by very low doses (2-3mGy acute dose) (Liu et al. 2006) and the binary nature of the response means that once activated further exposure does not lead to increased effect (Schettino et al. 2005). So while the direct dose effect will be influenced by dose rate, the NTE may not.

Overall however it is important to note that temperature and growth rates for these cell lines do reflect the temperature optima of fish and mammalian species in nature.

Considering the potential for sub-lethal doses from chronic exposure to radium and its daughters found in waste products, to remnants of historic commercial and medical usage of radium (ranging from self-luminous paints to cancer treatment), the unconventional behavior observed in both cell lines of this study have potential importance in radiological protection. Further, the presence of radium in waste reaching the ecosystem from mining and nuclear applications is important given the currently growing interest for non-human radiological protection.

The radioprotective quality of sub-lethal doses of alpha radiation that was observed in the HaCaT cell line, where cells displayed significantly lower lethality in the presence of alpha particles has not been noted previously in the literature. Current risk models for carcinogenesis and radiation protection at low doses do not consider effects like HRS/IRR and genomic instability phenomena, partly because the importance of such effects to the carcinogenic process have not been established. Research observing HRS/IRR behaviors suggest the activation of cell cycle checkpoints for increased cell repair, etc. as a possible mechanism for radioresistance (Fernet et al. 2010). Considering only radioresistance was observed in the presence of alpha radiation, the results suggest an ultra-low dose of alpha particles produces a sufficient level of genomic instability to activate the previously mentioned cell cycle checkpoints, inducing radioresistance. This effect was non-linear with dose with marked reduction as dose increased to the progenitor. The results observed in the CHSE-214 cell line on the other hand is in line with currently accepted descriptions on the effect of alpha particles at high doses. Here, the concentration of damage events is said to exceed a threshold at which effective repair becomes difficult (Blöcher 1988). Further lethality is seen in progeny as a de novo appearance of non-clonal lethal mutations, indicative of genomic instability. However, this decreases with subsequent generations suggesting the ability for existing damage repair mechanisms eventually to counteract the heritable susceptibility to lethal damage.

The results of the study support the need to consider dose-dependence when describing the effect of different radiation qualities. The results of the fish cell line experiments suggest a need to be aware of species differences and that protection for humans would not inherently protect ecosystems and non-human biota.

It should be noted that observed in vitro results cannot simply be translated to in vivo effects without much further research and discussion. For example, the evidence for heritable NTE through in vitro and non-human studies (Morgan 2012), is given appropriate context by epidemiological observations including germline mutation rates (Dubrova et al. 2002) and transgenerational genomic instability (Aghajanyan and Suskov 2009) observed in post-Chernobyl families. In addition, further research needs to be done to isolate the effect of dose rate during sub-lethal exposure to high-LET radiation, as differences in time for cell repair can affect the level of radioresistance observed.

Conclusion

At sub-lethal doses, survival greatly depends on repair mechanisms. Since the HaCaT cell line demonstrates hyper-radiosensitivity to acute gamma energy at low doses, high-LET alpha particle radiation may be able to produce sufficient genomic instability to induce radioresistance. In such instances, the ratio of biological damage caused by chronic alpha exposure is significantly lower compared with an equivalent dose of gamma energy alone, and as such a lower radiation weighting factor might be considered. However, while the CHSE-214 cell line demonstrates increased radioresistance to gamma energy, the concentrated nature of energy deposited causes increased lethality when exposed to alpha particles. These cases would suggest a higher radiation weighting factor, similar to what is currently recommended. Further study is required to isolate the effect of dose-rate at sub-lethal doses. In addition, further consideration is required to translate the observed in vitro results to in vivo effects.

In terms of the contributions of decay chain components of dose from the radium treatments, these appear to be negligible and below the threshold for inducing non-targeted effects except for the ^{214}Bi $\beta-$ decay for the highest applied ^{226}Ra activity.

Alpha-emitters are often found in environments and industrial applications related to the generation of nuclear energy. Further they can be dispersed during nuclear incidents and events of accidental release. As nuclear power grows as a potential carbon neutral solution to climate change, a shift from an over-simplistic model, to one that considers nuances at different exposure situations is crucial to provide meaningful risk management and radiation protection of human and non-human biota at low doses.

Disclosure statement

The authors declare that they have no known competing financial interests or personal relationships that could have appeared to influence the work reported in this paper.

Funding

This work was supported by the Natural Sciences and Engineering Research Council (NSERC) of Canada in the form of a Collaborative Research and Development Grant [Grant No. CRDPJ 484381-15]. We acknowledge continued support from the National Chronic Fatigue and Immunodeficiency Foundation Inc.

ORCID

Chandula Fernando http://orcid.org/0000-0001-9097-0358

References

Aghajanyan A, Suskov I. 2009. Transgenerational genomic instability in children of irradiated parents as a result of the Chernobyl Nuclear Accident. Mutat Res Mol Mech Mutagen. 671(1–2):52–57.

Amundson SA, Lee RA, Koch-Paiz CA, Bittner ML, Meltzer P, Trent JM, Fornace AJ. Jr. 2003. Differential responses of stress genes to low dose-rate irradiation. Mol Cancer Res. 1(6):445–452.

Belli M, Tabocchini MA. 2020. Ionizing radiation-induced epigenetic modifications and their relevance to radiation protection. IJMS. 21(17):5993.

Blöcher D. 1988. DNA double-strand break repair determines the RBE of alpha-particles. Int J Radiat Biol. 54(5):761–771.

Boukamp P, Petrussevska RT, Breitkreutz D, Hornung J, Markham A, Fusenig NE. 1988. Normal keratinization in a spontaneously immortalized aneuploid human keratinocyte cell line. J Cell Biol. 106(3): 761–771.

Browne E, Firestone RB, Shirly VS, editors. 1986. Radioactive isotopes. New York (NY): John Wiley & Sons.

Dubrova YE, Grant G, Chumak AA, Stezhka VA, Karakasian AN. 2002. Elevated minisatellite mutation rate in the post-chernobyl families from Ukraine. Am J Hum Genet. 71(4):801–809.

Eckert & Ziegler. 2010. Recommended nuclear decay data. Braunschweig: Eckert & Ziegler.

Fernet M, Mégnin-Chanet F, Hall J, Favaudon V. 2010. Control of the G2/M checkpoints after exposure to low doses of ionising radiation: implications for hyper-radiosensitivity. DNA Repair (Amst). 9(1): 48–57.

Fesenko S, Carvalho F, Martin P, Moore W, Yankovich T. 2014. Radium in the environment. In: The environmental behaviour of radium: revised edition technical reports series No 476. Vienna: IAEA; p. 33–105.

Goodhead DT. 1994. Initial events in the cellular effects of ionizing radiations: clustered damage in DNA. Int J Radiat Biol. 65(1):7–17.

Hoel DG. 2018. Nuclear epidemiologic studies and the estimation of DREF. Int J Radiat Biol. 94(4):307–314.

ICRP. 2007. 4.3 Dose quantities. In: The 2007 Recommendations of the international commission on radiological protection. Essen: Elsevier. p. 62–71.

Liu Z, Mothersill CE, McNeill FE, Lyng FM, Byun SH, Seymour CB, Prestwich WV. 2006. A dose threshold for a medium transfer bystander effect for a human skin cell line. Radiat Res. 166(1): 19–23.

Lyng FM, O'Reilly S, Cottell DC, Seymour CB, Mothersill C. 1996. Persistent expression of morphological abnormalities in the distant progeny of irradiated cells. Radiat Environ Biophys. 35(4):273–283.

MacLeod RA, Christie AF, Costa ND, Bryant PE. 1990. Repair kinetics in CHO cells of X-ray induced DNA damage and chromatid aberrations during a cell cycle extended by transient hypothermia. Mutagenesis. 5(3):279–283.

Marples B, Lambin P, Skov KA, Joiner MC. 1997. Low dose hyper-radiosensitivity and increased radioresistance in mammalian cells. Int J Radiat Biol. 71(6):721–735.

Mayer PJ, Bradley MO, Nichols WW. 1987. The effect of mild hypothermia (34 degrees C) and mild hyperthermia (39 degrees C) on DNA damage, repair and aging of human diploid fibroblasts. Mech Ageing Dev. 39(3):203–222.

Miller RC, Randers-Pehrson G, Geard CR, Hall EJ, Brenner DJ. 1999. The oncogenic transforming potential of the passage of single alpha particles through mammalian cell nuclei. Proc Natl Acad Sci U S A. 96(1):19–22.

Morgan WF. 2012. Non-targeted and delayed effects of exposure to ionizing radiation: I. Radiation-induced genomic instability and bystander effects in vitro. Radiat Res. 178(2):AV223–AV236.

Mothersill C, Kadhim MA, O'Reilly S, Papworth D, Marsden SJ, Seymour CB, Wright EG. 2000. Dose- and time-response relationships for lethal mutations and chromosomal instability induced by ionizing radiation in an immortalized human keratinocyte cell line. Int J Radiat Biol. 76(6):799–806.

Nagasawa H, Little JB. 1999. Unexpected sensitivity to the induction of mutations by very low doses of alpha-particle radiation: evidence for a bystander effect. Radiat Res. 152(5):552.

O'Reilly S, Mothersill C, Seymour CB. 1994. Postirradiation expression of lethal mutations in an immortalized human keratinocyte cell line. Int J Radiat Biol. 66(1):77–83.

Puck TT, Marcus PI. 1955. A rapid method for viable cell titration and clone production with HELA cells in tissue culture: the use of X-irradiated cells to supply conditioning factors. Proc Natl Acad Sci U S A. 41(7):432–437.

Rühm W, Azizova TV, Bouffler SD, Little MP, Shore RE, Walsh L, Woloschak GE. 2016. Dose-rate effects in radiation biology and radiation protection. Ann ICRP. 45(1_suppl):262–279.

Robinson R. 2010. Firing hot, firing cold: how poikilotherms compensate for temperature swings. PLoS Biol. 8(8):e1000470.

Ryan LA, Seymour CB, O'Neill-Mehlenbacher A, Mothersill CE. 2008. Radiation-induced adaptive response in fish cell lines. J Environ Radioact. 99(4):739–747.

Schettino G, Folkard M, Michael BD, Prise KM. 2005. Low-dose binary behavior of bystander cell killing after microbeam irradiation of a single cell with focused c(k) X rays. Radiat Res. 163(3):332–336.

Seymour CB, Mothersill C, Alper T. 1986. High yields of lethal mutations in somatic mammalian cells that survive ionizing radiation. Int J Radiat Biol Relat Stud Phys Chem Med. 50(1):167–179.

Thomas P, Tracy B, Ping T, Baweja A, Wickstrom M, Sidhu N, Hiebert L. 2007. Relative biological effectiveness (RBE) of alpha radiation in cultured porcine aortic endothelial cells. Int J Radiat Biol. 83(3):171–179.

Wheeler KT, Hickman R, Nelson GB, Moore SK, Wallen CA. 1992. Relationship between DNA damage, DNA repair, metabolic state and cell lethality. Radiat Environ Biophys. 31(2):101–115.

Investigation of radon sources, health hazard and risks assessment for children using analytical and geospatial techniques in District Bannu (Pakistan)

Huma Shakoor, Noor Jehan, Sardar Khan, and Nimat Ullah Khattak

ABSTRACT

Purpose: Radon (Rn) is a radioactive, odorless, and colorless gas which has a half-life of 3.83 days. One of the main sources of Rn which is directly consumed by the population is Groundwater (Tube well, Bore well, Hand pump). Rn gas is found naturally in rock, soil and water and can be considered as main health risk factor in terms of lung cancer, stomach diseases, leukemia and childhood cancer. The objective of this study was to determine the concentration of Rn in the drinking water sources, appraisal of health risk for children in District Bannu, Pakistan.
Material and method: Total of 98 drinking water samples were analyzed by using RAD-7 detector. The experimental data was statistically analyzed by using Pearson's test. The experimental and epidemiological data of the study area are shown on map using ArcGIS version 10.5.
Results: The analytical results show that Rn in drinking water was found varying from 10.1 Bq/l to 53.1 Bq/l with the average highest and lowest depth of 60 ft to 550 ft respectively. Pearson's test was used to show the concentration of Rn verses the depth of the water sources so +1 positive linear correlation was observed among the depth of water sources and the concentration of Rn. Out of 98 drinking water samples 40 sample were above the maximum contaminant level of 11.1 Bq/l (MCL) set by WHO, 2002. The effective doses (AED and DE_{ing}) for children ranges from 0.00001 to 3.792 mSv/y which exceeds the Permissible Exposure Limit (PEL) of Rn (0.1mSv/y) in 30 drinking water samples . On the basis of analytical results Rn high concentration areas are shown on the map using IDW model of interpolation and health risks were shown in areas where Rn content was above the maximum contaminant level. High correlations of diseases related to Rn were observed amongst the residence of the study area. Gastrointestinal diseases, brain tumor, lung cancer and kidney diseases were observed among the children of the study area.
Conclusion: From the overall analysis it was observed that high Rn concentration in drinking water may cause substantial health damage in children after long term exposure.

GRAPHICAL ABSTRACT

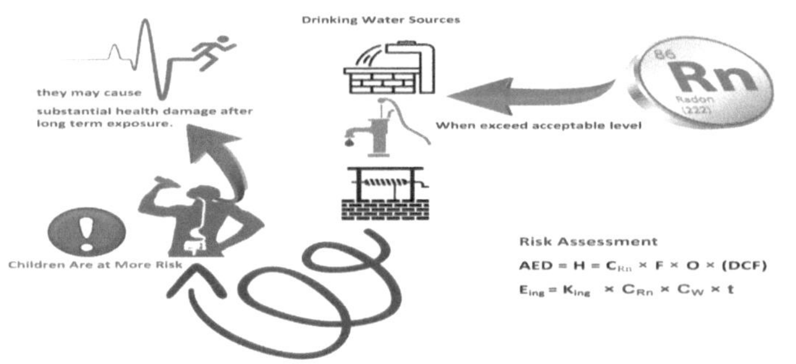

Introduction

In the early 1900s, Radon (Rn) was detected by a German scientist, Dorn. ^{219}Rn (actinon), ^{220}Rn (thoron), and ^{222}Rn are three naturally occurring isotopes of Rn (Al-Awadi and Khan 2019). It is a natural radioactive gas which is produced from the Radioactive decay of Radium (^{226}Ra), which in turn is the decay product of Uranium (^{238}U) Radioactive decay series (Partington 1957). Rn is a chemically inert gas without any color, odor or taste. Its nucleus is rich of neutron concentration due to which it becomes the densest Radioactive in nature (Goff and Becker 2010). It decays into Rn progenies by releasing particularly alpha particles that on inhalation or ingestion can pose great threat to a human's life (Alter and Oswald 1988). Transportation of Rn in soil is a mechanical flow of air and water (Chauhan et al. 2008). The bioaccumulation of Rn occurs in the lungs at higher concentration and at lesser degree other organ are affected through Rn gas dissolved in blood (Auvinen et al. 1996).

Stomach is the organ which receives the highest Rn dose, about the total of 90% dose (Kendall and Smith 2002). Due to the Radioactive and hazardous effects high concentration of Rn in indoor environment and drinking water poses a major threat to human health (National Research Council 1999; Leenhouts and Brugmans 2001; Keramati et al. 2018). Inhalation and ingestion are the two main ways by which human are exposed to Rn. Lung cancer due to the inhalation of Rn gas has got attention over the past few decades (Lubin 1994; Al-Awadi and Khan 2019). Lung cancer in late 1980s in the general population has been attributed to environmental Rn exposure in dwellings (Samet and Eradze 2000; Alberg and Samet 2003). The World Health Organization declared Rn as the second leading cause of lung cancer after tobacco smoke (Samet and Eradze 2000; Alina et al. 2012). When Rn gas is inhaled or ingested it dissolves in the body fat and its progeny moves within the human body and reaches the bone marrow (Alberg and Samet 2003). The accumulated dose of Rn in adult body is high, and gives way to the development of leukemia (Sandler et al. 2006). Rn and its progenies also produce melanoma and some childhood cancer (BEIR V 1990; National Research Council 1999). In the sampling areas, households obtained drinking water from tube wells, dug wells, bore wells, but in spite of this the exposure to Rn occurs primarily through inhalation.

This research shows the methodology used for the detection of Rn concentration of underground water samples collected from different locations of Bannu district (Pakistan). In the study area houses are mostly constructed from mud with a proper ventilation system. So indoor Rn inhalation risk is reduces as compared to the Rn ingestion due to directly consumption of ground water. There is no previous study concerning the Rn content in drinking water sources of district Bannu. The results are considered very important for public health. This study also draws a picture of detection of Rn concentration and its health effects on children.

Material and methods

Description of the study region

The Bannu district lies in the Bannu valley, which is a low structure valley that lies between 32°43″ and 33°06″ north latitudes and from 70°22″ to 70°57″ east longitude, divided into 4 tehsil and 49 union councils, in total area 2 major tehsil (Bannu and Domel) were selected for investigation, study area and sample location are shown in Figure 1. The total area of the district is 1227 km^2 with a population density of 552 persons per square kilometer. District Bannu is located in a hot and dry region and the water consumption per capital was 2 liter/day and 5–6 liters/day for cleaning and washing purpose. The only water resources in Bannu district are ground water sources (Tube well, bore well, Hand pump) with depth from 60 ft to 550ft. which is pumped out and distributed without a purification process to the community for consumption.

Water sample collection

A total of 98 drinking water samples were collected from different water sources ranging from a depth of the tube wells (TW- 230-550ft), Bore wells (BW- 60-230ft) and Hand pumps (HP- 60-120ft) for the determination of Rn concentration in order to know the environmental and health exposure from Rn in the study area. Number of samples and depth of water sources are shown in Table 1 and Figure 1.

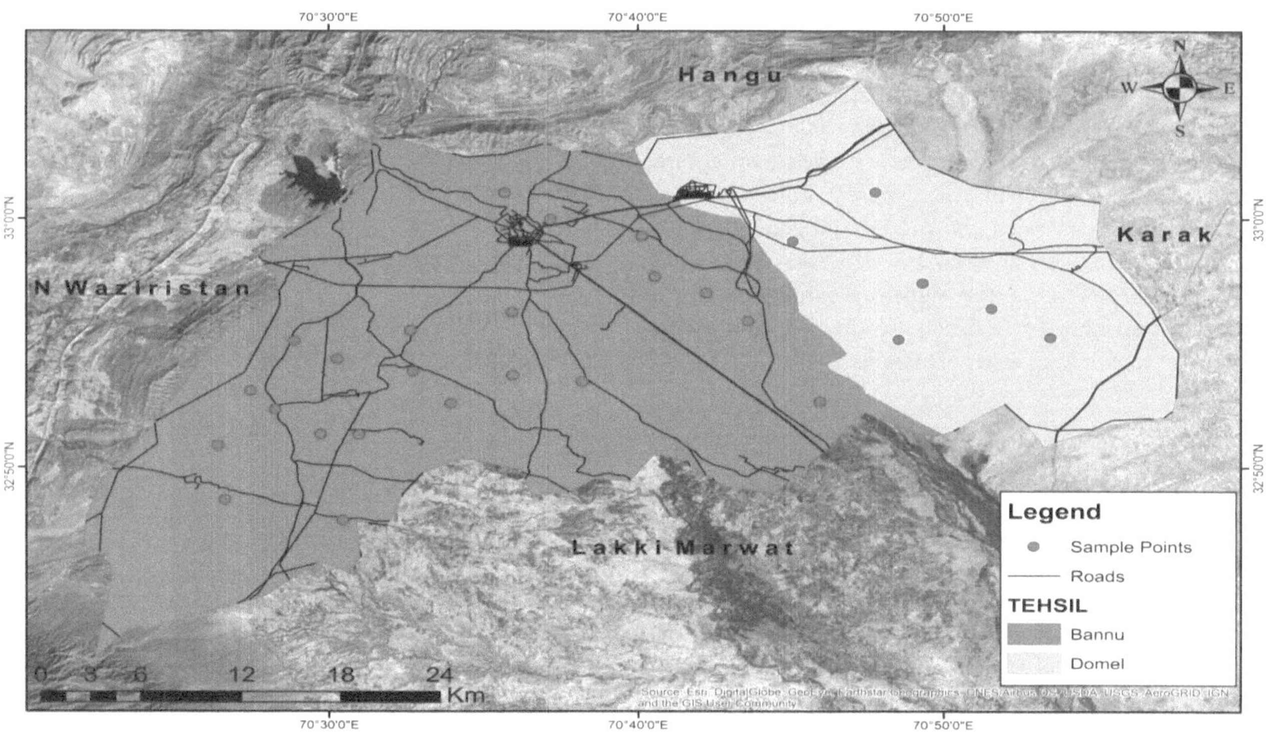

Figure 1. Location Map of District Bannu, Pakistan showing sampling sites.

Table 1. Drinking water samples collected from different ground water sources.

		Water Source			Depth of wells/ft		
S. no	Location	Tube well	Bore well	Hand pump	TW	BW	HP
1	District Bannu	40	29	29	230–550	60–230	60–120

Source: WSSP information

Rn concentration measurement

For the detection of Rn concentration in drinking water samples and assessment of annual effective doses and dose of ingestion for children were analyzed by using the following methodologies;

- Analytical
- Statistical
- Geostatistical

Analytical method: used for detection of Rn concentration in drinking water

For the determination of Rn concentration in drinking water of district Bannu RAD-7 radon detection equipment was used. The apparatus RAD-H_2O was developed by the manufacturer specifically to perform the detection and measurement of Rn gas in water (Avery et al. 2018). This apparatus presents a result after a 30 min analysis. After every sample analysis the apparatus is purged to remove the impurities and reduce the internal humidity of RAD-7 detector. The Rn content in water sample during the process of analysis ^{222}Rn nucleus that decay with in the cell and converted into a decay product positively charged ion ^{218}Po. The half-life of ^{218}Po is 3.1 min and emits alpha radiation which decays upon the detector's active surface, the alpha particles have a probability of 50% to enter the detector and produces an electrical signal proportional in strength, to the energy of the alpha particles (Faheem and Rahman 2008). The same alpha nucleus decays and produces beta particles, which are not detected in the detection cell. Different isotopes of Rn have different alpha energies, and produce different strength signals in the detector. The RAD-7 detectors RAD-H2O amplifies filters and sorts the signals according to their strength and uses only ^{218}Po signals for the detection of Rn concentration in drinking water (Durridge 2015).

Statistical analysis using Pearson's test: correlation of *Rn* concentration verses depth of water sources

Pearson's correlation coefficient is the test statistics that measures the statistical relationship, or association between two continuous variables (Rn concentration verses depth of water sources). It is known as the best method of measuring the association between variables of interest because it is based on the method of covariance. It gives information about the magnitude of the association, or correlation, as well as the direction of the relationship.

Geostatistical analysis: using spatial interpolation techniques: hazard mapping

Inverse Distance weighted (IDW) is a GIS based method of interpolation that estimate cell values by averaging the values of sample data points in the neighborhood of each processing cell. It is a process that uses random functions to assume that the surface or values created follow a pattern of spatial covariance. Random variables that vary in the same direction produce a positive correlation between two random values, or they may follow opposite directions which will produce a negative correlation between sample points (Mitas and Mitasova 1999). IDW predict the points closer to the predication location greater than those farther away. The general formula used for IDW is:

$$Z\ (S_0) = \sum_{i=1}^{N} \lambda i Z(Si)$$

here, the predication value is represented by Z (S_0) for location, N is the number of measured sample points surrounding the prediction location that will be used in the prediction. λ are the weights assigned to each measured point that used in analysis. These weights will decreases with distance. Z(s) is the observed value at the location

Health risk assessment

Calculation of annual effective doses

The concentration of Rn gas in drinking water is calculated by using Equation (1) used for the detection of annual effective doses introduced by the UNSCEAR report as

$$ADE = C_{Rn} \times F \times O \times (DCF) \quad (1)$$

where C_{Rn} is the Rn concentration measured in (Bqm^{-3}), the equilibrium factor between Rn and its progeny is F its value is (0.49 and 0.40 for smoke-bearing and non-smoke environment respectively), the average indoor occupancy time per person is 7000 ha^{-1} which is represented by O and the Dose conversion factor for Rn concentration is DCF and its value is 9nSv h^{-1}(Bq m^{-3})$^{-1}$ (Cross et al. 1985; UNSCEAR 1994).

Calculation of doses of ingestion for children

Rn concentration Dose ingestion through drinking water was detected in children by using Equation (2):

$$DE_{ing} = k_{ing} \times C_{Rn} \times C_w \times t \quad (2)$$

In this equation, DE_{ing} is the annual effective dose of Rn ingestion which is present in drinking water it is expressed in mSv unit, for children the conversion factor of ingestion is expressed as k_{ing} its value is in drinking water is (2×10^{-8}). Rn concentration is represented by C_{Rn}, Cw is the water consumption of 2 L/day. During 365 days it was considered and 't' is the time span of the consumption spend that 365 days (Whidden et al. 2010; Inácio et al. 2017).

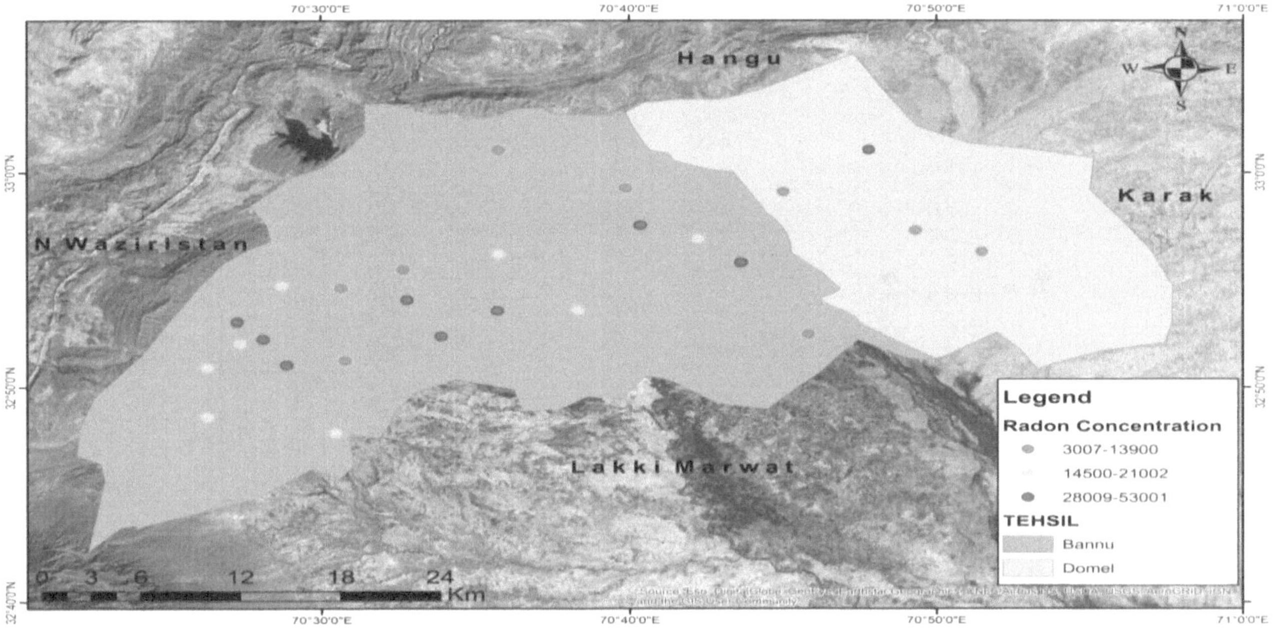

Figure 2. Rn concentration estimation in drinking water samples of district Bannu (Pakistan).

Investigation of epidemiological data

Sample size is 100 based on 95% prevalence and 5% precision according to the following formula:

$$n = \frac{(Z^1/_2)^2 pq}{d^2}$$

However, for maximum validity, we will take sample size of 100 from the selected area. A representative sample of children investigated using a questionnaire. The survey will be carried out on extensive scale hospitals, schools and exposed population such as houses and mosques. For investigation of epidemiological data, different age group children were selected for study. The occupant below 15 years is 48.08% of the total samples. Out of this 18.13% is below 5 years age including 2.56% infants. Both male and female children were selected for investigation of health related data.

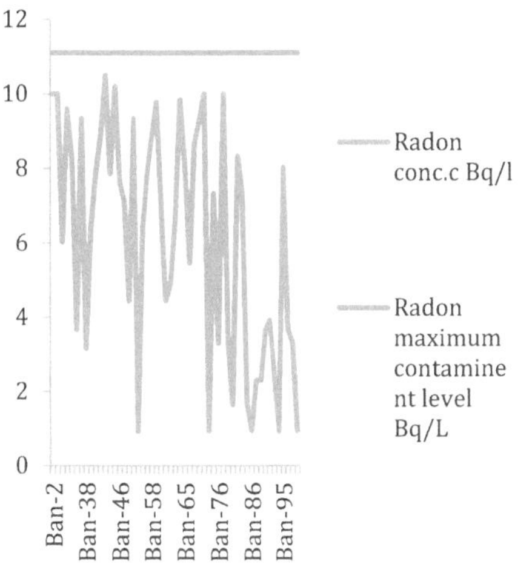

Figure 3. Drinking water Rn concentration with in the permissible limit of 11.1 Bq/l.

Results of Rn concentration estimation

Analytical results of Rn *concentration in drinking water*

The analytical results show that Rn concentration of drinking water samples varies from 10.1 Bq/L to 53.2 Bq/L. Figure 2 shows the variation of Rn concentration with different locations. In all the studied samples the higher Rn concentration in drinking water was observed in tube wells of district Bannu Pakistan. The USEPA has proposed that the allowed maximum contaminant level (MCL) for Rn concentration in drinking water is 11.1 Bq/l. This study reveals that out of 98 drinking water samples 40 samples have Rn levels higher than the maximum contaminant level of 11.1 Bq/L presents in Figures 2 and 3. Highest values of 53.1 Bq/l were found in the water samples from the tube well with depth of 550 ft of village Kallanger and Tur kakki. Lowest value of 3.007 Bq/L of Rn was observe in water sample collected from hand pump with an average depth of 60 ft of district Domel presented in Figure 2.

Statistical analysis and correlation of Rn *concentration verses depth of water sources*

Pearson correlation of Depth and Rn concentration is equal to 0.847 and p value is equal to 0.000, Pearson's test conclusion indicates that the strength of association between the depth of water sources and the concentration of Rn is very high ($r = 0.847$), and that the correlation coefficient is significantly different from zero ($p < .001$). We can also say that 94% (0.847^2) of the depth of the water sources effect the concentration of Rn in the study area shown in (Table 2).

Results of health hazards related to Rn content in drinking water

The World Health Organization (WHO) and United state Environmental protection agency (USEPA) recommended

Table 2. Pearson's Correlations of Rn concentration verses depth of water sources.

	Depth	Concentration
Depth		
Pearson Correlation	1	0.847**
Significance (2-tailed)		0.000
N	94	94
Conc.		
Pearson Correlation	0.847**	1
Significance (2-tailed)	0.000	
N	94	94

**Correlation is significant at the .001 level (2-tailed).

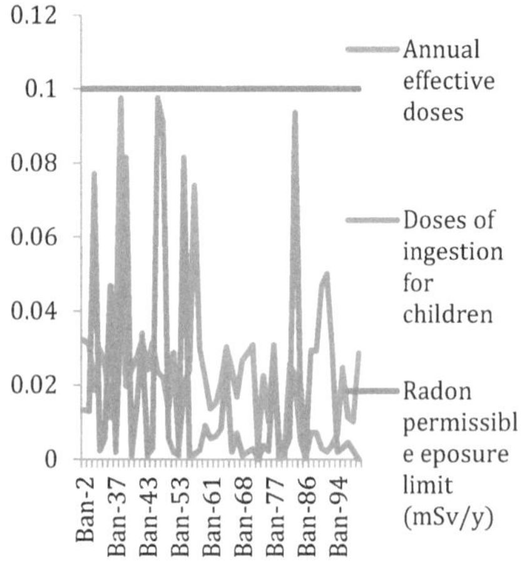

Figure 4. Drinking water Rn Doses of ingestion for stomach and annual effective doses below the exposure limit of WHO (0.01mSvy).

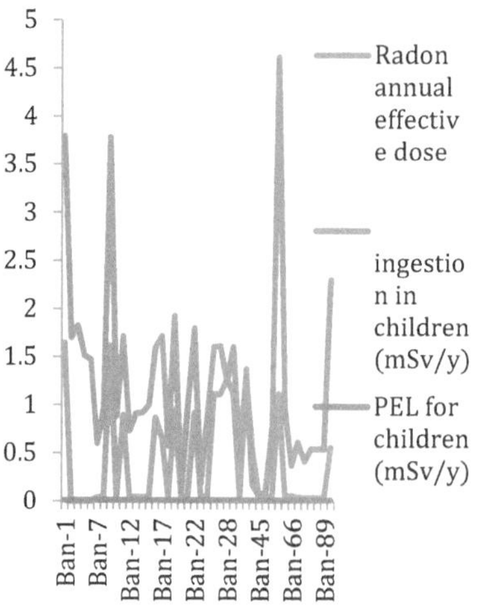

Figure 5. Drinking water Rn Doses of ingestion for stomach and annual effective doses above the exposure limit of WHO (0.01mSvy).

reference level of the annual effective dose received from drinking water is 0.1mSv/y. according to the WHO if annual effective dose and dose of ingestion is below the reference dose of 0.1mSv/y then the water is appropriate for drinking purposes and no additional action is required, if the calculated doses are above the reference value of WHO (0.1mSv/y), then remediation technology is required to reduce the risk.

Risk assessment of annual effective doses and doses of ingestion for children

In the present investigation, mean annual effective doses and ingestion dose for stomach received by children from the drinking water of the study area were calculated and compared with the permissible exposure limit of world health organization 0.01 mSv/y (WHO 2002). The effective dose received annually through drinking water by the children was 0.005–2.001 mSv/y, respectively. The effective doses received annually through ingestion of Rn in drinking water were 1.789–3.792 mSv/y, and 0.001–3.696 mSv/y presented in Figures 4 and 5. Doses of ingestion and annual effective doses presented in Figure 4 are quite below the reference level of 0.1mS/y of WHO and, therefore, do not pose any health problem from Rn dose received from drinking water in the study area. In case of children the conversion factor for dose of ingestion is 2×10^{-8}, children are more sensitive to Rn contaminated water. The effective dose received annually through drinking water from tube well in the children was 0.011 and 3.687 mSv/y, respectively. About 56% of samples collected from tube wells of different location of district Bannu, children receive high doses of Rn from these water sources. It will be a good practice to take precautionary measure to reduce exposure to Rn in drinking water. Rn doses received by the population of the study area above the reference dose of 0.1mSv/y are presented in Figure 5.

Hazard mapping for Rn concentration in drinking water

GIS application is one of the approaches which are used to explore the problem for detection of Rn concentration in District Bannu Pakistan. Inverse digital weighted model has been used for the prediction purpose of Rn concentration in ground water samples. The basic aim of this investigation is to predict the neighborhood values of the unmeasured areas from measured values. Results of the IDW prediction models are presented on map in which the red color shows high values of Rn in the ground water samples of selected study areas collected from different sources. While blue color presented low values of Rn in the analyzed drinking water samples. IDW modeling is analysis that attempts to locate 'hotspot' areas of increased Rn content in the study area from high to low concentrations. High concentration of Rn was detected in the following villages of District Bannu; Bharat, tur kaly, kalanger kally, Nurang which is presented as hot spot areas on (Figures 6 and 7). Inverse Distance weighted interpolation clearly implements the assumption that things that are close to one another are more alike than those that are far away. The purposes of this analysis are to predict the mean of Rn high concentration hot spot areas and its effects on the health of the inhabitant of the study area and prediction of neighborhood values from measured value shown in (Figure 6).

Discussion

Rn gas is present everywhere in the rocks of the earth's crust and is being released regularly through the ground water

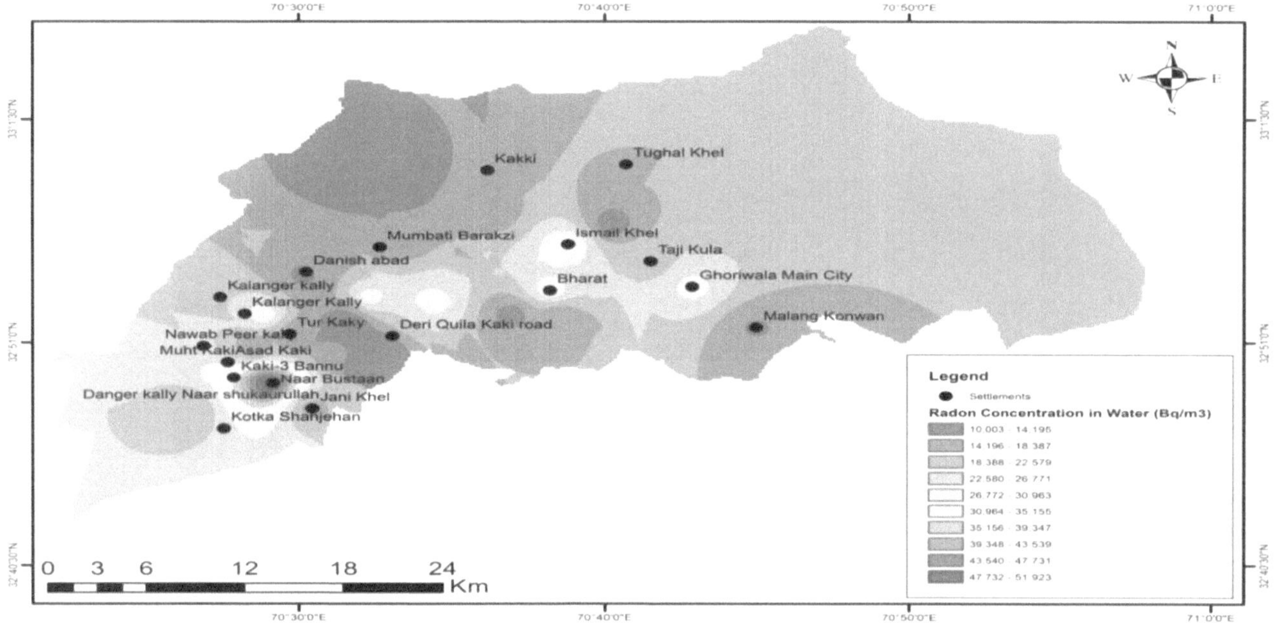

Figure 6. Evaluating the Potential of a Geospatial/Geostatistical Methodology for Locating Hot spot area and its effects on the neighborhoods unmeasured values of District Bannu.

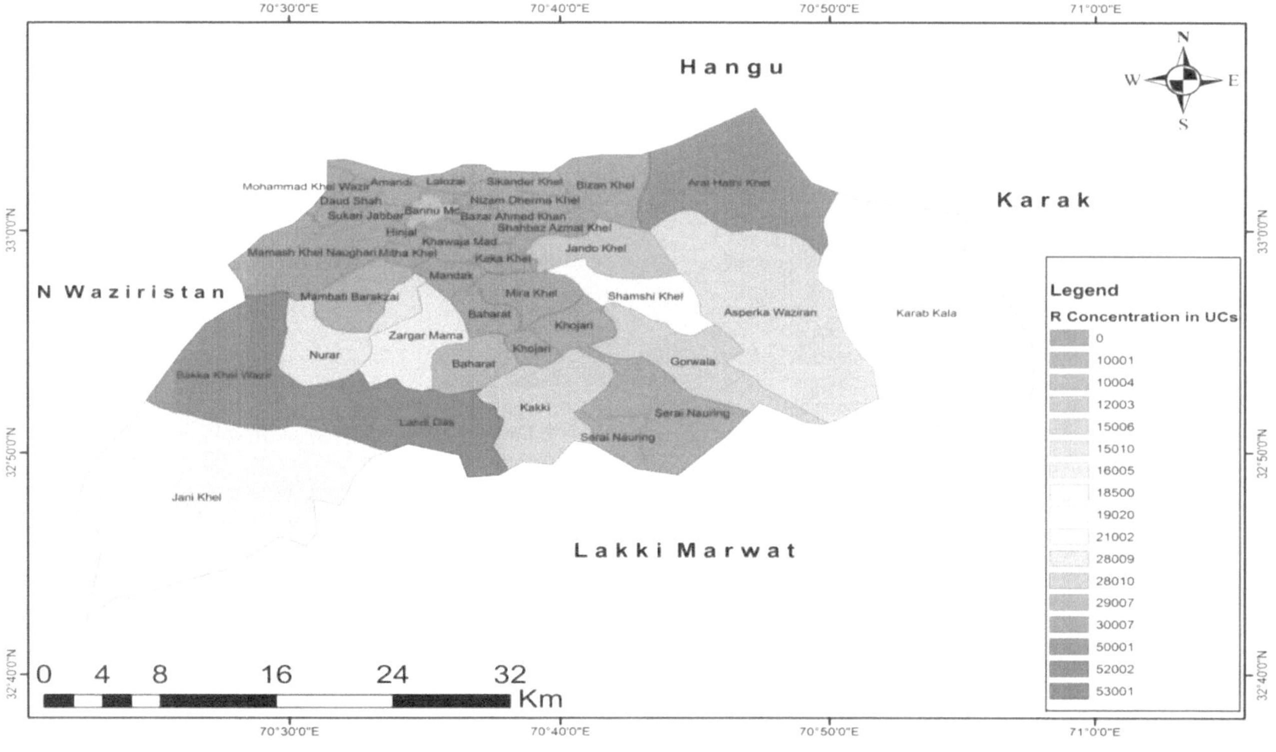

Figure 7. Hazard mapping for Rn concentration detection in drinking water of the study area.

present in the pore spaces of the strata, from the literature it was observed that high concentration of the Rn gas in ground water is due to the solubility property of the gas (Kearfott 1989). The observed analytical data of Rn in drinking water samples of the study area showed variation in their concentration. From the Geostatistical analysis of Rn content and its co-relation with diseases is observed as a significant co-relation in district Bannu (Figure 8). Maximum numbers of intestinal diseases, Brain disorder, neurological and brain tumor patients were witnessed (Figure 8). From the Pearson's test it was observed that high co-relation of Rn concentration with the depth of water sources were observed. The average effective doses ingested by the population through drinking water are above the permissible limits 0.01mSv/y set by (USEPA 2000). Study on Rn ingestion showed that multifold MCL inhalation can cause CNS brain tumors (Ruano-Ravina et al. 2017, 2018). From the epidemiological data of selected study area it was professed that

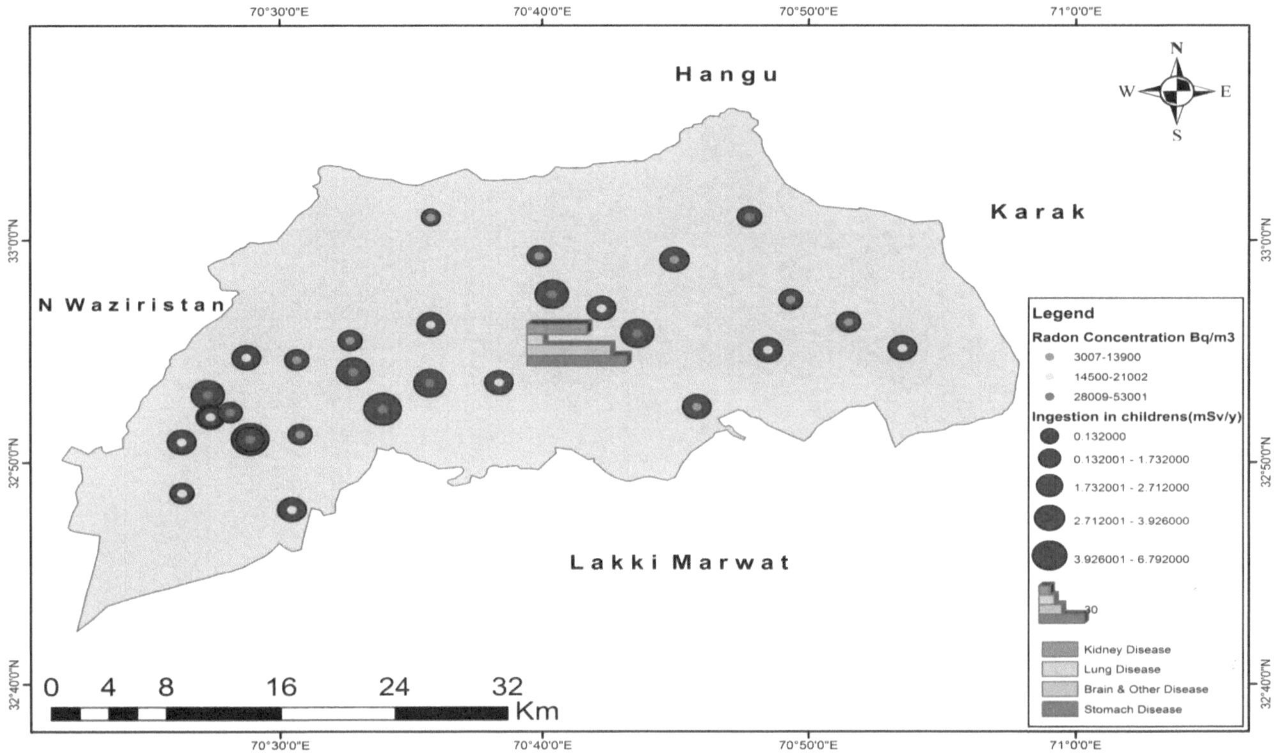

Figure 8. Rn concentration in drinking water doses of ingestion for children and risk of disease.

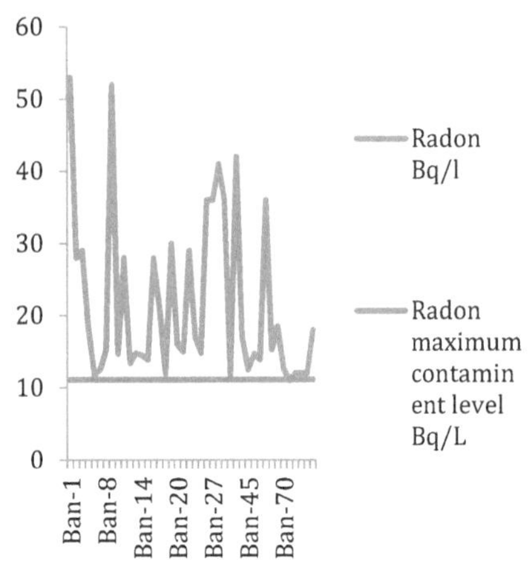

Figure 9. Drinking water Rn concentration above the permissible limit of 11.1 Bq/l.

maximum number of stomach diseases were observed among the occupants of the study area. Previous research work showed a high correlation between Rn ingestion in children regarding intestinal diseases, anemia and long term exposure causes childhood cancer (Bergdahl et al. 2010). The analyses have found statistically significant, positive associations between intestinal diseases and residential Rn levels among children, but the doses of ingestion in children was observed above the MCL (Figure 4) MCL of 0.1mSv (WHO 2011). From the experimental data it was observed that Rn rich drinking water causes gastrointestinal diseases (Figure 9) previous study showed that long term exposure to Rn gas ingestion can cause stomach cancer (Darby et al. 2005). Earlier ecological studies suggested positive associations between Rn levels in ground water sources and incidences of cancers, (Lorenzo-González et al. 2019) including lung cancer (López-Abente et al. 2018), all cancers combined (Mose et al. 1990) and childhood cancer, leukemia (Collman et al. 1990). Rn concentration doses of ingestion for children and its co-relation with diseases are shown in Figure 8 below.

Conclusions

On the earth 27% of the total population is effected from the radioactive exposure of Rn gas (Darby et al. 2005) and the highest level of Rn gas enters into the human body through ingestion (drinking Rn concentrated water) inhalation (breathing in from the air). From the health point of view, ^{222}Rn is the most important Rn isotope because its decay progenies, ^{218}Po and ^{214}Po can have a marked adverse effect in human tissues. Based on the detection of the concentration of Rn in drinking water samples portable device Rad-7 was used to measure the Rn concentration from the following drinking water sources (Tube well, bore well and hand pump). The detected concentrations of Rn in 40 drinking water samples were above the recommended level of WHO and UNSCEAR (11.1 Bq/l). Ingestion of Rn rich water annual effective dose of stomach for children in few isolated samples is multifold above the permissible values of 0.1 mSv/y (set by WHO 2011). Taking into account public health it is concluded that the measured drinking water sources are considered not safe for drinking purposes, according to the limits set by international organization for the protection of human from the exposure to radioactive doses. Dose should not be above the limit of 0.01 mSv/y and dose recommended by the United Nations Scientific Committee on the effect of atomic Radiation suggests that the contribution of Rn ingestion through drinking water for the total mean annual effective doses should be 0.002mSv (UNSC 2000; Inácio 2017). From the health risk assessment of Rn in drinking water for children, it is concluded that the children of the study area are more vulnerable to the Rn doses of ingestion from drinking water. On the basis of the conclusion recommendations are set out to protect the population of the study area from the health related issues due to high level Rn ingestion.

Recommendation

1. It is recommended that inhabitants of the District Bannu should rely on processed water by avoiding water from tube wells or direct supply lines, which are direct sources of Rn.
2. The inhabitants of vulnerable Rn high content zone should aerate or boil their water before use.
3. It is recommended to the WSSP of Khyber Pakhtunkhwa to install water treatment plants in the proposed study area, which will help in providing protection against mentioned diseases.
4. The area where the concentration of Rn is above the permissible limits, the inhabitants of the area should avoid smoking because smokers are more affected by Rn gas exposure as compared to nonsmokers.
5. In areas where the concentration of Rn is high, the population should be guided to construct well-ventilated homes to reduce the concentration of indoor Rn gas.
6. Observation and evaluation of toxic risks and phenomena in the community, their reporting to health authorities, and the promotion of actions aimed at reducing the risks.

Disclosure statement

No potential conflict of interest was reported by the author(s).

ORCID

Huma Shakoor http://orcid.org/0000-0001-5531-4400

References

Al-Awadi L, Khan AR. 2019. Indoor radon levels in schools and residential dwellings in Kuwait. Int J Environ Sci Technol. 16: 2627–2636.

Alberg AJ, Samet JM. 2003. Epidemiology of lung cancer. Chest. 123: 21S–49S.

Alter HW, Oswald RA. 1987. Nationwide distribution of indoor radon measurements: a preliminary data base. JAPCA. 37(3):227–231.

Arain MB, Kazi TG, Jamali MK, Afridi HI, Baig JA, Jalbani N, Shah AQ. 2008. Evaluation of physico-chemical parameters of Manchar Lake water and their comparison with other global published values. Pak J Anal Environ Chem. 9:9.

Auvinen A, Mäkeläinen I, Hakama M, Castrén O, Pukkala E, Reisbacka H, Rytömaa T. 1996. Indoor radon exposure and risk of lung cancer: a nested case-control study in Finland. JNCI. 88(14): 966–972.

Avery E, Bibby R, Visser A, Esser B, Moran J. 2018. Quantification of groundwater discharge in a subalpine stream using radon-222. Water. 10:100.

BEIR V. 1990. Health effects of exposure to low levels of ionizing radiation. National Research Council, Committee on the Biological Effects of Ionizing Radiations. Washington (DC): National Academy of Sciences.

Bergdahl L. 2010. Seeing otherwise: renegotiating religion and democracy as questions for education [Doctoral dissertation]. Stockholm: Department of Education, Stockholm University.

Bijwaard H, Brugmans MJP, Leenhouts HP. 2001. A consistent two-mutation model of lung cancer for different data sets of radon-exposed rats. Radiat Environ Biophys. 40:269–277.

Cardis E, Vrijheid M, Blettner M, Gilbert E, Hakama M, Hill C, Howe G, Kaldor J, Muirhead CR, Schubauer-Berigan M. 2005. Risk of cancer after low doses of ionising radiation: retrospective cohort study in 15 countries. BMJ. 331:77.

Chauhan RP, Nain M, Kant K. 2008. Radon diffusion studies through some building materials: effect of grain size. Radiat Meas. 43: S445–S448.

Ching-Chou F, Walia V, Yang TF, Lou-Chuang L, Liu TK, Cheng-Hong C, Kuo-Liang W. 2017. Preseismic anomalies in soil-gas radon associated with 2016 M 6.6 Meinong earthquake, Southern Taiwan. Terr Atmos Ocean Sci. 28:7.

Collman JP, Zhang X, Hembre RT, Brauman JI. 1990. Shape-selective olefin epoxidation catalyzed by manganese picnic basket porphyrins. J Am Chem Soc. 112(13):5356–5357.

Cook PG, Love AJ, Dighton JC. 1999. Inferring ground water flow in fractured rock from dissolved radon. Ground Water. 37:606–610.

Cornwell K. 1998. Quaternary break-out flood sediments in the Peshawar basin of northern Pakistan. Geomorphology. 25:225–248.

Cross FT, Harley NH, Hofmann W. 1985. Health effects and risks from 222Rn in drinking water. Health Phys. 48:649–670.

Darby S, Hill D, Auvinen A, Barros-Dios JM, Baysson H, Bochicchio F, Deo H, Falk R, Forastiere F, Hakama M. 2005. Radon in homes and risk of lung cancer: collaborative analysis of individual data from 13 European case-control studies. BMJ. 330:223.

Donovan TJ. 1974. Petroleum microseepage at Cement, Oklahoma: evidence and mechanism. AAPG Bull. 58:429–446.

Dörr H, Kromer B, Levin I, Münnich KO, Volpp HJ. 1983. CO2 and Radon 222 as tracers for atmospheric transport. J Geophys Res. 88: 1309–1313.

Durridge C. 2015. Radon in water accessory. Revision 2015-07-28 Edition.

Erickson BE. 2007. The therapeutic use of radon: a biomedical treatment in Europe; an "alternative" remedy in the United States. Dose Response. 5:48–62.

Faheem M, Rahman S. 2008. A review of radon measurement studies in Pakistan. J Radiol Prot. 28:283.

Fleischer RL, Giard WR, Mogro-Campero A, Turner LG, Alter HW, Gingrich JE. 1980. Dosimetry of environmental radon: methods and theory for low-dose, integrated measurements. Health Phys. 39: 957–962.

Fleischer RL, Turner LG. 1984. Correlations of radon and carbon isotopic measurements with petroleum and natural gas at Cement, Oklahoma. Geophysics. 49:810–817.

Genereux DP, Hemond HF. 1990. Naturally occurring radon 222 as a tracer for streamflow generation: steady state methodology and field example. Water Resour Res. 26:3065–3075.

Ghosh D, Deb A, Sengupta R. 2009. Anomalous radon emission as precursor of earthquake. J Appl Geophys. 69:67–81.

Goff K, Becker A. 2010. EnviRonmEnTAl PubliC HEAlTH. Public Health Foundations: Concepts and Practices, 197.

Hammood HA, Al-Khalifa IJM. 2011. Radon concentration measurement in water of Dhi-Qar Governorate (in Iraq) using emanometer. J Basrah Res Sci. 37:22–29.

Inácio M, Soares S, Almeida P. 2017. Radon concentration assessment in water sources of public drinking of Covilhã's county, Portugal. J Radiat Res Appl Sci. 10:135–139.

İnceöz M, Baykara O, Aksoy E, Doğru M. 2006. Measurements of soil gas radon in active fault systems: a case study along the North and East Anatolian fault systems in Turkey. Radiat Meas. 41:349–353.

Ismail AH. 2008. Analysis of radon concentrations in drinking water in Erbil governorate (Iraqi Kurdistan) and its health effects. Tikrit J Pure Sci. 13:1–8.

Jobbágy V, Altzitzoglou T, Malo P, Tanner V, Hult M. 2017. A brief overview on radon measurements in drinking water. J Environ Radioact. 173:18–24.

Kapała J, Mnich S, Karpińska M. 2018. Radon in human environment and carcinoma – part 2. Pol J Appl Sci. 4:5–11.

Kazi TG, Arain MB, Baig JA, Jamali MK, Afridi HI, Jalbani N, Niaz A. 2009. The correlation of arsenic levels in drinking water with the biological samples of skin disorders. Sci Total Environ. 407: 1019–1026.

Kearfott KJ. 1989. Preliminary experiences with 222Rn gas in Arizona homes. Health Phys. 56:169–179.

Kendall GM, Smith TJ. 2002. Doses to organs and tissues from radon and its decay products. J Radiol Prot. 22:389–406.

Keramati H, Ghorbani R, Fakhri Y, Mousavi Khaneghah A, Conti GO, Ferrante M, Ghaderpoori M, Taghavi M, Baninameh Z, Bay A. 2018. Radon 222 in drinking water resources of Iran: a systematic review, meta-analysis and probabilistic risk assessment (Monte Carlo simulation). Food Chem Toxicol. 115:460–469.

Kumar A, Kaur M, Sharma S, Mehra R, Sharma DK, Mishra R. 2016. Radiation dose due to radon and heavy metal analysis in drinking water samples of Jammu district, Jammu & Kashmir, India. Radiat Prot Dosimetry. 171:217–222.

Leenhouts HP, Brugmans MJP. 2001. Calculation of the 1995 lung cancer incidence in the Netherlands and Sweden caused by smoking and radon: risk implications for radon. Radiat Environ Biophys. 40: 11–21.

López-Abente G, Núñez O, Fernández-Navarro P, Barros-Dios JM, Martín-Méndez I, Bel-Lan A, Ruano-Ravina A. 2018. Residential radon and cancer mortality in Galicia, Spain. Sci Total Environ. 610: 1125–1132.

Lorenzo-González M, Torres-Durán M, Barbosa-Lorenzo R, Provencio-Pulla M, Barros-Dios JM, Ruano-Ravina A. 2019. Radon exposure: a major cause of lung cancer. Expert Rev Respir Med. 13:839–850.

Lubin JH. 1994. Radon and lung cancer risk: a joint analysis of 11 underground miners studies (No. 94). US Department of Health and Human Services, Public Health Service, National Institutes of Health.

Mitas L, Mitasova H. 1999. Spatial interpolation. In: Longley PA, Goodchild MF, Maguire DJ, Rhind DW, editors. Geographical information systems: principles, techniques, management and applications. Hoboken (NJ): Wiley.

Mose DG, Mushrush GW, Chrosniak C. 1990. Radioactive hazard of potable water in Virginia and Maryland. Bull Environ Contam Toxicol. 44(4):508–513.

Muhammad S, Khalid P. 2017. Hydrogeophysical investigations for assessing the groundwater potential in part of the Peshawar basin. Environ Earth Sci. 76:494.

Nain M, Chauhan RP, Chakarvarti SK. 2008. Alpha radioactivity in tobacco leaves: effect of fertilizers. Radiat Meas. 43:S515–S519.

National Research Council. 1999. Health effects of exposure to radon: BEIR VI. Vol. 6. Washington (DC): National Academies Press.

Partington JR. 1957. Discovery of radon. Nature. 179:912–912.

Price RR. 1987. Evaluation of occupational and environmental exposures to radon and radon daughters in the United States (NCRP Report No. 78). Med Phys. 14:427–428.

Rafique M, Rahman SU, Rahman S, Nasir T., Matiullah 2011. Radiation doses due to indoor radon exposure, before and after the 2005 earthquake, in dwellings of muzaffarabad and the jhelum valley, Azad Kashmir, Pakistan. Indoor Built Environ. 20:259–264.

Rangaswamy DR, Srinivasa E, Srilatha MC, Sannappa J. 2016. Measurement of radon concentration in drinking water of Shimoga district, Karnataka, India. J Radioanal Nucl Chem. 307:907–916.

Ruano-Ravina A, Aragonés N, Kelsey KT, Pérez-Ríos M, Piñeiro-Lamas M, López-Abente G, Barros-Dios JM. 2017. Residential radon exposure and brain cancer: an ecological study in a radon prone area (Galicia, Spain). Sci Rep. 7:1–7.

Ruano-Ravina A, Dacosta-Urbieta A, Barros-Dios JM, Kelsey KT. 2018. Radon exposure and tumors of the central nervous system. Gac Sanit. 32:567–575.

Samet JM, Avila-Tang E, Boffetta P, Hannan LM, Olivo-Marston S, Thun MJ, Rudin CM. 2009. Lung cancer in never smokers: clinical epidemiology and environmental risk factors. Clin Cancer Res. 15: 5626–5645.

Samet JM, Eradze GR. 2000. Radon and lung cancer risk: taking stock at the millenium. Environ Health Perspect. 108:635–641.

Sandler DP, Weinberg CR, Shore DL, Archer VE, Bishop Stone M, Lyon JL, Rothney-Kozlak L, Shepherd M, Stolwijk JAJ. 2006. Indoor radon and lung cancer risk in Connecticut and Utah. J Toxicol Environ Health A. 69:633–654.

Sextro RG. 1994. Radon and the natural environment. In: Nagda N, editor. Radon: prevalence, measurement, health risks and control. MNL15. West Conshohocken (PA): ASTM International. p. 9–32.

United Nations Scientific Committee (UNSC) on the Effects of Atomic Radiation. 1994. Sources and effects of ionizing radiation. UNSCEAR 1994 report to the General Assembly, with scientific annexes.

United Nations Scientific Committee (UNSC) on the effects of atomic radiation 2000. Annex a: exposures from natural sources of radiation. Tech Rep. United Nations Scientific Committee on the Effects of Atomic.

United States Environmental Protection Agency (USEPA). 2000. National Primary Drinking Water Regulations: Radionuclides. Final Rule. Environmental Protection Agency 40 CFR Parts 9, 141, and 142. Federal Register. Vol. 65, no. 236, 2000 Dec 7.

Whidden C, Beiko RG, Zeh N. 2010, May. Fast FPT algorithms for computing rooted agreement forests: theory and experiments. In: International Symposium on Experimental Algorithms. Springer, Berlin, Heidelberg. p. 141–153.

World Health Organization (WHO) 2002. Occupational health: a manual for primary health care workers (No. WHO-EM/OCH/85/E/L).

World Health Organization (WHO). 2011. Guidelines for drinking-water quality. WHO. 216:303–304.

Zalewski M, Karpiñska M, Mnich Z, Kapała J, Zalewski P. 2001. Study of 222Rn concentrations in drinking water in the north-eastern hydroregions of Poland. J Environ Radioact. 53:167–173.

From tangled banks to toxic bunnies; a reflection on the issues involved in developing an ecosystem approach for environmental radiation protection

Carmel E. Mothersill, Deborah H. Oughton, Paul N. Schofield, Michael Abend, Christelle Adam-Guillermin, Kentaro Ariyoshi, Nicholas A. Beresford, Andrea Bonisoli-Alquati, Jason Cohen, Yuri Dubrova, Stanislav A. Geras'kin, Tanya Helena Hevrøy, Kathryn A. Higley, Nele Horemans, Awadhesh N. Jha, Lawrence A. Kapustka, Juliann G. Kiang, Balázs G. Madas, Gibin Powathil, Elena I. Sarapultseva, Colin B. Seymour, Nguyen T. K. Vo, and Michael D. Wood

ABSTRACT

The objective of this paper is to present the results of discussions at a workshop held as part of the International Congress of Radiation Research (Environmental Health stream) in Manchester UK, 2019. The main objective of the workshop was to provide a platform for radioecologists to engage with radiobiologists to address major questions around developing an Ecosystem approach in radioecology and radiation protection of the environment. The aim was to establish a critical framework to guide research that would permit integration of a pan-ecosystem approach into radiation protection guidelines and regulation for the environment. The conclusions were that the interaction between radioecologists and radiobiologists is useful in particular in addressing field versus laboratory issues where there are issues and challenges in designing good field experiments and a need to cross validate field data against laboratory data and vice versa. Other main conclusions were that there is a need to appreciate wider issues in ecology to design good approaches for an ecosystems approach in radioecology and that with the capture of 'Big Data', novel tools such as machine learning can now be applied to help with the complex issues involved in developing an ecosystem approach.

General introduction

> It is interesting to contemplate an entangled bank, clothed with many plants of many kinds, with birds singing on the bushes, with various insects flitting about, and with worms crawling through the damp earth, and to reflect that these elaborately constructed forms, so different from each other, and dependent on each other in so complex a manner, have all been produced by laws acting around us.
>
> *Charles Darwin*
>
> *On The Origin of Species by Means of Natural Selection, or Preservation of Favoured Races in the Struggle for Life. London: John Murray, 1859.*

The final paragraph of Darwin's monumental monograph preempts our modern understanding of the interconnectedness of an ecosystem, its complexities and the variety of processes at play within it. Understanding the impact of environmental contamination on an ecosystem is a fearsome task, assessing damage still more so, given the subtlety of perturbations that might be later amplified to produce potentially catastrophic changes. In a series of papers, on topics where radiobiology and radioecology interact, of which this is the third, we have attempted to draw from the community a consensus view on how to capture the effects of radioactive contamination on an ecosystem, with the aim of defining ecological parameters that might be used for development of safety and regulatory guidelines, and help us to further understand the fundamental science of radioecology.

This discussion paper results from the 5th workshop organized by the International Union of Radioecologists (IUR) which was held in Manchester UK on 26th and 27th August 2019 during the International Congress of Radiation Research (ICRR). The aim was to stimulate the interaction of radiobiologists (who traditionally study mechanisms of radiation action at the physical, chemical and biological level) and radioecologists (who are more focused on uptake, transfer and effects of radiaoisotopes in ecosystems). The particular hope was to engage both groups in the on-going discussions about approaches to developing an ecosystem approach in environmental radiation protection. By an Ecosystem Approach we mean using methods and concepts within radiological environmental protection, which target populations and their interactions with other biota and abiotic components of ecological systems (Bradshaw et al. 2014). This allows a more holistic focus on the multiple facets that reflect the environment and the complexity of factors and interactions underlying the ultimate outcome for populations living in environments contaminated by radionuclides. The series of workshops stem from an initial consensus recognition that ecosystem approaches are better suited to fulfill environmental radiation protection goals (consensus statements from Miami Symposium, 2015; see Bréchignac et al. 2016). The 2019 workshop started with a discussion of three 'provocative statements' designed to be controversial. These served to determine what was impeding development of an ecosystem approach and what the concerns might be in formulating a conceptual roadmap. The discussion then moved to how we could progress toward an ecosystem approach in a practical sense including consideration of modeling needs, the role of big data, rigor in experimental design and data sharing, and the feasibility of tackling such a complex subject. Contributions from each attendee were sought and these have been edited to produce a summary for this paper as a prelude to the core discussions on how to identify and remove the blocks to progress in this field. Because the contributions from delegates provide a fascinating window to the multiple shades of opinion that were represented and aired, they will be published as submitted on the website of the International Union of Radioecologists (http://iur-uir.org). A key focus of the workshop was not only to reach areas of agreement, but also to: (a) better understand areas where consensus may not be reached; (b) understand what we disagree on and why; (c) identify what the knowledge gaps are and (d) propose what studies and experiments are required to fill those gaps. This paper summarizes the discussions at the workshop and does not necessarily represent agreement by all authors. Where required, some background to these discussions is given, but the paper is not intended to be a comprehensive review of the state of the art in the topic area.

Broad 'provocative statements'

Provocative Statement 1

Radiation is not a problem in the environment

Summary answer: The statement clearly begs the question 'in which context', e.g. do we mean in the context of regulated releases or at contaminated sites such as the Chernobyl Exclusion Zone (CEZ) and Fukushima Exclusion Zone FEZ) but discussions highlighted the different assumptions that might underlie such a point of view. For example, there could be disagreements as to whether effects have been demonstrated at all in contaminated areas, whether effects seen are due to radiation exposure, either past or present, or whether any changes seen have ecological relevance? While there was a general appreciation that 'radiation is not the most important problem in the environment' in light of other environmental threats, the question of when and how it might become a problem was more challenging.

The discussion topics covered a wide range of issues, from effects at the molecular level through to population and ecosystem impacts, from regulation to ethics. Some questioned whether a new system of assessment and management is needed if major effects on ecosystems are not immediately apparent. When rabbits were found to be burrowing close to Dounreay radioactive waste areas on the Dounreay nuclear site, the immediate reaction was to cull the rabbits to reduce risk to the human food chain (Ross 2003). This was followed by suggestions that it would be sensible to test other animals using the rabbits for food, such as cats and buzzards (Thomson 2003), and a retort by the neighboring Sandside estate owner that the answer to the problem was the secure containment of nuclear waste and not culling '*poor, innocent rabbits, cats or birds*' (Thomson 2003).

In this case, one might argue that radiation was at least indirectly damaging to that particular population.

If we are considering the release of radioactivity into the environment through current regulated activities then the statement ('Radiation is not a problem in the environment') is probably correct. However, if we are asking about the likelihood of observable effects of radiation in more highly contaminated areas around Chernobyl, Mayak, and Fukushima, then radiation doses in some areas are such that effects are likely and we do observe them (Geras'kin 2016; Beresford et al. 2020b, 2020c).

However, exactly what effects have been seen (or not) in contaminated regions, as well as their cause and ecological relevance are a matter of contention. Examples were given of papers that reported changes across a whole range of molecular and organism levels, from chromosome aberrations to cataracts, (e.g. Møller and Mousseau 2007a; Bonisoli-Alquati et al. 2010a, 2010b; Møller et al. 2012a, 2012b, 2013; Mousseau and Møller 2013; Baker et al. 2017). Other papers have reported little or no effect (e.g. Deryabina et al. 2015; Bonzom et al. 2016; Lerebours et al. 2018; Fuller et al. 2019; Goodman et al. 2019).

Other issues raised, with respect to interpreting results from contaminated areas (and especially studies in the CEZ) included the positive impacts from removing the human population, problems with the residual impact of high historic doses, and dose response relationships being driven by observations from areas with extreme high dose rates which may have additional confounding factors because of degraded habitat and slow ecosystem recovery (Beresford et al. 2020b, 2020c).

Some scientists believe changes seen in Chernobyl and Fukushima suggest that environmental risks of radiation are underestimated (Mousseau and Møller 2013). The intense discussion about this in the workshop and during the construction of this manuscript, captured very well the level of disagreement among scientists.

Related to studies conducted in the CEZ and FEZ one of the key points blocking development of an ecosystem approach (or indeed reaching any consensus on the effect of radiation on the environment), is a general lack of open access data (Lecomte-Pradines et al. 2020; Beresford et al. 2020b).

Potentially confounding factors arise if authorized or accidental releases of radionuclides take place in natural ecosystems that are already under pressure from habitat destruction, invasive species, or chemical pollution. For example, interactions between rapid climate change and radioactive contamination could compromise homeostasis and physiological responses, and potentially impair fitness, reproduction, and development (Noyes et al. 2009). Radiation exposure may further reduce the ability of organisms to acclimate and potentially make them more susceptible to infectious and vector-borne diseases (Dmitriev et al. 2011; Morley 2012).

Finally, harm to the environment caused by radioecologists related to studies conducted in the CEZ and FEZ was discussed but was not considered to be a serious issue. It was however emphasized that it is important to be aware of each other's work, as there is the potential to impact subsequent studies. Examples (known to have occurred) include trapping out small mammals and so impacting the population dynamics for ecologists coming later, and introducing unexposed individuals into contaminated sites.

To conclude, the type and the magnitude of either direct or indirect effects of ionizing radiation depend on ecosystem composition, and many ecological factors can be more important than radiation. On the other hand, ignoring population or higher-level effects, and focusing only on individual-level endpoints may lead to inaccurate risk assessments and errors in environmental management decisions. It should be noted here that population and higher-level protection are the stated aim of radiological protection – (ICRP 2008; IAEA 2014). Therefore, the build-up and use of ecological knowledge is essential for understanding responses of populations and ecosystems to radiation, including the potential for changes in interaction between species. In light of this, perhaps a more reasonable statement, than that proposed for this discussion, would have been that 'radiation is not the most important problem in the environment'

Provocative Statement 2

Not all change is bad, but no observable change may not indicate that nothing bad is going on. Focus on detection of negative effects can miss the detection of adaptive and protective effects operating at the ecosystem level.

Summary answer: This statement was generally agreed with, but it was not seen to advance the discussion of developing approaches in regulation where preventing harm is the main objective. In developing an ecosystem approach, the key should be, measuring change without assigning 'good' or 'bad' descriptors.

This statement was designed to address the role of adaptation and evolution. Biota can adapt in response to new stimuli and invoke protective mechanisms that are essentially heritable and conserved, and molecularly predicated, in the interest of survival. Such adaptation could occur in response to environmental exposure to low-dose radiation (Audette-Stuart et al. 2011; Mothersill et al. 2013; Lampe et al. 2017; Beresford et al. 2020b). These processes draw attention to the fact that the current environmental radiation protection regulation adapted from the human framework essentially measures or models dose and compares this to benchmarks below which impacts on individuals (mortality, morbidity, fecundity) likely to lead to population level effects are not anticipated (Howard et al. 2010). This can miss long-term processes playing out over generations in populations. It is difficult for humans to make the conceptual jump from individual-level short-term effects to very long-term ecosystem-level effects.

Discussions focused on the difficulties of extrapolating between different levels of effect (molecular, individual, population), as well as the types of mechanisms that could underlie these types of responses. Some changes at the ecosystem level might also be misinterpreted. For example, an increase in prey population density may be interpreted as a positive ecosystem level effect when it actually results from the eradication of the predator partner, perhaps signaling a decline in the condition of the ecosystem. Rabbit populations might increase if foxes or buzzards decrease, which might in turn bring about a decrease in abundance of a plant species, both of which would be indirect rather than directs effects of exposure.

Long-term field studies carried out on different plant species including winter rye (*Secale cereal* L.) and wheat (*Triticum aestivum* L.), spring barley (*Hordeum vulgare* L), oats (*Avena sativa* L.), Scots pine (*Pinus sylvestris* L.), Japanese red pine (*Pinus densiflora* Siebold & Zucc.), wild vetch (*Vicia cracca* L.), crested hairgrass (*Koeleria gracilis* Pers.) in various radioecological situations (nuclear weapon testing, the Chernobyl and Fukushima accidents, uranium and radium processing) have shown that in most cases strong effects at the molecular level (increased rate of mutations, changes in gene expressions) turn into moderate effects at the physiological level (enzyme activities, changes in the phytohormonal balance), and into slight effects at the organismal and population levels (morphological abnormalities, reproductive ability, radioadaptation) (Geras'kin et al. 2013; Boubriak et al. 2016). Evidence from field observations shows development of cytogenetic abnormalities in several generations of progenies distant from the initially irradiated surviving parental generation (Geraskin et al. 2003) and there is a rich literature in radiobiology concerning delayed *de novo* appearance of cellular effects such as lethal chromosomal aberrations or lethal mutations after many normal generations *in vitro* and *in vivo* (reviewed in Mothersill and Seymour 2019).

In a number of studies, it was noted that parental acute exposure to radiation affects both exposed organisms and their unexposed offspring (Streffer 2006; Sarapultseva and Dubrova 2016; Sarapultseva et al. 2019). At the same time, fertility and survival were restored by the second generation. Thus, it is important to use at least two generations of test design that analyze the long-term effects of ionizing radiation, as has been shown, for example, for the test organism *Daphnia magna* (Barata et al. 2017). However this raises the question of the long-term significance of genetic effects observed over generations; might there be long-term consequences of epigenetic changes in gene expression resulting in alterations in the genetic structure of populations (Nishikawa and Kinjo 2018)?

The Adverse Outcome Pathway (AOP) approach may help elucidate possible links and consequences across different levels of biological organization (Ankley et al. 2010; Villeneuve et al. 2014). This is a conceptual framework that gathers and organizes data from studies of organisms using a variety of endpoints. The data are then analyzed using pathway analysis methods to determine molecular initiating events and to link these to adverse outcomes seen at higher levels of complexity. By dissecting the problem into smaller organizational levels (from sub-cellular to organism), with each level potentially controlling the structure and function of the level above, may help us to reach a better understanding of how individual exposure responses may be expressed at higher levels of biological organization i.e. population and ecosystem.

However, it should also be appreciated that there is no reason why a dose-response relationship observed at one level of organization may be the same at other levels of biological organization (i.e. a dose response seen for an individual may not be mirrored at the population level). The responses, with latent periods, get attenuated at each level of organization and are influenced by many confounding factors in the complex environment. There is also the possibility of emergent properties as the system becomes more complex (Jha 2008). If the underlying mechanisms of adaptation were triggered by the responses at the molecular or cellular level, then adaptation would be strongly dependent on the dose/dose rate, and the radiosensitivity of the species at the life stage that experienced the exposure. Any population level adaptation therefore needs to be explained at molecular and cellular levels in different species. As an example, low dose hypersensitivity first seen as a response in radiotherapy is now thought to be a widely encountered integrated biological response to stress involving apoptosis, autophagy and mutations (Rodrigues-Moreira et al. 2017). Variation in stress response might underlie variation in low dose radiosensitivity within populations and among species.

Populations inhabiting radioactively contaminated territories may become physiologically adapted (i.e. acclimated) to chronic exposure through phenotypic plasticity or epigenetic changes, or even evolutionarily adapted through natural selection (e.g. Ruiz-González et al. 2016; Horemans et al. 2019). Investigations of environmental adaptation have mainly focused on single species, often overlooking the symbiotic context of the organisms under study (Exposito-Alonso et al. 2018; Schuman and Baldwin 2018; Song et al. 2019a). Symbiotic relationships, however, are ubiquitous in nature and it becomes increasingly clear that development, growth and health of macro-organisms is influenced by the complex microbial communities they host (Simon et al. 2019; Song et al. 2019a). Many examples of adapted mutualists conferring stress tolerance are available including rhizobia enhancing chickpea plant tolerance to desiccation (Bano et al. 2010), ectomycorrhizal (EcM) *Suilloid* fungi involved in pine tree metal tolerance (Krznaric et al. 2009) and common mycorrhizal network-connected *Nicotiana attenuate* plant communities showing an enhanced systemic defence responses to plant herbivory (Song et al. 2019a).

Sometimes we observe signs of adaptation to radiation exposure in affected populations (Galván et al. 2014), sometimes not (Geras'kin et al. 2013). While evidence of increased resistance is inconclusive, two arguments support this hypothesis. First, instances of rapid evolution – spanning a few to a few tens of generations – are increasingly documented in the eco-evolutionary literature (e.g. Pespeni et al. 2013; Oziolor et al. 2019). The only precondition seems to be that the drivers of selection are strong enough. Second, such rapid evolution is predicted to be even faster when the toxicant is responsible for generating new variants among which to select, as is the case for mutagenic ionizing radiation. However, an adapted populations' fitness might also decrease when its organisms move to uncontaminated territories (Hickey and McNeilly 1975; Levinton et al. 2003). So, what is the cost of adaptation? Such a fitness cost may be related to the use of metabolic energy to produce the adaptive trait (e.g. the DNA repair mechanism) even when that trait is not required (although it was disputed that the cell population would run such a tight energy budget). Moreover, the selection of genotypes for radio-resistance can lead to the loss of radiosensitive genotypes with valuable properties (Glazko 2001; van Straalen and Timmermans 2002; Hancock et al. 2019a). Since adaptation may play an important role in the response of populations to radiation exposure in natural settings, mechanistic information about population-level drivers of fitness and the role of adaptive responses, is directly applicable to predicting a radionuclide's impacts at the population level and hence for developing ecological risk assessment.

Even with a reduction in reproductive success of the organisms negatively affected by radiation, exposure may not result in population declines. Under ecologically realistic scenarios, reduced reproductive rate and survival of exposed organisms creates an opportunity for individuals of the impacted species to better exploit resources and hence reproduce. There is some evidence for compensation; embryonic mortality of fish was found to be accompanied by larger broods leading to no net effect on population numbers (Blaylock 1969). Ecological space left open by exposed and affected organisms, can also be filled by organisms immigrating from the outside, which would mask any deleterious effect on the resident population. Source populations can compensate for sinking numbers in that location, in a classic source-sink meta-population fashion.

Finally, some effects observed at affected sites could potentially result from non-targeted effects of acute exposure during the first period of the accident. In particular, historically-induced genomic instability leads to a phenotype with a greater tolerance for mutation than normal. This means that in addition to the mutation burden attributable directly to the ambient dose there may be a contribution due to the historically-induced genomic instability, which is not directly induced by the ambient dose but is a consequence of the increased mutation tolerance (genomic instability) in the population. This concept is discussed in Mothersill et al. (2017) and has been applied to datasets from CEZ and Fukushima (Omar-Nazir et al. 2018; Geras'kin et al. 2019; Hancock et al. 2019b, 2020). Therefore, a clear understanding of exposure history is of fundamental importance for interpreting field effects studies and subsequently for predicting how populations and ecosystems recover from radiation exposure.

Provocative Statement 3

'Mechanistic studies need to employ systems biology and be field-based to be useful (field v laboratory studies reveal discrepancies)'

Summary Answer: Both field and laboratory data have a role, and laboratory studies can be used to substantiate field observations under controlled conditions. But laboratory data should be validated in the field, whenever possible, before being used to develop ecosystem models. The corollary of this is that suspected field effects can and should be confirmed in the laboratory

The aim of this statement was to explore the suggestions that field-derived radiation sensitivities have been reported to be up to 10-fold greater than laboratory-based values, based on a comparison of between field and laboratory data (Garnier-Laplace et al. 2013). There are multiple reasons for the difference including stress, other pollutants, predation and disease, which could compromise survival of organisms from contaminated environments. However, the reason for the question was to discuss whether ONLY field data should be considered when developing an ecosystem approach or whether laboratory data has merit. A secondary issue is whether or not the difficulties in obtaining good field data, present a major block to developing an ecosystem approach?

The discussion centered mainly on the types of mechanistic analysis that might help in probing laboratory-field discrepancies, including the influence of neurological, immune and humoral responses and DNA damage. Although there is increasing evidence to suggest that ionizing radiation, like chemicals, can induce a variety of biological responses e.g. in the nervous, immune, endocrine or inflammatory systems (Jha 2004, 2008), these systems are relatively poorly understood in natural biota. Radiation is known to generate oxidative stress in humans, laboratory animals (Hurem et al. 2017; Maremonti et al. 2019) and wild populations (Einor et al. 2016; Volkova et al. 2017), by increasing oxidative damage and decreasing antioxidant defenses. If the key physiological systems mentioned above are impacted by redox imbalances under chronic exposure conditions, they could eventually impair the reproductive fitness of the organisms and the populations in a stressed ecosystem (Jha 2008). For example, sub-lethal effects can impair homeostatic or physiological conditions, and inflammation has been linked to a range of aging-related pathologies.

DNA damage is a recognized outcome of radiation exposure, together with a variety of key processes such as epigenetic modifications, including DNA methylation and transcriptomic (mRNA) and post-transcriptomic (small RNA and long non-coding RNA) measurements (e.g. Schofield and Kondratowicz 2018). However their ecological relevance is a matter of debate. DNA damage poses a direct threat as a precursor of mutation. Mutations are typically neutral to mildly deleterious. Most of the time they will cause no effect, - due to redundancy in the genetic code and the large proportion of the genome that does not perform any known function, - or have effects that (a) are non-lethal, and (b) only reduce performance of developmental, behavioral and physiological systems, rather than total impairment of those systems. DNA damage could also affect reproduction and survival because DNA repair requires a diversion of energy from other activities (Roff 2001). Epigenetic variation may however be expected to have a potentially more significant effect as epimutation is in its nature pleiotropic and affects the expression of many genes (Schofield and Kondratowicz 2018). Epigenetic changes have the potential to mediate toxicological, and transgenerational deleterious effects of exposure to ionizing radiation when they suppress the expression of genes otherwise useful for the organism. In principle, however, epigenetic changes can also favor a plastic response to ionizing radiation that can later be accommodated into an evolutionary one (Bossdorf et al. 2008). The translation of these organism-level effects to the population level is however complex. For example, higher mortality can be compensated by immigration from outside of the contaminated areas, and countered by reduced intraspecific competition and lower predation pressure. Time is also an important factor. In a laboratory, organisms are usually measured a short time after exposure, whereas field studies are often conducted years after the contamination event.

It is also difficult to relate or transfer the results of laboratory experiments on transgenerational effects and genomic instability to field conditions. While there is strong laboratory evidence for the manifestation of epigenetic transgenerational instability in the progeny of irradiated males, other studies suggest that low-dose, low-LET exposures do not destabilize F_1 (Mughal et al. 2012) or that lab-results are not always comparable to observations in the field (Horemans et al. 2019). In contaminated environments, both parents and their offspring/progeny of exposed populations are continuously exposed. This means that *in situ* analysis of mutation rates among the offspring cannot provide any evidence for the manifestation of transgenerational effects, since it would be impossible to distinguish between the direct effects of offspring exposure to ionizing radiation and transgenerational instability. However, the historically-induced genomic instability phenomenon referred to earlier, would suggest that there will be a component of the total mutational load induced by genomic instability. This is suggested by the dose reconstruction modeling done by Hancock et al. (2019b). To determine the extent of the additional mutational load experimentally, parents could be exposed in laboratory settings to the mix of radionuclides, similar to that on the contaminated territories. Subsequently, their offspring could be transferred to the clean environment, and the genome stability of non-exposed first- and second-generation offspring of irradiated parents could be analyzed. The results of such studies would provide a definitive evidence for the manifestation of transgenerational genomic instability in nature.

The major conclusions were that systems biology approaches to understanding the consequences of exposure to ionizing radiation at low doses under ecological conditions require thinking about in terms of the complex lives and demands of wild organisms. While laboratory experiments can yield useful insights into possible mechanisms, the population-level effects could be fully realized only when a systems biology approach is used within field-based studies. Mesocosm studies could also be useful in allowing 'more controlled' field observations.

How to make progress in developing an ecosystem approach

The second part of this paper deals with breakout discussions centered around practical approaches which need to be considered if progress is to be made. The three topics are

1. What is stopping the development of an ecosystem approach?
2. What can be gained from 'big data' and modeling approaches
3. How can we design 'good' experiments giving robust data?

What is stopping the development of an ecosystem approach?

The premise of the question is that ecosystem approaches are not being employed. Although this is correct from a regulatory perspective, it is incorrect for radiation ecology and ecology in general. Several examples of research that used ecosystem endpoints are available from Chernobyl and Fukushima accident sites as well as at nuclear facilities in the USA (Savannah River, Oak Ridge, Fernald, and Hanford), in Russia (Mayak), and microcosm studies (Bonzom et al. 2016; Geras'kin 2016; Fesenko 2019; Hevrøy et al. 2019).

To inform an ecosystem approach, a study needs one or more endpoints that pertain to ecosystem structure or function. Ecosystem endpoints that are gaining popularity are those enumerated in various lists of ecosystem services as defined in the US EPA document published in 2016 (US EPA 2016). For example litter degradation is an ecosystem function that ensures the service of recycling nutrients and protecting from fire. Ecosystem services can be synonymous with ecosystem functions and as such can become confusing if not defined explicitly (South Downs National Park Authority 2018). The definition in Wikipedia is as follows:

> They (Ecosystem services) include natural pollination of crops, clean air, extreme weather mitigation, human mental and physical well-being. Collectively, these benefits are becoming known as 'ecosystem services', and are often integral to the provisioning of clean drinking water, the decomposition of wastes, and resilience and productivity of food ecosystems.

Central to the definition of an ecosystem approach is that interactions among biotic entities must be considered. This derives from seminal work performed in ecology as far back as the 1960s exploring the regulation of populations within various communities (Slobodkin et al. 1967; May 1973; Pimm 1982; Brown and Munger 1985; Karr 1992; Matson and Hunter 1992). Emerging in these research efforts were two alternative mechanisms that control ecosystem dynamics, namely a bottom-up control (one that is driven by the producer community) and a top-down control (one that is regulated by top predators). A salient point of these works is that efforts that focus on a single taxon are likely to have limited predictive capacity because the various positive and negative feedback loops cannot be anticipated in isolation of interacting taxa.

Taking an ecosystem approach must begin with asking the questions – what are we trying to understand? What is our focus? Then, rather that attempting to assess or evaluate all possible components of the ecological system one looks one hierarchical level above the focal level for context and one hierarchical level below the focal level to explore mechanisms (Wu and Loucks 1995). This means that if one is interested in a population of a particular tree species, one would look at the dynamics of the forest association in which that population resides (context) and at autecological endpoints of the tree species for mechanisms governing the population. The focus could be on population structure of interacting taxa or guilds, productivity, energy flow, nutrient cycling, pollination services, etc. The complexity of ecological systems calls for multidisciplinary, even transdisciplinary teams that match the likely dynamics that will be encountered.

Although there are parallels that can be drawn between radiological stressors and other stressors, perhaps the most challenging problem is the accurate measurement or estimation of cumulative dose an organism might encounter in a complex environmental setting. The challenge relates to different emitters (alpha, beta, gamma) and determination of internal versus external dose. Beaugelin-Seiller et al. (2020) discusses approaches to improve the estimate of absorbed dose in field studies.

Many papers on radiological effects in biota pertain to alteration of DNA. These studies provide great insights into mechanisms of action at the molecular level. This knowledge helpful to understand cellular anomalies and has been shown to be important in epigenetic modifications that transmit to subsequent generations (Schofield and Kondratowicz 2018). However the importance of non-targeted effects such as genomic instability and bystander effects continues to grow as more mechanistic information becomes available (Mothersill and Seymour 2010) There is current enthusiasm about the utility of adverse outcome pathways (AOPs) first proposed by Ankley et al. (2010), in radioecological studies (e.g. Song et al. 2019b; Beresford et al. 2020b). The limitation that seems to be overlooked in this conceptual framework, is that there are currently no population models that can effectively forecast the population dynamics in a field setting. This imposes severe limitations on the use of AOPs in an ecosystem approach.

There has been some use made within radioecology of ecosystem endpoints that are common in mainstream ecology, such as ecosystem services, pollinators, and decomposers, level of resilience or objective measurements of interactions (e.g. Mousseau et al. 2014; Bonzom et al. 2016; Newbold et al. 2019; Beresford et al. 2020b). However, many of the concepts used in radiobiology are not integrated into radioecology and some of these might help progress toward developing an ecosystem approach. One suggestion was to have a 'dose and dose rate effectiveness factor' or 'DDREF' for the *organism resilience* -how sensitive is an organism and how adverse would it be for ecosystem instability if the organisms stopped functioning in a healthy way? A way to bridge from reference animals and plants to an ecosystem approach might be to develop reference values for organism resilience. This could be done by strategically combining mathematical models, laboratory and field study data of organisms' ability to deal with radiation stress and other pressures to establish a resilience factor (RF). The RF need only be based on a few species of each genus, and our knowledge of their role/interconnection to each other. However, an argument was put that an absolute resilience factor could not exist, as it would be context-dependent.

That said, there are biological features that make species more or less vulnerable, and more or less resilient. Among these are the reproductive strategies (see MacArthur and Wilson 1967) for K or r-selected species with r-selected species (e.g. mice) 'bouncing back' faster, and population connectivity (with more connected populations receiving immigrants that help the recovery).

Similarly, if you know and understand an animal by studying it in the laboratory you can better predict their interactions and contributions in the environment. This could help to model and inform our understanding of ecosystem resilience, which is based on empirical evidence of a particular ecosystem to resist change or to recover following a perturbation. This considers complexity of the system in terms of biodiversity, redundancy, and flows of information within the system that enable the system to retain basic functions. The concept of resilience and stability was developed by Holling (1973). Each ecosystem is unique and diverse, and to understand its response to a stress it should be mapped in a way which involves modeling the pathways and processes of organisms interacting with each other and their environment. This can be done by explaining the web of direct effects (express/measured at an individual level) and indirect effects (effects mediated/transmitted through interactions) in an ecosystem. If one organism is impacted, how will it ultimately affect the other organisms and the environment, for example if a predator becomes unable to hunt this may result in an imbalance and overpopulation of prey species.

Discussion on big data, machine learning, informatics and modeling as new approaches for radioecology

The effects of low dose radiation on ecosystems, such as in Chernobyl or Fukushima remains poorly understood. Although both field and laboratory studies synergistically contribute to understanding and evaluating the basis for environmental assessments, the complex nature of interactions on a population or ecosystem level make it harder to extrapolate the radiation effects observed in the laboratory-based studies. One approach to study this, at least qualitatively, is by using mathematical and computational modeling techniques that are motivated and formulated based on the insights gained from both laboratory and field experiments.

Mathematical and computational modeling approaches are widely used to understand and study ecological interactions. Novel concepts like multiscale modeling (Powathil et al. 2015), and bio-energetic model-based network analysis (Yodzis and Innes 1992) can be very beneficial for exploring the population dynamics and other ecological interactions under multi-stress scenarios, including radiation stress. Beyond these modeling approaches, current developments in Artificial Intelligence (AI) and Machine Learning (ML) techniques offer novel opportunities to gain insights from large data sets (see Jarry et al. (2003) or Feng et al. (2018) for an example in radiation oncology) and the availability of complex and large datasets in ecological studies permits the application of a large range of data-driven modeling approaches.

Machine learning in particular provides several main benefits: classifiers may be derived for data using supervised learning and requiring large training datasets, unsupervised learning can be used to find patterns in complex data and relationships between entities or measurements that had not been suspected. Convolutional artificial neural networks can find feature representations for datasets to facilitate, for example, the application of similarity metrics or both supervised and unsupervised learning for classification. These approaches are in many ways related to established techniques of data-driven model building as they provide functions for describing datasets. Both can use data collected in the field as training material or to derive functions in modeling, such as estimation of the rate of change of occupation of state spaces (position of a population in a multidimensional space such as that defined by population size, nutritional status and age distribution), but can also use synthetic data, discussed below.

A recent review (Christin et al. 2019) succinctly captures the range of recent data types and applications of machine learning in ecological studies which range from identifying individual bird calls from environmental recordings (Potamitis 2016) to wild animal counts (Norouzzadeh et al. 2018), tree defoliation (Kalin et al. 2019) and diversity assessment (Salamon et al. 2017). The use of drones to capture elements of the landscape (Richter et al. 2008; Gauci et al. 2018) is also amenable to ML analysis to search for changes which themselves can be used as inputs to ML problems.

All of these examples are potentially useful for radioecology but none to our knowledge have been applied at any scale with the exception of established population modeling methods (Vives I Batlle et al. 2012; Alonzo et al. 2016), or studies of Ra-226 characterization (Varley et al. 2015). These tools are only recently being applied to the area of nuclear science and radiation protection (Gomez-Fernandez et al. 2020)

There are several questions:

1. How much information is needed for effective application of machine learning? How large do populations need to be to provide sufficient information? For example in projects collecting meta-barcoding of earthworms for population diversity measures are complicated by poor numbers naturally occurring in Chernobyl (because of natural soil conditions) but there are high numbers of species and populations in Fukushima.
2. How many measurements are needed to make an estimation of time dependent changes? Collections may not be carried out at enough time points to do time series analysis
3. How rich, how defined and how reliable is the collection of defined phenotype data in existing databases?
4. Dosimetry is critical so that effects can be related to accurate estimates of dose. To quote Hansen et al. (2019)

> When effects data are presented versus absorbed dose rates and accumulated doses, and with information on the type of exposure, dose-response results from different experiments or situations can be compared. When this information is missing, it is difficult to interpret results from exposures, to compare results with literature data and to put these results into context.

5. How critical are the metadata on environmental parameters such as precipitation, temperature etc. when collecting complex data of a wide range of different types (high dimensional data)?

The advantages of applying machine learning or similar approaches are:

- Often no 'control' data are needed to determine patterns, just variation; highly dimensional data does not need 'clean' controls to be useful.
- Sonic and visual data are highly amenable to convolutional ANN (Artificial neural networks) representation for clustering and classifying
- For ML and model building there is the possibility of using synthetic data, developing the model and then iterating with data collection in the environment and rebuilding the model – bootstrap approach. As of yet this has not been explored in a radioecological context.

A key message here is that the collection of data, the types of data and the analytical methods available are not specific to radioecology, but are certainly all applicable to the assessment of the ecological impact of contamination. Dosimetry is probably the most important issue and this is discussed at length elsewhere in this commentary. Of particular interest in radiation ecology however is time series analysis and machine learning has been successfully applied to longitudinal time series to predict future events (e.g. Rammer and Seidl 2019 and review therein). Modeling applications such as these are potentially of great value when comparing historical and post-contamination time series as a way of identifying deviations from, for example, predicted changes in population size or complexity. This last example does raise another rather specific problem to the collection and use of big data in the context of radiological contamination, which is the likely lack of historical data on the contaminated site, requiring that a control site be used as a surrogate (Geras'kin et al. 2018). Identification of control sites is fraught with difficulty, a problem noted in several studies in Chernobyl.

Data scale and quality is important but there may be data available now such as birdsong and environmental soundscape recordings, which can be amenable to these approaches. The Internet of Things (IoT) is also being explored for its application to environmental monitoring (Lin and Liaw 2015; Muniraj et al. 2017).

Use of drones to capture elements of the landscape (Richter et al. 2008; Gauci et al. 2018) is also amenable to ML analysis to search for changes which themselves can be used as inputs to ML problems.

How can we use such large-scale data collection and analysis to define and quantify the damage caused to the ecosystem by any exposure (including ionizing radiation)?

A suggestion was to focus on diversity including not only the number of species but also intraspecies genetic variation. It has the advantage that it can be quantified by genome sequencing (which has become cheaper and cheaper). However, the main reason to focus on diversity is that it is characteristic of the ecosystem not its individual members, and that it is also a crucial property of the ecosystem determining its stability (McCann 2000), resilience and capability to cope with future stresses. A point was made that any small disturbance of the ecosystem can be harmful, because the physiological costs of the repair/adaptation might decrease the chances to reproduce or survive the next stress. However, it is important to recognize that organisms are almost always exposed to different stresses including for the resources within the population. From this point of view, exposures affecting the ecosystem homogeneously (causing similar and not too strong stress to all of its populations) do not cause big changes. However, if a population or a species is much more sensitive than others, it may lead to its elimination from the ecosystem or from the biosphere. Maybe, it is worth mentioning that if we focus on diversity, then the population size and the number of species in the ecosystem are crucial parameters describing its sensitivity. This also illustrates that, while useful, diversity estimates are univariate descriptors that cannot capture the complexity of biological communities. It is easy to imagine a community maintaining the same degree of species diversity while species composition drastically changes.

A much more ambitious aim would be to generate very large models from multiple species and non-biotic data using an agent-based approach. While widely used in ecology and socio-ecological contexts these have not yet found use in radioecology (DeAngelis and Grimm 2014) and have the potential to integrate the functions derived from ML for different components of an ecosystem to better define its structure and dynamics (DeAngelis and Diaz 2019). Such systems have the ability to develop stochastic rather than deterministic models of ecological interactions. These might have a rather specific sensitivity to the individual variation in response to radiation of species members and may approach the holistic ecological modeling for which we search.

Discussion on making experiments useful

Most scientists will have received formal training on research design, often as an early course within their undergraduate studies. Robust research design is fundamental to the scientific method. Without it, we cannot be sure that the conclusions drawn are valid. It is logical to assume that the professional scientific community, formally trained in research design principles, would draw on this in planning their own research and also when peer reviewing the work of others. Unfortunately, there are many examples where this is not the case. When we attempt to use research findings from radiation effects studies, either radiobiological or radioecological, to inform the development of benchmarks for use in environmental radiation protection, we often find that the research findings cannot be used in a meaningful way (cite benchmark derivation papers). This is especially true when we are considering the applicability of research findings to inform the implementation of an ecosystem approach.

A good research approach is to develop a hypothesis and then undertake studies to test that hypothesis. We may postulate a very general null hypothesis to direct research that could inform radiation protection in the context of an ecosystem approach: *Low dose radiation does not impact the environment*. This may appear to be a straightforward hypothesis and is certainly key to understanding radiation within the broader context of an ecosystem approach. However, it immediately raises three questions that must be considered if we are to ensure that research delivers meaningful results:

- What do we mean by 'low dose'? UNSCEAR (2010) defines low doses as 'those of 200 milligrays (mGy) or less and low dose rates as 0.1 mGy per minute (averaged over an hour or less) for radiations such as external Xrays and gamma rays' (UNSCEAR 2010). Research publications variously refer to low doses as being in the order of a few µGy through to cGy. There is a need to determine the relevant dose range over which we need to be targeting research effort from an environmental protection perspective and also to decide whether the focus is life-time accumulated dose or dose rate. Here we consider low dose to be environmentally relevant doses, including those resulting from authorized discharges and those likely to be encountered at contaminated sites. If the focus is accumulated dose then what constitutes low dose/low dose rate may also be species- and life stage- specific.
- What do we mean by 'environment'? – This is a question that has both a scientific and a social context. It requires a clear articulation of the goal(s) of protection and, for research findings to be widely applicable, the definition of environment in this context must encompass the varied goals and values of society. By viewing 'environment' through two lenses, population sustainability and ecosystem functioning, it is expected that species-specific through to ecosystem services goals would be addressed (assuming that habitat itself is also sufficiently protected).
- What do we mean by 'impact'? – To some researchers 'impact' is any measurable change resulting from radiation exposure, often at the sub-cellular level. Others view impacts as being changes that are observed at the whole organism level and above. Similarly, some researchers consider impacts as deleterious effects whereas others view impact as either beneficial or deleterious effects. Focusing on indicators of population sustainability and ecosystem functioning would enable radiation-induced changes, be they beneficial or deleterious, to be studied in relation to dose or dose rate.

Testing this hypothesis and understanding any apparent radiation effects will require a combination of laboratory and field-based studies. This is an approach that has been adopted in various projects over recent years, including the UK-based TREE project (https://tree.ceh.ac.uk/) and the EC COMET project (https://radioecology-exchange.org/), and there is a community consensus on the need to continue pairing laboratory and field studies into the future (Bréchignac et al. 2016; Beresford et al. 2020b). However, it is important to recognize some of the challenges in translating research findings between field and laboratory. For example, the response of organisms maintained under 'ideal' conditions within a laboratory to a particular radiation dose may be different to the response of organisms receiving that same dose in the field setting where radiation is but one of a range of stressors. Considering radiation in the context of other stressors is an essential step toward the potential integration of radiological assessments into an ecosystem approach. There are also differences in exposure in field situations. Chronic exposure conditions in a laboratory are generally stable over time, whereas chronic exposure in the field is variable both spatially (e.g. Aramrun et al. 2019) and temporally (e.g. due to seasonal changes in food sources and habitat utilization (Stark et al. 2017).

Both laboratory and field-based research on radiation effects require appropriate dosimetry. Where the absorbed dose is not quantified for the organisms under investigation, it is impossible to draw meaningful conclusions on the relationship between dose and effect. Many studies in both the Chernobyl and Fukushima Exclusion Zones have purportedly shown effects of radiation on wildlife but the effect has simply been related to an ambient air dose rate. Whilst this may provide a reasonable approximation of the above-ground external gamma dose to which an organism would be exposed at the location where the measurement is taken, the total absorbed radiation dose (considering both external and internal exposure) may be very different for some species (e.g. Aramrun et al. 2019; Beresford et al. 2020a).

A potential dosimetric challenge when pairing laboratory and field studies is the difference in exposure conditions. Laboratory exposures are often delivered using an external gamma source, whereas field exposures are a combination of both external and internal. Depending on the radionuclides within the organism, the dose deposition may be relatively homogeneous (e.g. radiocaesium) or highly localized (e.g. radioiodine). Consequently, for field studies focusing on radiation-induced changes at the sub-organism level, it will be important to consider whether a whole-organism dose or an organ-specific dose is the most appropriate measure against which to compare a response. For example, in an environment contaminated with radioiodine, relating whole-body dose to changes in thyroid cells may lead to quite different conclusions than in an environment where radiocaesium is the main contaminant. The ongoing development of voxel phantoms for various organisms provides the capability to determine doses for specific organs (e.g. Ruedig et al. 2015; Caffrey et al. 2017).

Given that our ultimate aim is protection of the environment, there is a need to further develop knowledge on the effects of radiation in complex systems. This may be both direct effects and indirect effects (e.g. a change in predator or prey abundance). For laboratory studies, the consensus view is that more use should be made of mesocosm studies (Haanes et al. 2020). For field studies, it is important to ensure that effects are not being masked by immigration of organisms from neighboring patches and also that other potential stressors and site history are considered.

Workshop participants were asked to suggest what field-based research should be prioritized if resources for research (both financial and personnel) were unlimited. There was a recognition that there was a need to effectively create mesocosms within the environment so that some of the confounding factors or the intra/interspecies interactions that influence interpretation of field-based radiation effects studies could be controlled. The CEZ has already been recognized as a radioecological observatory site (Steiner et al. 2013; Beresford et al. 2020b) although any CEZ observations need to be interpreted in the context of the contribution of absence of humans. The creation of a network of sub-observatories across the CEZ was proposed, covering a dose rate gradient and key habitats. Aquatic sub-observatories could be ponds which are geographically isolated from each other. In the terrestrial environment, each sub-observatory would be a large enclosure (perhaps 150 m × 150 m, with solid 1 m high walls extending to a depth of at least 0.5 m below ground). The walls of the enclosure should minimize the movement of smaller ground dwelling organisms (invertebrates and small mammals) between the enclosure and neighboring areas. Replicate sub-observatories would be required within each ambient dose rate band and habitat category; the number of replicates would need to be determined based on consideration of the statistical power of subsequent studies, mediated by the practicalities of locating suitable replicate locations to establish sub-observatories. There would need to be a community consensus on the sub-observatory siting. The sub-observatories should each be fully characterized from both an ecological and radiological perspective and instrumented to monitor climatic and soil parameters. These sub-observatories would be of particular interest to study smaller organisms like small mammals and their interactions (e.g. their food) but cannot be used to study e.g. large carnivores, birds or migratory animals.

The sub-observatories would become a focus for radioecological research in the CEZ, with multiple research groups using the same, well-characterized sites and openly sharing data (as required by many funders and recommended in the radioecology literature, e.g. Beresford et al. 2020b). Populations in these sub-observatories would then be monitored over generations, although it is acknowledged that by enclosing populations within sub-observatories this may change the dynamics of the populations over multiple generations. Of course one needs to take into account that the long-term follow-up of populations in these sub-observatories might be restricted by the fact that they live in a kind of confinement that potentially affects the natural course of ecological processes e.g. reduced ecological pressure on the population due to lack of big predators. Hence, it is as always essential to see if the experimental design matches with the hypothesis you want to test. Clearly there would be a need to ensure overall coordination of research in these sub-observatories to avoid studies by one research group impacting on studies by another group. An important aspect of such studies would be the ongoing development of non-lethal methods in radioecology (Wood et al. 2011), enabling quantification of internal radionuclide activity concentrations and biological responses without the need to kill the animal. The recent development of a field-portable radiation detector that enables quantification of Cs-137 and Sr-90 through live monitoring of animals (Fawkes 2019) is a significant step in this direction. For animals such as small mammals, this allows a robust internal dose estimation to be made for each individual studied and the animal can then be released back into the sub-observatory.

Whilst the sub-observatories would each be located within a specific ambient dose band, radionuclides will be heterogeneously distributed in three-dimensional space within each sub-observatory resulting in a complex external exposure situation. There will be spatial variation in radionuclides at the soil surface, differences in depth distribution profile and,. potentially, an additional above ground contribution to external dose field from Cs-137 that has transferred into vegetation. Developments in direct measurement of external exposure (e.g. Bonisoli-Alquati et al. 2015; Hinton et al. 2015; Aramrun et al. 2018, 2019) enable more accurate quantification of an individual's external dose within heterogeneously contaminated environments, especially for mobile organisms.

The act of creating sub-observatories, with physical boundaries constructed around them in the terrestrial environment, would be expected to have an influence on the dynamics of the system enclosed within them. For example, enclosure walls may lead to localized shading, micro-climatic variations, reduce grazing and predation, restrict access to resources and influence the movement patterns of more mobile organisms (e.g. small mammals) within the sub-observatory. However, there was broad agreement amongst workshop participants that the proposed sub-observatory approach would still enable more robust, coordinated field studies on radiation effects to be undertaken.

From this discussion the thoughts can be summarized as follows:

- Experiments should be 'hypothesis' based, with rationale and adequate planning, good design (including statistical approach).
- The experiments should consider the benefits to the stakeholders, societal benefit and consider the public confidence.
- Before performing the field studies, techniques/assays should be properly validated under laboratory conditions.
- For both laboratory and field studies - use appropriate controls defining the 'baseline'; generate 'historical control' data recorded before the event; select appropriate 'pristine' sites i.e. historic reference control sites having zero contamination as opposed to just not having radiation contamination while comparing the results from contaminated sites.
- Results should be compared with available previous studies- explore if laboratory and field studies support each other.
- Construct a 'dose-response' curve for different qualities of radiations, measuring different parameters that are sensitive, reproducible and reliable. Repeat the experiments if possible.
- Consider time and dose rate as important variables leading to differences between field studies and the more short term laboratory approaches.
- Consider development of techniques/assays which could be translated across species (e.g. DNA damage, oxidative stress).
- In the event it is not possible to conduct validation studies under laboratory conditions (e.g. experiments involving large mammals, trees etc.), studies aimed at identifying key mechanisms carried out in other model species could perhaps be of use to inform field studies.
- Identify the most sensitive species/life stages and most sensitive individuals (genotype) for the protection of environment. Know the biology of the species adequately.
- Consider different confounding factors which could influence the experimental outcomes (e.g. age, sex, seasonality, temperature, exposure to multiple stressors etc.).
- Consider appropriate measures while comparing the results of laboratory versus field studies (e.g. exposure to external radiations in laboratory exposures compared to exposure to radionuclides and diversity of radiation qualities in field conditions; define routes of exposures (e.g. water, food) and environmental realism).
- Consider ethical issues of using animals for experiments (i.e. 3Rs principles).
- Do not over-extrapolate the results, being aware of the limitations of the study and reporting them in unbiased manner.

Summary conclusions and potential way forward

The workshop covered an extensive range of topics but the overriding conclusion was that an interdisciplinary approach is needed for what was described as a 'wicked problem'. It was recognized that at this stage there is still a need for specific well-focused experiments studying ecological relations and ecosystem responses (where possible, drawing on risk assessment/pollutant/chemical studies for inspiration in set-up) because while there has been a lot of talk on ecosystem approach there are relatively few studies that fall under this term. Hence, we really need to encourage these ecosystem experiments – at both small and large scales.

In terms of a 'way forward' the conclusion was that while it is necessary to pursue development of an ecosystem approach to understand radiation action in ecosystems and to be able to detect early signs of issues, these are unlikely to be solely due to radiation. An ecosystem approach would be inappropriate for regulation as it would be too complex but it would be useful in providing the evidence on which regulation is based. At the practical level, the interaction of radioecologists and radiobiologists was seen as central to the development of useful biomarkers at all levels of organization. Time was seen as a key component that is missing in much of the existing analysis, which is very much concerned with ambient dose rates and effect, while ecosystem level effects develop over extended time periods and require change over time to be a central parameter.

Acknowledgements

We acknowledge support for the workshop from the International Union of Radioecologists and from the organisers of the International Congress of Radiation Research held in Manchester UK in August 2019.

Disclosure statement

No potential conflict of interest was reported by the author(s).

ORCID

Deborah H. Oughton http://orcid.org/0000-0002-5481-200X
Paul N. Schofield http://orcid.org/0000-0002-5111-7263
Kentaro Ariyoshi http://orcid.org/0000-0002-4639-6130
Andrea Bonisoli-Alquati http://orcid.org/0000-0002-9255-7556
Awadhesh N. Jha http://orcid.org/0000-0001-9660-4308
Nguyen T. K. Vo http://orcid.org/0000-0002-4709-3078

References

Alonzo F, Hertel-Aas T, Real A, Lance E, Garcia-Sanchez L, Bradshaw B, Vives I Batlle J, Oughton DH, Garnier-Laplace J. 2016. Population modelling to compare chronic external radiotoxicity between individual and population endpoints in four taxonomic groups. J Environ Radioact. 152:46–59.

Ankley GT, Bennett RS, Erickson RJ, Hoff DJ, Hornung MW, Johnson RD, Mount DR, Nichols JW, Russom CL, Schmieder PK, et al. 2010. Adverse outcome pathways: a conceptual framework to support ecotoxicology research and risk assessment. Environ Toxicol Chem. 29(3):730–741.

Aramrun K, Beresford NA, Skuterud L, Hevrøy TH, Drefvelin J, Bennett K, Yurosko C, Phruksarojanakun P, Esoa J, Yongprawat M, et al. 2019. Measuring the radiation exposure of Norwegian reindeer under field conditions. Sci Total Environ. 687:1337–1343.

Aramrun P, Beresford NA, Wood MD. 2018. Selecting passive dosimetry technologies for measuring the external dose of terrestrial wildlife. J Environ Radioact. 182:128–137.

Audette-Stuart M, Kim SB, McMullin D, Festarini A, Yankovich TL, Carr J, Mulpuru S. 2011. Adaptive response in frogs chronically exposed to low doses of ionizing radiation in the environment. J Environ Radioact. 102(6):566–573.

Baker R, Dickins B, Wickliffe J, Khan F, Gaschak S, Makova K, Phillips C. 2017. Elevated mitochondrial genome variation after 50 generations of radiation exposure in a wild rodent. Evol Appl. 10 (8): 784–791.

Bano A, Batool R, Dazzo F. 2010. Adaptation of chickpea to desiccation stress is enhanced by symbiotic rhizobia. Symbiosis. 50(3): 129–133.

Barata C, Campos B, Rivetti C, LeBlanc GA, Eytcheson S, McKnight S, Tobor-Kaplon M, de Vries Buitenweg S, Choi S, Choi J, et al. 2017. Validation of a two-generational reproduction test in Daphnia magna: an interlaboratory exercise. Sci Total Environ. 579: 1073–1083.

Beaugelin-Seiller K, Garnier-Laplace J, Beresford NA. 2020. Estimating radiological exposure of wildlife in the field. J Environ Radioact. 211:105830.

Beresford NA, Barnett CL, Gashchak S, Maksimenko A, Guliaichenko E, Wood MD, Izquierdo M. 2020a. Radionuclide transfer to wildlife at a 'Reference site' in the Chernobyl Exclusion Zone and resultant radiation exposures. J Environ Radioact. 211:105661

Beresford NA, Horemans N, Copplestone D, Raines KE, Orizaola G, Wood MD, Laanen P, Whitehead HC, Burrows JE, Tinsley MC, et al. 2020b. Towards solving a scientific controversy – The effects of ionising radiation on the environment. J Environ Radioact. 211: 106033

Beresford NA, Scott EM, Copplestone D. 2020c. Field effects studies in the Chernobyl Exclusion Zone: lessons to be learnt. J Environ Radioact. 211:105893

Blaylock BG. 1969. The fecundity of a gambusia affinis affinis population exposed to chronic environmental radiation. Radiat Res. 37(1): 108–117.

Bonisoli-Alquati A, Koyama K, Tedeschi D, Kitamura W, Sukuzi H, Ostermiller S, Arai E, Møller A, Mousseau T. 2015. Abundance and genetic damage of barn swallows from Fukushima. Sci Rep. 5(1): 9432

Bonisoli-Alquati A, Mousseau T, Møller A, Caprioli M, Saino N. 2010a. Increased oxidative stress in barn swallows from the Chernobyl region. Comp Biochem Physiol, Part A Mol Integr Physiol. 155(2):205–210.

Bonisoli-Alquati A, Voris A, Mousseau T, Møller A, Saino N, Wyatt M. 2010b. DNA damage in barn swallows (Hirundo rustica) from the Chernobyl region detected by use of the comet assay. Comp Biochem Physiol C: Pharmacol Toxicol Endocrinol. 151(3):271–277.

Bonzom JM, Hättenschwiler S, Lecomte-Pradines C, Chauvet E, Gaschak S, Beaugelin-Seiller K, Della-Vedova C, Dubourg N, Maksimenko A, Garnier-Laplace J, et al. 2016. Effects of radionuclide contamination on leaf litter decomposition in the Chernobyl exclusion zone. Sci Total Environ. 562:596–603.

Bossdorf O, Richards CL, Pigliucci M. 2008. Epigenetics for ecologists. Ecol Lett. 11(2):106–115.

Boubriak I, Akimkina T, Polischuk V, Dmitriev A, McCready S, Grodzinsky D. 2016. Long term effects of Chernobyl contamination on DNA repair function and plant resistance to different biotic and abiotic stress factors. Cytol Genet. 50(6):381–399.

Bradshaw C, Kapustka L, Barnthouse L, Brown J, Ciffroy P, Forbes V, Geras'kin S, Kautsky U, Bréchignac F. 2014. Using an Ecosystem Approach to complement protection schemes based on organism-level endpoints. J Environ Radioact. 136:98–104.

Bréchignac F, Oughton D, Mays C, Barnthouse L, Beasley JC, Bonisoli-Alquati A, Bradshaw C, Brown J, Dray S, Geras'kin S, et al. 2016. Addressing ecological effects of radiation on populations and ecosystems to improve protection of the environment against radiation: agreed statements from a Consensus Symposium. J Environ Radioact. 158-159:21–29.

Brown JH, Munger JC. 1985. Experimental manipulation of a desert rodent community: food addition and species removal. Ecology. 66(5):1545–1563.

Caffrey E, Johansen M, Caffrey J, Higley K. 2017. Comparison of homogeneous and particulate lung dose rates for small mammals. Health Phys. 112(6):526–532.

Christin S, Hervet É, Lecomte N. 2019. Applications for deep learning in ecology. Methods Ecol Evol. 10(10):1632–1644.

DeAngelis DL, Diaz SG. 2019. Decision-making in agent-based modeling: a current review and future prospectus. Front Ecol E. 6:237.

DeAngelis DL, Grimm V. 2014. Individual-based models in ecology after four decades. F1000Prime Rep. 6:39.

Deryabina TG, Kuchmel SV, Nagorskaya LL, Hinton TG, Beasley JC, Lerebours A, Smith JT. 2015. Long-term census data reveal abundant wildlife populations at Chernobyl. Curr Biol. 25(19):R824–826.

Dmitriev AP, Grodzinskii DM, Gushcha NI, Kryzhanovskaya MS. 2011. Effect of chronic irradiation on plant resistance to biotic stress in 30-km Chernobyl Nuclear Power Plant exclusion zone. Russ J Plant Physiol. 58(6):1062–1068.

Einor D, Bonisoli-Alquati A, Costantini D, Mousseau T, Møller A. 2016. Ionizing radiation, antioxidant response and oxidative damage: a meta-analysis. Sci Total Environ. 548–549:463–471.

Exposito-Alonso M, Vasseur F, Ding W, Wang G, Burbano HA, Weigel D. 2018. Genomic basis and evolutionary potential for extreme drought adaptation in Arabidopsis thaliana. Nat Ecol Evol. 2(2):352–358.

Fawkes R. 2019. An innovative portable detector for the live-monitoring of radionuclides in small terrestrial animals. PhD thesis, University of Salford. [accessed 2020 July 15]. http://usir.salford.ac.uk/id/eprint/47893/1/Thesis%20Ross%20Fawkes%20v18.07.20_Final.pdf.

Feng M, Valdes G, Dixit N, Solberg TD. 2018. Machine learning in radiation oncology: opportunities, requirements, and needs. Front Oncol. 8:110

Fesenko S. 2019. Review of radiation effects in non-human species in areas affected by the Kyshtym accident. J Radiol Prot. 39(1):R1–R17.

Fuller N, Ford AT, Lerebours A, Gudkov DI, Nagorskaya LL, Smith JT. 2019. Chronic radiation exposure at Chernobyl shows no effect on genetic diversity in the freshwater crustacean, Asellus aquaticus thirty years on. Ecol Evol. 9(18):10135–10144.

Galván I, Bonisoli-Alquati A, Jenkinson S, Ghanem G, Wakamatsu K, Mousseau T, Møller A. 2014. Chronic exposure to low-dose radiation at Chernobyl favours adaptation to oxidative stress in birds. Funct Ecol. 28(6):1387–1403.

Garnier-Laplace J, Geras'kin S, Della-Vedova C, Beaugelin-Seiller K, Hinton TG, Real A, Oudalova A. 2013. Are radiosensitivity data derived from natural field conditions consistent with data from controlled exposures? A case study of Chernobyl wildlife chronically exposed to low dose rates. J Environ Radioact. 121:12–21.

Gauci A, Abela J, Austad M, Cassar LF, Adami KZ. 2018. A machine learning approach for automatic land cover mapping from DSLR images over the Maltese Islands. Environ Modell Softw. 99:1–10.

Geraskin SA, Dikarev VG, Zyablitskaya YY, Oudalova AA, Spirin YV, Alexakhin RM. 2003. Genetic consequences of radioactive contamination by the Chernobyl fallout to agricultural crops. J Environ Radioact. 66(1–2):155–169.

Geras'kin S, Evseeva T, Oudalova A. 2013. Effects of long-term chronic exposure to radionuclides in plant populations. J Environ Radioact. 121:22–32.

Geras'kin S, Oudalova A, Kuzmenkov A, Vasiliyev D. 2018. Chronic radiation exposure modifies temporal dynamics of cytogenetic but not reproductive indicators in Scots pine populations. Environ Pollut. 239:399–407.

Geras'kin S, Volkova P, Vasiliyev D, Dikareva N, Oudalova A, Kazakova E, Makarenko E, Duarte G, Kuzmenkov A. 2019. Scots pine as a promising indicator organism for biomonitoring of the polluted environment: a case study on chronically irradiated populations. Mutat Res. 842:3–13.

Geras'kin S. 2016. Ecological effects of exposure to enhanced levels of ionizing radiation. J Environ Radioact. 162-163:347–357.

Glazko VI. 2001. A note on genetic structure of cattle breed within increased ionizing zone at the Chernobyl accident area. Anim Sci Pap Rep. 19:95–109.

Gomez-Fernandez M, Higley K, Tokuhiro A, Welter K, Wong WK, Yang H. 2020. Status of research and development of learning-based approaches in nuclear science and engineering: a review. Nucl Eng Des. 359:110479.

Goodman J, Copplestone D, Laptev GV, Gashchak S, Auld S. 2019. Variation in chronic radiation exposure does not drive life history divergence among Daphnia populations across the Chernobyl Exclusion Zone. Ecol Evol. 9(5):2640–2650.

Haanes H, Hansen EL, Hevrøy TH, Jensen LK, Gjelsvik R, Jaworska A, Bradshaw C. 2020. Realism and usefulness of multispecies experiment designs with regard to application in radioecology: a review. Sci Total Environ. 718:134485.

Hancock S, Vo NTK, Byun SH, Zainullin VG, Seymour CB, Mothersill C. 2019a. Effects of historic radiation dose on the frequency of sex-linked recessive lethals in Drosophila populations following the Chernobyl nuclear accident. Environ Res. 172:333–337.

Hancock S, Vo NTK, Goncharova RI, Seymour CB, Byun SH, Mothersill CE. 2020. One-decade-spanning transgenerational effects of historic radiation dose in wild populations of bank voles exposed to radioactive contamination following the Chernobyl nuclear disaster. Environ Res. 180:108816.

Hancock S, Vo NTK, Omar-Nazir L, Batlle JVI, Otaki JM, Hiyama A, Byun SH, Seymour CB, Mothersill C. 2019b. Transgenerational effects of historic radiation dose in pale grass blue butterflies around Fukushima following the Fukushima Dai-ichi Nuclear Power Plant meltdown accident. Environ Res. 168:230–240.

Hansen EL, Lind OC, Oughton DH, Salbu B. 2019. A framework for exposure characterization and gamma dosimetry at the NMBU FIGARO irradiation facility. Int J Radiat Biol. 95(1):82–89.

Hevrøy TH, Golz AL, Hansen EL, Xie L, Bradshaw C. 2019. Radiation effects and ecological processes in a freshwater microcosm. J Environ Radioact. 203:71–83.

Hickey DA, McNeilly T. 1975. Competition between metal tolerant and normal plant populations; a field experiment on normal soil. Evolution. 29(3):458–464.

Hinton TG, Byrne ME, Webster S, Beasley JC. 2015. Quantifying the spatial and temporal variation in dose from external exposure to radiation: a new tool for use on free-ranging wildlife. J Environ Radioact. 145:58–65.

Holling CS. 1973. Resilience and stability of ecological systems. Annu Rev Ecol Syst. 4(1):1–23.

Horemans N, Spurgeon DJ, Lecomte-Pradines C, Saenen E, Bradshaw C, Oughton D, Rasnaca I, Kamstra JH, Adam-Guillermin C. 2019. Current evidence for a role of epigenetic mechanisms in response to ionizing radiation in an ecotoxicological context. Environ Pollut. 251:469–483.

Howard BJ, Beresford NA, Andersson P, Brown JE, Copplestone D, Beaugelin-Seiller K, Garnier-Laplace J, Howe PD, Oughton D, Whitehouse P. 2010. Protection of the environment from ionising radiation in a regulatory context-an overview of the PROTECT coordinated action project. J Radiol Prot. 30(2):195–214.

Matson PA, Hunter MD. 1992. Playing chutes and ladders: heterogeneity and the relative roles of bottom-up and top-down forces in natural communities. Ecology. 73(3):723–732.

Hurem S, Gomes T, Brede DA, Lindbo Hansen E, Mutoloki S, Fernandez C, Mothersill C, Salbu B, Kassaye YA, Olsen A-K, et al. 2017. Parental gamma irradiation induces reprotoxic effects accompanied by genomic instability in zebrafish (Danio rerio) embryos. Environ Res. 159:564–578.

International Atomic Energy Agency (IAEA). 2014. Radiation protection and safety of radiation sources: international basic safety standards, General Safety Requirements No. GSR Part 3, IAEA, Vienna.

International Commission On Radiological Protection (ICRP). 2008. Environmental protection: the concept and use of reference animals and plants, ICRP publication 108. Ann. ICRP. 38:4-6.

Jarry G, DeMarco JJ, Beifuss U, Cagnon CH, McNitt-Gray MF. 2003. A monte carlo-based method to estimate radiation dose from spiral CT: from phantom testing to patient-specific models. Phys Med Biol. 48(16):2645–2663.

Jha AN. 2004. Genotoxicological studies in aquatic organisms: an overview. Mutat Res. 552(1–2):1–17.

Jha AN. 2008. Ecotoxicological applications and significance of the comet assay. Mutagenesis. 23(3):207–221.

Kalin U, Lang N, Hug C, Gessler A, Wegner JD. 2019. Defoliation estimation of forest trees from ground-level images. Remote Sens Environ. 223:143–153.

Karr JR. 1992. Bottom-up versus top-down regulation of vertebrate populations: lessons from birds and fish. In: Hunter MD, Ohgushi T, Price PW, editors. Effects of resource distribution on animal-plant interactions. San Diego (CA): Academic Press; p. 244–286..

Krznaric E, Verbruggen N, Wevers JHL, Carleer R, Vangronsveld J, Colpaert JV. 2009. Cd-tolerant Suillus luteus: a fungal insurance for pines exposed to Cd. Environ Pollut. 157(5):1581–1588.

Lampe N, Breton V, Sarramia D, Sime-Ngando T, Biron DG. 2017. Understanding low radiation background biology through controlled evolution experiments. Evol Appl. 10(7):658–666.

Lecomte-Pradines C, Adam-Guillermin C, Gashchak S, Bradshaw C, Copplestone D, Beresford NA. 2020. More than thirty years after the Chernobyl accident: what do we know about the effects of radiation on the environment? J Environ Radioact. 211:106108.

Lerebours A, Gudkov D, Nagorskaya L, Kaglyan A, Rizewski V, Leshchenko A, Bailey EH, Bakir A, Ovsyanikova S, Laptev G, et al. 2018. Impact of environmental radiation on the health and reproductive status of fish from chernobyl. Environ Sci Technol. 52(16): 9442–9450.

Levinton JS, Suatoni E, Wallace W, Junkins R, Kelaher B, Allen BJ. 2003. Rapid loss of genetically based resistance to metals after the cleanup of a Superfund site. Proc Natl Acad Sci Usa. 100(17): 9889–9891.

Lin TH, Liaw DC. 2015. Development of an intelligent disaster information-integrated platform for radiation monitoring. Nat Hazards. 76(3):1711–1725.

MacArthur RH, Wilson EO. 1967. The theory of island biogeography. Princeton (NJ): Princeton University Press.

Maremonti E, Eide DM, Rossbach LM, Lind OC, Salbu B, Brede DA. 2019. In vivo assessment of reactive oxygen species production and oxidative stress effects induced by chronic exposure to gamma radiation in Caenorhabditis elegans. Free Radic Biol Med. 152:583–596.

May RM. 1973. Stability and complexity in model ecosystems. Princeton (NJ): Princeton University Press.

McCann KS. 2000. The diversity-stability debate. Nature. 405(6783): 228–233.

Møller A, Bonisoli-Alquati A, Mousseau T. 2013. High frequency of albinism and tumors in free-living birds around Chernobyl. Mut Res Genet Toxicol Environ Mut. 757(1):52–59.

Møller A, Bonisoli-Alquati A, Rudolfsen G, Mousseau T. 2012b. Elevated mortality among birds in Chernobyl as judged from skewed age and sex ratios. PLoS One. 7(4):e35223

Møller A, Mousseau T. 2007a. Species richness and abundance of forest birds in relation to radiation at Chernobyl. Biol Lett. 3(5):483–486.

Møller AP, Barnier F, Mousseau TA. 2012a. Ecosystems effects twenty-five years after Chernobyl: pollinators, fruit set and recruitment. Oecologia. 170(4):1155–1165.

Morley NJ. 2012. The effects of radioactive pollution on the dynamics of infectious diseases in wildlife. J Environ Radioact. 106:81–97.

Mothersill C, Rusin A, Seymour C. 2017. Low doses and non-targeted effects in environmental radiation protection; where are we now and where should we go? Environ Res. 159:484–490.

Mothersill C, Seymour C. 2010. Eco-systems biology-from the gene to the stream. Mutat Res. 687(1–2):63–66.

Mothersill C, Seymour C. 2019. Targets, pools, shoulders, and communication – a reflection on the evolution of low-dose radiobiology. Int J Radiat Biol. 95(7):851–860.

Mothersill C, Smith R, Lariviere D, Seymour C. 2013. Chronic exposure by ingestion of environmentally relevant doses of (226)Ra leads to transient growth perturbations in fathead minnow (Pimephales promelas, Rafinesque, 1820). Int J Radiat Biol. 89(11):950–964.

Mousseau TA, Milinevsky G, Kenney-Hunt J, Møller AP. 2014. Highly reduced mass loss rates and increased litter layer in radioactively contaminated areas. Oecologia. 175(1):429–437.

Mousseau TA, Møller AP. 2013. Elevated frequency of cataracts in birds from Chernobyl. PLoS One. 8(7):e66939.

Mughal SK, Myazin AE, Zhavoronkov LP, Rubanovich AV, Dubrova YE. 2012. The dose and dose-rate effects of paternal irradiation on transgenerational instability in mice: a radiotherapy connection. PLoS One. 7(7):e41300.

Muniraj M, Qureshi AR, Vijayakumar D, Viswanathan AR, Bharathi N. 2017. Geo tagged internet of things (iot) device for radiation monitoring. In 2017 International Conference on Advances in Computing, Communications and Informatics (ICACCI), p. 431–436.

Newbold LK, Robinson A, Rasnaca I, Lahive E, Soon GH, Lapied E, Oughton DH, Gashchak S, Beresford NA, Spurgeon DJ. 2019. Genetic, epigenetic and microbiome characterisation of an earthworm species (Octolasion lacteum) along a radiation exposure gradient at Chernobyl. Environ Pollut. 255(Pt 1):113238.

Nishikawa K, Kinjo AR. 2018. Mechanism of evolution by genetic assimilation: equivalence and independence of genetic mutation and epigenetic modulation in phenotypic expression. Biophys Rev. 10(2): 667–676.

Norouzzadeh MS, Nguyen A, Kosmala M, Swanson A, Palmer MS, Packer C, Clune J. 2018. Automatically identifying, counting, and describing wild animals in camera-trap images with deep learning. Proc Natl Acad Sci USA. 115(25):E5716–E5725.

Noyes PD, McElwee MK, Miller HD, Clark BW, Van Tiem LA, Walcott KC, Erwin KN, Levin ED. 2009. The toxicology of climate change: environmental contaminants in a warming world. Environ Int. 35(6):971–986.

Omar-Nazir L, Shi X, Moller A, Mousseau T, Byun S, Hancock S, Seymour C, Mothersill C. 2018. Long-term effects of ionizing radiation after the Chernobyl accident: possible contribution of historic dose. Environ Res. 165:55–62.

Oziolor E, Reid N, Yair S, Lee K, VerPloeg S, Bruns P, Shaw J, Whitehead A, Matson C. 2019. Adaptive introgression enables evolutionary rescue from extreme environmental pollution. Science. 364(6439):455–457.

Pespeni MH, Sanford E, Gaylord B, Hill TM, Hosfelt JD, Jaris HK, LaVigne M, Lenz EA, Russell AD, Young MK, et al. 2013. Evolutionary change during experimental ocean acidification. Proc Nat Aca Sci. 110(17):6937–6942.

Pimm SL. 1982. Food webs. New York (NY): Chapman and Hall.

Potamitis I. 2016. Deep learning for detection of bird vocalisations. ArXiv:1609.08408 [Cs]. [accessed 2020 July 15]. http://arxiv.org/abs/1609.08408.

Powathil GG, Swat M, Chaplain MA. 2015. Systems oncology: towards patient-specific treatment regimes informed by multiscale mathematical modelling. Semin Cancer Biol. 30:13–20.

Rammer W, Seidl R. 2019. Harnessing deep learning in ecology: an example predicting bark beetle outbreaks. Front Plant Sci. 10:1327.

Richter N, Staenz K, Kaufmann H. 2008. Spectral unmixing of airborne hyperspectral data for baseline mapping of mine tailings areas. Int J Remote Sens. 29(13):3937–3956.

Rodrigues-Moreira S, Moreno SG, Ghinatti G, Lewandowski D, Hoffschir F, Ferri F, Gallouet A-S, Gay D, Motohashi H, Yamamoto M, et al. 2017. Low-dose irradiation promotes persistent oxidative stress and decreases self-renewal in hematopoietic stem cells. Cell Rep. 20(13):3199–3211.

Roff DA. 2001. Life history evolution, 1st ed. Sunderland: Sinauer Associates.

Ross D. 2003. Menace of the nuclear rabbits. The Herald, 25th June 2003.

Ruedig E, Beresford NA, Gomez Fernandez ME, Higley K. 2015. A comparison of the ellipsoidal and voxelized dosimetric methodologies for internal, heterogeneous radionuclide sources. J Environ Radioact. 140:70–77.

Ruiz-González M, Czirják G, Genevaux P, Møller A, Mousseau T, Heeb P. 2016. Resistance of feather-associated bacteria to intermediate levels of ionizing radiation near chernobyl. Sci Rep. 6(1):22969

Salamon J, Bello JP, Farnsworth A, Kelling S. 2017. Fusing shallow and deep learning for bioacoustic bird species classification. In 2017 IEEE International Conference on Acoustics, Speech and Signal Processing (ICASSP), p. 141–145.

Sarapultseva EI, Dubrova YE. 2016. The long-term effects of acute exposure to ionising radiation on survival and fertility in Daphnia magna. Environ Res. 150:138–143.

Sarapultseva EI, Ustenko KV, Dubrova YE. 2019. The combined effects of acute irradiation and food supply on survival and fertility in Daphnia magna. J Environ Radioact. 199–200:75–83.

Schofield PN, Kondratowicz M. 2018. Evolving paradigms for the biological response to low dose ionizing radiation; the role of epigenetics. Int J Radiat Biol. 94(8):769–781.

Schuman MC, Baldwin IT. 2018. Field studies reveal functions of chemical mediators in plant interactions. Chem Soc Rev. 47(14): 5338–5353.

Simon JC, Marchesi JR, Mougel C, Selosse MA. 2019. Host-microbiota interactions: from holobiont theory to analysis. Microbiome. 7(1):5.

Slobodkin LB, Smith FE, Hairston NG. 1967. Regulation in terrestrial ecosystems and the implied balance of nature. Am Nat. 101(918): 109–123.

Song Y, Xie L, Lee YK, Brede DA, Lyne F, Kassaye Y, Thaulow J, Caldwell G, Salbu B, Tollefsen KE. 2019b. Integrative assessment of low-dose gamma radiation effects on Daphnia magna reproduction: toxicity pathway assembly and AOP development. Sci Total Environ. 705:135912.

Song YY, Wang M, Zeng RS, Groten K, Baldwin IT. 2019a. Priming and filtering of antiherbivore defences among Nicotiana attenuata plants connected by mycorrhizal networks. Plant Cell Environ. 42(11):2945–2961.

South Downs National Park Authority. 2018. Ecosystem Services Background Paper South Downs Local Plan. [accessed 2020 July 15]. https://www.southdowns.gov.uk/wp-content/uploads/2018/04/Core-05-Ecosystem-Services-Background-Paper-April-2018.pdf.

Stark K, Goméz-Ros JM, Vives I Batlle J, Lindbo Hansen E, Beaugelin-Seiller K, Kapustka LA, Wood MD, Bradshaw C, Real A, McGuire C, et al. 2017. Dose assessment in environmental radiological protection: state of the art and perspectives. J Environ Radioact. 175–176:105–114.

Steiner M, Willrodt C, Wichterey K, Ikäheimonen T, Ioshchenko V, Hutri KL, Muikku M, Outola L, Beresford N, Bradshaw C. 2013. Observatories for radioecological research: Description. [accessed 2020 July 15]. https://radioecology-exchange.org/sites/default/files/STAR_Deliverable-2.3.pdf.

Streffer C. 2006. Transgenerational transmission of radiation damage: genomic instability and congenital malformation. J Radiat Res. 47(SupplementB):B19–24.

Thomson C. 2003. MSP calls for tests on potential atomic kittens, Caithness Courier 2nd July 2003.

U.S. Environmental Protection Agency (US EPA). 2016. Ecosystem Services as Assessment Endpoints in Ecological Risk Assessment

Technical Background Paper. [accessed 2020 July 15]. https://www.epa.gov/sites/production/files/2016-08/documents/ecosystem_services_technical_paper.pdf.

United Nations Scientific Committee on the Effects of Atomic Radiation (UNSCEAR). 2010. Summary of low-dose radiation effects on health. New York (NY): United Nations.

van Straalen NM, Timmermans M. 2002. Genetic variation in toxicant-stressed populations: an evaluation of the "genetic erosion" hypothesis. Hum Ecol Risk Assess. 8(5):983–1002.

Varley A, Tyler A, Smith L, Dale P, Davies M. 2015. Remediating radium contaminated legacy sites: advances made through machine learning in routine monitoring of "hot" particles. Sci Total Environ. 522:271–279.

Villeneuve DL, Crump D, Garcia-Reyero N, Hecker M, Hutchinson TH, LaLone CA, Landesmann B, Lettieri T, Munn S, Nepelska M, et al. 2014. Adverse outcome pathway (AOP) development I: strategies and principles. Toxicol Sci. 142(2):312–320.

Vives I Batlle J, Sazykina TG, Kryshev A, Monte L, Kawaguchi I. 2012. Inter-comparison of population models for the calculation of radiation dose effects on wildlife. Radiat Environ Biophys. 51(4):399–410.

Volkova PY, Geras'kin SA, Kazakova EA. 2017. Radiation exposure in the remote period after the Chernobyl accident caused oxidative stress and genetic effects in Scots pine populations. Sci Rep. 7:43009.

Wood MD, Beresford NA, Yankovich TL, Semenov DV, Copplestone D. 2011. Addressing current knowledge gaps on radionuclide transfer to reptiles. Radioprotection. 46(6):S521–S527.

Wu J, Loucks OL. 1995. From balance of nature to hierarchical patch dynamics: a paradigm shift in ecology. Q Rev Biol. 70(4):439–466.

Yodzis P, Innes S. 1992. Body size and consumer-resource dynamics. Am Nat. 139(6):1151–1175.

Index

Note: **Bold** page numbers refer to tables and *italic* page numbers refer to figures.